D0290233

APPLICATIONS OF DIGITAL SIGNAL PROCESSING

PRENTICE-HALL SIGNAL PROCESSING SERIES

Alan V. Oppenheim, Editor

ANDREWS and HUNT *Digital Image Restoration*
BRIGHAM *The Fast Fourier Transform*
HAMMING *Digital Filters*
OPPENHEIM and SCHAFER *Digital Signal Processing*
OPPENHEIM, et al *Applications of Digital Signal Processing*
RABINER and GOLD *Theory and Applications of Digital Signal Processing*

APPLICATIONS OF DIGITAL SIGNAL PROCESSING

ALAN V. OPPENHEIM, Editor

Massachusetts Institute of Technology Cambridge, Mass. 02139

PRENTICE-HALL, INC., *Englewood Cliffs, New Jersey 07632*

Library of Congress Cataloging in Publication Data
Main entry under title:
Applications of digital signal processing.
Includes bibliographical references and index.
1. Signal processing. I. Oppenheim, Alan V., 1937-
TK5102.5.A68 621.38'043 77-8547
ISBN 0-13-039115-8

Printed in the United States of America

10 9 8 7 6 5 4 3

PRENTICE-HALL INTERNATIONAL, INC., *London*
PRENTICE-HALL OF AUSTRALIA PTY. LIMITED, *Sydney*
PRENTICE-HALL OF CANADA, LTD., *Toronto*
PRENTICE-HALL OF INDIA PRIVATE LIMITED, *New Delhi*
PRENTICE-HALL OF JAPAN, INC., *Tokyo*
PRENTICE-HALL OF SOUTHEAST ASIA PTE. LTD., *Singapore*
WHITEHALL BOOKS LIMITED, *Wellington, New Zealand*

To
Phyllis
and
Justine

CONTRIBUTING AUTHORS

CHAPTER 1

Stanley L. Freeny *(Bell Telephone Laboratories, Inc.)*
James F. Kaiser *(Bell Telephone Laboratories, Inc.)*
Henry S. McDonald *(Bell Telephone Laboratories, Inc.)*

CHAPTER 2

Barry Blesser *(Massachusetts Institute of Technology)*
James M. Kates *(Teledyne Acoustic Research)*

CHAPTER 3

Alan V. Oppenheim *(Massachusetts Institute of Technology)*

CHAPTER 4

Bob R. Hunt *(University of Arizona)*

CHAPTER 5

James H. McClellan *(Massachusetts Institute of Technology)*
Robert J. Purdy *(M.I.T. Lincoln Laboratory)*

CHAPTER 6

Arthur B. Baggeroer *(Massachusetts Institute of Technology)*

CHAPTER 7

Enders A. Robinson *(Private Consultant)*
Sven Treitel *(Amoco Production Company)*

CONTENTS

PREFACE

Digital signal processing has been a growing and dynamic field for more than a decade. Throughout this period it has maintained a close coupling between the development of new algorithms and theoretical results, the exploitation of new technology, and the application of the techniques to new fields. The technology for signal processing has advanced from discrete semiconductor components to large scale integration (LSI) and grand scale integration (GSI) with densities above 100,000 components per chip. Furthermore, there is a growing emphasis on analog sampled-data devices such as charge transfer devices. The availability of low-cost, fast microprocessors and the growing potential for custom, high density integrated circuits promises the reduction to practice of increasingly complex and sophisticated algorithms. For example, because of the current CCD technology and its potential for implementing spectral analysis and finite impulse response filters, a custom LSI implementation of some formerly impractical speech processing systems is now possible, requiring a very small number of custom chips. Modern radar systems now rely almost totally on digital techniques for the signal processing, with the analog to digital interface at the receiver. Using parallel architecture and high-speed logic, fast, flexible digital radar processors are operating in the Megahertz range. These and similar examples suggest why the digital signal processing field has been and will probably continue to be characterized by an extremely optimistic attitude in terms of the way in which new technology will impact areas of application.

As the applications areas have developed there have been important similarities and differences in the ways in which the basic techniques and technology have been applied. The differences are partly due to the presence of different assumptions and constraints such as data rate, real-time requirements, etc. In some cases, however, the differences have arisen simply from a lack of interaction between these areas and it has often happened that an awareness of how digital signal processing techniques have been applied in one area has led to improvements in another area.

The goal of this book is to discuss a number of important applications which digital signal processing has found over the past decade. Some of these application areas, such as speech processing and seismic data processing, have utilized the techniques of digital signal processing from the very beginning and in fact have been catalysts for many of the developments in the field. Others, such as audio processing, have in the past relied principally on analog signal processing methods and only recently have begun to exploit digital signal processing techniques. The principal reason for this is that in contrast with other areas the audio field is heavily consumer oriented and highly sensitive to cost. With the current technology, however, it is increasingly clear that we are on the verge of a major revolution in audio signal processing.

In surveying the applications of digital signal processing, one can see a variety of motivations for exploiting techniques and technology associated with digital signal processing. In some cases, such as radar, sonar and seismic processing, the overriding consideration is the tremendous flexibility and accuracy available through digital signal processing. This motivation is clearly in evidence in many of the applications discussed in this book. In other applications, such as telephony and audio processing, it is the anticipated long term cost advantages that become a significant factor.

Each of the chapters in this book has been written with two principal aims in mind. The first is to provide an introduction and tutorial review of each specific field for readers having prior exposure to the basic techniques of digital signal processing, but whose detailed experiences are outside that applications area. As the reader will find, many of these application areas have many aspects in common, but also differ in a variety of ways and it is undeniable that a detailed understanding of how the techniques have influenced one area can contribute significantly to future developments in another. In addition to this heavily tutorial aspect, each chapter represents a state-of-the-art discussion of each field and will undoubtedly be of interest to those working in that field. Thus, it is hoped that each chapter and this book in general will have appeal to readers from a wide variety of areas and that through an exposure to new applications areas, a synthesis of techniques and new ideas will emerge.

I would like to express my thanks to all of the contributing authors for the care with which each of the chapters was prepared and for their sensitivity to the anticipated audience and the various deadlines. The preparation of a chapter for an edited book such as this invariably requires considerably more time and effort than anticipated when the commitment is originally made. Many of the contributing authors postponed other commitments to insure that publication of this book would not be delayed. In addition, I would like to thank Ms. Monica Edelman for her good nature and efficiency in handling the secretarial demands associated with my role as editor of this book.

A. V. OPPENHEIM

Massachusetts Institute of Technology
Cambridge, Mass.

1

SOME APPLICATIONS OF DIGITAL SIGNAL PROCESSING IN TELECOMMUNICATIONS

S. L. Freeny

Bell Telephone Laboratories, Inc.
Holmdel, N.J. 07733

and

J. F. Kaiser

Bell Telephone Laboratories, Inc.
Murray Hill, N.J. 07974

and

H. S. McDonald

Bell Telephone Laboratories, Inc.
Holmdel, N.J. 07733

1.1 INTRODUCTION

Signals are a very fundamental part of every telecommunication system, and such systems abound with examples of signal-processing techniques. In recent years more and more attention has been given to the use of digital signal processing in telecommunication systems. This is due in large part to the increasing availability of medium- and large-scale digital integrated circuits with their desirable properties of small size, low power, low cost, noise immunity, and reliability. For some time now, the authors have been associated with a number of exploratory applications of digital signal processing in the Bell System; in this chapter we propose to review a representative sampling of these in some detail. To help the reader make his way through a broad and complex subject, we have chosen to use a treatment that is more descriptive than mathematical. Furthermore, we have tried to keep to a minimum the use of the highly specialized, and sometimes arcane, terminology that pervades this field.

The discussion begins with a brief look at a number of different parts of the telecommunication system where digital techniques can be advantageously applied. These subsystems include digital transmission, digital switching, pulse-code modulation (PCM) and frequency-division multiplex (FDM) transmission terminals, supervision, and echo control. The discussion then centers on the design of programmable digital signal processors and the interplay of such design considerations as structure, coefficient sensitivity, nonlinear effects, and the role of design aids. The versatility and use of a programmable digital signal processor are described in detail. The chapter ends with a prognosis for the immediate future.

1.2 A VIEW OF TELECOMMUNICATIONS SYSTEMS

Since signals are a fundamental part of telecommunication systems, the role that signal processing plays in communications is a major one. To multiplex many signals on a high-speed digital line or a broadband analog radio channel, the signals must be filtered to restrict out-of-band components, which could cause distortions and interchannel interference. Telephone numbers are forwarded between major switching centers as coded tone signals, and many modern telephone sets use tone signaling to replace the sequential pulsing of rotary dials. Signal processing is required to generate and detect these signals. Testing to confirm proper operation or to locate faults involves substantial signal processing of a wide range of signals.

1.3 DIGITAL TRANSMISSION

As the need to process signals permeates all the telecommunications plant, there is a growing penetration of digital facilities into most parts of this same plant.(1) The shift to digital facilities started in the early 1960s when it was demonstrated that a pair of 22-gauge copper wires could support transmission of 1.5 megabits per second (Mbit/s) for about 6000 ft without significant errors. Six thousand feet just happened to be the usual spacing for inserting loading inductors into the line to equalize the high-frequency roll-off caused by shunt capacity. By replacing a loading coil unit with an active unit containing preamplifiers, clock retrieving circuits, and output pulse amplifiers, the incoming pulse train could be regenerated for another 6000-ft trip to the next repeater.

Thus an ordinary pair of wires that could normally support only one voice frequency signal can be made to carry a digital signal of greater than 1 Mbit/s. Taking advantage of this property, a new transmission system, known in telecommunications parlance as the T-carrier system, has been designed to transmit 24 voiceband channels simultaneously over a single wire pair with a 1.544-Mbit/s signal. In this system each voiceband signal is sampled 8000 times/s and each sample is converted to a digital number for transmission. This process is called pulse-code modulation or PCM. In ordinary PCM, each digital number represents the voltage (or current) amplitude of the analog signal at a particular instant. The type of PCM employed in T-carrier transmission is known as μ-255 companded (or logarithmic) PCM,[2] because each digital number represents (approximately) the logarithm of the instantaneous signal amplitude. This is done because in this type of PCM only 8-bit accuracy is required in each digital word to provide the desired signal-to-noise ratio, whereas 13-bit words would be required in ordinary linear PCM.

Introduced in 1961, the economic advantage of these PCM systems is so strong that a 40% per year growth rate has continued through the 1965–1975 decade. Over 65 million circuit miles of digital carrier are operating at the present time. Figure 1-1 shows a photograph of a line repeater for a 1.544 megabits/second system.

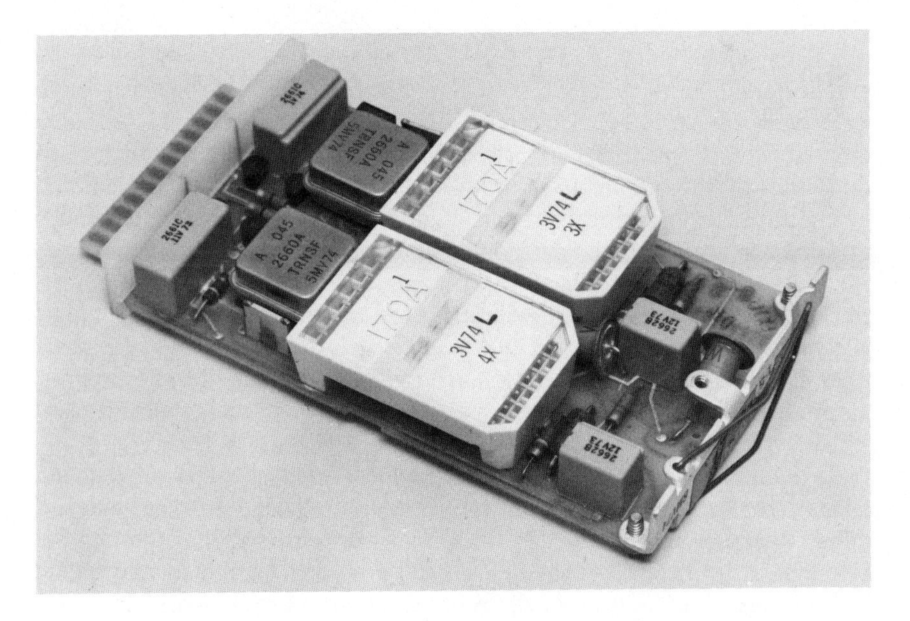

FIGURE 1-1. Tl Integrated circuit repeater.

Digital penetration into the transmission of telecommunication signals will also show up in many future systems.[3] As the ether fills with more and more signals, growth in radio transmission must either use higher and higher frequencies or systems must use confined (guided) energy to carry the information. Both techniques lead to the use of digital modulation because of inherent properties of the transmission medium. High-frequency line-of-sight radio systems have multipath problems because of reflections from the flat surfaces common to our man-made world. Waveguides and optical fibers have multipath-type effects when energy in different modes propagates at different velocities. Digital signal modulation, which allows distortion on discrete signals to be stripped off and the signal regenerated anew, solves the multipath problem. Before the end of the twentieth century, probably as much as half of the telecommunications signals will use digital transmission techniques. At the present time many metropolitan areas use T-carrier digital lines in at least one half of the trunk circuits between local central offices; Houston, Texas, is an example. But by the 1980s most growth areas will be over half digital.

1.4 DIGITAL SWITCHING

Digital technology has also penetrated digital switching. A very large toll (long distance) switch[4] is now operating in Chicago, Illinois. This switching system (called Number 4 Electronic Switching System or simply No. 4 ESS) is a very large switch, which can grow to over four times larger than previous toll switches (10,000 trunks). Digital time-division-multiplex (TDM) operation is the key to building very large switches. In ordinary switches there are practical limits to the size of a switching matrix used as an element to build large networks. A 16 by 16 matrix is typical in metallic networks. For very large networks, where any input may be connected to any one of tens of thousands of outputs, many stages of switching matrices are required and, owing to the very large fan-out and small switch size, the intrastage groups of connections are sparse. Sparse links lead to low occupancy and thus poor efficiency of utilization. Therefore, it is impractical to build very large networks out of a group of small switches. Time division solves this dilemma by allowing what might appear to be a small switch (say 16 by 16) to carry via time sharing as many as 120 simultaneous signals on each of its 16 input or output connections. So in order to meet the demand for a large toll switcher, a change to a digital technology was necessary.

In addition, the TDM switching solution turned up other advantages. Digital trunk circuits homing on the TDM switch need only be buffered, rather than converted to analog form, and can remain digital through the switch at a significant cost savings. Digital echo control circuits and digital tone receivers also appear to be more economical and reliable than their analog counterparts. TDM also saves intrasystem wires, since a single wire (a digital highway) serves hundreds of simultaneous conversations.

Future switches will most likely extend the TDM concept to include slow scanning and control signals along with talking channels within a hierarchical multiplex. Channels through the network could then also carry packets of information between control computers and between microprocessors that interface digital signal-processing equipment.

One view of the node in the local telephone system of the future[5] is a system that terminates cables and digitizes the signals as soon as they arrive in the building. Such a plan would have a plug-in interface unit[6] for each cable pair that terminated the circuit in a suitable fashion, depending on its function as a subscriber line, a trunk circuit, or some kind of special circuit such as a dedicated line. After termination, the signals would be filtered, sampled, and digitized. The data would be multiplexed with input–output data to sense and control the various functions[7] in the interface such as dc line current status or states of the test or ring relays. The digital interface of each interface unit would be standardized and would fit in any empty socket in an interface frame. All the intrasystems connections to this frame would be through a digital highway using several levels of multiplexing. Many properties of such a system, called a "Digital Wire Center", have been demonstrated in a laboratory experiment.[5] This experiment showed that for the local telephone office application more than a 200 to 1 savings in intrasystems wiring over that used in present offices is possible. Digital systems enable the designer to trade gates for wires. The prospects for the appearance of digitally implemented local switching systems in the last two decades of this century look very promising.

Thus, for reasons of both necessity and economics,[8] the telecommunication plant is undergoing a long-term metamorphosis toward digital transmission and digital switching. And since signal processing plays a fundamental role in telecommunications, digital signal processing will play a major role in future plant.

1.5 DIGITAL SIGNAL PROCESSING IN PULSE-CODE MODULATED TRANSMISSION TERMINALS

In transmission terminals, the basic bandlimiting functions now done using analog filters will be in part done in the future with digital filters both in time-division multiplex (TDM) and frequency-division multiplex (FDM) terminals. In TDM terminals, the basic D (for digital) channel bank serves to convert a group of analog voice-frequency signals to and from high-speed serial digital streams. The present generation of D banks uses either LC or active RC filters that cut off at about 3.2 kilohertz (kHz) in order to eliminate any higher frequencies that would cause aliasing about the 8-kHz sampling rate in the transmitting path or disturb other cable pairs in the same sheath in the receiving direction. A highpass filter to remove power line fundamental frequencies is also necessary in the transmitting direction. Usually, the highpass filter is second order, and the lowpass filters are fifth order and have several complex zero pairs near the band edge. In the conventional banks, the filtering operation is followed by a sampling gate, which connects to a shared PCM encoder. Figure 1-2(a) shows a block diagram of such a terminal.

An alternative approach,[8,35] using digital processing, would sample the analog signals at a high rate (say 32 kHz) and digitize them into a linear PCM or DPCM code. Digital filters operating at the high sampling rate would remove energy above 3.2 kHz and the hum below 200 hertz (Hz). Although this digital filter would do most of the work of removing unwanted out-of-band signal energy, it is still necessary to place a simple analog lowpass filter in front of the A/D converter. The sampling rate can then be safely reduced to 8 kHz by throwing away three of every four samples. The linear PCM representation necessary for the digital filtering must then be converted into the 15-segment μ law code used on PCM-TDM lines. In the reverse direction, a similar digital process would be used to linearize the code, interpolate new samples between the incoming data samples, and then convert back to analog for outward signal transmission. A block diagram of a digitally

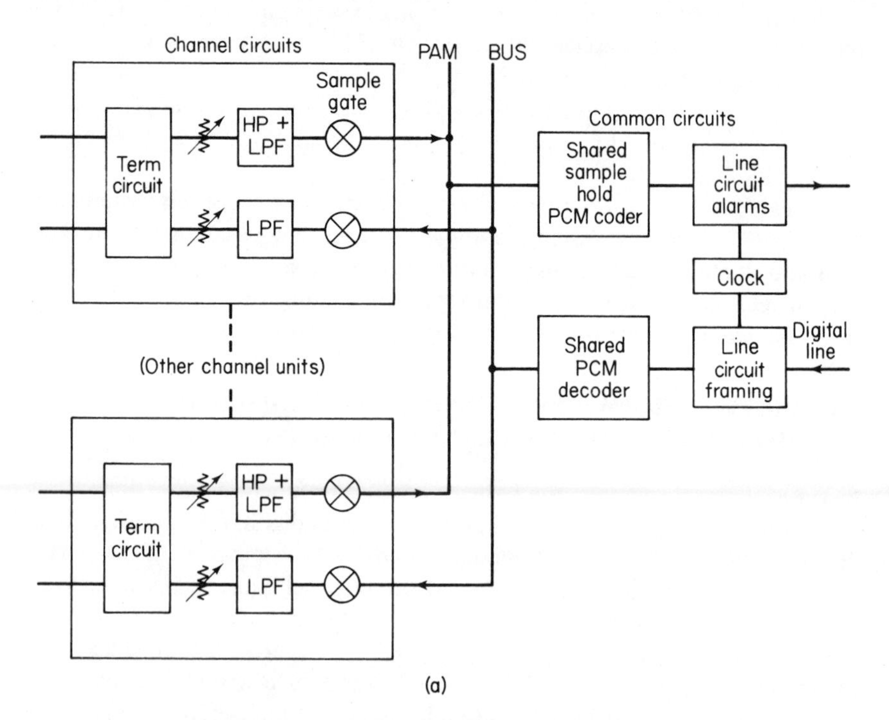

FIGURE 1-2a. Digital channel bank using analog filters.

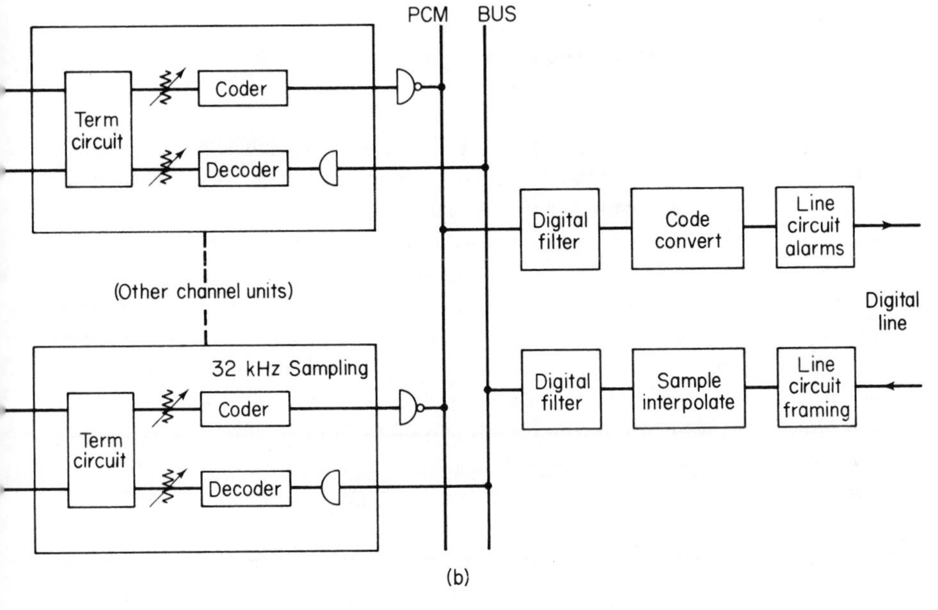

FIGURE 1-2b. Digital channel bank using digital filters.

implemented D bank is shown in Fig. 1-2(b). The advantages of digital implementation are as follows:

1. Multiplexed digital filters will probably be less expensive than their analog counterparts as integrated circuit costs come down.
2. There is no drift or tolerance problem with digital filters.
3. With a digital processor filtering a high sampling rate representation of the signal, a sample interpolator can be added at nominal extra cost, such that the line rate and the sampling rate of the channels need not be synchronized. Stated in a different way, the sampling frequency and desampling frequency can be synchronized even though the transmitting and receiving bit rates are slightly different.
4. The use of a per-line CODEC (coder–decoder) followed by a digital multiplex reduces the chances of interchannel intelligible crosstalk.

The disadvantages of the digitally implemented channel bank over the conventional analog bank are as follows:

1. At present integrated circuit costs, the digital filters are slightly more expensive.
2. Even though digital filters do not drift, they introduce round-off noise, so extra bits (greater precision) must be used to represent the signal internally.

1.6 DIGITAL SIGNAL PROCESSING IN FREQUENCY-DIVISION MULTIPLEX TRANSMISSION TERMINALS

With central offices such as the No. 4 ESS, which switch signals in digital form, it is necessary to digitize any analog signals that must pass through the office. This is true not only for individual voice lines but for any frequency-division multiplex (FDM) trunks that interface with the office, carrying groups of single sideband (SSB) modulated analog signals. An obvious course is to demodulate the SSB signals to baseband by conventional techniques and handle them from there on the same as individual voice lines. However, there exists the opportunity for introducing digital signal-processing techniques, and to do so at a higher level than baseband, by digitizing the composite SSB signal and then doing the demultiplexing digitally. (Only the receiving direction is being described here; it should be clear that in every case a symmetrical process exists for the other direction, i.e., converting digital voiceband signals to SSB.)

The first question to be answered is at what level in the FDM hierarchy should this conversion take place. The standard groupings of voiceband signals in the telephone network are the 12-channel group, the 60-channel supergroup (5 groups), and the 600-channel mastergroup (10 supergroups). The choice will depend on the relative expense and difficulty of encoding and decoding various-sized bundles of SSB signals, the relative cost of the digital processing in each case, and the reliability problem, which is generally made worse as more channels are handled by a single processing entity.

For an experimental system built at Bell Laboratories, the smallest bundle was chosen (i.e. the 12-channel group), although others have suggested using the 60-channel supergroup for this process.[9]

The next major question to be faced is what digital processing algorithm should be used in demultiplexing the SSB signals once they have been digitized? Over the course of time a large number of workable algorithms[9,10,11] have been examined. The final choice will depend strongly on the most efficient match between various algorithms and the digital IC technology available at the time the system is built. Since the technology is constantly evolving, it is difficult to make a prediction of the final outcome at the time of this writing. One approach that has proved very efficient, and that was used in the experimental system referred to above, is a variant to the SSB modulation/demodulation method first proposed by Weaver[12]; it is illustrated in Fig. 1-3.

It is perhaps easier to consider the analog version of Weaver modulation first (the method was, of course, invented for analog systems since Weaver did this work long before the advent of digital signal processing). Also, it is now most convenient to consider the interface in which baseband signals are

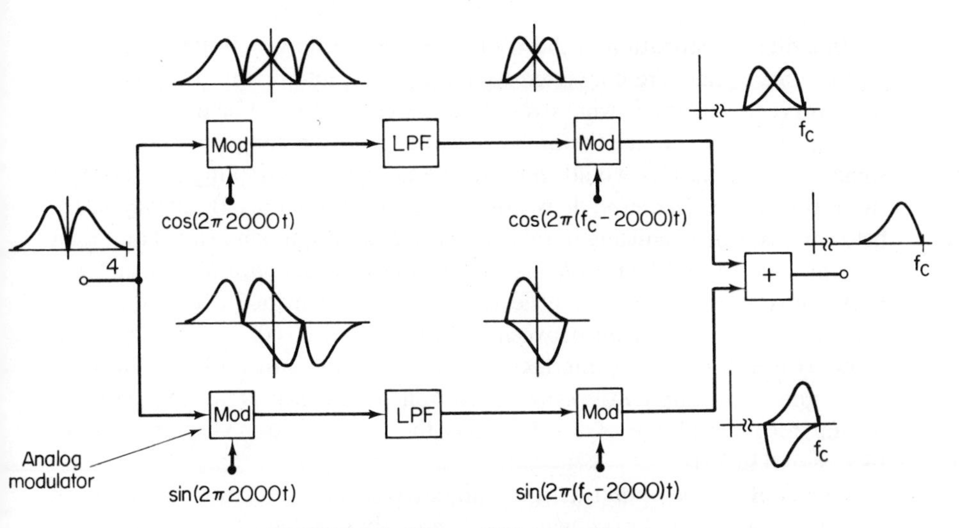

FIGURE 1-3. Weaver SSB generation.

converted to SSB; as before, an exactly symmetrical inverse process exists. We assume each voiceband signal to have a spectrum confined to the band from about 100 Hz to 4 kHz.

The signal is split into two bands and modulated by midchannel (2-kHz) carriers 90 degrees apart in phase, producing the output spectra shown. (These spectral plots are really pictorial representations of Fourier transforms in which overlapping sidebands are depicted as being distinct. This artifice helps one visualize how various sidebands cancel at the appropriate points.) Two identical lowpass filters (LPF) remove energy above 2 kHz. In a second modulation step, the sideband combinations produced thus far are translated to the desired channel position in the groupband (the standard groupband used in telephony consists of twelve 4-kHz slots occupying the frequency space from 60 to 108 kHz). If the two modulated signals are now added together, the unwanted sideband is canceled out. The degree to which this cancellation takes place depends, of course, on how closely the two versions of the rejected sideband underwent identical transmission in the two branches. As a matter of interest, this is precisely the characteristic that makes this type of SSB realization difficult with analog processing.

Stability and repeatability are fundamental attributes of digital circuits, however, and the necessary accuracy can be obtained in a digital version of the Weaver modulator by an appropriate choice of word length to represent the signals. It should be further observed that one advantage of the Weaver approach is that cancellation of the unwanted sideband takes place in the channel under consideration, rather than in an adjacent channel as in other SSB generating methods of this general type. This considerably relaxes the requirements on the degree of cancellation.

In a digital realization of the Weaver modulator in which standard 8-kHz PCM voice signals are converted to analog SSB, additional questions must be answered. The first is what overall sampling rate should be used for representing the signals. Since the highest frequency in the composite groupband signal is 108 kHz, this would indicate the need for a sampling rate at least twice this value. However, it is possible to use the empty frequency space below 60 kHz by producing in this band a pseudo groupband signal and then filtering out the first image of this signal, which occurs naturally in the final D/A conversion process. A little reflection will convince one that a basic sampling rate anywhere between 108 kHz and $2 \times 60 = 120$ kHz would work. However, we must interface initially with standard 8-kHz PCM sampled signals. To simplify the synchronization problem, it is advantageous to use an integer multiple of 8 kHz. Thus the choice is quickly narrowed to $14 \times 8 = 112$ kHz.

One could at this point design a digital Weaver modulator, operating at a sampling rate of 112 kHz, which looks exactly like that of Fig. 1-3 with the analog modulators and filters replaced by digital counterparts. It would, of course, be necessary at the input to change from the 8- to the 112-kHz sampling rate by using a sample interpolation method such as that described by Crochiere and Rabiner.[13] This, however, would entail operating the complicated digital lowpass filters at the 112-kHz sampling rate, when in fact they, and the 2-kHz modulators that precede them, can be operated just as well at the basic 8-kHz input rate. It is thus advantageous to move the 8- to 112-kHz sample rate change to a point just beyond the LPFs, since this produces a 14:1 reduction in the amount of hardware required for these filters. The resulting digital version of the Weaver modulator is shown in Fig. 1-4. Details of the design of this entire system are described in several papers.[10,14]

A rough measure of the hardware complexity of a digital processing system is the number of multiplications per second and the number of bits of memory required. It is instructive to compare these numbers for the various steps in the evolution of the Weaver modulator outlined above. The numbers are for processing a single channel.

Method	*Mult./Second*	*Memory (bits)*
216 kHz Weaver	6,048,000	388
112 kHz Weaver	3,136,000	388
112 kHz Weaver, interpolation after LPF	1,088,000	392
112 kHz Weaver, interpolation after LPF, combined interpolation and final modulation	640,000	400

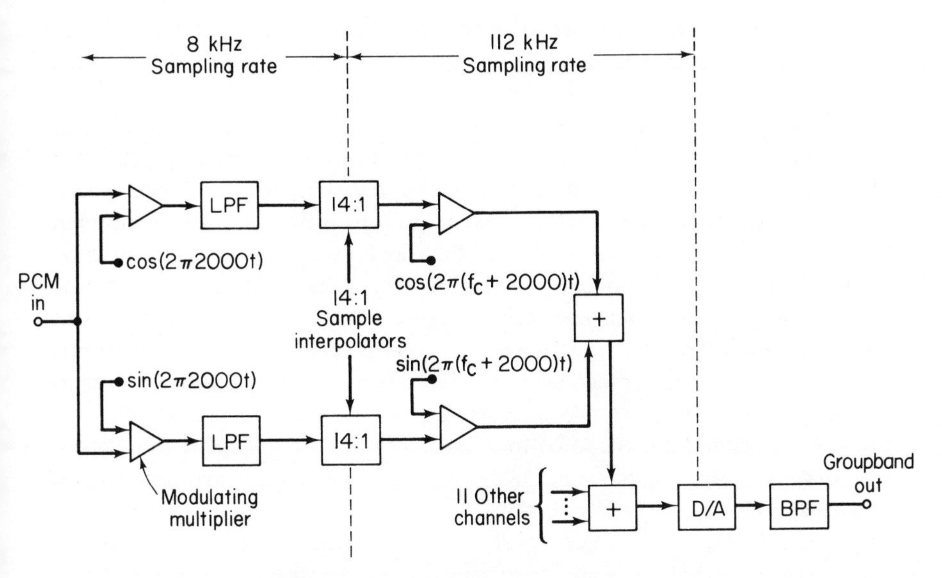

FIGURE 1-4. Digital Weaver modulator.

The fourth entry represents yet another reduction, too complicated to describe in this brief discussion, which takes advantage of the fact that 12 channels are to be added together at the output. It combines the interpolation and final modulation operations in a way mathematically reminiscent of the fast Fourier transform (FFT) algorithm.

1.7 DETECTION OF SIGNALING TONES

Another application of digital processing in the future communications plant is the detection and generation of signaling tones. Telephone central offices forward telephone numbers to each other by means of dc pulses or tone signals. At the receiving central office the signaling tones are detected or the pulses counted to derive the called number. In the case of a digitally implemented central office, it is most economical to use detectors that are implemented using digital techniques which operate on a digital version of the incoming signal. In the case of dial pulses, a simple third-order digital lowpass filter with about 40-Hz cutoff followed by a level detector with hysteresis serves as an adequate detector. A high-speed comparator with no prefiltering might at first seem to be acceptable, but the real world of bouncing relay contacts in the senders and hum pickup in the lines soon convinces the systems designer that digital filtering is necessary for robust operation. In fact, the relays of today's electromechanical telephone systems that detect signaling pulses are built with coil inductances, armature mass, and sometimes lag windings, which make the relay itself act as about a third-order lowpass filter followed by a detector with hysteresis.

Many systems signal with tones instead of dc pulses. TOUCH-TONE® calling telephones send the dialing information to the central offices in a "two tones present of eight possible" format. These tones are arranged such that the detector can make an almost perfect discrimination between speech signals and the tone pair. This is a very important property since, if a person were to dial a digit and then talk into the transmitter, a detector that would respond to speech would cause a dialing error. Such a property is called *talk-off* or *digit simulation* protection. Figure 1-5 shows a block diagram of such a detector. The initial highpass filter rejects both hum (primarily 60 Hz) and dial tone that is present when receiving the initial digit. Then the high-passed signal is fed into two band-rejection filters, which are arranged to exclude either the high group of four tones or the low group of four tones. The output of these band-reject filters is then limited and passed on to two groups of four bandpass filters, which select the eight tones.

If the signal is a valid digit signal composed of the sum of two pure tones, then through the action of the band-rejection filters each limiter is excited by a single tone and produces a symmetrical square wave at its output. If, on the other hand, a speech signal excites the detectors, the limiter inputs will be complicated signals, and the limiter output transitions will in general be very irregular in time. If the outputs of the tone-detection bandpass filters are matched to thresholds that are just a few decibels below the outputs expected with square wave out of the limiters, then the speech excitation that produces complicated signals out of the limiters will not produce overthreshold outputs in exactly two of the eight channels, which would simulate a digit. Thus the dynamic range and talk-off protection properties are the result of the limiter action and the rather exact threshold operation on the tone channels.

An experimental digital version of the TOUCH-TONE calling receiver has been implemented on a programmable digital signal processor constructed at Bell Laboratories. The processor is a 128-way multiplexed version of a second-order digital filter. In addition, it has scaling and nonlinear functions, such as limiting and rectification, that are selected by the program, which also sets the coefficients for the 128 filter sections. Figure 1-6 shows the second-order section configuration of this processor. Since the TOUCH-TONE calling receiver implementation requires only 24 sections of the 128 sections available, five TOUCH-TONE calling receivers can be simultaneously implemented by the processor. Figure 1-5 shows the section number in the circle next to the boxes representing second-order filter sections. This is another illustration of the power and flexibility of multiplexing.

It is in the areas of limiter performance and threshold stability that the digitally implemented detector shows differences from its analog counterpart.

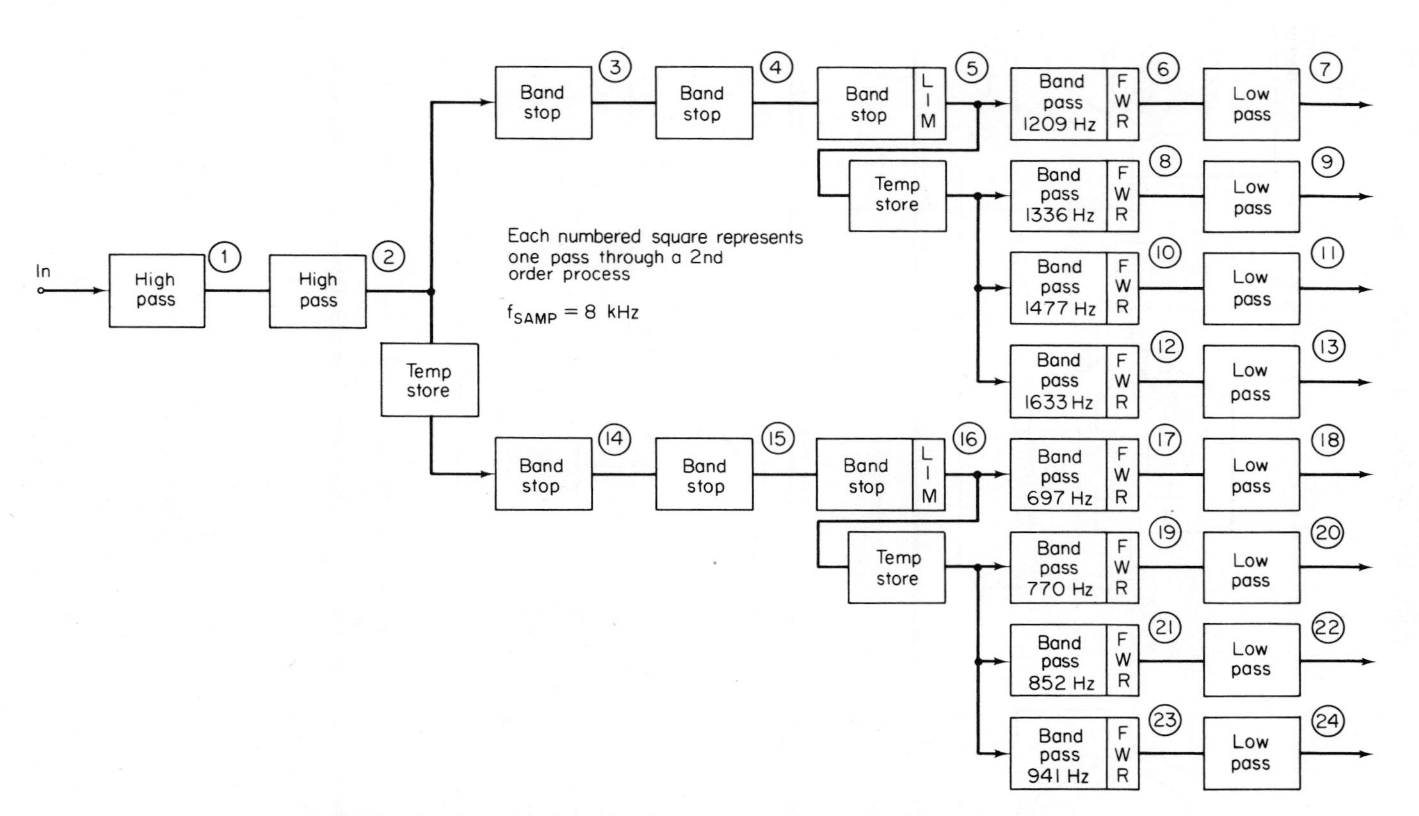

FIGURE 1-5. Block diagram of a Touchtone® receiver using 24 passes through a second-order digital filter processor.

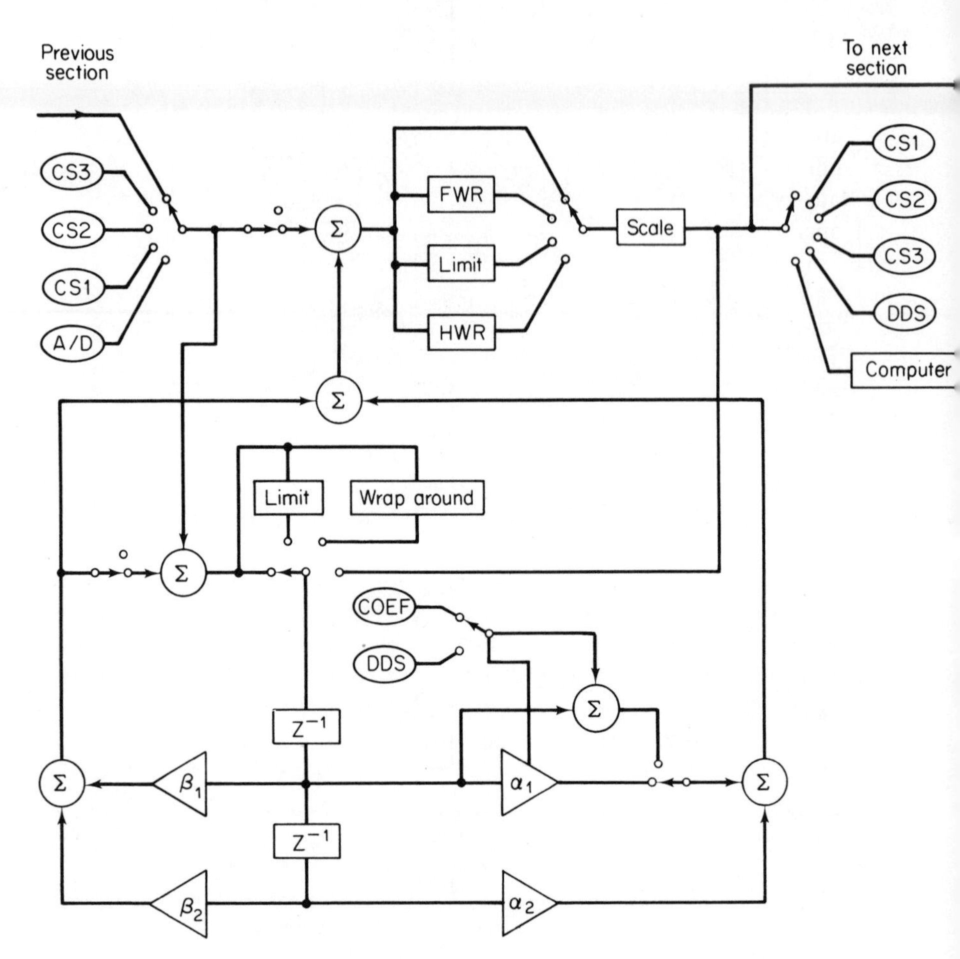

FIGURE 1-6. Block diagram of a programmable digital signal processor comprised of a multiplexed second-order digital filter.

Probably the most critical parameter in the system is the level adjustment of the threshold circuit after the tone detectors. Too low a level reduces the talk-off margin. Too high a level decreases the bandwidth of the tone filters, and therefore increases the sensitivity to tone frequency or filter response drift. In the digital implementation, there is no drift in the threshold levels or in the filter characteristics, since digital calculations are exact and very accurate. The limiter dynamic range is as good as the A/D converter in the system, since limiting is simply the selection of the sign bit. In general, digital implementation yields a system that is exceptionally stable and accurate.

1.8 ALIASED HARMONIC DISTORTION: A NONLINEAR PHENOMENON

There are, however, impairments in digitally implemented systems that make them different from their analog counterparts. Nonlinear operations in sampled data systems tend to generate new spectral components, just as nonlinear operations in continuous systems do. In the sampled data system, the images of the new spectral components reflected about the sampling frequency are also present. If the new frequency components are greater than half of the sampling rate, they are aliased back into the baseband. Typically, a nonlinearity such as a limiter or a full-wave rectifier will produce new spectral components that are harmonics of a periodic input signal. In the case of the TOUCH-TONE calling receiver, the 1336-Hz tone causes trouble in that the limiter produces energy at the fifth and seventh harmonics (6680 and 9352 Hz), which alias with the 8-kHz sampling rate to produce 1320- and 1352-Hz components. When these components are summed with the original 1336-Hz tone, they appear approximately as a 16-Hz amplitude modulation of about 17% of the carrier (1336 Hz) amplitude. To operate the receiver with a threshold suitable for both talk-off protection and tone bandwidth, the threshold for that channel must be reduced by 1.4 decibels (dB).

Distortions due to aliasing modulation products of nonlinear operations on the samples are fundamental to all sampled data systems, and it is vital that they be understood, particularly for detector designs. These effects, called *aliased harmonic distortion*, are very important when working with detectors for communications systems, because the signals are usually a large percentage of the Nyquist frequency (half of the sampling rate), and frequently the harmonics fall out of band.

A detection algorithm for the popular 103–113 type FSK (frequency shift keying) data set often used with computer terminals demonstrates how these distortion effects can be overcome. The 103–113 data set communicates both ways (full duplex) by using a high pair of tones (2025 and 2225 Hz) for one direction, and a low pair of tones (1070 and 1270 Hz) for communicating in the reverse direction.

The detection of the 2025-Hz tone presents a problem owing to the fourth harmonic (8100 Hz) causing a 100-Hz difference with an 8-kHz sampling rate. If the detector of Fig. 1-7, consisting of a pair of bandpass filters, full-wave rectifiers followed by a subtractor, and a lowpass filter, is used, then when a 2025-Hz sine wave is applied to the input, the lowpass filter output dc contains a 50% ripple at 100 Hz. The full-wave rectifier produces dc with 13.3% ripple owing to the aliased fourth harmonic, but the subtraction process in the discriminator reduces the net dc such that a 57% ripple results.

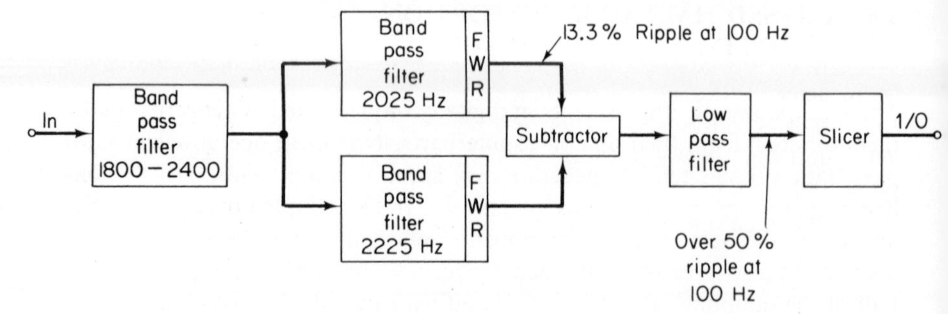

FIGURE 1-7. An FSK Detector based on a discriminator showing the effects of aliased harmonic distortion.

This makes the digital version of the detector very marginal and therefore unsuitable.

A second type of FSK detector is based on a zero crossing counting technique in which a standard area pulse is generated at each carrier zero crossing. These pulses are averaged by a lowpass filter whose output is sliced at an appropriate level ($f_m = 2125$ Hz) to produce a mark or space signal. A block diagram of such a detector is shown in Fig. 1-8. The 8100-Hz harmonic from

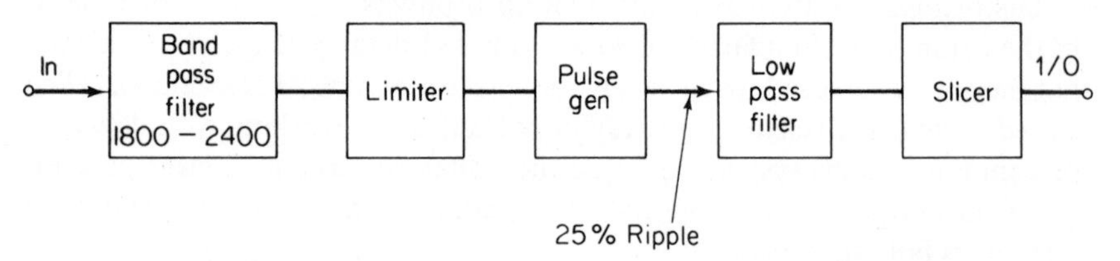

FIGURE 1-8. An FSK Detector based on a zero crossing and smoothing technique showing the effects of aliased harmonic distortion.

the limiter aliases about the 8000-Hz sampling frequency to produce a 100-Hz ripple in the lowpass filter output as before. Because of the additional nonlinearities in the pulse formation, it is difficult to calculate the ripple amplitude from a Fourier analysis of the limiter output. However, a time-domain analysis makes the calculation simpler. At 2025 Hz, the limiter should have zero crossings every 494 microseconds (μs), and if the sampling period is 125 μs, there should be 3.95 samples in each period of the carrier. What actually happens is that there are four samples in a period most of the time, but sometimes there are only three. The periods with three samples occur every $\frac{1}{100}$ s and cause a peak change of 25% in the input of the lowpass filter. This extra signal interferes with the detection process, and this type of FSK detection algorithm must also be rejected.

There is yet another way to detect FSK signals using a delay and a product demodulator, as shown in Fig. 1-9. If the delay is adjusted to be an odd multiple of 90 degrees at the center frequency between the mark and space frequencies (2125 Hz), the smoothed product of the input and the delayed signal will be positive for mark and negative for a space frequency. Since delay in a sampled data signal can be achieved easily by sample storage, a phase shift of 95.6 degrees at 2125 Hz results from a single sample storage. By inserting a single highpass section in the other input to the product demodulator, a 5.6-degree phase shift is easily obtained to trim the phase difference to 90 degrees at 2125 Hz.

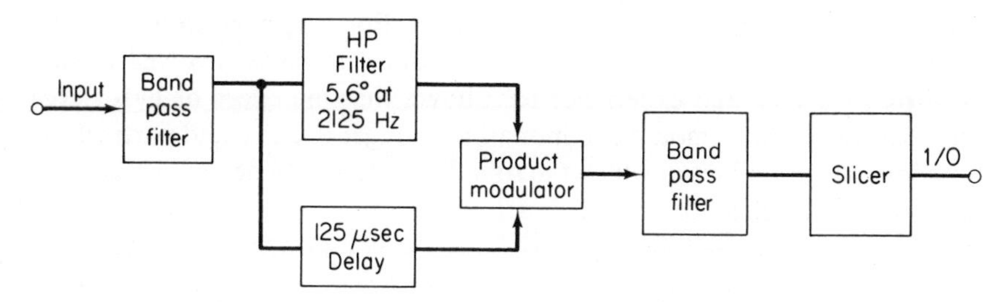

FIGURE 1-9. An FSK detector based on a product modulator type of detection which is unaffected by aliased harmonic distortion.

The product modulator is a nonlinear device and generates new spectral components which may alias back into baseband and interfere with the detection process. However, a product of a sinusoidal input with a delayed version of itself generates only a second harmonic, which in the case of the FSK data set falls at about 4050 to 4430 Hz, which alias into 3950 and 3550 Hz, respectively. These harmonics are removed by the lowpass filter prior to the slicer and thus do not interfere with the detection process. A product detector requires a larger dynamic range on the slicer than do the limiter-based or the rectifier-based algorithms, as an input change of 15 dB will produce a 30-dB change in the product modulator output. Fortunately, it is simple and relatively inexpensive to increase the internal dynamic range of digital systems by adding extra bits to the basic word length. This detection algorithm is therefore suitable for a digital implementation.

1.9 ECHO CONTROL

Another potential application of digital signal processing in the telephone industry is in the control of echos in long-distance transmission circuits. For economy, the subscriber loop plant (that part of the telephone network which connects each individual subscriber to the central office serving him) is two-wire (i.e., both the incoming and outgoing signals are carried by a single pair of wires). This is accomplished by making the subscriber's telephone set, the connecting wire pair, and the termination (hybrid coil) in the central office form a balanced bridge network such that the signals traveling in the two directions can be separated by the hybrid coil at the two- to four-wire conversion point in the central office (see Fig. 1-10). Ideally, this separation can be carried out to a high degree of precision. In practice, however, it is not economically feasible to dedicate a hybrid coil at the central office to each subscriber, and subscriber lines are not uniform in impedance. Thus in many cases the effective imbalance of the aforementioned bridge circuit can be quite large, causing a substantial portion of the distant talker's signal to be returned to him in the form of an echo. For telephone calls over distances of up to a few hundred miles, the round-trip delay involved is only a few milliseconds, and experience has shown that the effect of even large echos delayed by this amount is minimal. For longer circuits, and particularly those involving satellites in which the round-trip delay can be several hundred milliseconds, the effect of echos is much more severe.

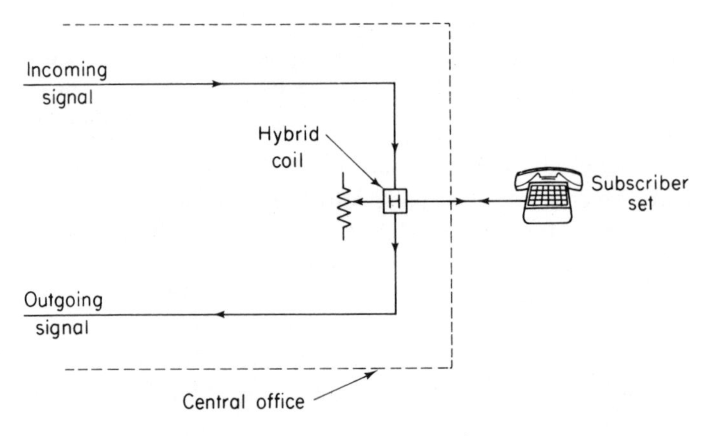

FIGURE 1-10. 2/4 wire conversion.

For circuits of up to a few thousand miles in length (a few tens of milliseconds of round-trip delay), the echo problem is customarily handled by devices known as *echo suppressors*. An echo suppressor is inserted in a circuit at any point where access to both directions of the conversation is available. It functions by detecting which direction of the conversation is active at any one time and inserting attenuation in the opposite (idle) direction, thus breaking the round-trip path and eliminating (or at least greatly attenuating) the echo. It should be clear that this method of echo control introduces its own form of distortion, that is, the clipping of parts of utterances when both parties talk at the same time. Nevertheless, echo suppressors are a very workable solution to the problem over a geographical area the size of the continental United States.

For longer circuits, however, the increased round-trip delay aggravates the double-talk problem to the point where more sophisticated means of controlling echos must be resorted to. The most promising solution to date involves the use of an *echo canceler*. Such a device and its arrangement in the network are depicted in Fig. 1-11. It consists basically of a transversal (FIR) filter connected between the two talking paths as shown. Using the incoming signal as input, the taps of the filter are adjusted to replicate as closely as possible the portion of the signal that leaks through the hybrid coil. Although echo cancelers can certainly be built with analog components, the digital approach is receiving considerable attention because it provides high stability, accurate tap weights that can be rapidly adjusted, and because the digital circuit can be time shared over many channels simultaneously.

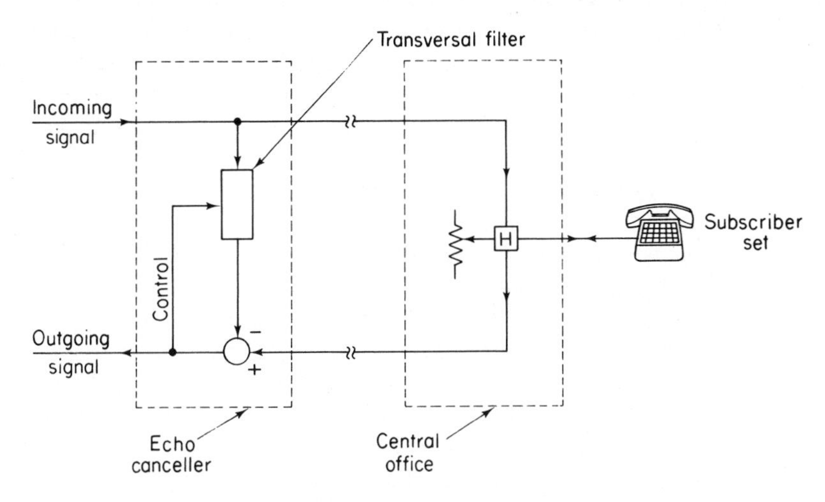

FIGURE 1-11. Echo canceler.

Whether realized in analog or digital form, however, an echo canceler is sufficiently expensive so that one cannot be dedicated to each voice circuit. Indeed, since the proportion of calls with round-trip delays long enough to require a canceler is quite small compared to the total number of calls, each canceler will be required to serve a large number of customers on a demand basis. This means in turn that the canceler's transversal filter must be complex enough (i.e., have enough taps) to cope with a myriad of different possible trans-hybrid transfer functions. Although an accurate measure of the necessary complexity is not yet available, it is estimated that a transversal filter of perhaps 200 taps having tap weight accuracies of the order of 10 bits will be required to provide the 35- to 40-dB echo attenuation deemed necessary for satisfactory performance. The estimate of the number of taps assumes that the canceler, although centrally located to serve many users, is nevertheless not too distant (within a few hundred miles) from the associated two- to four-wire conversion point. Otherwise, the required number of taps, a portion of which must be used to compensate for the round-trip delay between the canceler and the hybrid coil, would be prohibitively large. This implies, of course, that two cancelers must be used, one near each end of the conversation.

When operated in the manner described, the canceler must be capable of adjusting its tap weights quickly (on the order of a few hundred milliseconds) as each new talking path is set up, using only the signals that occur naturally during this time. The requirement of a special initial training period would unduly complicate the existing long-distance telephone network and, furthermore, would preclude the canceler's being able to follow slow changes in the circuit throughout the course of a conversation. Fortunately, a method of rapidly adjusting the tap weights in the desired manner exists, the adaptive algorithm first described by Sondhi.(15) This method, in effect, performs a running correlation between the incoming and outgoing signals of Fig. 1-11, and moves the tap values in an efficient manner to drive the mean-square value of the latter to a minimum. This adaptive scheme is very similar to the ones used by Lucky(16) for setting the taps of adaptive equalizers for data transmission.

Unfortunately, the adaptive tap-setting algorithm works only when the incoming signal is present and the outgoing signal, except for the echo, is absent. Energy generated by the near-end user severely compromises the algorithm, and, in fact, will often cause a properly adjusted filter to diverge markedly from its optimum setting. The most straightforward solution to this problem is simply to disable the tap-adjusting mechanism, and freeze the tap values during periods when near-end generated signal is present. This necessitates quickly recognizing the difference between energy that is correlated, albeit in a complicated manner, with the incoming signal and energy that is uncorrelated. This task is in no way simple since information about the near-end signal alone is normally not available to the canceler. A number of promising approaches have been suggested, but this is still an area under active investigation.

1.10 DESIGN CONSIDERATIONS FOR DIGITAL SIGNAL-PROCESSING HARDWARE ELEMENTS

The examples in this chapter are all drawn from voice-frequency applications; there are no examples of applying digital signal processing to video- or carrier-frequency systems. There are several reasons for this choice of applications. First, there are more opportunities in voice-frequency circuits, because there are so many more of them than any other by a large factor owing to the nature of the telecommunication systems. Second, because of the relatively low bandwidth signals to be processed compared to the present-day logic speeds of the integrated circuits, the digital clocks controlling the signal processors can be made thousands of times faster than the per-channel sampling rate. But multiplexes of thousands of channels are rare in the telephone network, and opportunities to exploit these large multiplexing factors are limited. Fortunately, this discrepancy does not lead to a dilemma, because serial-data, as opposed to parallel-data, implementations(17) can lead to trade offs that use more clock periods for calculations at significant savings in power and chips in the arithmetic processing units. For these reasons, the hardware discussions center on highly multiplexed (32× to 128×) processors that are usually implemented using serial arithmetic.(18)

1.10.1 Choosing a Structure

When faced with the design of signal processor systems, the designer usually concentrates on the digital filter aspects of the system first, because most of the hardware, including storage and arithmetic processing, is dominated by the filtering portion of the signal-processing algorithm. The nonlinear processes usually require very simple hardware for implementation compared to the multipliers and delay elements of the digital filters. In the detectors considered for the FSK demodulator, the limiting, full-wave rectification or product demodulation functions are all implemented with a fraction of the number of gates used to implement the digital filters.

There are several structures (cascade, direct form, parallel, etc.) of digital filters to choose from. The choice depends on the parameters of the filter, the type of arithmetic unit implementation, and the type of control or programming desired. In the case of fixed sets of identical filter sections, such as found in the PCM channel bank example, a read only memory (ROM) type multiplier(19,20) implementing a cascaded set of second-order sections seems promising. Since each filter is identical for each time slice, the coefficients are fixed, and the arithmetic processing can be broken down into two ROM-driven sums of products per second-order section.

At first it might appear attractive to implement higher-order filters in the direct form, since the ROM realization of sums of products works even more efficiently for four or five terms than for three. However, for most digital filters that have significant gain, a third order or higher direct realization can, with the proper initial conditions, sustain large-amplitude limit cycles(21) even with saturation arithmetic. High gain can result from either high Q poles or poles that are located at a small fraction of the sampling rate. For many systems where the initial conditions necessary for such limit cycles could come about during power turn on, such a filter would be unacceptable.

If the filter is not needed in multiple identical copies, but is instead required to be programmable as in the test equipment example, a two-input multiplier must be used. Although there are structures other than cascaded second-order sections, for programmable filters the second-order section seems appropriate. In the SSB terminal example,[14] both recursive second-order sections and finite impulse response (FIR) transversal filters are used. In that example the transversal filter implementation could capitalize on the fact that 13 of each 14 input samples are zero. In general, it is our experience that the FIR filters will find most applications in the cases where a change of sampling rate is desired such that simplifications can be exploited in implementation.[22]

There are other cases when the cascaded second-order sections are not adequate. In cases where a section has very high Q poles (to achieve sharp cutoff), or a very low pole frequency, the gain per second-order section can be very large. In one example of a lowpass filter designed to shape dial pulses, the 40-Hz second-order lowpass filter operated at a 32-kHz sampling rate had a gain of over 15,600, or about 14 bits. Thus, for an input data word length of 16 bits, the output would be quantized to only 2 bits (16 minus 14)! To prevent such a coarse quantization of the output, a structure was chosen that could be scaled after each order of filtering (i.e., a pair of coupled first-order sections). In the example of the dial pulse filter, each stage had 7 bits of gain, so the output signal was quantized to 9 bits instead of 2 bits.

Scaling to prevent undue coarseness of signal quantizing is a very important aspect of digital filter implementation.[23-25] It is related to the choice of data word length and to structure, as is seen in the preceding example. Although the scaling does not influence the characteristics of the filter, it plays a major role in determining the signal-to-noise ratio of the filter and the related idle-channel noise performance.

A related problem is the assignment of pole and zero factors to the various stages of a cascaded filter. There are several published search procedures[25-32] for finding minimum noise assignments, and the reader should consult the references for them. Proper section ordering is an important step in the implementation as a chance or casual assignment can result in the loss of many bits of signal accuracy.

There are other structures possible; second-order sections can be paralleled, series paralleled, or combined in many ways that all ensure real coefficients. However, there does not seem to be any major advantage to combining second-order sections in other than a cascade form. Another structure is the wave structure,[33,34] which is recommended because of low coefficient sensitivity properties. However, the wave implementations proposed thus far seem to be of limited practical usefulness because they are difficult to multiplex.

1.10.2 Coefficient Sensitivity

The questions of coefficient sensitivity turn out to be relatively unimportant in the filters used as examples for this chapter. In general, the length of the data word used to ensure suitable output signal-to-noise ratio and output idle-channel noise exceeded the length of the coefficients required for the desired accuracy in the filter characteristic. For example, the lowpass filters in the PCM terminal example required only 8 bits of coefficient accuracy to ensure a frequency characteristic within ± 0.2 dB of ideal specifications.

1.10.3 Design Aids

The designer of a digital processing system is faced with many choices and decisions about sampling rates, word lengths, structures, types of logic, and trades of gates of arithmetic logic for memory.(18,35) The better designs are not too difficult to obtain, however, because of the simple property of a digital system that it can be simulated exactly on a digital computer. In fact, the simulation is a perfect prediction of how each and every copy of the final system will perform. If good design programs for designing and testing algorithms are coupled with good simulation programs, the digital system can be made with low development costs and low risk of poor performance. Good machine aids are the key to good digital processor design, and, of course, good design yields good systems.

1.11 SOME USES OF THE PROGRAMMABLE DIGITAL SIGNAL PROCESSOR

The flexibility of a programmable digital signal processor, which can be programmed to any of a large number of tasks, has significant implications on test instrumentation for communications systems. One instrument can perform all the voice-frequency tests now done by over a dozen different types of test sets. It can also serve as a computer terminal to access a time-sharing computer for additional information, for logging test results, and for retrieving new trouble information.

The digital filter processor shown in block diagram form in Fig. 1-6 has been programmed to perform the following tasks:

1. C-message weight noise meter.
2. Tone generator with very high accuracy.
3. Tone sequence generator for testing signaling receivers.
4. Sweep oscillator.

5. TOUCH-TONE calling receiver.
6. MF tone receiver.
7. Ringer current filter and meter.
8. Data sets.
9. Impulse noise generator.
10. Impedance measuring set.
11. Voice synthesizer.
12. Music generator.[36,37]

The list will grow as time passes, since the limits to what a programmable machine can do are immense. By combining the filter with a more conventional processor organization, such as is found in a microprocessor, the fast fine-structured calculations on individual samples are done in a specialized arithmetic unit, while the sequential control and reconfiguration on particular subtests are done in a more conventional computing sense. Figure 1-12 shows a photograph of a specialized arithmetic unit for realizing 1,024,000 second-

FIGURE 1-12. A hardware implementation of a multiplexed second-order digital filter capable of realizing 1,024,000 second-order section calculations per second.

order section calculations per second. The unit is designed to be directly connected to the memory bus of a commercially available microprocessor system. This processor was designed by H.G. Alles of Bell Laboratories.

Imagine an attaché case with a small cathode-ray tube (CRT) that swings, up, a keyboard, a small tape cassette, and a few test leads that connect up to the equipment to be tested, as shown in Fig. 1-13. There are several impor-

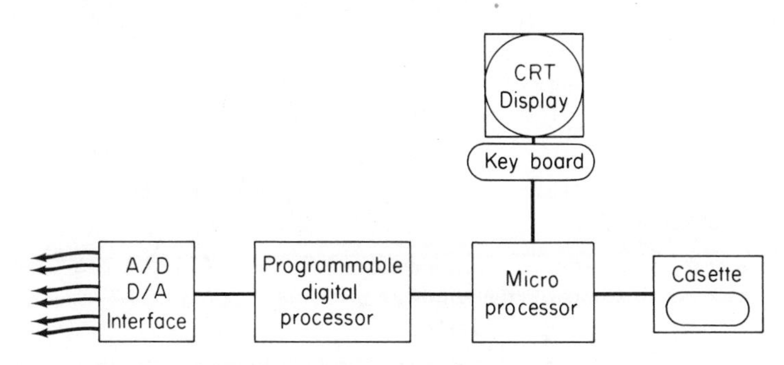

FIGURE 1-13. A programmable test set.

tant tasks that such a system could do in a far superior manner than a craftsman with ordinary instruments. Balancing a hybrid on a two- to four-wire conversion is a good example. It is presently done by adjusting taps on a capacitor block until minimum return signal at a single frequency is observed. In a programmable system, return measurements at many frequencies could be made in a few seconds, and an optimum tap choice could be determined without a lot of trial and error by the operator. In addition, such parameters as hum and noise and impedance ranges could be checked. New tests for new equipment are added by changing only the program. The system can be made self-calibrating. Such a system has been demonstrated in the laboratory; as the cost of the terminal-oriented large-scale integrated circuits (LSI) comes down, the appearance of such a set in the field is only a matter of time.

1.12 SUMMARY

The move toward increased digital transmission facilities and digital switching equipment in telecommunication systems has opened up many new opportunities for digital signal processing. Digital systems have the desirable properties of small size, low power, noise immunity, increased reliability, and a reduced number of parts and interconnection wiring. In this chapter a brief look was taken at a number of potential applications of digital signal-processing techniques in the communications plant. These applications included PCM transmission terminals, FDM transmission terminals, detection of signaling tones, and echo control. Considerations in both the design and the use of a multiplexed programmable digital signal processor were discussed in detail.

With the rapid development of large-scale integrated circuits and now microprocessors, one can expect their appearance as an integral part of the communications plant of the 1980s.

REFERENCES

1. W. E. DANIELSON, "Exchange Area and Local Loop Transmission," *Bell Lab. Record*, vol. 53, no. 1, Jan. 1975, pp. 40–49.
2. H. KANEKO, "A Unified Formulation of Segment Companding Laws and Synthesis of Codecs and Digital Compandors," *Bell System Tech. J.*, vol. 49, no. 7, Sept. 1970, pp. 1555–1588.
3. E. F. O'NEILL, "Radio and Long Haul Transmission," *Bell Lab. Record*, vol. 53, no. 1, Jan. 1975, pp. 50–59.
4. H. E. VAUGHAN, "An Introduction to No. 4 ESS," *Proc. 1972 Intern. Switching Symp.*, Cambridge, Mass., 1972, pp. 19–25.
5. H. S. MCDONALD, "An Experimental Digital Local System," *Proc. 1974 Intern. Switching Symp.*, Munich, Germany, 1974, pp. 212.1–212.5.
6. J. H. CONDON and H. T. BREECE, III, "Low Cost Analog–Digital Interface for Telephone Switching," *Proc. 1974 Intern. Conf. on Comm.*, Minneapolis, Minn., 1974, pp. 13.B.1–13.B.4.
7. H. G. ALLES, J. H. CONDON, W. C. FISCHER, and H. S. MCDONALD, "Digital Signal Processing in Telephone Switching," *Proc. 1974 Intern. Conf. on Comm.*, Minneapolis, Minn., 1974, pp. 18.E.1–18.E.2.
8. H. S. MCDONALD, "Impact of Large-Scale Integrated Circuits on Communications Equipment," *Proc. Natl. Electron. Conf.*, vol. 24, 1968, pp. 569–572.
9. M. G. BELLANGER and J. L. DAGUET, "TDM–FDM Transmultiplexer: Digital Polyphase and FFT," *IEEE Trans. Comm. Tech.*, vol. COM-22, no. 9, Sept. 1974, pp. 1199–1205.
10. S. L. FREENY, R. B. KIEBURTZ, K. V. MINA, and S. K. TEWKSBURY, "Systems Analysis of a TDM–FDM Translator/Digital A-Type Channel Bank," *IEEE Trans. Comm. Tech.*, vol. COM-19, no. 6, Dec. 1971, pp. 1050–1059.
11. C. F. KURTH, "SSB/FDM Utilizing TDM Digital Filters," *IEEE Trans. Comm. Tech.*, vol. COM-19, no. 1, Feb. 1971, pp. 63–71.
12. D. K. WEAVER, JR., "A Third Method of Generation and Detection of Single-Sideband Signals," *Proc. IRE*, vol. 44, no. 12, Dec. 1956, pp. 1703–1705.
13. R. E. CROCHIERE and L. R. RABINER, "Optimum FIR Digital Filter Implemen-

tations for Decimation, Interpolation and Narrow Band Filtering," *IEEE Trans. Acoust. Speech Signal Processing*, vol. ASSP-23, no. 5, Oct. 1975, pp. 444–456.

14. S. L. Freeny, R. B. Kieburtz, K. V. Mina, and S. K. Tewksbury, "Design of Digital Filters for an All Digital Frequency Division Multiplex–Time Division Multiplex Translator," *IEEE Trans. Circuit Theory*, vol. CT-18, no. 6, Nov. 1971, pp. 702–711.

15. M. M. Sondhi, "An Adaptive Echo Canceller," *Bell System Tech. J.*, vol. 46, no. 3, Mar. 1967, pp. 497–511.

16. R. W. Lucky, "Techniques for Adaptive Equalization of Digital Communication Systems," *Bell System Tech. J.*, vol. 45, no. 2, Feb. 1966, pp. 255–286.

17. L. B. Jackson, J. F. Kaiser, and H. S. McDonald, "An Approach to the Implementation of Digital Filters," *IEEE Trans. Audio Electroacoustics*, vol. AU-16, no. 3, Sept. 1968, pp. 413–421; also in *Digital Signal Processing*, Rabiner and Rader, eds., IEEE Press, New York, 1972, pp. 210–218.

18. S. L. Freeny, "Special-Purpose Hardware for Digital Filtering," *Proc. IEEE*, vol. 63, no. 4, Apr. 1974, pp. 633–648.

19. A. Croisier, D. J. Esteban, M. E. Levilon, and V. Riso, "Digital Filter for PCM Encoded Signals," U.S. Patent No. 3777130, Dec. 4, 1973.

20. A. Peled and B. Liu, "A New Hardware Realization of Digital Filters," *IEEE Trans. Acoust. Speech Signal Processing*, vol. ASSP-22, no. 6, Dec. 1974, pp. 456–462.

21. Debasis Mitra, "Large Amplitude, Self-Sustained Oscillations in Difference Equations Which Describe Digital Filter Sections Using Saturation Arithmetic," *IEEE Trans. Acoust. Speech Signal Processing*, vol. ASSP-25, no. 2, Apr. 1977, pp. 134–143.

22. R. E. Crochiere and L. R. Rabiner, "Further Considerations in the Design of Decimators and Interpolators," *IEEE Trans. Acoust. Speech Signal Processing*, vol. ASSP-24, no. 4, Aug. 1976, pp. 296–311.

23. L. B. Jackson, "On the Interaction of Round-off Noise and Dynamic Range in Digital Filters," *Bell System Tech. J.*, vol. 48, no. 2, Feb. 1970, pp. 159–184.

24. L. B. Jackson, "Roundoff-Noise Analysis for Fixed-Point Digital Filters Realized in Cascade or Parallel Form," *IEEE Trans. Audio Electroacoustics*, vol. AU-18, no. 2, June 1970, pp. 107–122.

25. L. B. Jackson, "Roundoff Noise Bounds Derived from Coefficient Sensitivities for Digital Filters," *IEEE Trans. Circuits Systems*, vol. CAS-23, no. 8, Aug. 1976, pp. 481–485.

26. S. Y. Hwang, "On Optimization of Cascade Fixed-Point Digital Filters," *IEEE Trans. Circuits Systems*, vol. CAS-21, no. 2, Mar. 1974, pp. 163–166.

27. W. S. Lee, "Optimization of Digital Filters for Low Roundoff Noise," *IEEE Trans. Circuits Systems*, vol. CAS-21, no. 3, May 1974, pp. 424–431.

28. E. Lueder, H. Hug, and W. Wolf, "Minimizing the Roundoff Noise in Digital Filters by Dynamic Programming," *Frequenz*, vol. 29, no. 7, July 1975, pp. 211–214.

29. B. Liu and A. Peled, "Heuristic Optimization of the Cascade Realization of Fixed Point Digital Filters," *IEEE Trans. Acoust. Speech Signal Processing*, vol. ASSP-23, no. 5, Oct. 1975, pp. 464–473.

30. K. Hirano, H. Sakaguchi, and B. Liu, "Optimization of Recursive Cascade Filters," *Proc. 1976 IEEE Intern. Conf. on Acoust. Speech Signal Processing*, Philadelphia, Pa., 1976, pp. 513–516.

31. T. R. Lapp and R. A. Gabel, "An Algorithm for Optimally Ordering the Sections of a Cascade Digital Filter," *Proc. 1976 IEEE Intern. Conf. on Acoust. Speech Signal Processing*, Philadelphia, Pa., 1976, pp. 517–520.

32. K. Steiglitz and B. Liu, "An Improved Algorithm for Ordering Poles and Zeros of Fixed-Point Recursive Digital Filters," *IEEE Trans. Acoust. Speech Signal Processing*, vol. ASSP-24, no, 4, Aug. 1976, pp. 341–343.

33. A. Fettweis, "Digital Filter Structures Related to Classical Filter Networks," *Arch. Elek. Ubertragung.*, vol. 25, no. 2, Feb. 1971, pp. 79–89.

34. A. Sedlmeyer and A. Fettweis, "Digital Filters with True Ladder Configuration," *Intern. J. Circuit Theory Appl.*, vol. 1, no. 1, Mar. 1973, pp. 5–10.

35. J. D. Heightley, "Partitioning of Digital Filters for Integrated Circuit Realization," *IEEE Trans. Comm. Tech.*, vol. COM-19, no. 6, Dec. 1971, pp. 1059–1063.

36. H. G. Alles, "A Hardware Digital Music Synthesizer," *EASCON'75 Record*, 1975, pp. 217-A to 217-C.

37. H. G. Alles, "The Teaching Laboratory General Purpose Digital Filter Music Box," *IEEE Proc. 1975 ISCAS*, 1975, pp. 387–389.

2

DIGITAL PROCESSING IN AUDIO SIGNALS

B. Blesser

Massachusetts Institute of Technology
Cambridge, Mass. 02139

and

J. M. Kates

Teledyne Acoustic Research
Norwood, Mass. 02062

2.1 INTRODUCTION

Audio engineering encompasses the recording, storage, transmission, and reproduction of signals to which people listen. In practice, such signals are predominantly natural music, although such signals as electronic music, bird calls, theatrical performances, and underwater sounds must be also included. Unlike digitally processed speech, which must be subjected to an intelligibility test, much digitized audio is required to meet fidelity criteria as well. These fidelity criteria are necessarily subjective, since the final test of quality is based on the perception of listeners. For this reason, human perception is frequently considered in this chapter; the translation of its limits into technical parameters is an overriding concern of the audio engineer. Because of the commercial importance of music reproduction, much of the professional interest in digitized audio systems concerns music. Throughout the following chapter, we shall treat digitized music as representative of the broader class of signals classified as audio.

Throughout its history, audio technology has been interdisciplinary, drawing from chemistry and physics, especially electronics, magnetics, and acoustics. Digital signal processing, perhaps nearest mathematics in principle, is the newest discipline to be absorbed into the audio family; many engineers expect it to produce a quantum jump in audio system performance. Although digital signal-processing techniques are just beginning to be applied to audio, many inherent advantages are already apparent. Nevertheless, at the time of writing, this field is still in its infancy; many of the more sophisticated digital-processing techniques have not yet found applications in audio systems. This will, no doubt, change in the near future.

The need for digital processing in audio may not be readily apparent until one has considered some of the difficulties encountered in bringing a musical performance into the listener's living room. The reproduction chain from the microphone to the loudspeaker is extremely long. There may be as many as 100 separate systems in the chain, each performing a necessary function, but also introducing its own form of degradation. Not uncommonly, each musical instrument in an ensemble is recorded on a single track of a multitrack tape recorder with as many as 24 tracks. This process gives the sound engineer a high degree of flexibility, such as allowing the rerecording of a single instrument part when necessary, and also aids in isolating each performer from acoustic background noise. However, the sound of such a recording is somewhat unnatural, when compared to a performance in an auditorium, in that it has no reverberation and may contain considerable spectral bias, depending on microphone location. These deficiencies can often be corrected by manipulation of the signals during the mix-down phase. The mix-down console allows the sound engineer to subject each track of the master recording to individual processing: artificial reverberation and other special effects, spectral equalization, dynamic range compression, noise reduction, and limiting, to name the techniques most frequently used. The complexity of this process, and the apparatus invoked to carry it out, parallels that of a NASA mission control center.

After a highly trained sound engineer has mixed the processed signals to form a stereo or quadraphonic submaster, additional processing takes place to make the signal suitable for recording on disc or cartridge tape. The resulting submaster tape is used to feed a precision record-cutting lathe or a tape duplicator. In recent times, these lathes have also acquired their own complex signal processors, with dynamic control of the cutting mechanism and compensation and predistortion for nonlinear effects employed in both the cutting and the playback of the record. Furthermore the record master produced on the lathe is only the first step in a complicated process that results in the manufacture of the record played at home or in the broadcast studio. An equally long chain can occur in broadcasting. The home audio system and its loudspeakers provide a critical last stage in the reproduction chain. We can thus view the reproduction chain as being composed of three basic parts:

1. Generation and recording of the original signals.
2. Storage and transmission of these signals.
3. Reproduction of the signals in the form of an acoustic wave.

Although it might appear that some of the complexity of the reproduction chain is unnecessary, each stage has been proved important, often as a way of dealing with technical defects introduced in another stage. For example, compression during the original recording phase is necessary because of dynamic range limitations of the storage devices.

Much work on digitized audio has been aimed at replacing the weak links of the transmission or recording chain. Digital magnetic tape recorders and digital long lines transmission of audio are two important examples. Although conceptually simple, these systems are difficult to implement. However, they do offer a dramatic improvement in quality. The control of the mix-down console has also been digitized to relieve the sound engineer of the burden of actively controlling hundreds of adjustments in real time. All-electronic digital reverberation has been introduced as a replacement for mechanically based systems. Home-ambience synthesizers are available to alleviate the need for more than a stereo pair of signals in reproducing certain acoustic fields.

More advanced techniques have been employed in the laboratory in the restoration of old acoustic recordings. Turn-of-the-century Caruso performances are thus currently available in which the primitive sound of acoustic recording has been transformed quite remarkably. Digital processing is also being used as a research tool for improving electroacoustic transducer design. The loudspeaker is one of the weakest and least understood parts of the chain, producing complicated amplitude, phase, and directional effects, as well as complex forms of distortion. Signal-processing techniques are used in experiments to determine the physical properties of transducers and to suggest the perceptual consequences of measured irregularities.

All these systems share a common element: A–D and D–A conversion, a subject that is here treated separately because of its fundamental character. Any degradation introduced in these stages can substantially defeat the benefits of the signal processing. The need to match the performance of the converters to perception is motivated by several factors:

1. Excessive quality produces an economic burden and can result in unnecessary computational power in later stages to handle the excessive input data rate.

2. Measured degradation does not necessarily correspond to perceived degradation.

The issue is further complicated by subtle design factors that can have a profound influence on performance. There are therefore quite different approaches to conversion, depending on the intended application of the final system.

The relation between the physical and electrical properties of the system and the quality perceived must be clear to the engineer. For example, the classical definition of signal-to-noise ratio is based on the ratio of maximum signal energy to the noise energy in the absence of signal. Yet the perception of the noise depends on its spectral similarity or dissimilarity to the signal, probability density function, and temporal variation. Thus two noise mechanisms that measure 20 decibels (dB) apart on an energy scale may be equivalent in terms of perceived degradation.

Such examples indicate that underlying theory for audio systems must come from psychoacoustic studies rather than from systems theory. Systems theory deals with how to solve a problem; psychoacoustics, in this case, indicates the nature of the desired goal. Thus, in the above example, the goal is to render noise inaudible, not necessarily to remove it. The economic consequences of choosing the wrong goal can be severe. A 16-bit A–D converter has neither perceptual nor measurable noise, in general; however, it is two orders of magnitude more expensive than a 12-bit converter. The unifying concern in audio should therefore be the exploitation of both technology and the human auditory system to optimize the perceived quality of the sound reproduction.

2.2 FUNDAMENTALS OF ANALOG–DIGITAL CONVERSION

The A–D and D–A conversion circuits that precede and follow the digital processing are common to all digitized audio systems. The design issues are reviewed in detail here, because the converters can have a much more significant effect on the system performance than would normally be expected. In some respects, audio applications impose different constraints on conversion than other digital applications. Particular attention must be given to the bit rate, since minimizing it has strong economic and design consequences in the storage, transmission, and processing areas. As has been pointed out above, choice of implementation strategies can have a profound influence on perceived degradation. The trade-off finally chosen amid opposing considerations must be carefully selected because of the economic consequencies of errors in commercial audio systems. A laboratory prototype is not subject to these difficulties, since arbitrarily high quality conversion systems can be acquired, if needed, from specialized manufacturers.

The operation of an A–D converter is simple enough in principle: it transforms a sampled sequence of analog signals into a corresponding sequence of binary numbers. There are, however, many different ways of doing this, including linear pulse-code modulation (PCM), differential pulse-code modulation (DPCM), delta modulation (DM), adaptive delta modulation (ADM), and other methods.

2.2.1 Linear Pulse-Code Modulation

A block diagram of a digitized audio system using a PCM converter is shown in Fig. 2-1 with the A–D converter at the input and the D–A converter at the output. The input signal is sent through an anti-aliasing lowpass filter and sampled, with the analog value held constant for the complete conversion cycle. Digital logic seeks the digital word that, when converted to an analog signal, approximates the held value of the input signal. These digital words, appearing one per sampling period, are the digitized input. The output voltage of the converter is obtained from the setting of an array of switching transistors, which operate as a kind of binary potentiometer. The number of bits represented by the array determines the number of different voltage levels available and, therefore, the resolution of the converter. The digital logic determines the best approximation to the analog value by a series of successive approximations, beginning with the array's most significant bit (MSB) and ending with its least significant bit (LSB). The MSB is turned on, and a comparator determines if the resulting voltage exceeds that of the input. If it does, the bit is turned off; if it does not, the bit is left on. The process continues, one bit at a time, until the LSB is determined.

At the output, the reverse process takes place. A digital word is transferred to the D–A output register, an array similar to that in the A–D converter; the resulting successive samples are stored by a sample-and-hold. A lowpass filter removes any aliasing components above the Nyquist frequency, and an aperture filter corrects the high-frequency attenuation caused by the output sample-and-hold. The reader is referred elsewhere for a more complete discussion of such converters[8,75] and for examples of circuit designs for the D–A component.[2] Because the design of a high-accuracy D–A component is an art that changes continuously with evolving semiconductor technology, it is assumed that most users will purchase complete units rather than build their own. Nevertheless, the specification[76] of a conversion system for audio applications must take into account the particular requirements of this field.

As stated earlier, the number of bits in the conversion process determines the number of discrete levels available to represent the analog signal. A linear

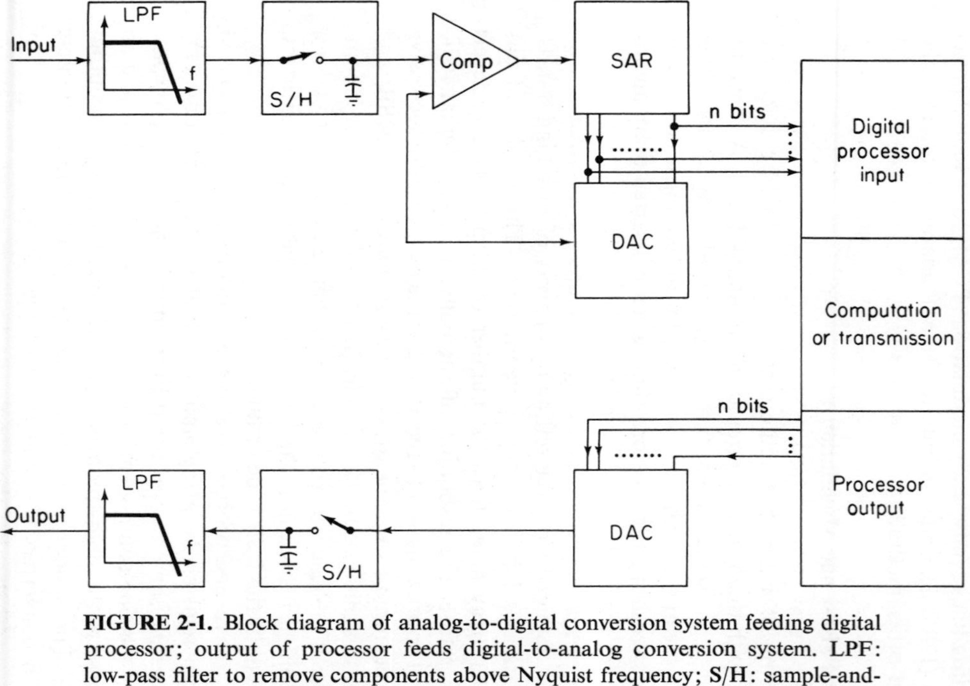

FIGURE 2-1. Block diagram of analog-to-digital conversion system feeding digital processor; output of processor feeds digital-to-analog conversion system. LPF: low-pass filter to remove components above Nyquist frequency; S/H: sample-and-hold to keep sampled value constant during conversion; Comp: analog comparator to compare input signal with approximation from digital-to-analog converter; DAC: digital-to-analog converter module with digital word input and analog signal output; SAR: successive approximation register which performs the conversion (standard LSI integrated circuit); *n* bits is the digital word corresponding to the analog signal. The processor transforms the input digital word to the output word. In the case of delay, the processor would be a memory; in the case of reverberation, the processor would be a computer.

n-bit converter has 2^n levels, or a peak signal of 2^{n-1} levels. The peak quantization error is one half of a level. In the presence of high-level wideband signals, the quantization error can be modeled as a random variable with a rectangular probability density function, statistically independent from sample to sample. Under these conditions, the quantization error is constant white noise[1] equivalent to the usual noise found in analog systems.

A figure of merit commonly used to describe audio systems is the dynamic range. This is the ratio of the largest signal, before overload distortion, to the smallest signal, before the latter is masked by the noise. The dynamic range is thus synonymous with the maximum signal-to-noise ratio (SNR) when the noise is additive and stationary.

For a converter, this is approximately

$$\text{SNR (dB)} = 6n$$

where n is the number of bits employed in conversion. If the signal is low level or narrowband, the quantization error produced by digital conversion is more like a complex form of distortion than noise. Consider, for example, the conversion of a low-level sine wave, centered precisely in level at the transition between two converter values, and with amplitude equal to one level. The quantized signal is then a square wave, which contains an infinite series of odd harmonics. In this case, the quantization error produces a result analogous to the harmonic distortion of an analog system. The relation of this distortion to signal level is unlike that in an analog system, however. In analog systems, the percentage of harmonic distortion increases with increasing signal level, while the system can be assumed to be perfectly linear for infinitesimally small signals. In the above example, the switching of the single quantization level is like the action of a hard limiter that captures the highest amplitude components. The degradation increases with decreasing signal.

If the low-level signal is also high in frequency, the distortion mechanisms become still more complex because of the sampling process that is part of conversion. The harmonics of the square wave produced by quantization error form aliasing products with the sampling frequency.[7] Had these harmonics existed in the original signal, they would have been filtered out by the input lowpass filter, which removes all signal components above the Nyquist frequency. When such frequencies are injected after the lowpass filter, their effects cannot be removed. Figure 2-2 illustrates this process. With a 31-kilohertz (kHz) sampling frequency, the fifth harmonic of a 6-kHz input signal will alias to produce a 1-kHz error component. The presence of quantizing harmonics and the aliasing frequencies they cause give rise to a very unpleasant sound, which is sometimes referred to as *granulation noise*. As the size of an input sine wave increases, the signal-to-error correlation factor decreases from 0.5 to less than 0.01[4]; however, the energy in each harmonic

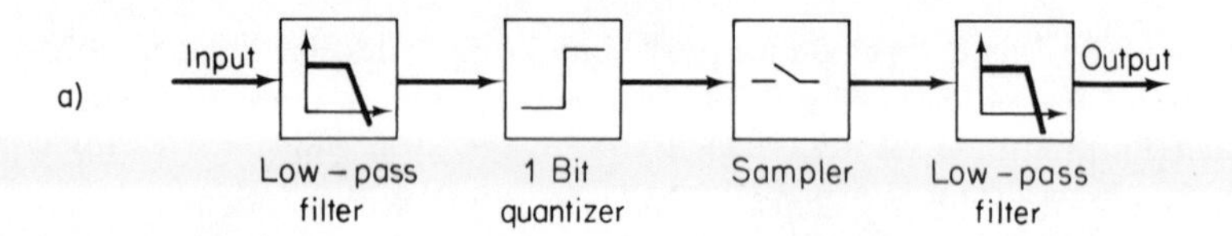

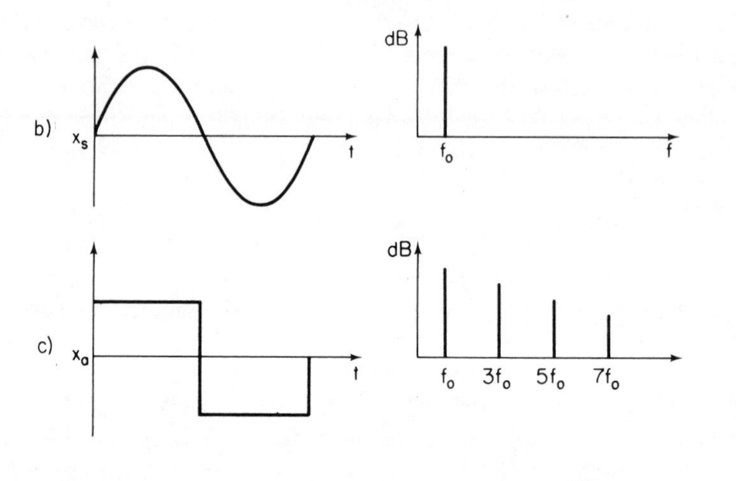

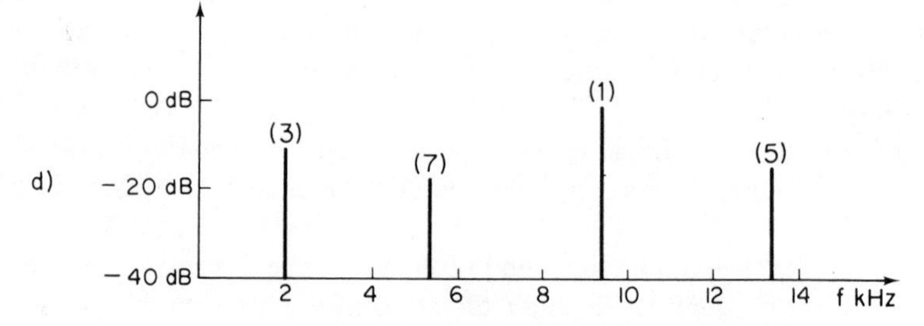

FIGURE 2-2. Illustration of the process by which a low-level sine wave produces granulation noise. 9.33 kHz input signal with sampling frequency of 30 kHz. The low-pass filters have a cut-off frequency of 15 kHz. (a) Model of the A/D conversion system when the sine wave has a peak value of less than 1 level (LSB) and is centered at a level crossing. The quantizer operates as a 1-bit A/D converter; (b) Input sine wave at 9.33 kHz with its spectrum; (c) Signal after quantization with its spectrum; (d) After sampling and low-pass filtering, the output spectrum contains aliased harmonics at frequencies which correspond to beats with the sampling frequency. For example, the 3rd harmonic (3) of the signal is 28 kHz which produces a 2 kHz beat. (Reprinted with permission of the Audio Engineering Society, Copyright 1974, from Blesser, Volume 22, No. 1, pp. 20–22.)

in the granulation remains approximately the same.[1] The aliasing components disappear completely as the input signal becomes broadband.

There are several ways to avoid the undesirable results of granulation noise. The most direct approach is to use a number of bits sufficient to reduce the harmonics generated by quantization error to a value below the threshold of perception. An alternative approach is the use of an additive *dither noise*,[4] which makes the quantization error vary more or less randomly from sample to sample. A simple version of dither with rectangular probability density function is additive Gaussian noise. In addition, conversion techniques can be used in which an inherent limit cycle oscillation produces a dither mechanism.[6] As long as the coherent nature of the quantization error is removed, the error can be treated as if it were an analog noise.

A similar, and equally disturbing, effect occurs when the input signal exceeds the maximum level of the converter or sample-and-hold amplifier. The use of high-frequency preemphasis,[11] which is common practice in various audio systems, makes this a particularly frequent problem. The resulting clipping produces distortion components above the Nyquist frequency. When this occurs, energy in the region near one third of the sampling frequency will be increased, as will the probability of overload. Frequencies near one third the sampling frequencies are most critical for symmetric distortions, since the third harmonic is the sampling frequency. Because the aliasing components which then appear are below the frequency of the signal that gave rise to them, they will not usually be masked by a music signal. These components are sometimes referred to as *birds* because their sound is similar to that of a bird's song. Technically, the noise produced in this way is identical to granulation noise. To avoid this problem, the converter's analog input stage is normally designed to limit the maximum signal to a value 3 dB below the maximum level that can be converted. The extra margin is required since a low-passed square wave will have a higher peak value than the original square wave. The use of compression or limiting amplifiers can also be recommended, with the control voltage taken from the sample-and-hold.

A more subtle form of clipping or quantizing effect occurs when the levels of either the A–D or D–A converter are not uniformly spaced owing to manufacturing imperfections, or if the sample-and-hold amplifier has some droop in the hold mode. In both cases, the resulting nonlinearity is introduced after lowpass filtering. To make matters worse, the effect is most pronounced in the center of the conversion range, which is the active region for low-level signals.

Still another mechanism produces birds at the output D–A converter. The output sample-and-hold amplifier usually has some slew-rate limit,[3] which produces a constant slope ramp between successive sampling values. The error between the ideal transition and the constant slope transition is nonlinear and has been shown[9] to be

$$\text{Error} = \operatorname{sgn}(x_n - x_{n-1})\cdot(x_n - x_{n-1})^2$$

where x_n is the nth time sample analog signal from the D–A converter. This error can be avoided by designing the sample-and-hold to integrate during the sample period, producing linear transitions.

2.2.2 Floating-Point PCM

The question of how many bits to use in the converter is a difficult one and needs to be examined carefully. The dynamic range of the undamaged human auditory system has been measured in the laboratory as about 130 dB; a more practical range is about 90 dB. The lower part of the range measured in the laboratory is usually masked by background noise in the real world; the upper part of the range produces discomfort or pain. In concert performances of symphonic music, peaks of between 100 and 110 dB have at times been observed,[77] with the background noise less than 20 dB at high frequencies.[78] These observations suggest a design goal of 90 dB for full dynamic range reproduction of such program material. If the music to be digitized has already been recorded or processed by conventional analog means, the dynamic range will be considerably less, for example, 75 dB with professional tape recorders.[79]

Based on the simple relation of dynamic range to digital word size cited earlier, a 90-dB dynamic range and 17.5-kHz signal bandwidth would require a 15-bit word and at least a 35-kHz sampling rate. Unfortunately, implementing a 15-bit A–D converter that can operate at a 35-kHz sampling frequency is at the limit of the present state of the art, and is therefore very expensive. This is not difficult to understand if one considers that such a converter must have an accuracy of resolution of 300 microvolts (μV) over a 10-volt (V) range, and that the MSB must settle level in less than 2 microseconds (μs).

It should be noted, however, that the wide dynamic range requirement is not in itself an accuracy requirement, but is the range of program levels that must be representable in the digitized word. Because the measurement definition assumes that the noise was stationary and additive, the minimum signal was defined as equal to the noise. This noise is only important when the signal level is very low, not when it is high. The difficulty of A–D conversion is considerably reduced if the noise level is allowed to increase with increasing signal level. We can therefore define a new figure of merit, signal-to-noise ratio when the noise is measured in the presence of the signal. The notation $\text{SNR}_{ws}(S)$ is used, where the subscript *ws* indicates that the noise is measured "with signal" and S indicates the amplitude of the signal. This definition indicates that the $\text{SNR}_{ws}(S)$ is a function of both channel and signal properties. This differs from the usual definition of signal-to-noise ratio, which is the ratio of the maximum signal before overload distortion to the constant additive noise. Since the conventional signal-to-noise ratio is usually measured in the absence of signal, the notation SNR_{ns} is adopted, where *ns* indicates that the noise is measured with *no signal*. SNR_{ns} is equivalent to dynamic range, and the terms are used interchangeably. For the conventional A–D converter,

$$SNR_{ws}(S = \text{Max}) = SNR_{ns}$$

and

$$SNR_{ws}(S) \propto S$$

To keep the noise below the threshold of audibility for almost all program material, the $SNR_{ws}(S)$ should be about 60 dB,[10] which implies a converter resolution of about 10 bits. The SNR_{ns} is still about 90 dB. A hybrid A–D converter technique using a floating-point approach appears capable of giving this performance. The converter shown in Fig. 2-3 has a switchable prescale amplifier preceding the actual A–D converter. The switch state is controlled by a logic algorithm, which attempts to keep the signal level at the converter input in the upper half of the converter range. As the input music program decreases in amplitude, the gain of the amplifier is increased. The resulting digital word is composed of two parts, the scale factor and the mantissa. The converter in Fig. 2-3 has four prescale states: 0, +6, +12, and +18 dB, which are encoded as a 2-bit word. This is followed by a conventional 12-bit A–D converter. For input signals between 0 and −24 dB, the converter is operating in the upper half of its range, and the SNR_{ws} is given by

$$SNR_{ws}(S) = 66 + S/\text{modulo } 6 \text{ dB} \qquad 0 > S(\text{dB}) > -24$$

As the input signal decreases further, the prescale amplifier remains at maximum gain, and the SNR_{ws} becomes

$$SNR_{ws}(S) = 90 + S \qquad -24 > S(\text{dB})$$

The total dynamic range of this conversion technique is the sum of the converter dynamic range and the prescale range. This is given by

$$SNR_{ns} = 90$$

The floating-point representation can readily be converted to standard binary format, since the prescale information represents the number of right shifts to be applied to the mantissa. The dynamic range of the floating-point conversion technique can be increased further by adding a third bit to the prescale amplifier, which gives eight gain states. This exceeds any present practical requirement and is not normally done.

The control of the prescale factor can be either instantaneous or syllabic. When it is instantaneous, the optimum value of gain is determined at each sampling interval. The syllabic algorithm decreases the gain whenever the converter would have been overloaded, but it does not increase the gain until after the signal has remained below half-scale for a predetermined waiting period. Typical waiting periods are on the order of 100 to 300 milliseconds (ms), which corresponds to a speech syllable or a musical note. The digital

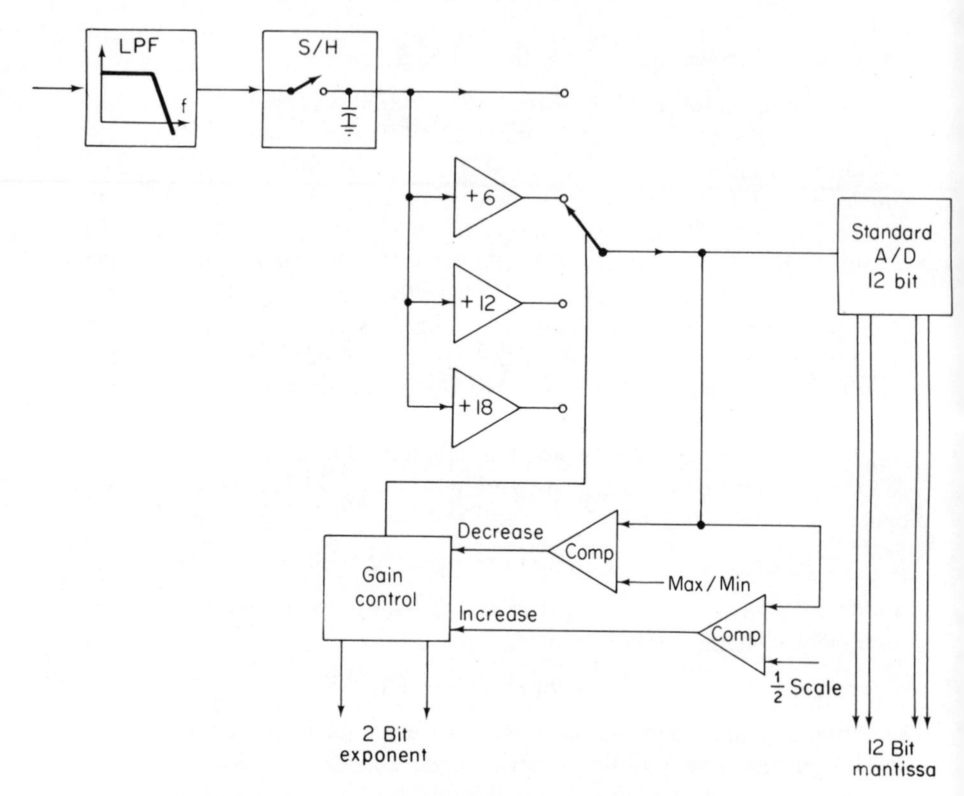

FIGURE 2-3. Floating-point converter technique. Following the normal S/H and lowpass, a switchable gain stage is inserted. The stage shown has four gain settings, 0 dB, +6 dB, +12 dB, and +18 dB. The gain control seeks that gain-state which maximizes the signal level at the input of the A/D converter without producing overload. A comparator tests the signal to determine if it exceeds the range of the converter. If so, the control logic decreases the gain until overload no longer exists. Correspondingly, the control increases the gain if the signal is less than half scale. In some systems, there is a delay (e.g., 200 ms) before the gain is increased. This minimizes switching and makes the accuracy requirements for the system less severe without introducing any degradation. The final digital word is composed of two parts: a 12-bit mantissa and a 2-bit exponent, thus giving rise to the name *floating-point conversion*. By using a gain of 6 dB, the resulting word can easily be reconverted to standard binary formats.

output, when using the syllabic algorithm, is equivalent to block floating point (see page 81).

The choice of instantaneous or syllabic control is one of implementation. The instantaneous algorithm requires that the accuracy of the prescale amplifiers be of the same order as that of the converter which follows; the syllabic algorithm is much more tolerant of offset and gain inaccuracies. Gain switching normally takes place only a few times per second, and inaccuracies are inaudible since they are synchronous with changes in the input program. Because the syllabic floating-point converter usually does not noticeably degrade the program and is easy to implement, it is generally preferred. The output of the system can be designed in a corresponding fashion with a D–A converter followed by a postscale amplifier.

This approach to conversion has become typical, with variations of the technique also used.[3,12,80] In certain applications, to be discussed later, the reduction in the number of bits to a minimum is of prime concern. Because compression is easier to accomplish digitally, a high-quality conversion system is often used as the input to a circuit using a bit-compression algorithm.

2.2.3 Delta Modulation

In delta modulation (DM), it is the level difference between samples that is represented digitally. This class of converter has as its primary asset extremely simple hardware design, which, unlike PCM, dispenses with successive-approximation logic, sample-and-hold amplifiers, anti-aliasing filters, prescaling, and precision components. Nevertheless, such a system can achieve a high SNR_{ns} of about 65 dB. A disadvantage of delta modulation is that the required bit transmission rate is usually much higher than that of PCM for a given quality. For these reasons, DM is attractive for systems that cannot have a high cost associated with storage or processing. One such example, discussed later, is in a delay line used in an ambience enhancement system. A comprehensive survey of DM systems can be found elsewhere,[47] but a brief introduction is presented here.

The simplest form of A–D converter is the 1-bit delta modulator shown in Fig. 2-4. At each clock interval, a binary decision is made by comparing the input signal to a stored approximation of the previous sample. If the input signal is more positive than the stored approximation, a fixed positive increment is added to the approximation. Conversely, if the input signal is more negative than the previous sample, a negative increment is added. The process is repeated at successive sampling periods, bringing the approximation continuously closer to the value of the input signal. The accuracy of the approximation is a direct result of the value of increment used. The 1-bit digital data used to create the approximation at the encoder can also be

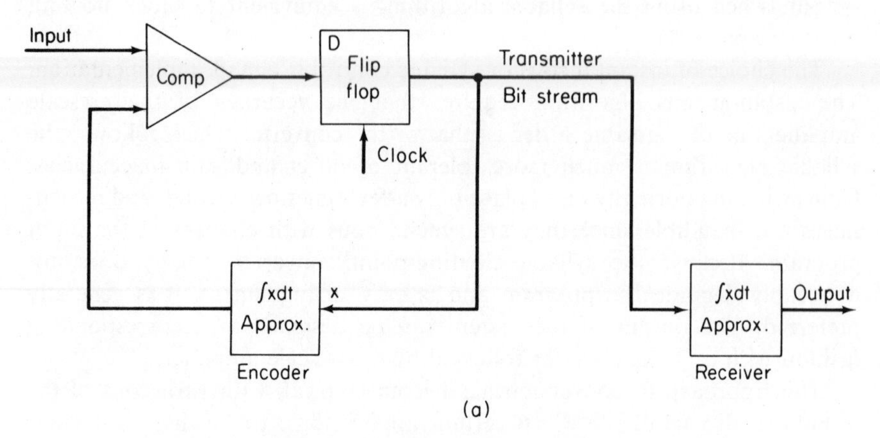

FIGURE 2-4a. One-bit delta modulation system. The input is compared to the output of the approximation module, shown as an integrator. The output of the comparator is a binary one or zero; this value is strobed into the D-flip flop at the clock pulse. If the error between the input and the approximation is positive, a binary one is generated; if the error is negative a binary zero is produced. In the approximation circuits, a binary 1 produces a positive current, a binary 0 a negative current. During normal operation, the approximation output will be reasonably similar to the input. Since this circuit derives all its information from the transmitted bit stream, the receiver will create the same approximation.

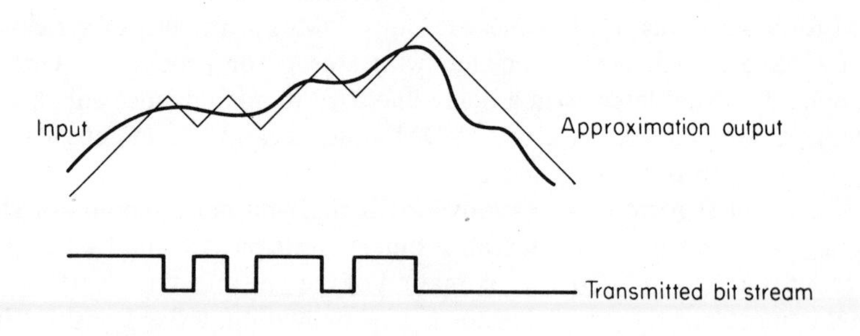

FIGURE 2-4b. Signals in delta modulation: input, approximation output, and binary bit stream from which approximation signal is derived. When the bit stream stays in one state, slew-rate limiting is occurring since the approximation cannot match the rate of change of the signal. A series of alternating states means that the approximation signal is oscillating about the correct value.

transmitted and used to regenerate the same approximation at a remote decoder.

Single-bit conversion exhibits two kinds of signal degradation. When the input signal is increasing rapidly, the approximation cannot follow the input, because the maximum rate of change at the output is only one increment per sampling periods. This produces *slope overload*, otherwise known as slew-rate limiting or clipping of the first derivative. The second kind of degradation consists of a gross error for signals smaller than a single positive or negative step. This produces *hunting* or oscillation about the correct value, and corresponds to the quantization noise of a PCM converter; it is also referred to as *granulation noise*.

The quality of the 1-bit converter is effectively determined by the sampling frequency, and hence by the bit rate. For a given step size, which produces a given granulation noise, the maximum signal is a function of both the slew rate and the signal frequency. Doubling the sampling frequency doubles the slew rate and doubles the maximum signal at a given frequency. Alternatively, the slew rate could be kept fixed and the step size decreased by a factor of 2, thus halving the granulation noise. Depending on the statistics of the input signal assumed, an optimum step size can be found that maximizes the signal-to-error ratio.(51,55) However, in this type of calculation, slope overload and granulation are combined as a single error measure. The perception of slope overload appears to be more closely related to the energy in the derivative error(53) for speech signals, rather than energy in the signal error. Moreover, slope overload occurs in the presence of high-frequency energy, which tends to mask the audibility of the distortion products. Granulation noise, on the other hand, is present whenever the signal becomes small.

One criterion of a music system is its dynamic range, usually the SNR_{ns} measured at 1 kHz. This frequency is often used to represent the maximum steady-state fundamental of a music note. Higher-frequency components are present as the lower-amplitude harmonics of a fundamental, or in the transients of percussive instruments. In can be shown that the SNR_{ns} for a simple 1-bit delta modulator is given by

$$\mathrm{SNR}_{ns} = \frac{0.2 f_0^{1.5}}{f_s W^{0.5}}$$

where f_0 is the sampling frequency, f_s the sine wave signal frequency, and W the bandwidth over which the noise is measured.(66) A 500-kHz sampling frequency, which produces a 500-kbit/s data rate, will therefore encode a 1-kHz sine wave with an SNR_{ns} of 50 dB over a 14-kHz bandwidth. We note that this performance is considerably worse than that of the PCM converters discussed earlier. The cause of the difference is easily found. With DM, a doubling of the bit rate (sampling frequency) produces only a 9-dB increase in SNR_{ns}, whereas a doubling of the number of bits in a PCM converter produces exponential increase in SNR_{ns} (doubling on a decibel scale). Thus, where high SNR_{ns} may be required, the delta modulator is not an optimal choice.

This can be understood by noting that the doubling of sampling frequency only allows for a step size reduction by a factor of 2 (6 dB) and a doubling of bandwidth over which the quantization noise is distributed (3-dB decrease in energy density). The sampling frequency of such a DM system would have to be several megahertz to produce acceptable audio quality, with an accompanying high bit rate. At low sampling rates, however, and hence low SNR_{ns}, the linear delta modulator is at least as good as the classical PCM converter.(67) It is therefore more appropriate for telephone-quality speech transmission than high-quality audio.

Rather than use a single integrator or lowpass filter for the approximation circuit in Fig. 2-4, a second-order integrator can be used. In this case, the SNR_{ns} is given by

$$SNR_{ns} = \frac{0.026 f_0^{2.5}}{f_s W^{1.5}}$$

where f_0 is the sampling frequency, f_s the signal frequency, and W the bandwidth over which the quantization noise is measured.(66) Using the same reference frequency of 1 kHz and a 14-kHz bandwidth, the SNR_{ns} is increased from about 50 dB to about 65 dB.(67) This improvement is not obtained, however, without incurring certain new problems. One is that the noise behavior is now strongly influenced by potential instabilities. This is because the delay between samples plus the phase shift introduced by the second-order integration can result in more than 180 degrees of phase shift unless special care is exercised in design.(69,47)

The overload behavior of the second-order system consists of a clipping in the second derivative rather than the first. Thus the allowable maximum signal must be decreased 12 dB/octave to avoid distortion. The tendency to produce audible distortion is substantially increased, since the high-frequency energy in music signals does not decrease very rapidly with increasing frequency. Some of the extra dynamic range gained by the second-order integrator is thus lost, because the input signal must be held at a lower level to avoid a high probability of second-derivative overload.

Some additional improvement can be achieved by a careful shaping of the loop characteristics. A loop compensation filter that operates on the error signal can be added directly before the 1-bit decision.(52) Although the transient behavior of the system is then somewhat corrupted, the noise is reduced by about 8 dB by this technique.

The second-order delta modulator just described would probably be just acceptable for many consumer audio applications. Even further improvement can be obtained by using an adaptive strategy.

2.2.4 Adaptive Delta Modulation

The linear delta modulator produces a constant noise, which results in an SNR_{ws} that is maximum only for maximum signals. As the signal decreases, the SNR_{ws} also decreases by the same factor, as is the case for all uncompanded A–D converters. The delta modulator can, however, incorporate a version of the floating-point algorithm, which increases the SNR_{ns} dramatically without need to increase the sampling frequency or bit rate.

In the system shown in Fig. 2-5, the step size is increased or decreased as a function of signal behavior. For signals rapidly increasing in level, the step

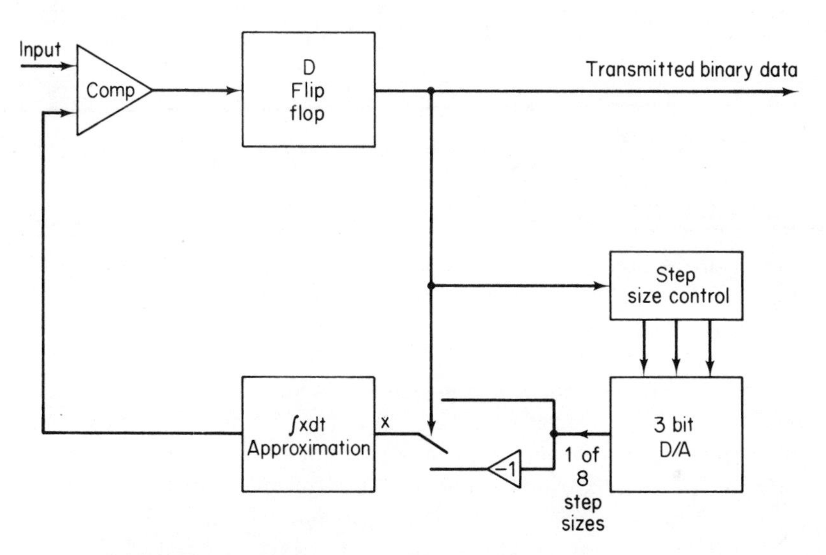

FIGURE 2-5. Block diagram of adaptive delta modulation encoder. The binary value at the output of the flip flop controls the sign of the correction step; the step size is variable. In the version shown, the 3-bit D/A converter provides 8 step sizes. The control logic determines the step size based on the previous binary values. For example, a series of binary 1's, indicating slew-rate limiting, results in increases in step size to allow the approximation circuit to catch the signal. A series of alternating 1's and 0's, indicating oscillation, results in a decrease in step size.

size is increased to avoid slew-rate limiting; for small signals, the step size is correspondingly decreased to minimize granulation noise. The performance is determined by the specific algorithm for step-size increase and decrease, as well as the range of the step-size changes. For the decoder to recover the signal correctly, the step adaptation algorithm must be derived from the digital bit stream it receives, and not from the input signal. Logically, the decisions as to when and by how much to change the step size must therefore be based on the results of previous decisions.

In the simplest embodiment, such a decision algorithm may use only the current and previous values.[50,70] If both have the same sign, the step size is increased by a factor P. If not (i.e., if consecutive signs alternate between $+$ and $-$), the step size is decreased by a factor Q. The condition for stability in such a system is given by[70]

$$PQ \leq 1$$

As PQ increases beyond 1, the instability results in a very rapid increase in noise. For PQ between 0.8 and 1, the SNR_{ws} remains approximately constant. Using speech signals, it has been claimed that optimum values are

$$P = 1.5, \qquad PQ = 1$$

when the optimization is based on an error energy measure[70]; however, a set of measurements based on listening tests[68] suggests an optimum of $P = 1.2$. This apparently arises because the subjects prefer overload distortion to granulation noise. Fortunately, a selection of $P = 2$ and $Q = 0.5$ does not produce serious degradation, since this pair of values is particularly convenient for implementation.[71]

This algorithm might be considered rather primitive in that slope overload is considered to be present when two consecutive decisions have had the same sign, and granulation noise is considered present when they have alternate signs. Further improvement in performance can in fact be obtained by using more than one previous decision in the computation of the new step size. One such algorithm, using 6 previous bits of information instead of 1, has been applied to increase the SNR_{ws} by about 8 dB.[49] The number of previous decisions that should be used is a strong function of the relation between signal and sampling frequency. The higher the sampling frequency, the more stationary the signal behavior. The above data were taken from work done with speech at a low sampling frequency, about 14 times the highest signal frequency. For music, with a 500-kHz sampling frequency, the signal can generally be approximated by a constant ramp or, at worst, a parabola.

In one variation of the adaptive DM technique, a basic direction is derived from the bit stream; a binary 1 then represents continuing in the same direction, and a 0 signifies no change.[72] Although not much work has been done with this type of converter for high-quality audio, the results thus far reported suggest that an inexpensive DM system could be built having a bit rate no greater than that of the classical PCM converter. Experiments carried out by one of the authors (B.B.) suggest that a 70-dB SNR_{ws} at 1 kHz may be attainable over a 15-kHz bandwidth. The SNR_{ns}, which is determined by the range of step sizes, was 96 dB in one case.

A serial bit stream is very difficult to use for applications other than storage or transmission. To process a signal in this form, the delta-modulated signal must be decoded into the standard binary format. The nonadaptive delta modulator, by contrast, can be processed directly. It is normally lowpass filtered digitally to remove the noise above the audio band and then resampled.(74) These samples are equivalent to an ordinary converter output.

2.2.5 Adaptive Differential Pulse-Code Modulation (ADPCM)

A hybrid form of converter combining the differential properties of the delta modulator and the binary representation of PCM can be formed by replacing the 1-bit error decision with a multibit representation.(54,93) Consider, for example, the use of a 3-bit code. The error information is then at one of eight possible levels, rather than simply negative or positive, providing much higher signal resolution. However, the sampling frequency must be reduced by a factor of 3 if the bit rate is to be held constant.

The primary advantage of ADPCM is that more information is available for the adaptation algorithm during each sampling period. For the example above, the magnitude of the error is available in four ranges (plus sign). Each of these ranges can then be assigned a scale factor for adaptively changing the step size. Typical usable values found for speech are 0.9, 0.9, 1.25, and 2. Thus, if the previous decision were 101, the step size for the next sampling interval would be reduced by a factor of 0.9. In contrast, if the decision were 111, which suggests slew-rate limiting, the step size would be increased by a factor of 2. Depending on the statistics of the input, other scale factors may be preferred. The use of memory of more than one word has apparently not yet been considered.

The ADPCM converter has been extensively evaluated for speech signals using an instantaneous companded PCM for comparison. In terms of energy in the quantization error, the ADPCM offers an improvement of about 1.5 bits over PCM. Subjective judgments, however, suggest an improvement of about 2.5 bits. A 4-bit ADPCM has been judged better than a 6-bit log-PCM, but worse than a 7-bit log-PCM.(56) For speech signals, at least, the saving in bit rate is about 40%.

Although no ADPCM data for music systems are available, certain inferences can be made from sine-wave measurements. With a 4-bit ADPCM converter measured over a 2.8-kHz bandwidth, the SNR_{ws} is given as 20 dB for an 800-Hz sine wave. These data can be extended to music by using the 9-dB/octave factor for increases in the sampling rate and the $W^{0.5}$ for increases in bandwidth. An 8-bit ADPCM at 50-kHz sampling frequency should then yield a 64-dB SNR_{ws} at 800 Hz over a 15-kHz bandwidth. The bit rate, which is 0.4 Mbit/s, is comparable to or even better than floating-point 10-bit PCM, sampled at 35 kHz, having 2 bits of level scale. One might expect still better performance if the adaptation algorithm used more than one past sample. Because of the additional logic complexity, this may or may not prove attractive in application.

2.2.6 Psychoacoustic Factors

In view of the difficulty of comparing the performance of different types of converters analytically, it is helpful to make such comparisons from a psychoacoustic standpoint. By determining the required bit rate of a hypothetical converter matched to the auditory system, we can evaluate the degree to which conventional converters deviate from an ideal value. This provides a figure of merit that has been frequently applied in evaluating speech systems. We can ask, for example, the following question: What would be the bit rate of a converter that was perfectly matched to the human auditory system? We can consider this question with respect to all audio signals covering the complete dynamic range and hearing bandwidth or with respect to the smaller class of signals referred to as natural music. In both cases, the optimum converter will be one with the minimum perceptual error. Audio signals in the natural music class are somewhat restricted, since natural music originates by the excitation of mechanical resonance or a temporally limited percussive sound. Although evaluation of a converter for the broader class of all audio signals might provide a more universal result, commercial systems for natural music are usually designed to keep costs within reason.

An optimum converter is one with minimum bit rate such that further minimization of the error between the converted and unconverted audio signal is inaudible. Attainment of this goal requires a good understanding of the psychoacoustic effects of various conversion errors. Unfortunately, there is no complete model of auditory perception, but only a large literature documenting various phenomena.

The studies most relevant to audio encoding are those that concern detection under various conditions.[16,82] In general, acoustic energy at a given frequency below a certain minimum level cannot be detected; and, in the presence of another signal, this threshold increases greatly. The first case involves the measurement of an absolute auditory threshold. For any given frequency, a sound below this threshold is completely inaudible. Figure 2-6 shows the threshold for people with normal hearing.[124] From these curves one can see that audition is most sensitive in the region from 1 to 5 kHz. At higher frequencies sensitivity decreases, being about 20 dB less sensitive at 15 kHz. Similarly, at low frequencies the threshold of detectability increases. The second case concerns masking, where the presence of one sound inhibits the perception of another sound. This is equivalent to a shift in the perceptual threshold due to the presence of signal. In the steady state, a given

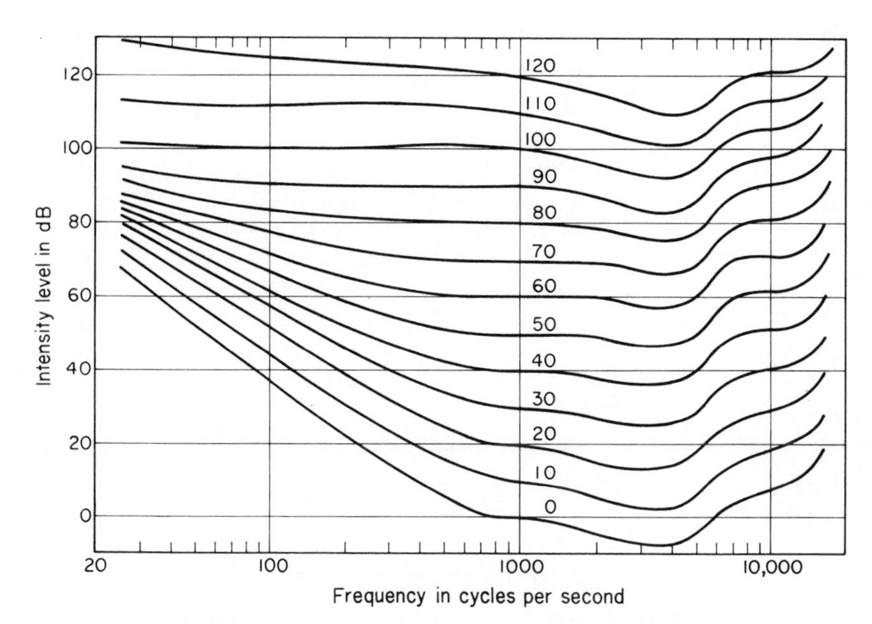

FIGURE 2-6. Contours of equal loudness. Each curve connects levels of equal loudness as perceived by normal adults for sinusoidal signals. The lowest curve (0 phons) is the threshold of audibility (after Fletcher and Munson).

frequency will mask signals at neighboring frequencies, rendering them completely inaudible. A transient sound of limited duration will mask signals occurring before and after the transient. In each of these cases, the audio signal can make both the quantization noise and other signals completely inaudible. This allows the converter to ignore signals that would be masked and to generate quantization noise that will also be masked.

Figure 2-7 shows the results of a frequency-domain masking experiment.(18) An 80-dB sound pressure level (spl) sine wave at 1200 Hz was presented to subjects, and a secondary tone was added. The frequency and amplitude of this secondary test tone were varied to determine the region over which it remained inaudible. The region in the figure labeled "primary only" indicates complete masking, where only the 1200-Hz signal was heard. Outside of this region, the secondary tone was heard either as a tone or a beat.

If the masking tone is replaced by narrowband noise,(19) a slightly different pattern emerges, as shown in Fig. 2-8. This type of signal is probably more typical of music than is a single-frequency tone. This set of curves shows that the spectral region in which masking occurs is a strong function of the amplitude of the masker. The higher its amplitude, the greater is the band-

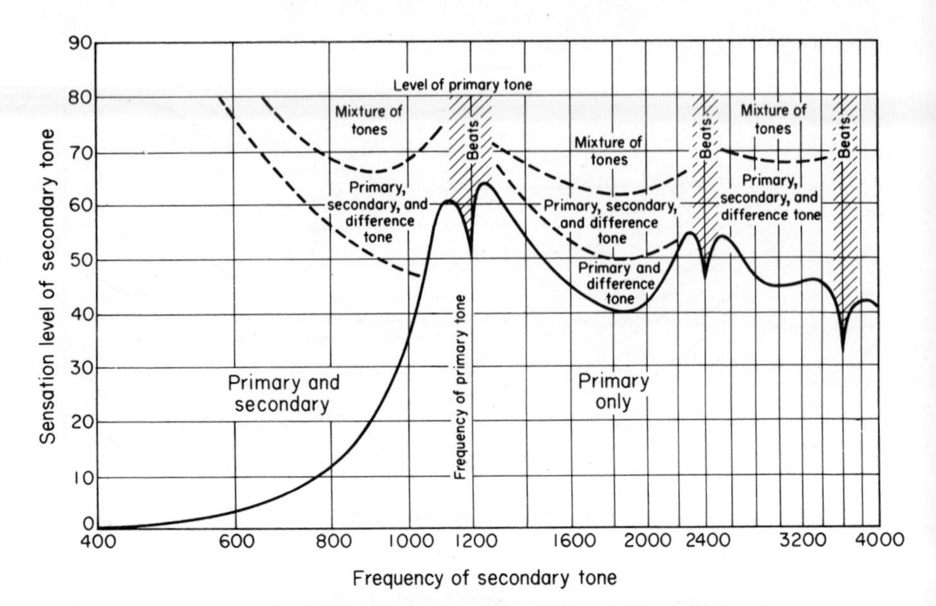

FIGURE 2-7. Masking of a secondary sinusoid of varying amplitude and frequency by a primary sinusoid at 1200 Hz, 80 dB above threshold. When the secondary tone falls below the solid curve, it has no effect on the perception of the primary tone. Above the solid curve, the secondary tone produces a complex effect which is perceived as either a beat, a mixture of tones, or both. (From Licklider, 1951, after Fletcher, 1929, from Wegel and Lane, 1924. By permission from John Wiley and Sons, Inc., from Speech and Hearing by Harvey Fletcher, Copyright, 1929, by D. Van Nostrand Company, by permission of Van Nostrand Reinhold Company; and from Bell Telephone Laboratories.)

width of the masking region. Extensive data on masking have been accumulated for various combinations of masking and test signals.[82]

Temporal masking is also an important phenomenon. In this case the primary signal can inhibit the perception of less intense signals that are near it in time. When the louder sound masks a sound that follows it, the effect is *forward masking*. The temporal range for this effect is about 250 ms, and the masking contour is approximately linear on a logarithmic scale.[17] A sound can also mask a sound that precedes it, called *backward masking*, but the effect operates only over 20 ms.[125] See Figure 2-9.

The most obvious implication of the above data is that the SNR_{ws} requirement needs to be only about 30 dB if we consider narrowband signals only. By way of example, assume that the input signal is passed through a series of bandpass filters, and the output of each filter is digitized separately. The SNR_{ws} specification, and hence the number of bits in the A–D converter, is based on the audibility of noise present in each band. From Fig. 2-8,

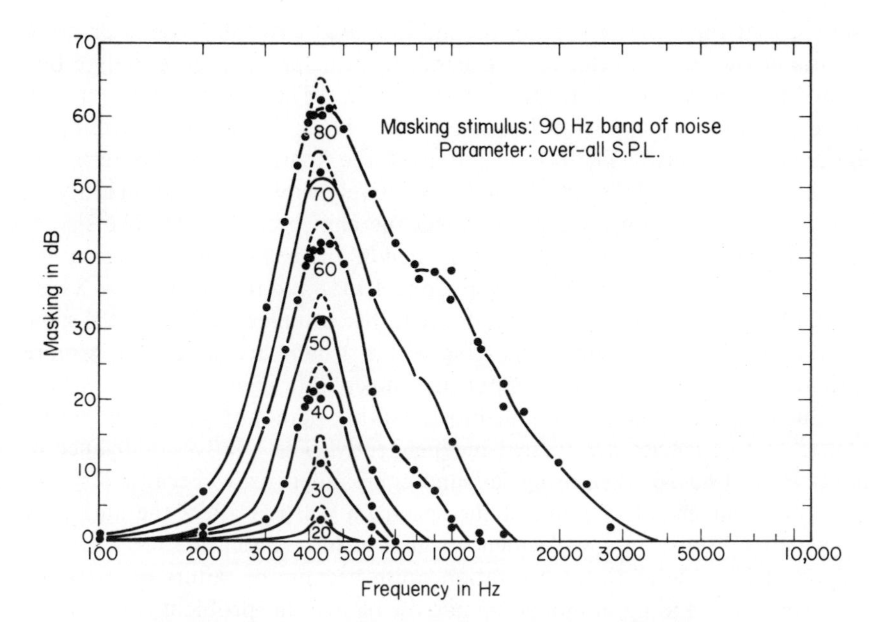

FIGURE 2-8. Masking region produced by narrow-band noise at 410 Hz and a bandwidth of 90 Hz, at various amplitudes. A test signal below a given curve will remain inaudible. The bandwidth of masking increases as the intensity of the noise is increased. (Reprinted with permission of the Acoustical Society of America, from Egan and Hake, 1950.)

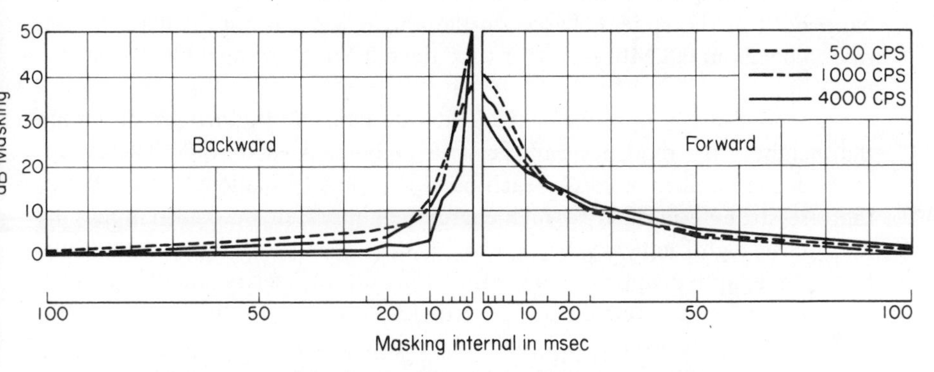

FIGURE 2-9. Forward and backward masking of white noise with a probe tone. A white noise burst is presented and a test tone at a reduced level is presented later (forward masking) or earlier (backward masking). The threshold shift of the tone as a function of the time difference to the noise masker is shown. The effect is even stronger if the masker is also a tone rather than broadband white noise. (Reprinted with permission of the Acoustical Society of America, Volume 34, No. 8, 1962.)

we see that for an 80-dB signal the masking will cover any signal or noise that is below 50 dB in this narrow band. In principle, therefore, only 6 bits would be required to achieve the required SNR_{ws}. The dynamic range of each band could be handled with the same kind of block floating-point technique described previously. The response time of the floating-point is determined by the forward-masking time, which is relatively slow. Thus the high SNR_{ns} specification does not appreciably affect the choice of bit rate. The size of these frequency bands, called *critical bands*, increases with frequency.[87] Critical bands are about 100 Hz wide up to 1 kHz and then increase to 2 kHz at 10 kHz. The data for critical bands are rather complex but lead to the same general view: signal components lying within a band are handled as a more primitive unit compared to components in separate bands.

Such a system differs from the block floating-point approach only in that bands of frequencies are treated independently. The relative annoyance of an SNR_{ws} of 60 dB when using fullband companders arises because a strong signal component at one end of the spectrum will not affect the audibility of wideband noise. A low-frequency signal at high amplitude will cause an apparent increase in the quantization noise because of failure of masking. By using companding in individual narrow bands, this problem is alleviated. Similar observations have been made in the design of classical analog noise-reduction equipment,[38,88] although these systems used bands that were much wider than the critical bands.

Information in each bandpass filter is sampled at a rate twice its bandwidth, and the resulting outputs are multiplexed. The effective sampling rate is not affected, since the sum of individual rates is still equal to twice the total bandwidth of the system. Based on the above, we can approximate the bit rate to be about 0.2 Mbit/s, rather than the 0.3 Mbit/s using a block floating-point approach.

A second-order improvement in bit rate can be obtained by observing that neighboring frequency bands can strongly mask each other. The assignment of the number of bits to each band can thus be made adaptive. Bands that are strongly masked by high-energy neighbors, or masked remotely by an intense high-frequency band, can be squelched, or assigned fewer bits. Such an adaptive approach is rather difficult to specify and may prove extremely difficult to implement; nevertheless, one might estimate that another 20 to 30% reduction in bits could be achieved in this way.

Using this preliminary analysis, we might conclude that an optimum converter based on psychoacoustic matching would probably have a bit rate on the order of 0.15 Mbit/s for a single monophonic program. This is an improvement by a factor of 3 over a 15-bit uncompanded word at 35-kHz sampling frequency, and it is a factor of 2 improvement over a floating-point compander.

Whether such additional complexity would be worth the additional compression rate is an open question. In certain applications where the transmission or storage cost was high, the benefits would be likely to outweigh the cost. An archival music library with 10,000 rolls of tape might find a factor of 3 decrease in tape volume a worthwhile compensation. Similarly, the cost of long lines transmission could easily absorb the terminal cost of such companders, since a 3000-mile high-fidelity link is extremely expensive.

2.3 RECORDING

Much, perhaps most, of contemporary musical experience reaches listeners by way of recordings. Even broadcast music consists almost entirely of commercial disc recordings. Recording techniques are therefore an important application of audio engineering, since recording technology is a major factor in determining the quality of the signal finally reproduced at home.

2.3.1 Digital Tape Recorders

Virtually all professional music recording is now made in multiple tracks, each track containing the contribution of a musician or group of musicians. In addition, different sections of a work of music may be recorded at different times. To assemble a complete performance requires the cutting apart, splicing together, and rerecording of the original material from many separate pieces of tape. With a conventional analog tape recorder the signal is degraded on each pass, the final signal including the sum of all the noise and distortion introduced at each stage of editing and mixing.

The advantages of a digital tape recording system can be understood by examining the conventional analog system. Professional tape recorders are based on the same principles as tape recorders intended for home use, but at a much higher level of reliability. A thin tape coated with magnetic particles is drawn, at constant speed, past a recording head, an electromagnet designed to concentrate its flux across a very small gap placed in contact with the tape. The recording head produces a magnetic field, modulated by the music signal, that orients the particles on the tape. The resultant tape contains a relatively stable physical pattern that represents the original music. Upon playback, the magnetic properties of this pattern are sensed by induction and thus reconverted into an electrical signal. Degradation mechanisms arise from two primary sources: (1) characteristics of the magnetic material and recorder heads, and (2) mechanical imperfections in the constant-speed tape transport mechanism.

The magnetic materials used for recording tapes are susceptible to additive noise and also have very nonlinear transfer characteristics, even with optimal adjustment of the recording equipment.[31] At best, a SNR_{ns} of about 70 dB can be obtained,† even when special noise-reduction processors are added externally.[38] At high frequencies, the dynamic range is even more severely limited unless the recorded spatial wavelengths are increased by running the tape at higher speeds (e.g., 30 inches per second (ips)). The noise mechanism is not static, but is modulated by the signal, giving rise to a complex form of modulation noise.[29] A reproduced sine wave will generally have noise sidebands. This kind of noise is very perceptible when the signal is around 1 kHz, where the SNR_{ws} deteriorates to less than 40 dB.[30] Manufacturing imperfections in the tape and accumulated dirt particles result in short-term variations in amplitude and frequency response, notably at high frequencies.[33] These drop-outs produce a rough quality that can be heard clearly on pure musical notes such as organ or flute.

Any eccentricities in the rotating parts of the tape recorder, such as the capstan or idlers, result in speed variations of the tape during playback and recording. This produces a frequency modulation to which the auditory system is particularly sensitive. Designers of professional machines generally try to limit this wow and flutter to less than 0.1 %, especially at rates near 4 Hz.[34] There are subtle problems in keeping the tape speed constant. Nonlinear friction between the tape and the heads can result in microscopic starts and stops. These can be accentuated by mechanical resonances in the unsupported tape. This scrape flutter has to be minimized by careful location of supporting idlers.[30]

Because of these difficulties, as well as the ultimate instability and unreliability of analog recording, digitally based tape recorders would find a welcome place in the audio profession. This is especially true for record mastering and archival storage. Several digital tape recording systems have been built on an experimental basis.[20,22,32] Nippon Columbia has already issued recordings that have been mastered with digital tape recorders.[35] Commercial digital recorders do not seem far off.

Conceptually, the digital tape recorder is straightforward, as shown in Fig. 2-10. Each input channel is lowpass filtered, sampled, and converted to digital words using the techniques discussed earlier. The bit streams are multiplexed, and additional control bits for timing, error correction, parity,

† The measurement definition of SNR_{ns} varies depending on the standard being used. These differ from country to country and also within the profession itself. For example, the maximum signal can be defined as the root-mean-square (RMS) sine wave at 1 kHz, which produces 0.3, 1, or 3 % harmonic distortion. The frequency can be other than 1 kHz. Similarly, the noise can be spectrally weighted, measured as peak, average, RMS. Sometimes the noise is measured with virgin tape or bulk erased tape. All these variables must be considered in evaluating the measured value of SNR_{ns}; the comparison of one machine to another is still extremely difficult unless the measurement conditions are the same.

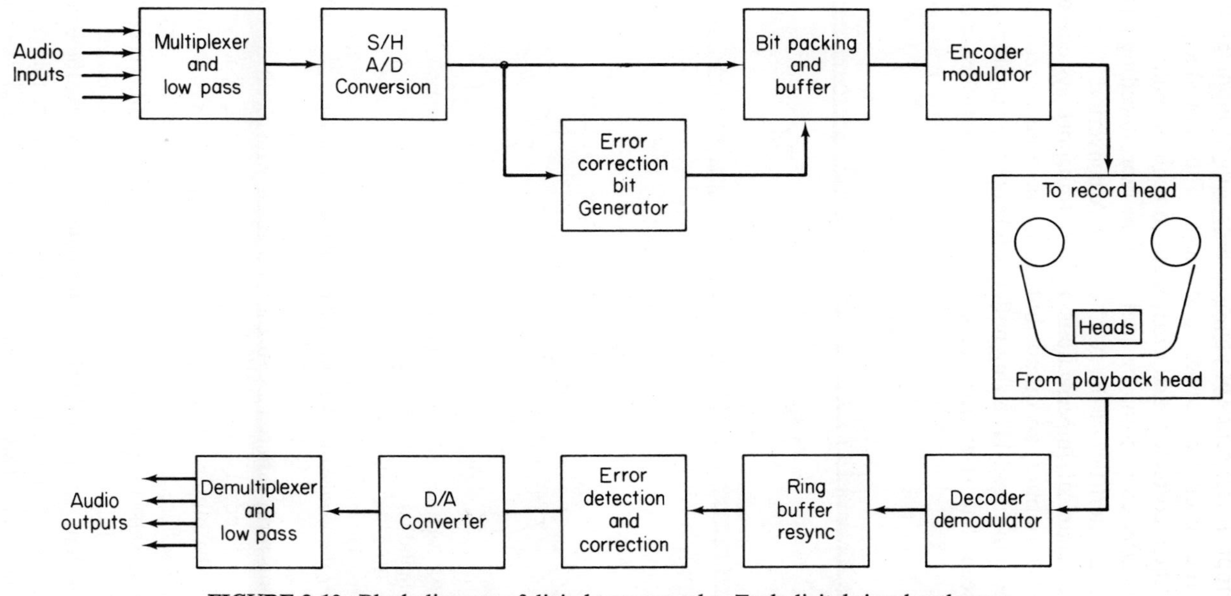

FIGURE 2-10. Block diagram of digital tape recorder. Each digital signal path may transmit bits either in series or in parallel. At the record and playback head, there will be one line for each recording track.

and block scaling are added. The resulting composite bit stream is transformed to an appropriate series of analog pulses using a modulator. During reproduction, the reverse processes take place. The reproduced signal is decoded, unpacked, corrected, and distributed to each output. If we assume no digital errors after correction, the quality of the reproduced signal is determined by the characteristics of the A–D and D–A converters.

Although mechanical imperfections in the transport system will still introduce speed variations, an output buffer can be used to reclock the bit stream. As shown in Fig. 2-11, a ring buffer is filled at the variable rate equal to the tape speed, and is emptied at a fixed crystal-controlled clock rate.

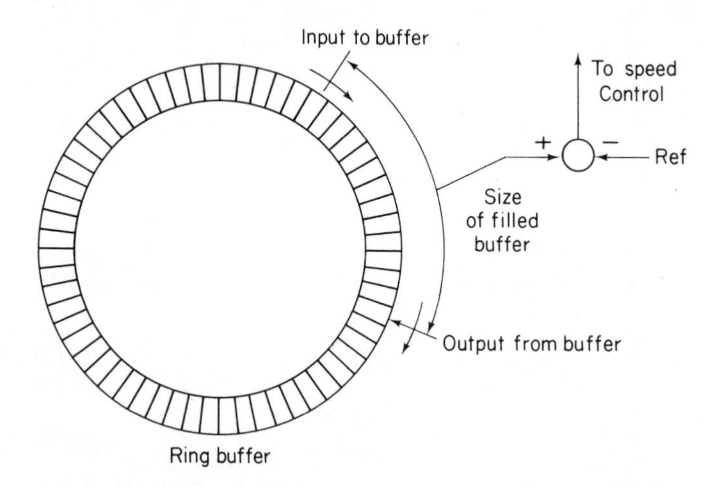

FIGURE 2-11. Ring-buffer storage system to synchronize playback from digital tape recorder. The inputs to the buffer are the digital words from the playback head; words are placed in successive registers. The output from the buffer is fed to the D/A conversion system; the words are taken from the buffer consecutively.

Clearly, any long-term differences in the two rates will result in the buffer being emptied or filled. To avoid this, an auxiliary signal, representing the degree to which the buffer is filled, is generated to control the capstan servomechanism. The difference between the input and output address, modulo the buffer size, is sent to a simple D–A converter. Any difference between this value and half-full is an error signal, which is used to control the main capstan servomechanism. Thus, as the buffer begins to empty, the error signal causes the tape to speed up; conversely, if the buffer fills, the error signal reduces the tape speed. The buffer need only be as large as the time constant of the capstan servomechanism. A buffer of 100 words has proved adequate.[32]

A more difficult design consideration is the physical nature of the tape recording process and its effect on both bit density and bit errors. A low bit density on the tape will make the reproduction more immune to errors, but the recording process will then consume a large amount of tape. Alternatively, a high bit density uses tape economically but introduces a higher probability of error. Because the digital tape recorder is intended for highest-quality applications, any bit errors must be corrected by some kind of error-correction algorithm. High-density recording places a substantial burden on the type of error-correction process being used.

To displace existing professional equipment already high in cost, a commercially viable digital tape recorder would need to provide substantially better quality. Thus we can assume that each channel will have to carry at least 0.3 Mbit/s until higher efficiency compression techniques are discovered. A four-channel system would thus require 1.2 Mbits/s; but eight channels would be preferred, and sixteen are already commonplace in professional practice. These are very high information rates for tape speeds of 15 to 30 ips. A laboratory digital audio tape recorder, described by Sato,(22) deals with this problem by basing the design on a video tape recorder, which has a much higher bandwidth than the conventional audio machine. Two audio signals with an information rate of 1.28 Mbits/s were recorded on 1-in. tape at 7.5 ips. The effective head speed of the helical scan mechanism, however, was about 700 ips; the bit density was in reality only 500 bits/in. A more sophisticated modulation scheme, however, was used to record eight audio signals on the same type of video recorder using a tape speed of 15 ips.(21) Engineers at the BBC Research Laboratories based an experimental model on a 16-track instrumentation recorder.(32) Each bit of a 16-bit digital word was recorded on a separate signal recording track with a density of about 500 bits/in.

The subject of modulation techniques is complex and cannot be discussed fully here. Additional information can be found in the references.(28,39) Modulation schemes form two distinct classes: self-clocking and non-self-clocking. The latter class requires an auxiliary timing signal to locate each bit. The former class contains all the information to determine the temporal location of each bit. For this application, the self-clocking codes are clearly preferred, since the recording of a time track would waste unnecessary bandwidth. If we assume that some kind of error-correction system will be incorporated in the digital tape recorder logic, the choice of codes should be based solely on the maximization of the information rate. The limit is then set by the channel capacity of the tape medium. This kind of analysis requires a detailed investigation of both the bandwidth and the SNR_{ws}, but such data are difficult to obtain and cannot be incorporated into the standard linear model since the tape medium is far from ideal. The transfer characteristics of tape contain a strong nonlinearity and hysteresis, which changes with different kinds—even different lots—of tape, recording heads, etc. The SNR_{ws} is equally complex in origin.

For a given tape width, multiple tracks require that each track be smaller in width. Although this decreases the SNR_{ns}, the bit density measured in bit per square inch is increased, because the SNR_{ns} often remains better than the required minimum. These issues have not been fully explored at the time of writing, and the performance of the prototype machine is probably far from optimum. Experience with the design of standard digital tape recorders for computer data gives very little insight into the limiting performance for the audio application. Data storage machines are very conservatively designed for minimum error. Typically, they use non-self-clocking codes with an auxiliary timing track. Densities are only about 800 bits/in./track. Experimental machines can, however, be built with densities ranging up to 30,000 bits/in; and densities of 80,000 bits/in. are thought possible.[131]

Error-correction techniques might seem an attractive alternative to a reduction in the recorded bit density, since the information density can often be made substantially higher without danger of irrevocable information loss. A simple form of error correction for digital tape recorders is a 1-bit parity check of the five most significant bits, followed by use of a concealment technique when an error is detected.[22,32] The concealment can consist of either a first-order interpolation between neighboring words or a zero-order hold of the previous word, as has been used for error concealment on long lines transmission[23] (see Sec. 2.4.2). Such techniques are, however, inadequate for digital tape recorders because the errors usually occur in bursts. Concealment is perceptually adequate only under the assumptions that the errors are random and occur at rates on the order of 1 in 10^5.[25]

Errors in tape recorders most often originate in drop-outs caused by tape imperfections and embedded dust particles. Empirical data indicate that these "holes" in the reproduced signal last for about 100 μs at speeds of 30 ips.[33] For a recording density of 10,000 bits/in., such a drop-out would create a burst of errors lasting upward of 50 bits. A simple approach is not likely to overcome so high a concentration of bit errors. Instead, an error-correction algorithm specifically designed for burst errors must be used. None of the prototype digital tape recorders discussed earlier had such error correction; they might therefore be considered somewhat primitive.

Because error correction is a broad and complex subject beyond the scope of this discussion, the reader is referred elsewhere for details[26,83] beyond the basic review of the design issues provided here. Error correction is based on the notion of adding redundant bits to the information bits that create the message sequence. The redundancy then forms the basis for error correction at the receiver. The resulting degree of redundancy can be measured using the Hamming distance: the number of bits in a message sequence that must be changed to create another legal sequence. A message sequence without any redundancy bits, that is, without error protection, has a Hamming distance of 1, because changing 1 bit will create another legal sequence. In contrast, if we consider the case in which 2 information bits are represented by the sequences 111000 and 000111, the Hamming distance is 6, because 6-bit errors must occur for the first message to be transformed into the second. This degree of redundancy would allow 2-bit errors to be corrected, while 3-bit errors could be detected but not corrected. In this example the redundancy has a very high cost since each information bit is encoded into 6 message bits.

One of the earliest 1-bit error-correcting codes is the *Hamming block code*.[84] The incoming data are divided into blocks of k information bits. From these k bits, $n - k$ redundancy bits are generated and become part of the block to be transmitted. Since the existence of a single error in any of the n bits of the block must be represented by the information of the $n - k$ redundancy bits, the following inequality must hold:

$$n \leq 2^{n-k} - 1$$

For example, 4 redundancy bits indicate 1 of 16 possible situations: a single error in any of 15 bits of the block or no errors in the block. Similarly, 5 redundancy bits are needed for 1-bit error correction in a block of 31 bits. Another measure of the code is its efficiency in terms of the ratio of the message-transmission rate to the actual information-transmission rate. The 31-bit block, for example, has an efficiency ratio of 0.84. The error-correcting effectiveness is inversely proportional to the efficiency. A 255-bit block has an efficiency of 0.97, but only 1 error in every 255 bits can be corrected.

The random error assumption is removed in another class of codes called *burst error-correcting codes*. There is a wide variety of types, which differ both in the assumptions that they represent about the nature of the burst errors and their respective difficulty in implementation. The reader may consult ref. 85 for a review. The simplest form of burst-error code can be derived directly from the random codes, although they tend to be either impractical or inefficient. Conceptually, the Hamming code can be transformed into a burst-error code by interleaving the bits from multiple blocks. Imagine that the information bits are divided into 63 blocks of 57 bits each. A 6-bit redundancy code is then added to each block as discussed previously. Each block is protected against 1-bit error at the receiver. Instead of transmitting the bits from each block sequentially, the first bit from each block is transmitted, then the second bit from each block, and so forth. Since the successive bits from a particular block are separated from each other during transmission by 63 other bits, a burst-error sequence of less than 63 bits would affect no more than 1 bit from each block.

In practice, errors in digital tape recorders originate from two sources:

(1) burst errors caused by drop-outs, and (2) random errors caused by background noise. The interleaved Hamming code will correct for either type of error since its original random error-correcting property is still present. Also, burst errors that originate from discrete particles and imperfections are unlikely to occur at a high rate. A burst every 100,000 bits would probably indicate tape of poor quality. However, the combination of a burst error and the background random error would probably result in a significant uncorrected error. A burst error of 30 bits would mean that 30 of the original blocks could no longer tolerate any random error. There are more complex codes that can correct both random and burst errors. See ref. 85 for an example of a recursive code. Among its other features, this type of code is relatively easy to implement in hardware. Unfortunately, there is almost no literature describing empirical results with these error-correcting codes in high-density digital tape recording.

Because its inherent performance capability is so much better than the analog equivalent, we may expect that the recording industry will eventually adopt digital recording. Among its advantages are the following:

1. An arbitrarily high SNR_{ns} and SNR_{ws} limited only by the A–D converter and packing density.
2. No wow–flutter speed variation because of the precision of a crystal clock.
3. No harmonic distortion near the upper signal range.
4. No interchannel crosstalk† for multiprogram recording.
5. No amplitude variations due to slight changes in magnetization of the tape.
6. Full bandwidth up to the Nyquist frequency.
7. No complex modulation-noise sidebands near the signal frequency.
8. No print-through between layers of the tape.‡

Such performance specifications are totally beyond any analog means and are extremely attractive even at additional cost.

† Crosstalk is the leakage of one signal into another because of magnetic leakage across the heads or through the amplifiers. A digital system could tolerate such leakage at −20 dB, whereas the analog specification is on the order of −60 dB.

‡ Print-through is a kind of crosstalk between two layers of tape. The close proximity of the tightly wound layers induces some magnetization. Like crosstalk, the specification for an allowable level is −60 dB for an analog machine, but only −20 dB for the digital equivalent.

2.3.2 Editing and Mix Down

Once the various musical segments have been recorded, they must be combined into an aesthetic whole. A selection might be recorded many times, and the final recorded performance made up of sections of each rendering. In a multiple-track recording, the separate tracks must be edited and combined to form the actual performance. Filtering, level changes, and other signal processing will take place during editing and mix down at the discretion of the recording engineer and producer.

The mix-down process can be extremely complicated. Assume, for example, that a two-channel master tape is being made from 16-track source material. One of the first decisions to be made is which instrumentalist or combination will appear in which of the two or four channels of the final recording. This implies that different amounts of each track must be summed and directed to each of the two or four outputs; this mix can also change with time.

In addition, each track is subject to separate signal processing. This can be as simple as level changes, or can involve filtering or special effects. The filtering, to effect timbre changes, might be constant for the selection or might be time dependent, being designed to treat different sections of the music differently. Special effects, such as artificial reverberation, are also used extensively. The result of this manipulation is a sound environment that exists only in the recording; the performance as finally heard is constrained only by the limits of electronics.

The mixing console used for this process can have hundreds of settings and adjustments. The engineer must determine the settings required during each stage of the program, and then effect changes in these settings as the music proceeds. Adjustments can be so complicated that they are virtually impossible to remember between editing sessions. Making changes in real time can also be quite difficult.

To facilitate mixing, several manufacturers have developed automated mix-down equipment. These systems are essentially small special-purpose computers that control the console settings. The engineer can program the initial settings as well as precisely timed changes. The signal processing, however, is still analog.

Conventional editing and mix down, even with automated equipment, have many problems. One is equipment redundancy, with its resultant expense. An analog circuit can handle just one input at a time. Sixteen or more channels must be controlled from the console, and each channel has its own level, filtering, and other controls, although all channels do not use all the controls available at any one time. Additional separate signal-processing hardware can be quite expensive. A studio will have a limited number of such devices, which are moved from room to room as needed. Their use requires remaking the signal connections every time a special effect is desired on a different channel. The simpler signal-processing and switching functions are therefore duplicated on each channel, a convenient but expensive measure, and the specialized hardware, whether built into the console or not, is available for only a limited number of channels, which is both inconvenient and expensive.

A second set of problems arises from the limitations of analog circuitry. The mixing console can perform only those functions that have been designed into it. Even external special-effects devices are constrained in function or range by the actual circuits and control functions supplied by the manufacturer. Any changes or modifications in the signal processing require either hardware changes in existing equipment, if possible, or purchase of new equipment. Despite the complexity of the equipment now available, the flexibility of the signal processing is in many ways quite limited.

A system incorporating digital signal processing as well as programmed operation overcomes many of the problems of conventional mix down. A small computer can be programmed to carry out the digital signal processing of each channel, and also to control the timing and synchronization of the mix down as it is made. Unlike the analog console, a digital system offers unlimited flexibility in the signal processing. Digital filters can be programmed and changed at will, and can be designed for specific applications and program material. Changes in filtering and level can be synchronized with steps in the digital sequence representing the program material, and timing can be specified to the exact digital word. Since these operations are stored as a sequence of program instructions, the signal-processing steps can be added, modified, or recalled as desired, allowing various effects to be tried without destroying previous control programs. In addition, an archival copy of the exact processing used can be stored along with the resulting music record.

The digital system also provides the engineer with a powerful display and analysis tool. A graphic display terminal, if part of the system, can show both processing and signal parameters simultaneously, making the effects of the signal processing immediately apparent. Spectral analysis of the signal is also possible, allowing frequency-domain filter design for special equalization effects. Control settings can be displayed, showing the signal processing under way and indicating changes as they occur. One major problem in any complex system is the man–machine interface, and a computer-controlled display can allow for far better information transfer than can an analog console.

A further advantage of a digital mix-down system is the ability to process digital recordings directly. Conventional consoles would require conversion of the digital sequence to an analog signal for processing and then back to a digital sequence for storage on each processing pass. The signal degradation caused by repeated decoding and encoding can be avoided if all the processing is digital. In fact, high-quality, high-density digital tape recording is necessary for the success of digital mix down.

For all these benefits, there are some clear-cut difficulties caused by the extensive computation power required to process music. Real-time processing requires computational speeds several orders of magnitude faster than commercial minicomputers. Non-real-time processing is also problematic, since this requires longer processing times and an intermediate storage system. A typical mass-storage device on a small computer is a disc with a capacity of about 10^6 words of storage, or 30 s of wideband program material. This essentially limits the processing system possible today.

Special-purpose computers with an architecture specially designed for audio have been built (see Sec. 2.3.3), but these can be extremely expensive. Moreover, such simple tasks as compression or spectral equalization, can consume most of the computational power, indicating that processing for multitrack mix down would still have to be performed in separate time segments. Nevertheless, a digital mix-down system is being developed by Stockham[126,127] using a Digital Equipment PDP 11/45 and an 800-Mbit disc storage peripheral. This is enough for storage of about 26 min of single-channel music at 32-kHz sampling rate using 16-bit words. General-purpose programming is being developed to implement such mix-down functions as level control, panning, and equalization. Stockham's system is intended for use with a digital tape recorder, with sections of music being read into the computer prior to processing. The results are recorded on digital tape for storage. Because the basic processing is digital, repeated operations do not produce the usual degradations. From a quality standpoint, a digital mix down is clearly superior to analog techniques, although this advantage imposes a substantial economic penalty.

2.3.3 Artificial Reverberation

Much of the signal processing performed during mix down requires special hardware (e.g., systems to generate artificial reverberation, dynamic range compressors, and noise reduction systems). Although this processing is usually carried out with analog equipment, digital techniques offer advantages both in quality and flexibility. Because most mixing is still being performed in the analog domain, and will remain analog in the immediate future, digital hardware to perform specialized tasks must be interfaced to analog parts of the system using standard converters. Such digital systems must therefore be capable of real-time operation. Not only are presently available minicomputers much too slow, but even most larger machines are also lacking in sufficient speed. This has stimulated the design of specialized high-speed processors whose architecture is oriented toward optimum audio signal flow.

The user will need to be provided with a control interface that corresponds to natural parameters. It is thus not desirable to provide a sound engineer with a typewriter-style keyboard console for programming; he must be given controls that correspond to familiar physical or psychoacoustic variables, the latter being preferred. This requirement places a considerable burden on the equipment designer, since he must have a good understanding of both the physical processing required and its perceived effects. These issues are considered in the following discussion on reverberation.

The sounds we hear are strongly affected by the acoustics of the listening environment. Acoustic energy injected into a space is reflected from the various boundary surfaces, so that the listener experiences the results of a multitude of reflections from different directions, which die away as the energy is gradually absorbed. An anechoic chamber makes listeners uncomfortable because of the complete absence of reverberation, whereas a large cathedral gives a satisfying sensation of overwhelming space because of the very high reverberation energy. In less extreme cases, the listener's awareness of the acoustic space remains only partially conscious. Nevertheless, the reverberation is a very important aspect of the auditory experience; the difference between an excellent concert hall and a mediocre auditorium is almost exclusively related to reverberation.

The listener's sense of space is dependent on the time of arrival of reflected energy, the gradual decay of the energy, and the directional properties of the individual reflections. All are necessary for the perception of being inside a hall, and one without the others gives an incomplete impression. A home living room is typically quite "dry" and does not produce the intense reverberant field present in a concert hall, so that the home music listener experiences the recorded reverberation as coming from the direction of the loudspeakers alone. Because domestic music systems are based on the use of two discrete channels, the recorded reverberation can only come from these two directions. In a later section, we discuss how home ambience enhancement systems can be used to provide the additional delayed and directional reflections normally missing from conventional two-channel systems, to supply a sensation of being within a large hall. In the present section, we shall discuss how reverberant energy can be added to the recorded signal.

The need for artificial reverberation normally arises because of the close microphone placement used in recording the original performance. The use of multitrack recording, with each instrument or group of instruments having its own microphone, allows for an adequate separation between the signals and yields an improvement in signal-to-noise ratio, but also prevents the natural room reverberation from being recorded. Popular music offers another motivation for multitrack recording, since different performers can record their parts at different times. However, the small recording studios most often used to record such music have little or no reverberation. A given performer can rerecord only his contribution without the other performers. Not only is this economically advantageous, but the sound engineer has better control over the final result. As a result, with most popular music, no "original performance" ever exists. Many special effects are added after all the performers have recorded their respective sections. In more than one example, a performer has recorded a number of tracks sequentially, creating a full-fledged "one-person band."

Classical music is often recorded with artificial reverberation. In this case the purpose is as much control of the reverberation in the final mix down as it is reduction in background noise. If the microphones were placed in the audience seating section of the concert hall, the sound would be natural; but it would also contain the background noise produced by ventilation systems, passing airplanes, trucks, elevator motors, and, at a live concert, the audience. Close-microphone techniques minimize background noise, because the signal intensity in the region of the instruments is as much as 20 dB greater than elsewhere in the hall. Aside from the issue of background noise, the reverberation in a concert hall varies with the size of the audience. An empty hall has a reverberation time that is as much as 20 to 30% higher than a filled hall. Only at actual performances do the acoustic conditions exist that might permit a "natural" result, but these are also conditions that make high-quality recording impossible because of the added background noise present.

This problem of background noise is even more acute during sound recording for motion pictures and television. The equipment used and movement of personnel generate noise that can only be minimized by the use of highly directional microphones, or, in the case of films, by dialogue "dubbing" after photography. Often the visual illusion required of the stage set will produce undesirable acoustic characteristics.

Three methods of producing artificial reverberation are used currently:

1. Acoustic chamber.[98]
2. Mechanical plates.[96]
3. Mechanical springs.[97]

In facilities where the space is available, a reflective room, sometimes filled with randomly shaped objects, is used to provide reverberation. The "dry" signal is played through loudspeakers in the reverberant room; this sound, picked up by microphones in another part of the room, is mixed with the dry signal before recording. The objects in the room are used to break up the regular modes of a rectangular room. If well designed, the reverberation chamber produces satisfactory reverberation. However, the characteristics of the chamber are almost impossible to change. A more common system uses a mechanical plate of gold alloy composition.[94,95] A transducer excites the plate at one point, and a pickup located at another point detects the energy transmitted through the plate along various paths. A similar system uses a nonuniform spring. All these techniques have quality limitations.

The design of such mechanical systems is very difficult, amounting almost to an art form, with an admixture of alchemy, in which the designer attempts to overcome the pronounced limitations of the physical device. A spring, for example, must be nonuniformly etched to create a high density of resonances and an irregular echo pattern. The plate is constructed from a complex and costly alloy, with characteristics that must be extremely carefully controlled. Even with very sophisticated manufacturing technology, these mechanical systems exhibit severe limitations at high frequencies, where short wavelength and high transmission velocities result in excessive attenuation and limited dynamic range.

During the last 15 years efforts have been aimed at using digital signal-processing techniques to produce all-electronic artificial reverberation. To understand the techniques by which such a system is designed, an understanding of natural reverberation is desirable.

Sound emitted in a room will be partially reflected from the room boundaries until all the energy has been absorbed. Because the reflection coefficient of these surfaces is always less than 1, the energy in the acoustic wave is reduced with each reflection. At high frequencies, most materials used in building construction have lower reflectance, and additional losses are caused by air absorption. In theory, the reverberation process could be completely described by the echo pattern or impulse response of the room. The process is indeed sufficiently linear so that one can view the performing stage as the input and a listener's seat as the output. The Fourier transform of the impulse response provides an alternative description of the reverberation process.

However, the detailed echo pattern is sufficiently complex and difficult to measure that a complete description is not possible. Although empirical investigations have tried to measure room impulse response using spark excitation (actually a doublet) and recording the echo pattern at a particular seat, the spark intensity cannot be made large enough to overcome the background noise without introducing nonlinearities in the air medium. In addition, the echo density is so great that any air currents and normal dispersion make the result vary from trial to trial, which prevents the use of averaging techniques. Approximations for the initial part of the reverberation process have been made using scale models of Boston's Symphony Hall;[60] computer studies of ray tracing have produced similar results.[99]

However, the perception of the reverberation process does not appear to depend on the detailed fine structure of the process but only on its general properties. Thus we need not simulate a specific reverberation process, but only those physical properties that have important correlation to perception.

Data describing the perception of reverberation are not available except for a few specific aspects. For this reason, assumptions must often be made during design, which are later tested empirically.

Calculations from physical acoustics show that the average echo density in a rectangular room is given by

$$\frac{dN}{dt} = \frac{4\pi c^3}{V} t^2$$

where V is the room volume and c is the speed of sound. In an irregularly shaped room, the equation changes form, but remains similar once the statistical aspect of the process becomes dominant. The initial echo pattern is dependent on the particular room shape. In his study of concert halls around the world,[(65)] Beranek has observed that all halls rated A+ by expert listeners had a delay time between the direct sound and the first reflection of between 10 to 20 ms. This measure appears to correlate with the listeners' sense of "intimacy."

Reverberation systems are sometimes evaluated by measuring the time for the echo density to reach 1/ms. For quality systems, this time is on the order of 100 ms. Although this measure is used with synthetic reverberation systems,[(100,64)] it has not been used for predicting the quality of concert halls. Rather, Meyer[(101)] proposed that the "clearness" should be defined as the ratio of the reverberation energy arriving in the first 50 ms to the total reverberant energy. This measure discards the geometry of the initial echo pattern and focuses instead on the energy returned from room boundaries in the initial process. For New York's Avery Fisher Hall, this ratio was -3 dB;[(102)] it is felt that change of a few decibels in this variable is critical.[(63)]

As the reverberation process continues, the energy should normally decay exponentially. Sabine's[(107)] early exploration of reverberation in an irregularly shaped room showed that the process was indeed exponential; however, for certain room shapes nonexponential decay can occur. The time for the energy to decay to -60 dB is defined as the reverberation time (RT). This variable is probably one of the most important in determining the perceived character of the acoustic environment. A cathedral, for example, might typically have a reverberation time of about 5 s; for a typical home living room the value is usually less than 1 s. Beranek[(60)] reported that the average mid-frequency reverberation time for the best concert halls was 1.9 s.

Perceptually, however, it has been found that the sense of reverberation time is essentially determined by the energy decay in the first 160 ms[(114)] with continuous music; the early part of the reverberation process dominates. It is necessary to distinguish between *running reverberance* and *stopped reverberance*. For continuous music, only the early part of the reverberation process (running) is audible between music notes. On the other hand, if the music abruptly stops, the complete process (stopped) is audible.

In contrast to time-domain analysis, the frequency domain offers a different set of insights into reverberation, which bears on the coloration of the music. The acoustic environment acts as a linear filter, attenuating certain frequencies and augmenting others. Although it is generally accepted that flat frequency response is most desirable for music reproduction, the wide variations in frequency response of a concert hall are not necessarily audible, because the irregularities in response are narrow enough so that even the purest musical notes cover many peaks and valleys.

The degree of spectral irregularity can be determined statistically. The real and imaginary parts of the response are Gaussian, and the probability density function of the logarithm of the energy (decibels) is given by

$$W(z) = \exp[z - \exp(z)], \qquad z = \frac{\ln (p^2)}{p^2}$$

where W is the density function and p the pressure random variable.(106) The standard deviation of the level about the mean has been found to be 11 dB. Thus 70% of the frequencies have less than 11 dB difference from the mean transmission gain. Other empirical data from 19 concert halls show that the mean difference between successive minima and maxima is about 9 dB, independent of room volume and reverberation time.(105) The spacing between successive maxima was found to be 6.73/RT, where RT is the reverberation time.

Since these concert halls do not produce a sensation of coloration, we can conclude that the roughness of the frequency response is not undesirable and is a natural part of the reverberation process. The origin of the variation lies in the high number of room resonances or eigentones (poles) in a large hall. For a regular-shaped room, this is given by

$$\frac{dN}{df} = \frac{4\pi V}{c^3} f^2$$

where V is the volume, c the speed of sound, and f the frequency in hertz. On the average, the resonances have a high Q, which is proportional to the reverberation time. If the eigentone density is very low, in a room with a long reverberation time, the individual resonances will be very strong and will produce a strong degree of coloration. The acoustics of a bathroom illustrate the strong coloration resulting from low eigentone density with a high reverberation time.

Experimental investigations have been carried out to determine the required eigentone density in an artificial reverberation system.(103) In one experiment, octave band white noise pulses were reverberated. The direct sound was removed, and the reverberation signal was compressed to keep its level constant; subjects were asked if they could hear any difference between this signal and the original. The results indicated that an eigentone density of 0.1/Hz was required at low frequencies, and a density of 3/Hz was required at 1 kHz. This is perhaps the strictest test of required density. A less stringent test excited comb filters with steady-state white noise, with results suggesting that a density of 0.14/Hz was required to achieve a minimum coloration.[104] This test technique is more closely related to the coloration for running reverberance, where the first test was more like stopped reverberance. It will become clear that these data are rather inadequate to specify design criteria for electronic reverberation.

An early study of the simulation of natural reverberation using a digital computer was reported by Schroeder and Logan.[109] They observed that a delay line with feedback behaved like a reverberation process, as shown in Fig. 2-12. The impulse response of such a system is a series of echoes of decreasing amplitude separated by the delay time. The frequency response is a comb filter, with comb spaces equal to the reciprocal of the time delay. The reverberation time is given by

$$\mathrm{RT} = \frac{3T}{\log_{10}(g)}$$

where g is the feedback coefficient and T the delay line length. This simple structure is not adequate as a reverberation simulator because the high echo density is accompanied by a low eigentone density. However, a series of such

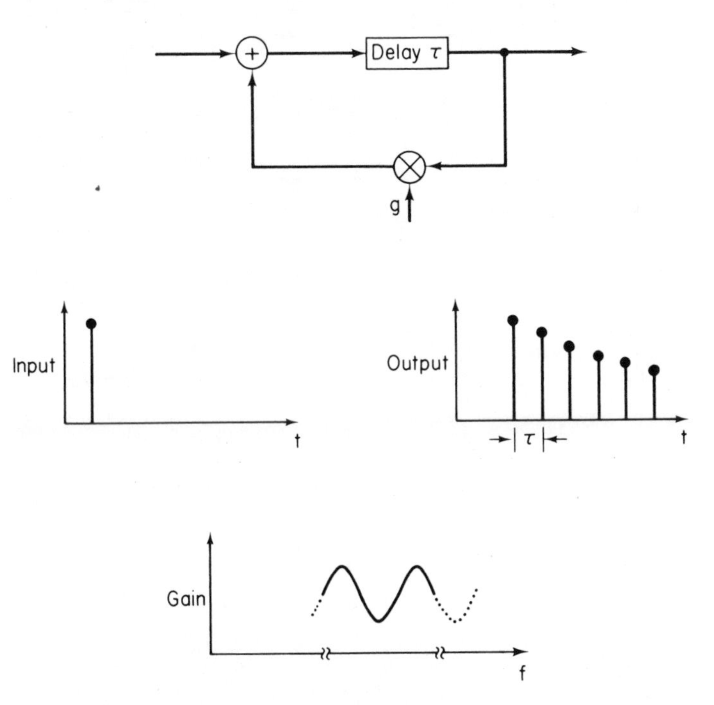

FIGURE 2-12. A delay line with feedback to simulate reverberation. A single impulse at the input produces a train of impulses separated by the delay time. The amplitude of the pulses decreases exponentially with the factor g.

networks can be cascaded. In this case, the echo density increases in accordance with the expression $d = t^{n-1}$, where n is the number of sections cascaded; the total eigentone density is the reciprocal of the total delay in the system.

Such a cascaded system is, in principle, a good approximation to real reverberation; however, the number of discrete sections means that some of the eigentones from different sections will match to produce very strong resonances at some frequencies. Other frequencies will be repeatedly attenuated. Schroeder observed that the individual sections could be made into all-pass networks by adding some of the unreverberated signal to the output of the feedback delay line, as shown in Fig. 2-13.

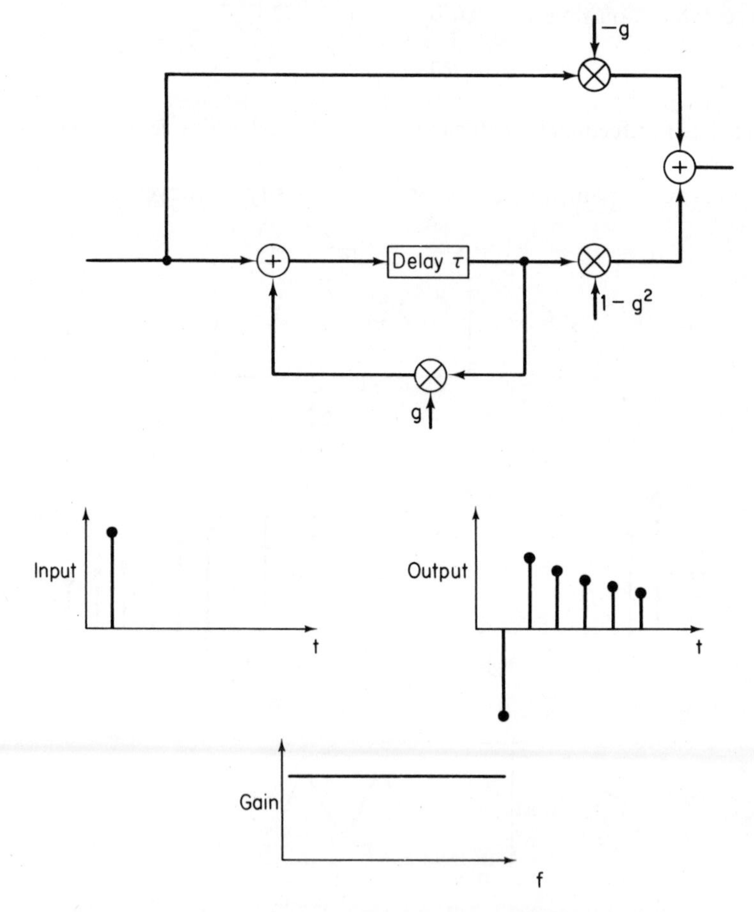

FIGURE 2-13. An all-pass delay structure with flat frequency response. It differs from Figure 12 in that the direct path is also fed forward to the output.

The transfer function of the all-pass is given by

$$H(s) = e^{-sT}\frac{1 - ge^{sT}}{1 - ge^{-sT}}$$

where g is the feedback coefficient and T the delay time. The all-pass configuration has the advantage that frequencies that are not strongly reverberated, by virtue of being far from a resonance, will be passed unattenuated to the next all-pass section, where they might be reverberated. The effect is to allow each resonance to be excited at the same intensity. Without the feed-forward paths to produce the all-pass, a few resonances would dominate all others. After much experimentation, Schroeder chose five all-pass sections with delay times of 100, 68, 60, 19.7, and 5.85 ms, with a feedback coefficient of 0.7. The resulting eigentone density is 0.25 Hz, which satisfies the running reverberance coloration criterion. The echo density of the system increases as t^4, which is faster than natural reverberation.

Schroeder(111,112) later modified his original configuration to that shown in Fig. 2-14. The main reverberation process no longer contained all-pass

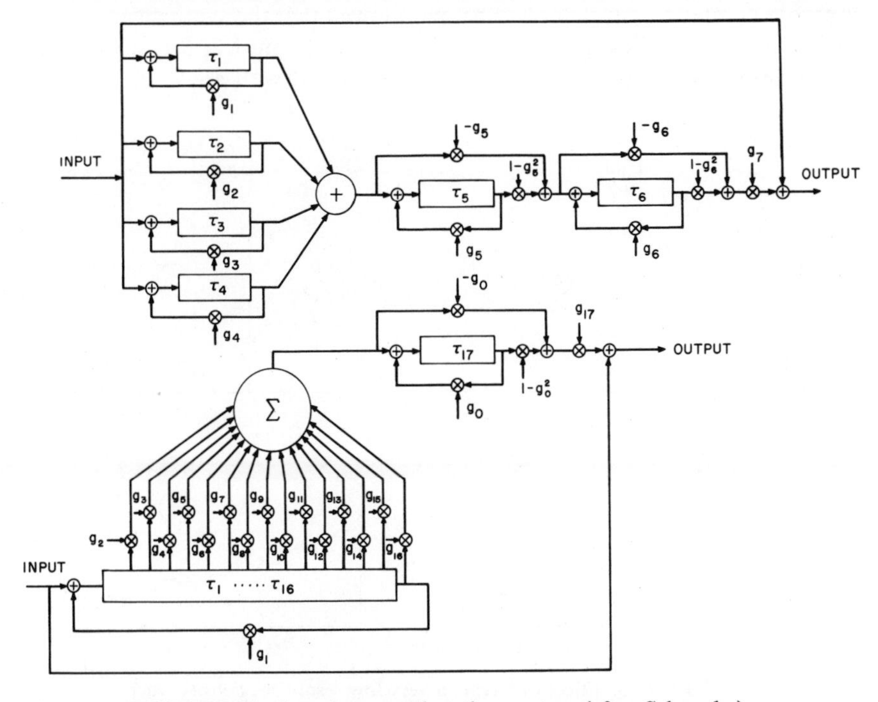

FIGURE 2-14. Complete reverberation system (after Schroeder). The four main loops provide long-term reverberation and the two all-pass networks following provide the desired short-term echo density. (Reprinted with permission of the Audio Engineering Society, Copyright 1975, Volume 23, No. 9.)

networks, since all four main loops were excited in parallel. Only the two sections that were used to boost the echo density were configured as all-pass networks. Delay times were relatively short for high echo density. The all-pass technique prevented the strong comb shape from affecting the steady-state frequency response. The delay times for the four main loops were chosen in the region of 30 to 45 ms, with values that were "maximally incommensurate." Unlike the first simulation, this configuration allowed the reverberation time to be readily adjusted, and the reverberation time could be made a function of frequency.(113)

All these simulations were programmed on an IBM 7090 computer, which is extremely slow by today's standards. The program could not be run in real time; a short segment of music was stored digitally and then reverberated. This limited the amount of experimentation.(110) From his own listening experience, Schroeder indicated that this synthetic reverberation was indistinguishable from the natural kind, even with such difficult signals as clicks and white noise.

A repetition of the Logan and Schroeder experiments using a special-purpose computer,(108) described below, showed that the synthetic reverberation obtained in this way was inadequate for pure musical tones. Consider, for example, a flute note at the frequency marked with an $\times$ in Fig. 2-15.

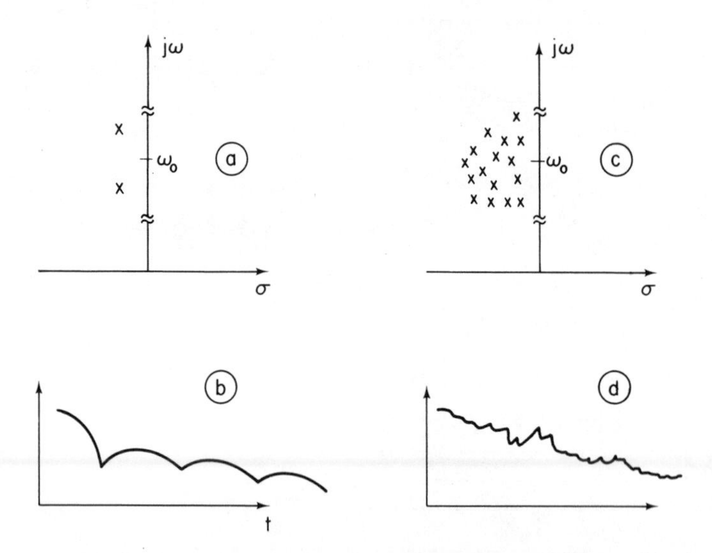

FIGURE 2-15. Effect of having a very low resonance density when simulating reverberation. (a) *s*-plane showing two dominant resonances near a tone frequency of ω_0. (b) Envelope response of a tone burst at ω_0 with an envelope ripple equal to the frequency difference of the two resonances. (c) Natural reverberation with very high resonance density. (d) Envelope response with minor higher-frequency perturbations from true exponential shape.

In this frequency region the reverberation might produce two dominant resonances indicated by the two poles. The reverberation of the flute then has a characteristic envelope modulation, which is equal to the frequency separation between the two poles; this is called *flutter echo*, a sound like that generated in some corridors or empty rooms with hard walls. Because the eigentone density is much higher in real reverberation, such dominance by a pair of poles does not occur in a quality concert hall. Instead, the envelope modulation appears to be random, with a bandwidth on the order of 10 to 30 Hz. The Schroeder–Logan simulation contains many frequencies at which the envelope is periodic at 1 to 4 Hz, which is clearly perceptible and musically disagreeable.

A need for extensive investigation of artificial reverberation in the development of an all-electronic system prompted the development of a special-purpose computer that can process music in real time.(115)

One major difficulty in researching artificial reverberation is in building digital systems that are both flexible and yet fast enough to process music in real time. The simulation just described could have been built in "dedicated" hardware so that it would run in real time; however, such rigid logical architecture would be difficult to change. The general-purpose computer is much more suitable for experimentation because of its programming flexibility, but the high data throughput rate is not compatible with the capabilities of such a computer. To process full-bandwidth music, the sampling period needs to be on the order of 30 μs. Thus the entire signal-processing program must run its course in this time in order to be ready to process the next music sample. The simulation described has 6 delay lines, and carries out 10 multiplications, 13 additions, and miscellaneous instructions for data transfer during each sample cycle. A typical computer that uses memory reference for each instruction has an internal cycle time of about 1 μs, but a hardware multiplication can take from 30 to 300 μs. Most computers have an architecture that is more appropriate for complex numerical calculations or data manipulation than for signal processing.

For this reason, a special-purpose minicomputer was designed(115) for music experimentation. The design goal was to have twice the computational power of the simple Schroeder–Logan reverberation system. This extra computational power would be needed to implement filters in the delay line feedback loop to make the reverberation time a function of frequency. Computational power, however, is not related to speed alone, since it also depends upon the number of instructions needed to perform certain types of operations. For example, if seven instructions are needed to implement a delay line with each instruction taking 100 nanoseconds (ns), the total time for the operation is 700 ns. An alternative architecture that only required three instructions of 200 ns each would be computationally more powerful. An optimization for music led to the architecture shown in Fig. 2-16. Only the basic signal flow paths are shown.

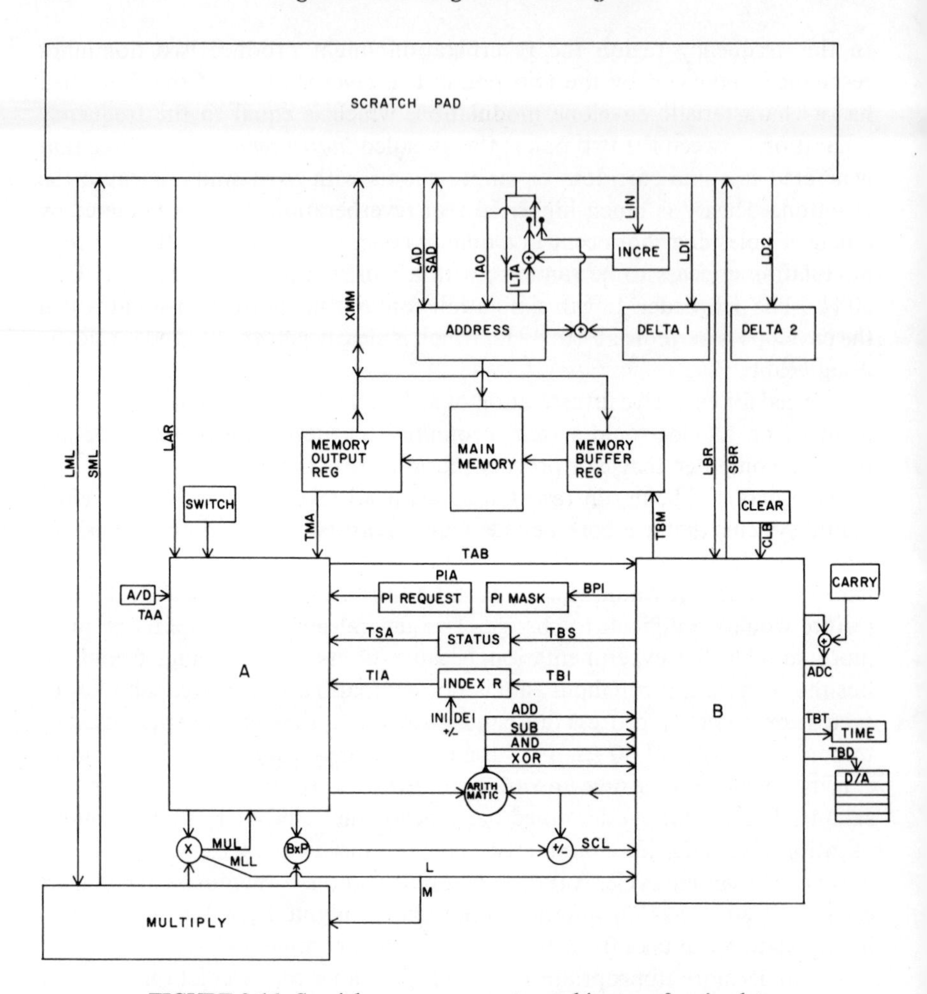

FIGURE 2-16. Special-purpose computer architecture for simulating reverberation systems. Only the main signal flow paths are shown. (Reprinted with permission of the Audio Engineering Society, Copyright 1975, Volume 23, No. 9.)

The computer contains two different types of memory, bulk storage of 16 K words to implement the delay lines, and a scratch pad memory for storage of temporary results and the program. The bulk storage memory provides a total delay time of over 400 ms, but its cycle time is rather slow. However, the slow speed does not reduce the computational power since the architecture separates the address load from the data transfer. Once the memory cycle has been initiated, the main program can carry out other operations while waiting for the data to become available. The scratch pad, on the other hand, contains only 512 words, but is built with high-speed transistor–transistor logic (TTL).

All arithmetic operations are performed in the A and B registers, with the M register being used for multiplication. The instruction sequence could be to load data into these registers and then to execute the required arithmetic operations. The effective speed is increased, however, since the instruction word is composed of two parts, each part consisting of a separate instruction. During the arithmetic operation, for example, new data can be loaded into a source register. By using such a split instruction word and a partial pipeline structure, the effective computational power is increased by a factor of 2.

A multiplication operation is usually very slow when it involves a full 16- by 16-bit product. For audio, the resolution of a 16-bit word is excessive; usually resolution of a fraction of a decibel is adequate as long as the range is very large. The MUL instruction is a 4-bit multiplicand in the range from 0.97 to 0.5. An additional instruction SCL (scale) produces a scaling by factors of 2^n. The SCL and MUL instructions can each be completed in one machine cycle of 200 ns. Moreover, since each instruction is only half of an instruction word, its execution can be combined with other operations.

Operation for a simple structure is shown in Fig. 2-17. The addition of two attenuated signals *Za* and *Zb* takes a total of 1.6 μs. The program is shown in Fig. 2-17.

Note that several half-instructions are available for other operations. Assuming that these instructions can be used for other purposes, the effective execution time is only 1 μs. This compares very favorably with even the fastest current computers.

The instruction set contains a special group for the operation of the delay lines. A hardware ring buffer is established by the use of three special instructions, which organize the main memory as a delay line. The addition of a tap to a delay line takes only an extra 200 ns. The main memory is divisible into an arbitrary number of delay lines. The program uses four half-instructions per delay line.

Using such a computer, a simulation with 15 delay lines, 35 multiplications, 48 additions, and a few miscellaneous control operations has been implemented in real time (i.e., each signal sample is fully processed in 30 μs). By application of this kind of research tool, various complex reverberation structures can be tested. Results from this kind of research have led to a commercial reverberation system.[80] This system has a quadraphonic output with each output being uncorrelated, an adjustable reverberation time from 0.3 to 4.5 s, a frequency-dependent reverberation time at high and low frequencies, a variable delay between the main sound and the first reflection, and a signal-to-noise ratio of over 70 dB. Such performance easily exceeds that possible with mechanical systems. As the cost of digital components decreases, we can expect that such systems will eventually replace mechanical reverberation devices.

The inherent advantages of control of perceptual parameters, such as reverberation time as a function of frequency, make this kind of system

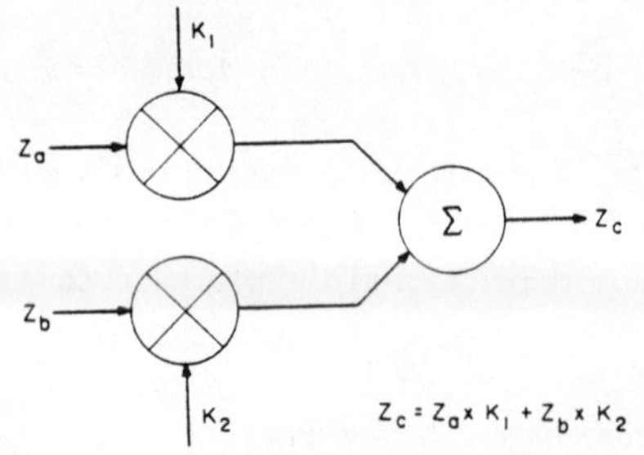

1. CLB LAR *a* — Clear *B* register, load *A* register with the contents of scratch pad location *a*. In this location the program (earlier part) placed the music sample Z_a.

2. --- LML *b* — Load the *M* register with the contents of scratch pad location *b*. In this location the program (earlier part) placed the coefficient K_1.

3. MUL --- --- — Execute a multiply of the music sample which was placed into the *A* register (machine cycle 1) by the lower four bits of the *M* register.

4. SCL LAR *c* — Execute a scaling of the *A* register by the next four bits of the *M* register and add the result to the contents of the *B* register. Since the contents of the *B* register was 0 by virtue of the CLB in machine cycle 1, the *B* register contains $Z_a \cdot K_1$. Also, simultaneously, load the *A* register with the contents of scratch pad location *c*. In this location the program (earlier part) placed the music sample Z_b.

5. --- LML *d* — Load the *M* register with a new coefficient which is found in scratch pad location *d*. In this location the program (earlier part) placed the coefficient K_2.

6. MUL --- — Execute a multiply of the music sample which was placed into the *A* register (machine cycle 4) by the lower four bits of the *M* register.

7. SCL --- — Execute a scaling of the *A* register by the next four bits of the *M* register and add the results to the contents of the *B* register. Since the old contents of the *B* register was the first attenuated signal, the *B* register now has the complete summation.

8. --- SBR *e* — Store the *B* register into scratch pad location *e*.

FIGURE 2-17. An example illustrating the programming of a special-purpose computer for reverberation. The flow chart (top) shows two signals being attenuated and mixed. The program (bottom) implements this structure in 8 machine cycles. Note, however, that there are 6 half-instructions available which can be used to simultaneously implement another operation. The effective computational time is about 1 μsec. (Reprinted with permission of the Audio Engineering Society, Copyright 1975, Volume 23, No. 9.)

highly adaptable for simulation of various acoustic environments. The user can control variables that correspond to the size of the hall, absorption of the walls, and the distance between listener and performer. Mechanical systems, on the other hand, have a single adjustment that damps the mechanical waves.

2.3.4 Chorus and Additional Effects

The additional kinds of processing that can be implemented with such a system are perhaps best illustrated by a technique that can transform a soloist into a chorus of as many performers as one wishes. Such a system is currently available(80) and is considered here both because of its simplicity, in technical terms, and because of its useful subjective effect in music recording.

When a group of performers play in unison, they cannot play each note at exactly the same time, and they do not achieve identical pitch. In a general way, the music of the performers is synchronized, but the fine structure shows a more or less random variation of the individual attack times and pitches. As a result, each musical note really consists of a spectral and time-domain distribution, which listeners identify as the sound of many performers. Figure 2-18 shows a block diagram of a system that transforms one voice into a chorus of five voices.

Each delay element contributes to the multiplicity of musical notes corresponding to the errors in the synchronization of a real performance. This delay, however, is driven by an external source. This has two advantages: (1) the actual delay times vary from note to note, and (2) the change in delay corresponds to a random pitch variation. In the steady state, the random interaction of phase variations corresponds to the effect produced by a real chorus. The nonstationary time delay of transients is also natural sounding.

In the block diagram of Fig. 2-18, the input signal is applied to the input of each of the four delay lines. Typical delays are 10 to 50 ms, corresponding to the range of asynchronism. The drive for the variable delay time is provided by low-frequency pseudorandom noise,(123) producing a noise bandwidth of about 5 Hz. Peak pitch variation needs to be limited to 0.1 to 1 Hz, depending on the energy distribution of the noise signal. Slightly different effects can be achieved by even lower frequencies (i.e., less than 1 Hz). Although the system is simple, the effect is dramatic and surprisingly convincing.

One of the most important advantages of digital systems is that the type of processing implemented can be changed without additional hardware if the architecture of the machine is sufficiently general. Each type of processing is simply another program that can be stored in a read-only memory. Thus the digital system can be all devices to all people, whereas a corresponding analog system must be dedicated to a single function. This advantage should not be overlooked, since the deceptively low cost of the analog equipment actually far exceeds the digital equivalent when the potential range of pro-

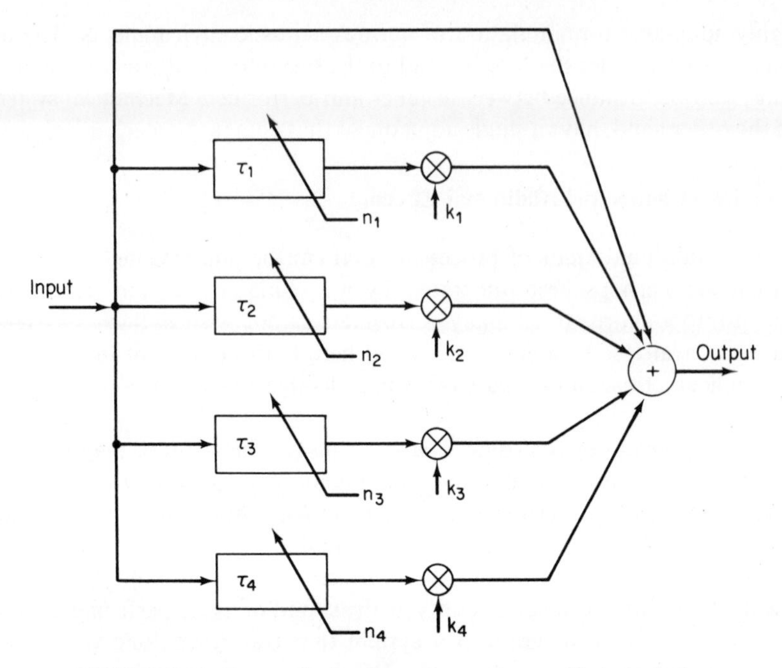

FIGURE 2-18. System for implementing a chorus effect. A single "dry" instrument voice is used to feed four delay lines of differing length. Each of the lines has a variable time delay which is controlled by an independent noise source with dominant energy in the region from 0–20 Hz. The outputs are summed to form a synthetic chorus.

grammed signal processing is considered. For this reason alone, digital systems will progressively invade the analog domain. At present, a determining variable is the cost factor. If history is any guide, digital systems will become more powerful at lower cost.

2.4 TRANSMISSION AND BROADCAST DISTRIBUTION

Once the final processed signal is available, it must be transferred to the domain of the listener through a physical medium, such as phonograph records, recorded tape, or broadcasting. Even though broadcasting can be considered an extra link in the reproduction chain, it is one of the most important means for transferring the audio signal into the home. The potential loss in quality is compensated by the vast range of program material that resides in the music library of broadcast houses. A typical broadcast station specializing in music might have many thousands of phonograph discs and innumerable transcription tape recordings. Moreover, highly efficient distribution networks allow programs to originate at locations remote from the broadcast studio.

The broadcast systems of the United Kingdom and European countries are considered reasonable subjects of national pride. In the United Kingdom, the BBC is organized as a quasi-independent government activity whose responsibility is to distribute several daily broadcast schedules, or programs, throughout England, Wales, Scotland, and Northern Ireland, where these programs are expected to reach virtually every household. Listeners owning receivers pay an annual license fee to support the activities of the BBC. Until recently, these broadcast signals were interconnected to local repeating transmitters using the standard techniques for distribution of telephone signals: long lines, modulated high-frequency carriers, microwave, and so on. The resulting quality at points distant from the originating facility was sometimes rather poor. As a simple matter of technical pride, it was desired to have higher-quality reception; politically, it was equally undesirable for listeners paying their license fees to have poor reception.

Several years ago a major research effort was initiated in Great Britain to consider the feasibility of changing the method of program distribution to a digital network.(36) Because of a lack of existing high-frequency distribution networks, it was initially thought that the audio part of a television program could be digitized and inserted into a blank part of the picture frame,(44) and such a system was developed. Today, however, the Post Office, which handles transmission and broadcasting, has installed standard 2.048-Mbit/s digital transmission lines, allowing audio programs to be distributed digitally. With this network, the broadcast system has raised the SNR_{ns} in the outlying regions from 40 dB to a minimum of 60 dB with full bandwidth. Many European broadcasting systems are now converting their transmissions to the same type of system in a standardized format.

The number of digitized audio programs that can be multiplexed using the Post Office transmission system is determined solely by the bit rate of the A–D converters. This is given by the product of the bits per word (dynamic range) and sampling frequency (bandwidth). Using the converters described previously, a single audio channel would generate 0.5 Mbit/s. Since this would only allow four monophonic or two stereophonic programs to be multiplexed onto the transmission system, further research into compression techniques was initiated about 1970. This effort was aimed at developing compression systems that allow use of a minimum bit rate without producing any perceptual degradation of the program.

2.4.1 Bit Compression

An initial experiment was conducted at the BBC Research Laboratories to determine the required dynamic range of the A–D converters. Based on subjective listening tests, a range of 84 dB (14 bits) was found necessary for granulation noise to remain inaudible to 90% of the listeners, using piano music.[37] Independent investigations in Germany later came to similar conclusions.[40]

Granulation noise could be made equally inaudible, however, if white noise, equal in level to twice that of the granulation noise, was combined with the program before quantization. It was also observed that this added noise could be reduced by 6 dB without loss of effect if a square wave at half the sampling frequency was added. The combination resulted in a noise degradation of only 2 dB referred to the granulation noise. These results suggested that a 78-dB (13-bit) dynamic range would be more than adequate when such a dither type of signal augmentation was used. It should be noted that this limited dynamic range was found acceptable mainly because the usual program sources at broadcast studios are recordings. Tape recorders, in particular, are used extensively, and these have a dynamic range of no more than 75 dB.[79] Thus there was no need to provide a transmission system better than the source material.

The specification of the SNR_{ns} (dynamic range) does not, however, indicate the noise permissible in the presence of the signal. Only by increasing the noise in the presence of signal can bit compression be achieved. A *compander* is one such system; a compression transfer characteristic precedes the digital conversion, and a complementary expansion characteristic follows reconversion. Since the compression is a direct function of signal level, it also is a function of the increase in quantization error.

These techniques, which have been studied for a long time for speech applications,[45,46] consist of a well-defined nonlinearity preceding the A–D converter and a corresponding nonlinearity following the D–A converter. Usually, these nonlinearities are of the lin–log form, as for example the μ law,

$$F(x) = \operatorname{sgn}(x)\frac{\ln(1+\mu|x|)}{\ln(1+\mu)} \qquad 1 \geq x \geq -1$$

where x is the input signal, $F(x)$ is the nonlinearized signal feeding the A–D converter, and the factor μ controls the degree of compression. As μ becomes larger, the SNR_{ws} remains constant over a wider signal variation, but takes on a lower value. The nonlinearity is a logarithmic characteristic for large signals and a linear one for small signals. The factor μ controls the signal level at which the transition from linear to logarithmic takes place.

Such a system was investigated by the German Post Office for transmission of music.[40] They used a slightly different compander characteristic, the A law, which is given by

$$\begin{aligned} F(x) &= \operatorname{sgn}(x)\frac{1+\ln(A|x|)}{1+\ln(A)} \qquad & 1 \geq x \geq \frac{1}{A} \\ &= \operatorname{sgn}(x)\frac{A|x|}{1+\ln(A)} \qquad & \frac{1}{A} \geq x \geq 0 \end{aligned}$$

where A, the compression factor, was set to 87.7. The companding advantage in this example is 24 dB, which is the incremental gain difference for low- and high-level signals. This allows a 14-bit dynamic range to be represented by a 10-bit word; however, the SNR_{ws} was only 50 dB for signals ranging from full scale to −24 dB. For signals below −24 dB, the SNR_{ws} decreased linearly. This performance level is only marginally adequate for high-quality transmission. Psychoacoustic experiments have shown that tone bursts between 500 and 5 kHz require a SNR_{ws} that exceeds 50 dB for high-level signals(10); however, low-frequency organ notes, and certain steady-state signals such as English horn notes, require a high SNR_{ws} if the signal is to mask the noise. Low-frequency organ notes are particularly problematic since they have high energy but little subjective loudness. Moreover, quantization noise is broadband and is not effectively masked at mid-frequencies.

To avoid these difficulties, pre- and deemphasis filtering is used before compression and after expansion. The European standard (CCITT) calls for a reduction in level at frequencies below 2 kHz of 13 dB, and a boost in frequencies above 5 kHz by 4 dB. The reverse filter is placed at the output of the D–A converter. This produces two benefits. First, high-amplitude low-frequency energy in the music is attenuated so that the compressor–converter combination is no longer operating in the region of maximum change, and therefore maximum error, thus reducing quantization error. Second, high-frequency attenuation at the output attenuates the noise that is generated. However, the program level must be somewhat decreased to avoid overload of the converter. Such filters raise the SNR_{ws} from 50 to 58 dB.

Companders have only recently gained widespread acceptance in the audio profession because the nonlinear characteristics introduced various errors, some actually emphasizing noise. Some circuits were unstable and difficult to adjust. A compander characteristic could, however, be implemented digitally to form a piecewise approximation to the desired curve. Figure 2·19 shows that a standard 14-bit A–D converter can be controlled by the compander characteristic directly. Each decision during the successive approximation does not correspond to a single bit, but to a fixed increment of the piecewise nonlinearity.

Researchers at the BBC Research Laboratories, examining the same problem,(42) came to a different conclusion than the German Post Office investigators. After extensive listening tests with a varied selection of music and listeners, they chose a block floating-point-type compression system, also called *syllabic companding*. The output of a 13-bit A–D converter was transformed to a floating-point representation having 2-bit scale information and a 10-bit mantissa. The four scale ranges provided 18-dB gain variation to achieve a 78-dB SNR_{ns}; the 10-bit mantissa provided a constant SNR_{ws} of 57 dB (average) for signals between full scale and −24 dB. With the addition of CCITT pre- and deemphasis, the SNR_{ws} was increased to about 65 dB.

The 2 scale bits of information did not accompany each 10-bit mantissa; rather, a block floating-point representation was used.[41] The digital word sequence from the 13-bit A–D converter was stored temporarily in a 1-ms buffer. All the words in the buffer were scaled such that the largest sample still remained below overflow. The scale factor is then transmitted at the relatively slow rate of once per block, while the mantissa transmission rate is the same as the sampling frequency. The effective rate is thus 10.125 bits/word.

Extensive listening tests shows that this form of encoding could not be distinguished from the original music samples. Both 9-bit mantissa and the instantaneous companding proposed by the German Post Office were judged inferior to the original.[11]

Such block floating-point companding does not produce any perceptible noise of its own, since the noise amplitude follows the signal amplitude with a time constant of about 1 ms. Since this is several orders of magnitude faster than the response of the human auditory system, the delay can be ignored. Although the *A* law instantaneous compander and the block floating-point compander both have the same information transmission rate, the latter performs better because the SNR_{ws} is a critical 6 dB higher. Seen analytically, however, the advantage is not inherent in the system, but is a function of the statistics of the music signal. The floating-point compander assumes that the music signal has a low peak-to-average ratio for short periods of time. This average is extracted in the scale factor and transmitted at a slow rate. In contrast, the instantaneous compander can respond to a rapid level change of a single sample without affecting the neighboring samples. Such a system would perform better if the music signal contained very

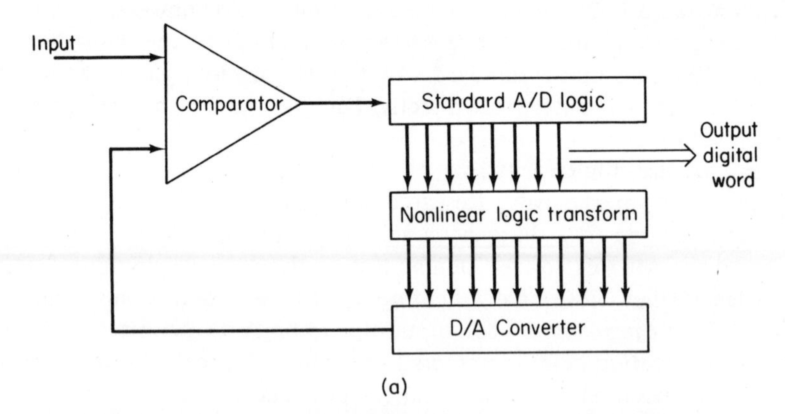

FIGURE 2-19a. A successive-approximation A/D converter incorporating a non-linear compander characteristic. A standard successive-approximation logic circuit drives a non-linear transformation network so that the bit tests correspond to the desired non-linear curve. The output is a non-linear version of the input.

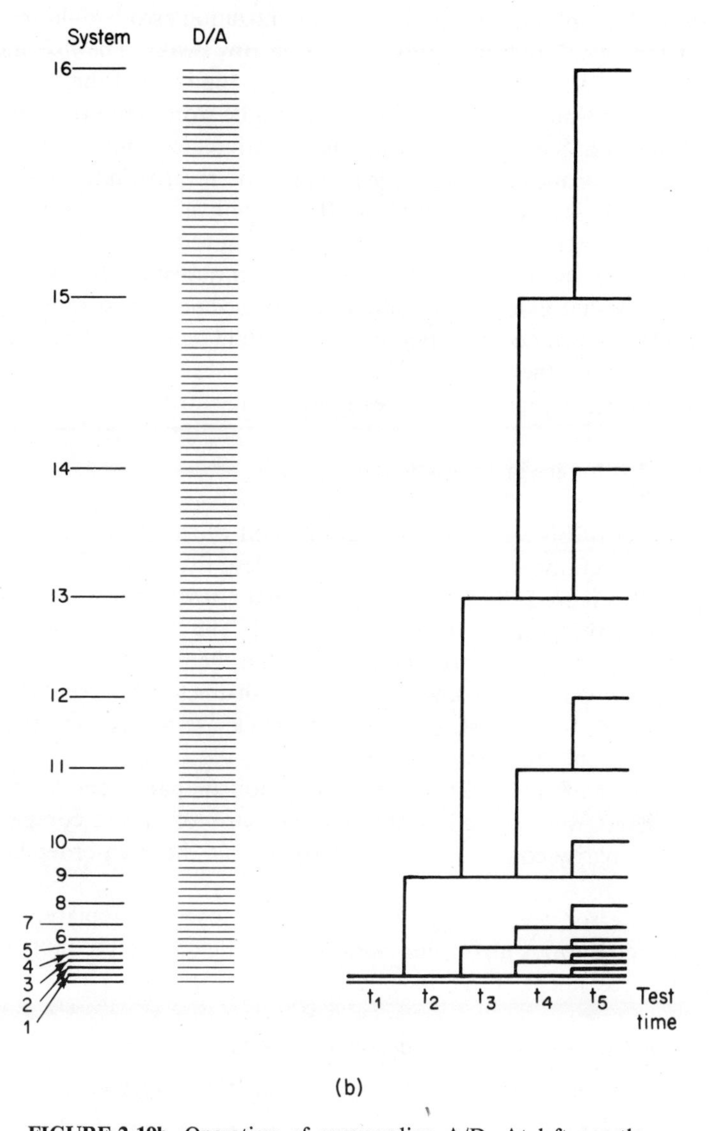

(b)

FIGURE 2-19b. Operation of companding A/D. At left are the logarithmically spaced quantization levels for a 5-bit process (4 bits plus sign), and the linear characteristic of the actual D/A converter. The logarithmic quantization uses many fewer levels than are available. At right is the decision sequence of the system showing the order in which the bit levels are tested. The non-linear logic produces an increment depending on the current value. In the example, an 8-bit D/A is the basis for a 5-bit conversion. At low levels, the companded and straight conversion both have the same resolution; at high levels, the companded version lacks most of the quantization levels.

narrow, high-amplitude peaks, since the quantization error would be low for most of the signal and high only for the narrow peaks. The floating-point system must minimize its quantization error on the basis of the highest peak during the 1-ms interval. In principle, it would be worse than the instantaneous compander. Since most music sounds originate from an acoustic or mechanical resonance and are, at least partially, reverberant, the low peak-to-average ratio is a good assumption. If electronic music were used as a test sample, however, the performance advantage might go to the other system, as electronic music tends to have few a priori statistics. In this example, we see clearly that hidden assumptions about the source program can sometimes lead to difficulties. A better approach would be to match the converter characteristics to the human auditory system, since this represents the invariant to which signal properties must be matched.

2.4.2 Bit Errors in Transmission

The effect of bit errors during transmission of the digital data was also considered carefully, since error rates much less than 1 in 10^6 were not to be expected. Figure 2-20 shows the perceptual consequence of bit errors as a function of the position in the digital word. The data were taken during silent intervals in a music program. Notice that the high-order bit errors are always objectionable since they appear as a distinct click, even at rates as low as 1 in 10^6. In contrast, the lower-order bits are barely perceptible even at relatively high error rates of 1 in 10^4.[23]

The choice of using either error correction or parity checking for the high-order bits was decided mainly on the basis of circuit complexity. A simple 5-bit parity code was used to determine if the high-order bits were correct.

Once an error was detected, a synthetically generated replacement word was inserted. Various algorithms were tried to determine the most effective form of concealment. These included the following:

1. Muting: choosing a word equal to zero.
2. Zero-order extrapolation: holding the previous received word.
3. First-order interpolation: average of the two neighboring words.
4. First-order extrapolation: continuing the slope represented by the two previous words.

The effectiveness of the various concealment approaches is strongly dependent on the program at the time of the error. For a silent passage, the approaches are equivalent, since they all indicate a value of zero. In the presence of low frequencies, the first-order interpolation best approximates the

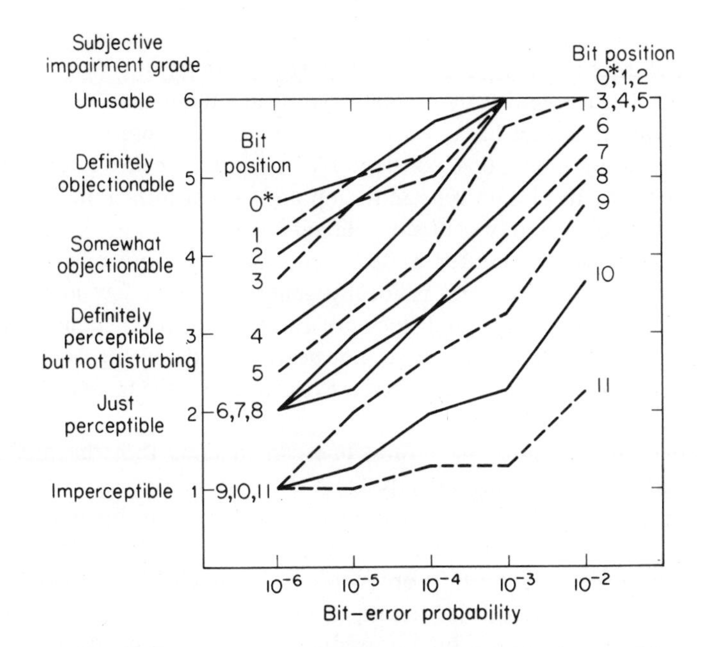

FIGURE 2-20. The subjective evaluation of bit errors in a straight PCM system during silent portion of program. Bit numbers correspond to significance: 0 is MSB. (Reprinted with permission from the BBC Technical Monograph 1972/18; "Pulse Code Modulation for High Quality Sound-Signal Distribution: Protection Against Digital Errors.")

actual data. Zero-order extrapolation is somewhat poorer than first-order interpolation, except at high frequencies, where they become equivalent.[23] First-order extrapolation is always inferior to interpolation.

Listening trials showed that zero-order extrapolation was adequate. For a given audible degradation, the zero-order extrapolation allowed the error rate to increase by 1.5 orders of magnitude.[25] These results were derived with a 13-bit uncompanded digital word. The use of block floating point makes the situation more complicated, since any error in the scale factor will affect 30 sample words. On the other hand, the block scale information appears at a very low rate, which reduces the frequency of scale errors. Nevertheless, it was recommended that the 2 scale bits have their own parity check.[43] The extra parity bit has no appreciable effect on the information rate.

2.4.3 Future Developments

Based on the successful experience in Great Britain, we can expect that most countries will eventually convert to digital transmission systems for broadcast distribution. The quality level, even in remote corners of the country, was dramatically increased. The post office departments in Germany, Switzerland, France, and Italy are all engaged in similar developments with systems at varying stages of completion. Owing to the interconnected nature of European communications, considerable attention is being given to using a standardized form of data representation and synchronization. This, however, often can be a highly charged political issue.

The situation in the United States is considerably more primitive since much radio program material is locally generated, with few and notable exceptions. Standardized digital lines are currently being installed between some major U.S. cities, but these are not being considered for use in broadcasting. Among other factors, the lines are organized on the basis of 50 kbits/s transmission.

Looking very far into the future, one can imagine that the digital bit stream could be transmitted directly from the microphone and be decoded only in the receiver. This would guarantee the highest possible quality. This step awaits two major developments: (1) the bit rate of the digitized audio must be substantially reduced in order to occupy much less bandwidth, and (2) the cost of decoding the bit stream must be brought within reach of the home listener. The required bit rate reduction will need a new concept in A–D conversion for the encoding to be efficiently matched to the properties of speech and music. The cost of such equipment will depend on the growth of integrated-circuit technology and improved production yields.

2.5 RECEPTION AND TRANSDUCTION

At this point in the audio chain the sound has been recorded, processed, and distributed to the listener's home sound reproduction system. Signal processing at this final stage is limited, since the listener is already provided with the signal to be converted into an acoustic waveform. Since this signal generally consists of two channels, the home equipment might be designed to re-create the extra dimensions of reverberation. Frontal reverberation is generally included in the main channel signals, but reflections coming from the sides and back must be synthesized. The resulting effect is referred to as *ambience enhancement*, since the actual reverberation from the rear has been lost. Quadraphonic systems try to fill this gap by providing a pair of back channels, but usually the basic information is still two channel. Only discrete four channel, which is mostly limited to special four-track home tape systems, can offer the correct signals for the back channels. Matrix quadraphony is limited in ambience enhancement capability since the extra channels are primarily sums and differences of the two main signals, and delayed reflections are not provided.

The sensation of being within an enclosed space depends on hearing reflections that are displaced from the source in both azimuth and time. Experiments[128] with electronically generated reflections show that adding delayed laterally located versions to a frontal source results in an increase in the apparent width of the source or an increase in the apparent size of the listening space. As a result of these experiments, a measure of the *spatial impression* of a room is proposed as being the ratio of the lateral to frontal energy over the first 80 ms after the reception of the direct sound. Although the sound produced by the frontal main speakers in a stereophonic system includes prerecorded reverberation and is subject to additional reverberation in the living room, the amount of lateral reverberant energy is still relatively low. The surfaces of the living room readily absorb the acoustic energy, so the echoes that do exist are quickly attenuated. Thus no matter how much reverberation has been added to the main channels, a strong spatial impression cannot be created in the living room without some additional mechanism. Stereophonic sound reproduction can thus provide only a restricted sense of space.

Regardless of the original direction of a reflection recorded in the studio or concert hall, the reproduction of the reflection comes from the front loudspeakers. Reverberation is necessary to give the sensation of a performance taking place in an appropriate space, but it is not sufficient to give the listener the sensation of being within the space. The listener is effectively looking into, or listening in on, the concert hall, and is not inside it. Perceptually, the stereophonic listening environment is equivalent to a small anteroom connected to the concert hall through an open window. This effect is most observable with liturgical music and least noticeable with small chamber music groups.

The spatial and temporal properties of the perceived sound pattern are quite complex. Differences in arrival time and intensity of the direct sound from the front loudspeakers provide localization cues.[58] The early reflections from different directions provide a sense of being within an enclosed space;[62] the later reverberations provide a sense of the total size of the environment. In most acceptable concert halls, the reflections begin to arrive from the lateral direction within 10 to 20 ms after the arrival of the direct sound.[65] After about 100 ms, the reflections are quite dense and arrive at a rate greater than 1/ms.[64] The time for the reverberation energy to decay to 60 dB, defined as the *reverberation time*, is related to the size of the space and the absorptive properties of the room boundaries and contents.

One attempt at increasing the spatial impression is to generate lateral reflections acoustically. Some loudspeakers direct sound to the sides and rear of the loudspeaker as well as to the front. The energy radiated to the side bounces from the walls of the room, giving rise to a reasonably strong image at the side of the loudspeaker. This will often increase the spatial impression or apparent source width in a most satisfying way. The increase is not great, however, since generally only one strong reflection is generated, and this reflection is still well forward in the listening area. The effect is also somewhat dependent on speaker and furniture placement.

To generate an adequate spatial impression, the sound reproduction system must provide lateral reflections. Since these reflections are essentially delayed versions of the original program material, they need not be recorded or transmitted. Rather, signal processing upon playback can provide the desired delays, and then direct the delayed signals to appropriate loudspeakers to simulate reflections. In this manner, a room can be made to sound much larger than it actually is.

A concert-hall simulation system was built at Bolt, Beranek, and Newman[60,61] using a 12-channel system to recreate the sound field that a listener would experience if he were sitting in a hall. Using a scale model of the Boston Symphony Hall, the directional impulse response from the stage to the listener was measured along 12 axes. Figure 2-21 shows the measurement procedure; a spark, serving as an impulse (actually a doublet), was the excitation, and 12 directional microphones recorded the signal at a particular seat

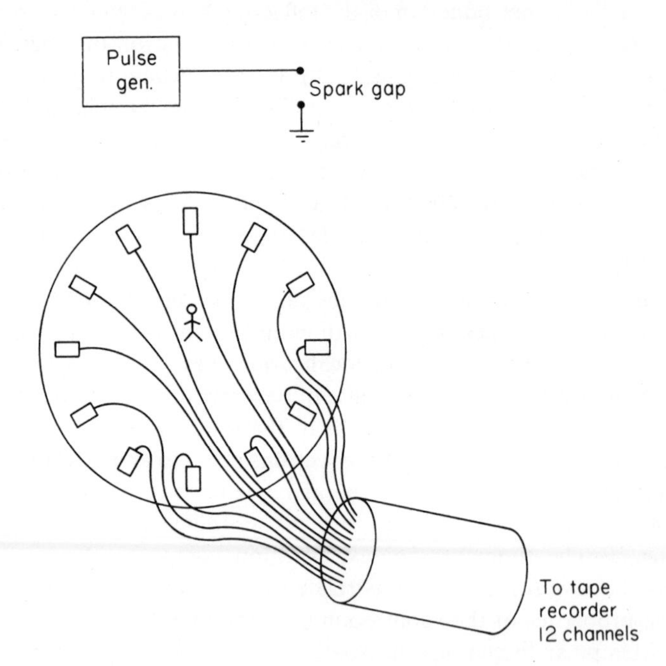

FIGURE 2-21. Method for determining the directional impulse response of a concert hall. In this example, 12 highly directional microphones are arranged in a circle around a particular seat; an impulse source in the form of a spark gap excites the hall. The signals from the microphone are recorded for later processing and analysis.

in the hall. This impulse response was then simulated by using a tapped delay line to generate the early reflection pattern. The delay line was effectively 150 ms long with outputs in steps of 10 ms. Each discrete output feeds all of the 12 channels with amplitudes adjusted to approximate the measured data. The stopped reverberation is supplied by adding the outputs from a reverberation plate or reverberation chamber.

An ordinary stereophonic recording can be used to demonstrate the simulation. The left and right channels feed two speakers in front of the listener to provide directionality. The left and right channels are then added together to form the source for the 12 transversal filters, each filter feeding its appropriate loudspeaker. This simulation has been called "the ultimate hi-fi system," since the sound field around the listener is perceptually matched to that which would have occurred in the live experience. It is reported that listeners familiar with the Boston Symphony Hall could identify the particular seat.(61)

A digital delay system for a similar purpose has been constructed at Teledyne Acoustic Research.(129) This system uses 12-bit PCM to encode a monophonic audio signal. The digitized signal is stored in a random-access memory of 8192 words at a 32-kHz sampling rate, providing 256 ms of delay. Sixteen output channels are available, each adjustable in delay in 1-ms increments and in amplitude. Every time a new sample is read in, each of 16 samples corresponding to the desired delays is converted to an analog voltage sample by the D–A converter and read out through a 16-channel demultiplexer.

The delay settings are based on the early reflections calculated for a rectangular room the size of a concert hall. Each delay feeds one loudspeaker. The loudspeakers are arranged in two tiers around the listener, and each delay corresponds to the first reflection that would occur in the vicinity of the associated loudspeaker. This system differs from the system of Bolt, Beranek, and Newman in that each loudspeaker receives only one reflection, not 12, and in that a general space is being simulated rather than a specific concert hall. A reverberant tail is not provided, as it is assumed that most prerecorded music already contains appropriate reverberation. Listener reaction is quite favorable in that the system successfully generates a strong and realistic spatial impression that can be varied by changing delay and amplitude values.

Both systems described here were designed as research tools. The technology used is still too expensive for the consumer, but such systems should be the music reproduction systems of the future. Simpler systems for home use, however, have been designed.

Madsen(57) has shown that two 12-ms delays are sufficient to achieve a high degree of spatial ambience. Admittedly, such a crude approximation is not as compelling, but it is still an improvement over stereophonic reproduction. It is also better, at least for some listeners, than proposed matrix quadraphonic systems.(91) The use of only two discrete echoes, as shown in

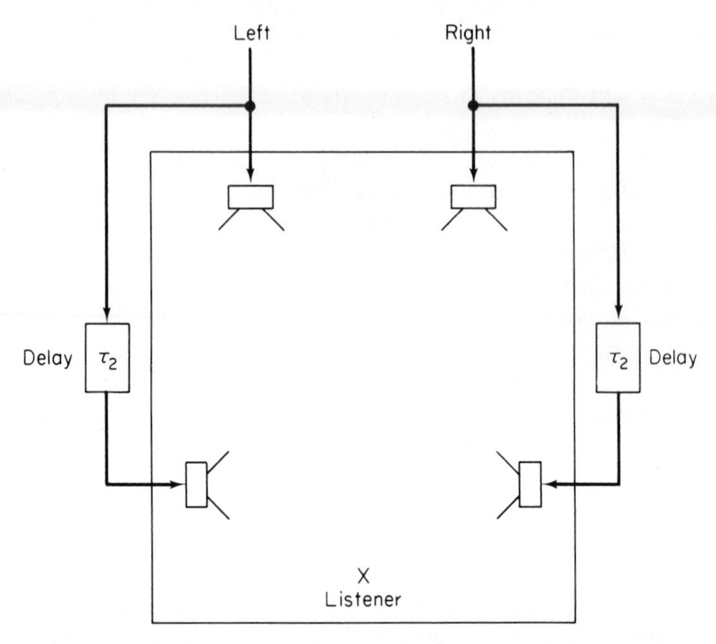

FIGURE 2-22. Technique for increasing spatial ambience in the home listening environment (after Madsen). Two side speakers are fed with delayed version of the front signal to produce two strong echoes.

Fig. 2-22, rather than a continuum, still results in a clear broadening of the image and a sense of space.

As long as they occur within an appropriate interval, the discrete echoes will fuse with the direct sound and will not be heard as an echo.(62) Because the system proposed by Madsen uses a very short delay, the memory requirement is minimal even at relatively high bit rates.

An ambience system more sophisticated than that of Madsen has been commercially realized by Hybrid Systems and is shown in Fig. 2-23.(92) Unlike the Madsen system, this approach regenerates reverberation to be played through additional rear or side channels using the inherent reverberation present in the stereo signals played through the front speakers. The two front signals are added together and fed to a delay line; the output is recirculated in a feedback loop around the delay line. The operation can be understood by observing that the impulse response samples the main front signals. Since the reverberation in the front signals is incoherent between samples, the output signals for the rear are incoherent. The front signals themselves are composed of a direct signal plus the sum of echoes in the reverberation. The rear signals are thus the direct signal plus a sum of echoes, where the echo pattern of the rear is the convolution of the original reverberation pro-

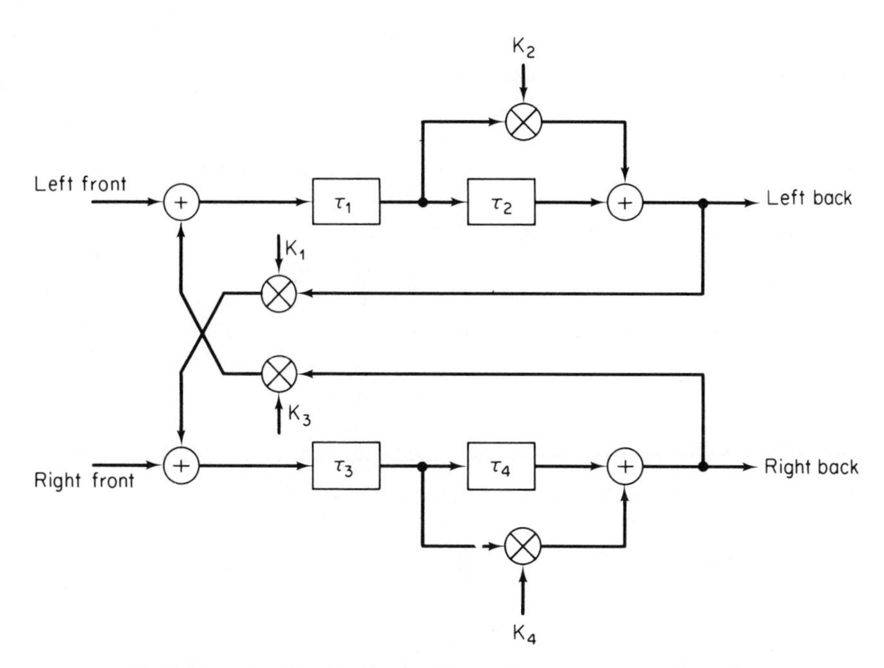

FIGURE 2-23. "Audiopulse" ambience enhancement system. This system contains a primitive reverberation generation algorithm in that the input audio signal is recirculated by the internal loops. The system takes the main left/right front signals as inputs and creates a pair of secondary signals which are used to drive the back pair or side pair.

cess and the synthetic process. To the listener, the reverberation from the four directions is substantially incoherent.

The success of the various ambience enhancement systems is based on the designer's understanding of psychoacoustics. These systems do not duplicate the events within a concert hall, but instead help generate the perceptual correlates necessary for a spatial impression. By bypassing architecture, a delay system will allow the listener to select the perceived environment he desires for a particular piece of music and to vary this environment at will. The effects chosen might be impossible to realize in any actual performance space. Signal processing therefore offers the opportunity of creating new spatial effects as well as duplicating existing ones, and will be an important aspect of sound reproduction in the future.

2.6 TRANSDUCER DESIGN

From this point on in our discussion of digital audio, we shall consider applications that take place primarily in the laboratory, for example, measurement techniques during the design of digital audio systems. Audio measuring equipment, like most of the reproduction chain, has been the exclusive domain of analog signal-processing equipment. This equipment has been designed to measure such traditional parameters as frequency response, harmonic or intermodulation distortion, phase response, transient response, attack time (compressors), and decay time (reverberation). This equipment has and does serve the needs of the audio engineer quite adequately, with one noticeable exception—loudspeaker design. Because of the inherent complexity of this transducer, its design and manufacture have remained something close to an art form, although not nearly as much as some manufacturers suggest.

Most electronic equipment can be measured in a relatively straightforward way. An amplifier, for example, can be constructed so that it is virtually free of distortion and frequency irregularities. As a result, its presence in the audio chain is said to be *transparent*, having essentially no measurable or perceptible effect on the signal. In contrast, the loudspeaker is often judged subjectively, since almost all loudspeakers sound different from each other, and engineers disagree widely about operational descriptions of good or excellent loudspeakers.

Unlike the amplifier, the loudspeaker is an energy transducer that must convert electrical power into acoustic power over a frequency range of ten octaves. The acoustic wavelengths range from 10 meters (m) to 2 centimeters (cm), that is, from about 30 times larger than the device to an order of magnitude smaller. Moreover, loudspeakers typically have complex distortion and response irregularities that are much greater than any of the associated electrical equipment. These difficulties are further intensified by the fact that the loudspeaker is handling high power, sometimes on the order of hundreds of watts, unlike a small signal transducer such as a microphone. In addition, the acoustic energy must be considered in terms of its distribution in three dimensions. The inability of engineers to make loudpseakers transparent forces them to evaluate the relative importance of different perceptual variables, since the optimization of design often reverts to the listening test. Ideally, only those degradation mechanisms that produce audible differences need be considered. However, there is little agreement over what effects are truly inaudible to some, if not all, listeners.

A typical audio loudspeaker system is composed of an enclosure and one or more loudspeakers. The loudspeaker is in itself a complicated electromechanical system. A moving-coil loudspeaker, for example, consists of a cone- or dome-shaped diaphragm to which is attached a coil positioned in a magnetic field. Electric currents through the coil produce electromagnetic forces, which move the coil and diaphragm to produce acoustic radiation. At low frequencies the diaphragm moves as a piston, displacing relatively large volumes of air, but at high frequencies the diaphragm no longer behaves as a rigid object. As the radiated wavelengths approach the size of the diaphragm, normal modes are excited in the diaphragm and different parts of the diaphragm move in different directions. These modal vibrations can become quite pronounced, causing severe irregularities in the measured response of the driver.

This problem can be alleviated by using several loudspeakers in the system, each covering a different frequency range. Electrical dividing networks, called *crossovers*, distribute low-frequency energy to the large-dimensioned woofer, midrange energy to a midrange driver, and high-frequency energy to tweeters. The design of such a system must take into account the fact that different parts of the spectrum are being radiated from different locations, since interference patterns are potential difficulties. The directional properties of the drivers present an interesting problem: should they be designed so that the frequency response is flat as measured by the intensity along the main axis, or should the total acoustic energy be used as the criterion? Again, respectable authorities are diametrically opposed.

The enclosure in which the loudspeaker drivers are mounted can also affect the overall system performance. Cabinet moldings, mounting hardware, and the other drivers present can all generate reflections that give rise to irregularities in measured frequency response and further complicate the directional patterns of the radiation. The drivers also radiate into the cabinet from the back side of the diaphragms; this radiation can bounce off the cabinet walls and emerge into the listening space. All these effects produce a complicated phase and frequency response, but it is difficult to be certain of the effect of these irregularities on perception.

In addition to these linear effects, the loudspeaker system is subject to many nonlinear distortion mechanisms. The magnetic field in which the voice coil moves is not uniform, so the electromotive force depends on the actual position of the diaphragm. For large input signals, the mechanical suspension limits the excursion of the driver; if a wide range of frequencies is present, a low-frequency signal can cause Doppler shifts in the high-frequency components.

To design a loudspeaker or system, adequate measurement equipment must be available to provide a complete description of its physical properties, and an analysis technique must be available to interpret the results of these measurements. Digital signal processing is well suited to this purpose.

2.6.1 Measurement Techniques

Loudspeaker measurements can be made in the frequency domain or the time domain. Conventional techniques use swept-frequency measurements, and require a special anechoic chamber to remove the effects of room reflections on the measured frequency response. Phase measurements, especially at high frequencies, are normally ignored as being unimportant. Thus most loudspeaker design proceeds on the basis of incomplete measurements of the transducer, and the importance of the quantities not measured is difficult to determine.

An alternative to frequency-domain measurements is to use the loudspeaker impulse response, since the impulse response is a complete description of a linear system. The excitation pulse, however, must be of low intensity to avoid overloading the loudspeaker. At low energies, however, the ambient noise can mask the response. One solution to this problem is to use a digital computer to average the response over many trials.[121] An experimental setup is shown in Fig. 2-24.

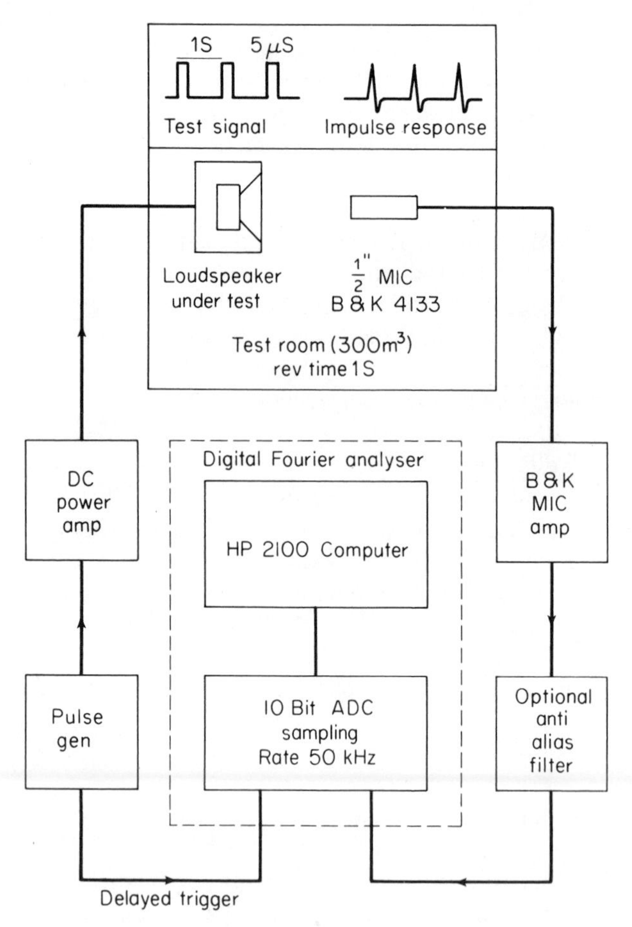

FIGURE 2-24. Experimental set-up used by Berman to produce an impulse response measurement of a loudspeaker. Repeated measurements are stored and averaged in the computer. (Reprinted with permission of the Audio Engineering Society, Copyright 1975.)

The loudspeaker under test is excited with successive low-level impulses, and a microphone of instrumentation quality is used to pick up the acoustic signal. This signal is digitized and stored in the computer. The process is repeated numerous times so that the resulting impulse response is the average of many independent tests. Averaging 1000 such impulses will reduce the noise by a factor of 30 dB. In principle, there is no limit to the accuracy specification, if there is time to acquire enough impulses.

The impulse reaches the microphone long before its reflection from a wall. The need for a purely anechoic environment is therefore removed, since the effect of the testing environment on the impulse response will only appear at a time much later than the arrival of the impulse from the loudspeaker under test. For example, if the loudspeaker and test microphone are in the center of a room which is 4 m^2 with the microphone 1 m from the speaker, the first reflection from the wall will be received about 9 ms after the direct signal. The impulse response can be truncated before this time. An example of such an impulse response is shown in Fig. 2-25, which is taken from the results reported by Berman.[121] The extra magnification shows that the radiation from the enclosure is present for about 7 ms. The problem that such a picture presents is that no interpretation of the psychoacoustic quality of the loudspeaker can be made without further analysis.

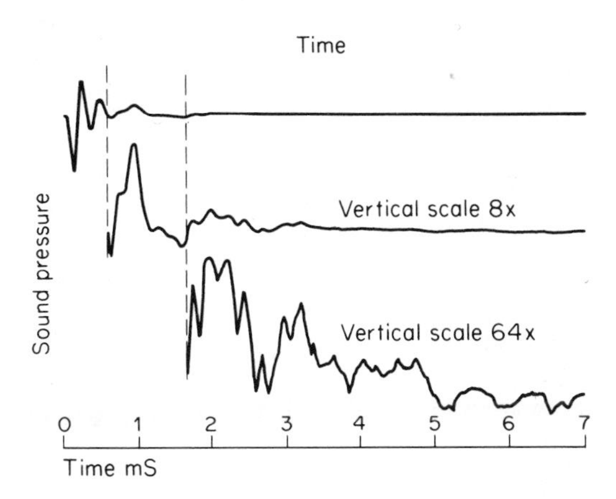

FIGURE 2-25. Impulse response of a loudspeaker as derived from repeated averaging. Second and third curves are magnified versions of the top trace showing the details of the impulse response (after Berman). (Reprinted with permission of the Audio Engineering Society, Copyright 1975.)

Nevertheless, this is true of most measured loudspeaker data, and the impulse data are much more accurate than the data usually obtained by analog measuring techniques. Unlike with the swept-frequency method, the room has no influence, and details of the system magnitude and phase response are not degraded by the sweeping process.

Using a standard FFT algorithm, the impulse response can be converted to the standard amplitude and phase response, as shown in Fig. 2-26. Again, however, the significance of the various irregularities in the response is not obvious. The two main dips in the frequency response at 1.1 and 2.2 kHz could be compensated with an equalizer filter matched to this frequency response. The behavior above 6 kHz is less important, since this loudspeaker is a bass driver, which does not receive any high-frequency energy. Still, it is difficult to interpret the data. As a reference, the minimum-phase Hilbert transform is shown for comparison to the actual phase. The difference suggests a degree of extra phase, which implies that some right-hand plane zeros exist. In such cases as these, the phase data may be significant.

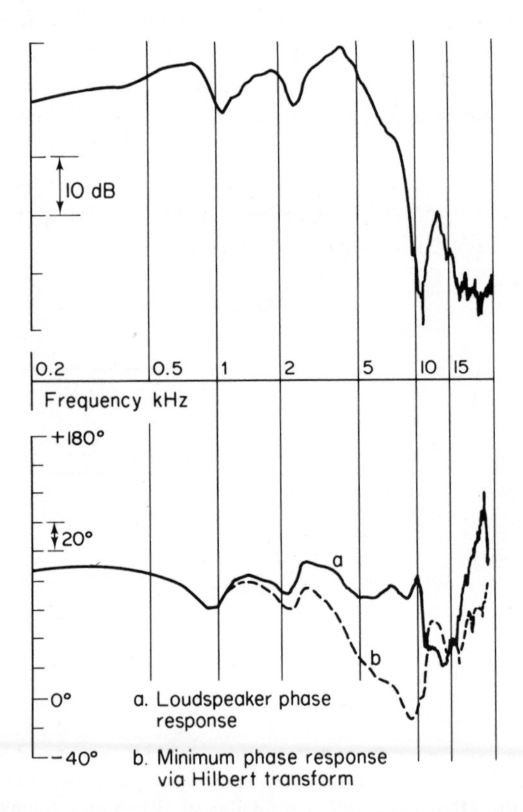

FIGURE 2-26. Amplitude and phase response as a function of frequency for loudspeaker under test. This data is the result of taking a Fourier transform of the impulse response in the previous figure. The phase response on the bottom shows the actual response and the minimum-phase Hilbert transform of the amplitude response (after Berman). (Reprinted with permission of the Audio Engineering Society, Copyright 1975.)

Because of the computational flexibility of the FFT, alternative display formats can be used for the data. Figure 2-27 shows what is currently called *cumulative decay spectra.* This display attempts to show the history of each frequency component. Conceptually, it is equivalent to exciting the loudspeaker with a steady sine wave and measuring the decay process when the sine wave is abruptly terminated. This process is repeated for each frequency. The resulting picture shows that the envelope of each frequency behaves quite differently. The distribution of spectral energies after the excitation has stopped shows the idiosyncratic properties much better than either the amplitude–phase response or the impulse response. The computational process for deriving the cumulative decay spectra is based directly on the FFT of the impulse response, but is also equivalent to the tone burst decay.

Examination of the cumulative spectra shows well-defined patterns. The peak in the response at 0.8 ms after excitation has ceased is presumably a

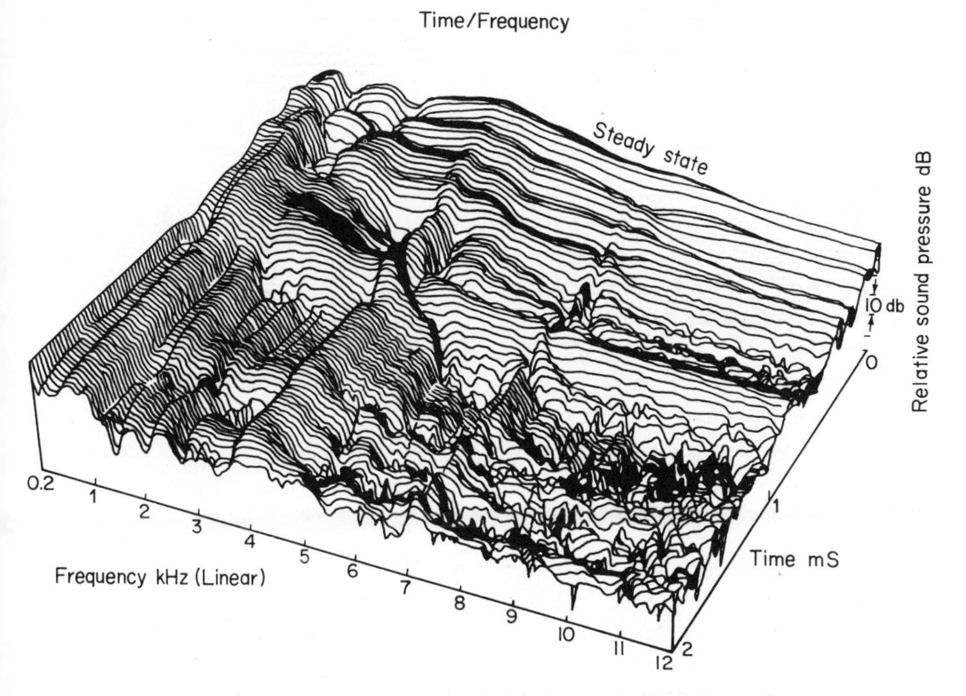

FIGURE 2-27. Cumulative decay spectra for loudspeaker under test. This format allows both the spectral and temporal properties to be viewed simultaneously. Ridges along a given time value suggest a coherent reflection; ridges along a spectral dimension suggest resonances which can be masked in the steady-state analysis (after Berman). (Reprinted with permission of the Audio Engineering Society, Copyright 1975.)

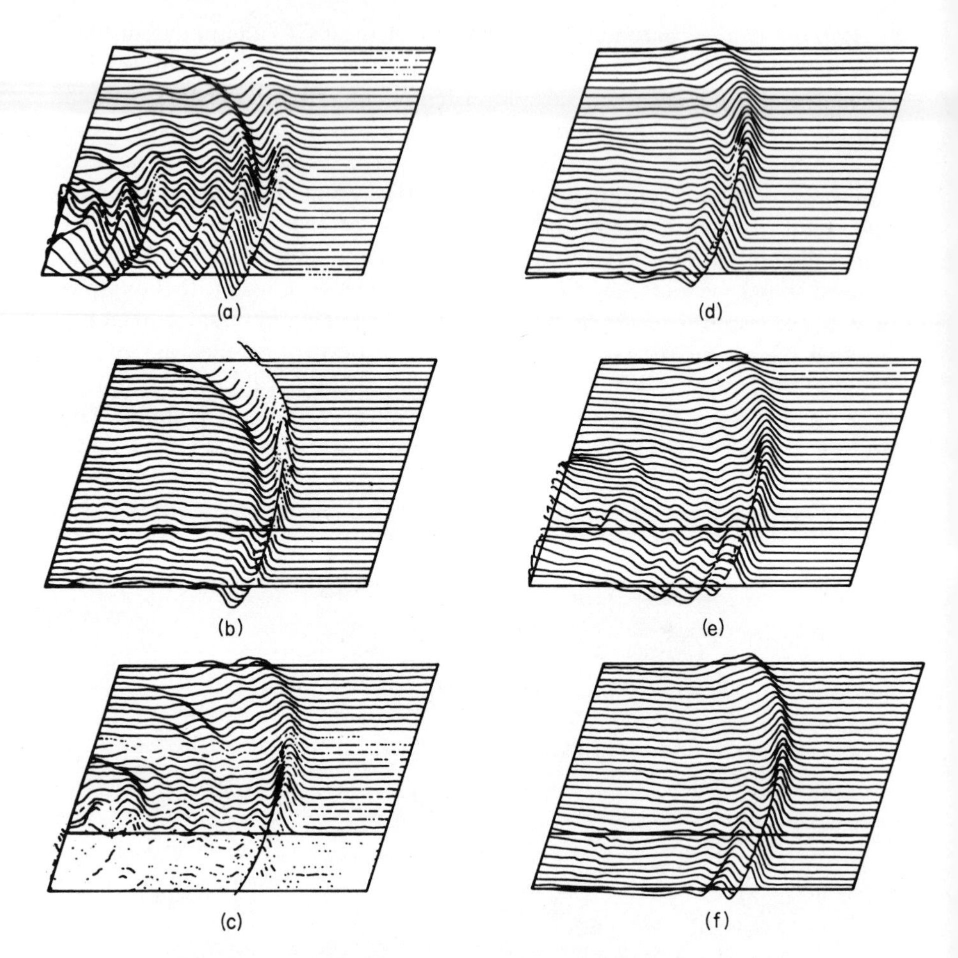

FIGURE 2-28. Wave-front patterns from 6 different tweeters when excited by a raised cosine pulse [130]. (Reprinted with permission of the Audio Engineering Society, Copyright 1975.)

reflection from the rear wall of the enclosure. The dominance of energy at 3.8 kHz after 2 ms has elapsed suggests a strong cone resonance at this frequency. Notice that this resonance existed in the steady-state signal, but was only a few decibels above the main response. In the delayed response, this resonance predominates. This kind of display seems appropriate, since the ear is a spectral resolver only over a limited time interval. The normal frequency response does not distinguish the time dependency of this energy addition, but the cumulative display does.

The value of the above-mentioned measurement procedures is yet to be demonstrated. One way to test the validity of interpretations of data is through computer simulation. The loudspeaker response can be modified by using the computer as a digital filter, or the system impulse response can be used to process digitized and stored music to compare the effects of various changes.

Because this approach has only recently been introduced, and perhaps because of the proprietary nature of the results, no data have been reported. However, this technique appears promising. A computer can also be used as a graphical technique to display information which has been gathered. In Figure 2-28, the wavefronts from six different tweeters are displayed when excited by a raised cosine pulse(130). The differences are quite apparent although somewhat difficult to interpret.

2.7 ADVANCED SIGNAL-PROCESSING STRATEGIES

The applications of digital signal processing discussed so far have been of a very practical nature. Digital technology is being pursued in audio in order to improve upon existing systems, as in the case of digital tape recorders, or to provide signal processing for predetermined perceptual effects, as in the case of digital ambience enhancement systems. The signal processing needed for these applications is relatively straightforward and is based upon the classical lines of linear filtering.

More advanced signal-processing strategies have also been developed, but have not been applied to commercial audio systems owing to their complexity and expense in implementation. These techniques are, in a sense, solutions looking for a problem. The mathematics involved has been thoroughly developed on a theoretical basis, however, and some laboratory examples have been successfully investigated.

One very interesting class of signal processing is called *generalized linearity*, or *homomorphic*, signal processing, and is based on theories developed by Oppenheim.(116) In brief, signals that are convolved or multiplied together are subjected to a nonlinear operation to produce signals that are added instead. These additive representations are then processed by classical linear filters, and returned to the original domain by the inverse of the nonlinear operation. This general sequence of operations can be the basis of a compandor system, can be used to separate sound from its reflections, and can be used to restore old acoustic recordings. A discussion of these techniques follows.

2.7.1 Compandor System

A music signal can be considered to be the product of two signals. The first is the slowly varying envelope, $e(t)$, and the second the much faster vibrations, $v(t)$, giving the signal $s(t)$ as

$$s(t) = e(t)v(t) \tag{2.1}$$

The envelope is always positive, whereas the vibration can be both positive and negative.

This multiplicative signal can be mapped into an additive signal through the operation of the logarithm, yielding

$$\log s(t) = \log e(t) + \log v(t) \tag{2.2}$$

Since $v(t)$ can be positive or negative, the complex logarithm must be used. The phase of $v(t)$ is either 0 or π, and this angle forms the imaginary part of the logarithm, while the log of the magnitude forms the real part. After the signal is operated on in this manner, it is subjected to linear processing. The filtered signal can then be exponentiated to give a new music signal, $s'(t) = e'(t)v'(t)$. A block diagram of the total system is shown in Fig. 2-29.

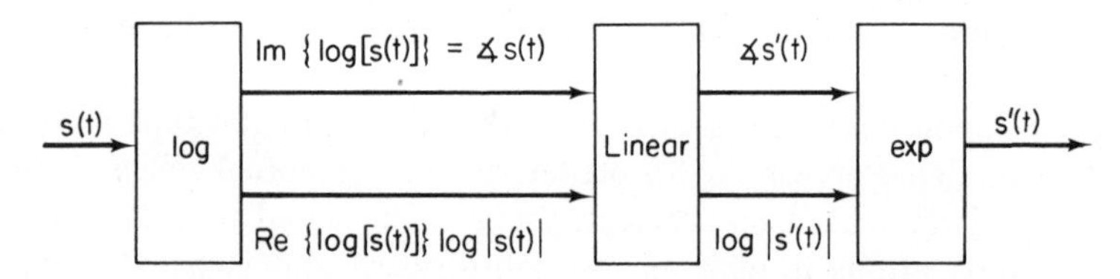

FIGURE 2-29. Block diagram for homomorphic filtering of multiplied signals. The linear system is a filtering operation of the log-magnitude of the input signal (after Stockham). (Reprinted with permission of the Institute of Electrical and Electronic Engineers from IEEE Transactions on Audio and Electro-acoustics, 1968.)

The value of this kind of processing is realized if $\log e(t)$ and $\log |v(t)|$ can be separated by classical filtering. If these two components occupy separate frequency bands, for example, the filtering can perform different operations on each. For most music and speech material, this is the case, as shown in Fig. 2-30. The envelope is concentrated below 16 Hz, while the vibration extends above that frequency.

Let the linear filter have unity gain above 16 Hz and a gain of K below 16 Hz. Thus $\log |v(t)|$ is kept the same, but $\log e(t)$ becomes $K \log e(t)$ or, equivalently, $\log [e^K(t)]$, giving the system output of $s'(t) = e^K(t)v(t)$. If K is less than 1, the amplitude of the envelope is reduced and the dynamic range compressed. When $K = 0$, the system becomes an automatic gain control, since only the vibration components are passed and these with equal amplitude. The system with $K > 1$ becomes an expandor, increasing the dynamic range of the program material.

The system described is not necessarily a digital one. High-quality audio applications, however, would require that the nonlinearities be extremely

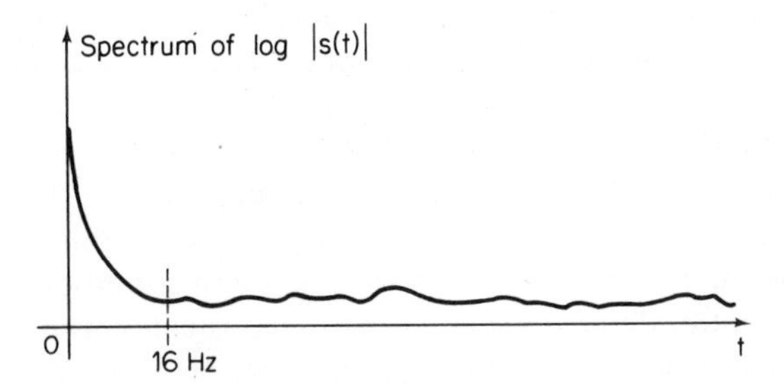

FIGURE 2-30. Spectrum of the log-magnitude of a typical audio signal. The dominant energy at low frequencies represents the wide dynamic range; the high-frequency components represent the details of the signal modulation. Each can be filtered separately (after Stockham). (Reprinted with permission of the Institute of Electrical and Electronic Engineers from IEEE Transactions on Audio and Electro-acoustics, 1968.)

stable and their transfer characteristics be very accurately defined. This suggests that the best quality would be realized with a digital system. Compression or expansion for recording would therefore be a logical part of a digital mix-down system.

2.7.2 Echo Removal

A signal that has been corrupted by an unwanted echo can potentially be restored by using homomorphic signal processing. Here the two signals have been convolved rather than multiplied, which means that the operations must take a different form. The basic technique, however, is the same as for the multiplied signals.

Consider a system with a single echo. The impulse response is then

$$h(t) = \delta(t) + a\,\delta(t - T) \tag{2.3}$$

If the original signal is $s(t)$, the output of this system is

$$\begin{aligned} x(t) &= s(t)*h(t) \\ &= s(t)*[\delta(t) + a\,\delta(t - T)] \end{aligned} \tag{2.4}$$

In the frequency domain, this is represented as

$$X(F) = S(F)(1 + ae^{-j2\pi FT}) \tag{2.5}$$

The goal is to separate the signal from its echo. The first step is to take the logarithm of Eq. (2.5), yielding

$$\log X(F) = \log S(F) + \log (1 + ae^{-j2\pi FT}) \tag{2.6}$$

Since $(1 + ae^{-j2\pi FT})$ is periodic in frequency, its logarithm will be periodic as well, and this gives an additive periodic component in Eq. (2.6). Thus the effects of the echo can be minimized by filtering out those components in log $X(F)$ that are periodic with a period of $2\pi/T$.

The transform of log $X(F)$ is the spectrum of the log spectrum and is defined to be the *cepstrum*. Since log $X(F)$ is the sum of two components, and since the Fourier transform is a linear operation, the cepstrum will also be the sum of two cepstra, one related to the signal and the other related to the echo. The filtering proposed in the above paragraph corresponds to multiplication in the ceptral domain. Thus, if the signal is concentrated near $t = 0$, the echo can be removed by "low-time" filtering.[117] This assumes that the signal is slowly varying in frequency and that the echo is quickly varying, so that multiplying by a function that is unity near the origin and zero elsewhere will remove the echo.

Alternatively, a series of notches can be placed over the impulses in the cepstrum once their location has been determined. Figure 2-31 shows the two forms of filters that can be used. These filters are represented as a multiplication in the cepstrum domain.

Although these techniques have been exclusively applied to speech and seismic processing, they are equally applicable to music. The computational complexity combined with the high data rate of music, however, makes the task somewhat formidable. Nevertheless, if the application required the

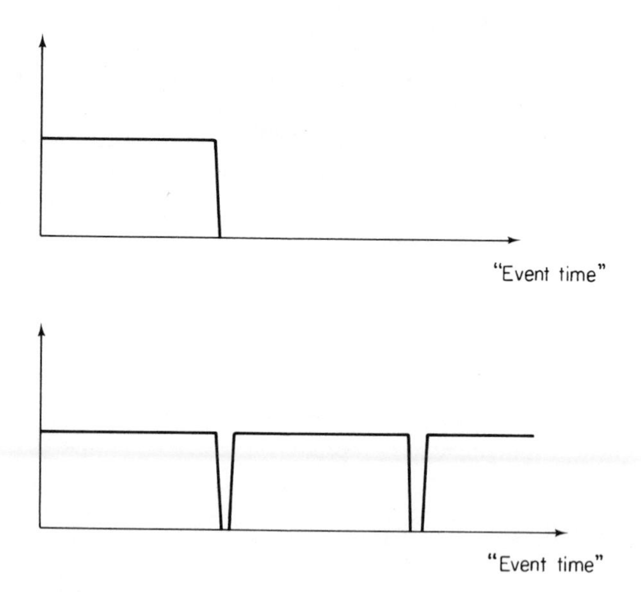

FIGURE 2-31. Two filters for removing components in the cepstrum of a signal which has been subjected to a single echo. The top filter is a "low event-time" filter; the lower filter is a "periodic-event rejection" filter.

removal of an unwanted echo, the digital signal-processing approach would be the only available one.

Filtering in the cepstral domain is somewhat difficult to visualize physically. The domain is a kind of event time, such that the notch filter centered at the impulses is a periodic-event reject filter with a rejection interval equal to the echo time. This filter removes all periodic events occurring at a rate equal to the echo time. The filter that passes components near the origin and rejects all others is a short-time pass filter; all events that occur over a short time are passed. The reverse filter would be a long-event pass filter.

In the case of echo removal from music, the periodic-event filter would produce a minimum effect on the music, except when there was a repetition in the music that was at the same rate as the echo. If the echo time is stable, however, the width of the event reject could be made arbitrarily small, and one would expect a minimum effect. Unfortunately, there are no empirical data to indicate just how well such a system would work in practice.

2.7.3 Blind Deconvolution

In the case of echo removal, the processing was made possible by having knowledge of one of the two signals being separated. The temporal distribution of the music cepstrum or the single echo period was used as the basis of the filtering. Often, however, both signals are less characterized or unknown, but it is still desired to separate them. This task is called *blind deconvolution*, which is the process of removing the effect of an unknown filter upon an unknown signal by observing the output of the filtering.

Blind deconvolution techniques have been used by Stockham[119] to restore old acoustic recordings. In restoration, the actual music input is unknown, as is the exact transfer function of the recording device; all that is available is the output of the recording process in the form of a shellac record. Stockham observes that these old recordings have a unique sound, which is due to the sharp resonances of the mechanical recording system. Each element in the recording chain, from the acoustic horn to the diaphragm stylus mount, exhibits strong resonances; a detailed discussion of the recording technique can be found elsewhere.[118] These resonances probably derive from the need to reduce any frictional losses in order to get as high an output signal as possible. Although flatter frequency responses could probably have been achieved, the main concern was to keep the signal above the noise level.

The effect of these resonances is to produce very sharp peaks and dips in the frequency response, especially in the region near 1 kHz. These variations can be up to 20 dB and are sufficiently broad to produce very strong spectral coloration. The nonuniform frequency response, in itself, is not particularly noticeable if the bandwidth of these peaks is small. Reverberation produces very sharp peaks and dips, but their bandwidth can be as small as 1 Hz or less. The resonance widths of the acoustic recording can be on the order of 10 to 50 Hz. As the pitch of a singer's voice crosses one of these resonances, there is a very strong change in the signal level. Thus, in addition to the spectral coloring, there is a very complex change in the dynamics of the music. It is this effect, rather than the excessive noise or limited bandwidth, that gives the old recordings their characteristic sound.

To remove the effects of an unknown filter, some assumptions must be made about the original signal, the filter, or both. In the case of restored acoustic recordings, two basic assumptions must be made. First, it must be assumed that the filtering process remained stable and linear throughout the recording session. This assumption is probably valid, except that some change in the frequency response is produced by movement of the singer's mouth relative to the recording horn. Nevertheless, the spectral balance of the singer's voice is reasonably constant throughout the musical passage, and the average spectra of a male signer on a given selection is probably the same today as in 1906. The second assumption is that the musical selection is much longer than the filter impulse response. This allows averaging to be used as part of the signal separation process. Deconvolution always requires some knowledge of the signal and filter. If, for example, a new type of musical instrument had been recorded only once, it would not be possible to produce a deconvolution.

The efforts of Stockham in the restoration of old Caruso recordings from 1906 clearly illustrate this process. The deconvolution technique is based on homomorphic signal processing, and uses operations similar to those used in echo removal. Let the recorded signal be $v(t)$, the singing input be $s(t)$, and the recording process impulse response be $h(t)$, giving

$$v(t) = s(t)*h(t) \tag{2.7}$$

The Fourier transform of both sides of this equation then gives

$$V(F) = S(F) \cdot H(F) \tag{2.8}$$

Taking the complex logarithm of both sides of Eq. (2.8) yields

$$\log V(F) = \log S(F) + \log H(F) \tag{2.9}$$

which has the desired additive property. It is desired to subject this equation to a linear operation that produces an accurate description of $H(F)$. Once found, this function can be used to generate the inverse filter to remove the coloration produced by the recording process. If there were a clear spectral difference between the two right-hand terms of Eq. (2.9), this separation could be performed by linear filtering, as was done in the echo-removal example. In this case, however, the interesting behavior of $H(F)$ occurs in the same spectral region as $S(F)$, between 100 and 1000 Hz.

The term $H(F)$ is, however, constant throughout the recording, while $S(F)$ is continuously varying. This suggests that the entire recording could be chopped into many segments, each longer than the recording system response, and these segments averaged to give an estimate of the filter. The term log $H(F)$ would be constant for each segment, but the terms log $S(F)$ would differ. Only one recording session can be used, since the recording apparatus and its adjustments differed from session to session.

For the ith interval, Eq. (2.9) can be represented as

$$\log V_i(F) = \log S_i(F) + \log H(F) \tag{2.10}$$

Care must be exercised in performing the averaging operation, since Eq. (2.10) contains a magnitude and angle part that must be treated separately. This gives the average magnitude

$$\frac{1}{N}\sum_{i=1}^{N} \log |V_i(f)| = \log |H(f)| + \frac{1}{N}\sum_{i=1}^{N} \log |S_i(f)| \tag{2.11}$$

and the average angle

$$\frac{1}{N}\sum_{i=1}^{N} \measuredangle V_i(f) = \measuredangle H(f) + \frac{1}{N}\sum_{i=1}^{N} \measuredangle S_i(f) \tag{2.12}$$

The left part of Eq. (2.11), which is empirically determined, would specify $\log |H(f)|$ if some estimate of the second terms on the right could be derived. This term converges to a function that is determined by the average energy density spectrum of the singer's voice. This can be approximated by subjecting a modern recording to the same process, which yields the equivalent Eq. (2.11) given by

$$\frac{1}{N}\sum_{j=1}^{N} \log |V_j(f)| = \log |H'(f)| + \frac{1}{N}\sum_{j=1}^{N} \log |S_j(f)| \tag{2.13}$$

except that $H'(f) = 1$ since these recordings have no coloration. The difference between Eqs. (2.11) and (2.13) yields

$$\log |H(f)| = \frac{1}{N}\sum_{i=1}^{N} \log |V_i(f)| - \frac{1}{N}\sum_{j=1}^{N} \log |V_j(f)| \tag{2.14}$$

which is an estimate for the magnitude of $H(f)$. The assumption that the average spectra of two male singers singing the same operatic passage will produce similar energy density spectra is probably a good one. Moreover, any errors will tend to produce the kind of gentle coloration similar to a misadjusted tone control. The empirical data derived from Eq. (2.11) are shown in Fig. 2-32, and the smoothed data from the modern recording are shown in Fig. 2-33.

The difference between them is taken as the $\log |H(f)|$. Actually, this process is only an approximation, since the actual data from the modern recording have significant peaks and valleys, which can neither be attributed to the singer's average spectrum nor to coloration in the recording process. Part of the spectrum contains power-line frequency components, presumably owing to imperfections in the power supplies used in the modern recording process. Nevertheless, the technique is sufficiently accurate to give a dramatic improvement in the perceived quality of the restored singing voice. Stockham

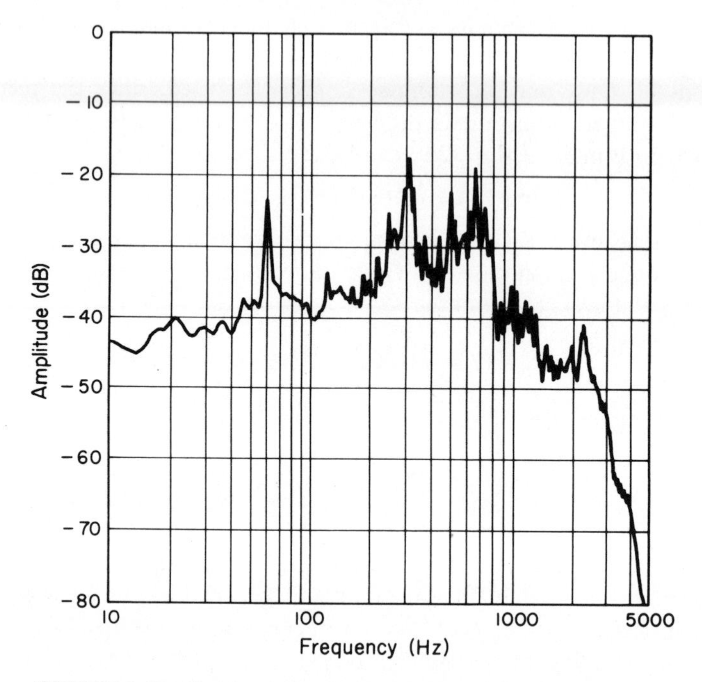

FIGURE 2-32. Graph of the average produced by equation (11) for 1907 acoustic recording by Enrico Caruso of "Vesti la Giubba" (after Stockham). (Reprinted with permission of the Institute of Electrical and Electronic Engineers, 1975.)

reports that the restored Enrico Caruso "retains some of the acoustic flavor but the clarity of expression, the texture of the voice, and the artistic interest are dramatically changed." The changes in level as the pitch excites a resonance are removed such that the megaphone quality no longer appears.

The compensation filter is derived from the inverse magnitude of $H(f)$, but no form of phase information is added because of mathematical difficulties in averaging Eq. (2.12). These difficulties arise because the averaging of phase requires an absolute phase rather than a principal value modulo 2π. The principal value of the average phase is not the same as the average of the principal values. For example, the average phase of 0 and 370 degrees is 185 degrees; but if the 370 degrees is represented as 10 degrees principal value, the average is 5 degrees. Because of computational difficulties, Stockham was not able to produce the phase function of $H(f)$, although aspects of this problem have been considered elsewhere.[116,117] The justification for ignoring phase compensation is based on the limited perceptibility of phase errors. The actual limits of phase perception are not well understood, however, and some of the remaining coloration in the restored recordings could be due to this effect.

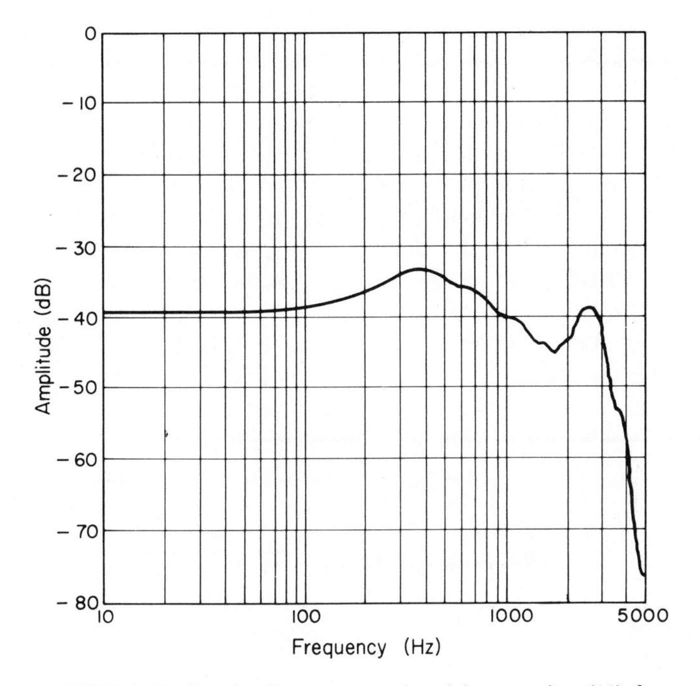

FIGURE 2-33. Graph of average produced by equation (11) for modern recording by Jussi Bjoerling of "Vesti la Giubba"; data is smoothed (after Stockham). (Reprinted with permission of the Institute of Electrical and Electronic Engineers.)

The computational process of generating the restoration is extremely time consuming even on the highest-speed computers. A few minutes of music can take hours to process. The music signal must be segmented, and each segment transformed into the frequency domain using an FFT. Approximately 500 segments are used, so the logarithms of 500 FFT signals must be averaged to derive the filter approximation. This filter is so complex in its detail that the music must again be subjected to a similar transformation into the frequency domain, and multiplied by the coefficients of the derived inverse filter. The extent of computation required alone prevents such techniques as these from leaving the laboratory. Nevertheless, the nature of the results are impressive. Favorable listener reaction is an indication of the effectiveness of the restoration.

2.8 CONCLUSIONS

It is clear that digital technology has already made significant inroads into the traditionally analog domain of audio engineering. Applications of digital signal processing are being developed for every part of the sound reproduction chain, from recording to distribution to home sound reproduction. In each area, digital techniques offer increased quality or greater signal-processing flexibility than the standard analog components.

It is not too difficult to imagine a totally digital audio chain. The A–D converter can be placed in the microphone or the associated preamplifier, and the D–A converter built into the loudspeakers. In between, the entire chain will be digital. The recording will be made with a digital tape recorder, and the editing and mixing will be done on a digital computer that is programmed to perform all the signal processing. Reverberation, compression, equalization, noise reduction, channel selection, and level setting will all be accomplished in the digital domain. The distribution will also be digital, either by digital transmission or by a variation of commercial video discs. The bit rate in the encoding process will become less of an issue as new techniques are developed and higher-density media evolve.

As a result of this technology, listeners will enjoy much more accurate and much more realistic sound reproduction at home. Traditional controls will be augmented to permit variation of the apparent size of the listening space, giving the listener the impression of actually sitting in the performance hall. Expanders and compressors will give the listener control over the dynamic range of the music, allowing for compression for background listening and expansion for the restoration of natural dynamic levels. Transducer systems incorporating signal processing will allow extremely accurate sound reproduction, completing the system.

This vision is far from realization, yet every one of these applications of digital technology is currently under investigation. The benefits of digital signal processing in audio are clearly recognized, and it is only a matter of time before an all-digital system is actually implemented.

ACKNOWLEDGEMENT

The authors wish to express their sincere appreciation for the helpful comments provided by Robert Berkovitz of Teledyne Acoustic Research.

REFERENCES

1. W. Bennett, "Spectra of Quantized Signals," *Bell System Tech. J.*, vol. 27, no. 7, 1948, pp. 446–472.

2. D. Hoeschele, *Analog-to-Digital/Digital-to-Analog Conversion Techniques*, Wiley, New York, 1968.

3. J. Kriz, "A 16 Bit A–D–A Conversion System for High-Fidelity Audio Research," *IEEE Trans. Acoust. Speech Signal Processing*, vol. ASSP-23, no. 1, 1975, pp. 146–149.
4. N. Jayant and L. Rabiner, "The Application of Dither to the Quantization of Speech Signals," *Bell System Tech. J.*, vol. 51, no. 6, 1972, pp. 1293–1304.
5. W. Manson, "Digital Sound Signals: Subjective Effect of Timing Jitter," *BBC Monograph*, Great Britain, 1974/11, Research Department, Engineering Division, 1974.
6. J. Candy, "A Use of Limit Cycle Oscillations to Obtain Robust Analog-to-Digital Conversion," *IEEE Trans. Commun.*, vol. COM-22, no. 3, 1974, pp. 298–305.
7. B. Blesser, "An Investigation of Quantization Noise," *J. Audio Eng. Soc.*, vol. 22, no. 1, 1974, pp. 20–21.
8. B. Blesser and F. Lee, "An Audio Delay System Using Digital Technology," *J. Audio Eng. Soc.*, vol. 19, no. 5, 1971, pp. 393–397.
9. D. Freeman, "Digital to Analog Conversion Distortion," MIT internal memorandum, Lebel Laboratory, RLE, Massachusetts Institute of Technology, Cambridge, Mass., 1975.
10. B. Blesser and F. Ives, "A Re-examination of the S/N Question for Systems with Time-Varying Gain or Frequency Response," *J. Audio Eng. Soc.*, vol. 20, no. 8, 1972, pp. 638–641.
11. D. Osborne and M. Croll, "Digital Sound Signals: Bit-Rate Reduction Using an Experimental Digital Compandor," *BBC Monograph*, Great Britain, 1973/41, Research Department, Engineering Division, 1973.
12. Lexicon Delay System: Delta T 102, Lexicon Corp., Waltham, Mass.
13. A. Zverev, *Handbook of Filter Synthesis*, Wiley, New York, 1967.
14. L. Huelsman, *Active Filters: Lumped, Distributed, Integrated, Digital and Parametric*, Chapter 2, McGraw-Hill, New York, 1970.
15. T. Stockham, "A–D and D–A Converters: Their Effect on Digital Audio Fidelity," in *Digital Signal Processing*, L. Rabiner and Rader, eds., IEEE Press, New York, 1972.
16. L. Jeffress, "Masking," in *Foundation of Modern Auditory Theory*, J. Tobias, ed., Academic Press, New York, 1970.
17. E. Kuescher and J. Zwislocki, "Adaptation of the Ear to Sound Stimuli," *J. Acoust. Soc. Amer.*, vol. 21, 1949, pp. 135–139.
18. R. Wegel and C. Lane, "The Auditory Masking of One Pure Tone by Another and Its Probable Relation to the Dynamics of the Inner Ear," *Phys. Rev.*, vol. 23, series 2, 1924, pp. 266–289.
19. J. Egan and H. Hake, "On the Masking Pattern of Simple Auditory Stimulus," *J. Acoust. Soc. Amer.*, vol. 22, 1950, pp. 622–630.

20. J. MYERS and A. FEINBERG, "High-Quality Professional Recording Using Digital Techniques," *J. Audio Eng. Soc.*, vol. 20, no. 8, 1972, pp. 622–628.

21. H. IWAMURA, H. HAYASHI, A. MIYASHITA, and T. ANAZAWA, "Pulse-Code Modulation Recording System," *J. Audio Eng. Soc.*, vol. 21, no. 7, 1973, pp. 535–541.

22. N. SATO, "PCM Recorder—A New Type of Audio Magnetic Tape Recorder," *J. Audio Eng. Soc.*, vol. 21, no. 7, 1973, pp. 542–458.

23. J. CHEW and M. MOFFAT, "Pulse Code Modulation for High-Quality Sound-Signal Distribution: Protection Against Digital Errors," *BBC Monograph*, Great Britain, 1972/18, Research Department, Engineering Division, 1972.

24. A. JONES and J. CHAMBERS, "Digital Magnetic Recording: Conventional Saturation Techniques," *BBC Monograph*, Great Britain, 1972/9, Research Department, Engineering Division, 1972.

25. G. MITCHELL and M. MOFFAT, "Pulse Code Modulation for High-Quality Sound-Signal Distribution: Subjective Effect of Digital Error," *BBC Monograph*, Great Britain, 1972/40, Research Department, Engineering Division, 1972.

26. W. PETERSON, *Error Correcting Codes*, The MIT Press, Cambridge, Mass., 1961.

27. M. HECHT and A. GUIDA, "Delay Modulation," *Proc. IEEE*, vol. 57, no. 1, 1969, pp. 1314–1316.

28. H. KOBAYASHI, "A Survey of Coding Schemes for Transmission or Recording of Digital Data," *IEEE Trans. Commun. Technol.*, Dec. 1971, pp. 1087–1100.

29. J. MELIS, "A Visual Display of Tape Modulation Noise with Audio Recording," *J. Audio Eng. Soc.*, vol. 19, no. 11, 1971, pp. 939–941.

30. E. TRENDELL, "The Measurement and Subjective Assessment of Modulation Noise in Magnetic Recording," *J. Audio Eng. Soc.*, vol. 17, no. 6, 1969, pp. 644–653.

31. K. NAUMANN and E. DANIEL, "Audio Cassette Chromium Dioxide Tape," *J. Audio Eng. Soc.*, vol. 19, no. 10, 1971, pp. 822–828.

32. F. BELLIS and M. SMITH, "A Stereo Digital Sound Recorder," *BBC Monograph*, Great Britain, 1974/39, Research Department, Engineering Division, 1974.

33. D. COMSTOCK, R. MISNER, E. ROBERTS, and R. ZENNER, "Dropout Identification and Cleaning Methods for Magnetic Tape," *J. Audio Eng. Soc.*, vol. 22, no. 7, 1974, pp. 511–520.

34. J. MCKNIGHT, "Development of a Subjective Flutter Measurement Standard," presented at the 40th Audio Engineering Society Convention, preprint no. 790, Apr. 1971.

35. Nippon Columbia, Tokyo, Japan, Issued PCM Recordings: NCB 8001, NCB 8002, NCB 8003.

36. D. Shorter and B. Chew, "Application of Pulse-Code Modulation to Sound-Signal Distribution in a Broadcast Network," *Proc. Inst. Elec. Eng.*, vol. 119, no. 10, 1972, pp. 1442–1448.

37. M. Croll, "Pulse-Code Modulation for High Quality Sound Distribution: Quantization Distortion at Very Low Signal Levels," *BBC Monograph*, Great Britain, 1970/18, Research Department, Engineering Division, 1970.

38. R. Dolby, "An Audio Noise Reduction System," *J. Audio Eng. Soc.*, vol. 15, no. 4, 1967, pp. 383–388.

39. W. Bennett and J. Davey, *Data Transmission*, McGraw-Hill, New York, 1965.

40. H. Hessenmueller, "The Transmission of Broadcast Programs in a Digital Integrated Network," *IEEE Trans. Audio Electroacoustics*, vol. AU-21, no. 1, 1973, pp. 17–20.

41. M. Croll, D. Osborne, and D. Reid, "Digital Sound Signals: Multiplexing Six High-Quality Sound Channels for Transmission at a Bit-Rate of 2.048 mbs," *BBC Monograph*, Great Britain, 1973/42, Research Department, Engineering Division, 1973.

42. D. Osborne, "Digital Sound Signals: Further Investigation of Instantaneous and Other Rapid Companding Systems," *BBC Monograph*, Great Britain, 1972/31, Research Department, Engineering Division, 1972.

43. D. Reid and M. Croll, "Digital Sound Signals: The Effect of Transmission Errors in a New Instantaneous Digitally Companded System," *BBC Monograph*, Great Britain, 1974/24, Research Department, Engineering Division, 1974.

44. D. Shorter, J. Chew, D. Howorth, and J. Sanders, "Pulse-Code Modulation for High Quality Sound-Signal Distribution," *BBC Monograph*, Great Britain, 1968/75, Research Department, Engineering Division, 1968.

45. Bell Telephone Laboratories Staff, *Transmission Systems for Communications*, Western Electric Co. Technical Publications, Winston-Salem, N.C., 1971.

46. B. Smith, "Instantaneous Companding of Quantized Signals," *Bell System Tech. J.*, vol. 36, May 1957, pp. 653–709.

47. R. Steele, *Delta Modulation Systems*, Wiley, New York, 1975.

48. C. Dalton, "Delta Modulation for Sound-Signal Distribution: A General Survey," *BBC Eng.*, July 1972, pp. 4–14.

49. L. Zetterberg and J. Uddenfeldt, "Adaptive Dalta Modulation with Delayed Decision," *IEEE Trans. Commun.*, vol. COM-22, no. 9, 1974, pp. 1195–1198.

50. C. Song, J. Garodnick, and D. Schilling, "A Variable Step Size Robust Delta Modulator," *IEEE Trans. Commun.*, vol. COM-19, no. 6, 1971, pp. 1033–1044.

51. D. Slepian, "On Delta Modulation," *Bell System Tech. J.*, vol. 51, no. 10, 1972, pp. 2101–2136.

52. H. Schindler, "Linear, Nonlinear, and Adaptive Delta Modulation," *IEEE Trans. Commun.*, vol. COM-22, no. 11, 1974, pp. 1807–1823.

53. H. Levitt, C. McGonegal, and L. Cherry, "Perception of Slope Overload Distortion in Delta Modulated Speech Signals," *IEEE Trans. Audio Electroacoustics*, vol. AU-18, no. 3, 1970, pp. 240–247.

54. N. Jayant, "Digital Coding of Speech Waveforms: PCM, DPCM, and DM Quantizers," *Proc. IEEE*, vol. 62, no. 5, 1964, pp. 611–631.

55. L. Greenstein, "Slope Overload Noise in Linear Delta Modulators with Gaussian Inputs, "*Bell System Tech. J.*, vol. 52, no. 3, 1973, pp. 387–421.

56. P. Cummisky, N. Jayant, and J. Flanagan, "Adaptive Quantization in Differential PCM Coding of Speech," *Bell System Tech. J.*, vol. 52, no. 7, 1973, pp. 115–118.

57. E. Madsen, "Extraction of Ambience Information from Ordinary Recordings," *J. Audio Eng. Soc.*, vol. 18, no. 5, 1970, pp. 490–496.

58. H. Wallach, E. Newman, and M. Rosenzweig, "The Precedence Effect in Sound Localization," *Amer. J. Psychol.*, vol. 62, 1949, 315–336.

59. H. Haas, "Ueber den Einfluss eines Einfachechos auf die Hoersamkeit von Sprache," *Acustica*, vol. 1, 1951, pp. 49–58 (also in English, "The Influence of a Single Echo on the Audibility of Speech," *J. Audio Eng. Soc.*, vol. 20, no. 2, 1972, pp. 146–159).

60. T. Horral, B. Blanchard, and B. Watters, "An Auditorium Acoustic Simulator," *79th Meeting Acoust. Soc. Amer.*, Apr. 21, 1970.

61. T. Horral, private communications, 1972.

62. M. Gardner, "Image Fusion, Broadening, and Displacement in Sound Localization," *J. Acoust. Soc. Amer.*, vol. 46, no. 2 (part 2), 1969, pp. 339–348.

63. T. Schultz, "Acoustics of the Concert Hall," *IEEE Spectrum*, vol. 2, 1965, pp. 56–67.

64. M. Schroeder, "Natural Sounding Artificial Reverberation," *J. Audio Eng. Soc.*, vol. 10, no. 3, 1962, pp. 219–223.

65. L. Beranek, "Rating of Acoustic Quality of Concert Halls and Opera Houses," *4th Intern. Congr. Acoust.*, vol. 2, Copenhagen, 1972, pp. 15–29.

66. F. DeJager, "Delta Modulation, a Method for PCM Transmission Using the 1-unit Code," *Philips Res. Rept.*, vol. 7, 1952, pp. 442, 466.

67. C. J. Dalton, "Delta Modulation for Sound-Signal Distribution: A General Survey," *BBC Eng. Rev.*, July 1972, pp. 4–14.

68. N. Jayant and A. Rosenberg, "The Preference of Slope Overload to Granularity in Delta Modulation of Speech," *Bell System Tech. J.*, vol. 50, no. 10, 1971, pp. 3117–3125.

69. P. Nielsen, "On the Stability of Double Integration Delta Modulators," *IEEE Trans. Commun. Technol.*, vol. COM-19, 1971, pp. 364–366.

70. N. Jayant, "Adaptive Delta Modulation with a One-Bit Memory," *Bell System Tech. J.*, vol. 49, no. 3, 1970, pp. 321–342.

71. P. Cummisky, "Adaptive Differential Pulse-Code Modulation for Speech Processing," Ph.D. Dissertation, Newark College of Engineering, Newark, N.J., 1973.

72. R. Rouquette, "Audio Delta Modulator Design and Evaluation," S.M. Dissertation, Massachusetts Institute of Technology, Cambridge, Mass., 1975.

73. B. Blesser, unpublished experiments, 1975.

74. D. Goodman and L. Greenstein, "Quantizing Noise of Delta Modulation PCM Encoders," *Bell System Tech. J.*, vol. 52, no. 2, 1973, pp. 183–204.

75. H. Schmid, *Electronic Analog/Digital Conversions*, Van Nostrand Reinhold, New York, 1970.

76. D. Sheingold, *Analog–Digital Conversion Handbook*, Analog Devices, Inc., Norwood, Mass., 1972.

77. S. Ehara, "Amplitude Distribution of Orchestral Music," *Proc. 6th Intern. Congr. Acoust.*, vol. D, Tokyo, Japan, 1968, pp. 133–136.

78. L. Beranek, *Acoustics*, Chapter 31, McGraw-Hill, New York, 1954.

79. D. Gravereaux, A. Gust, and B. Bauer, "The Dynamic Range of Disc and Tape Records," *J. Audio Eng. Soc.*, vol. 18, no. 5, 1970, pp. 530–535.

80. Electronic Reverberation: EMT 250, Franz Vertriebsgesellschaft m.b.H., Lahr, West Germany.

81. D. Kahr, *The Codebreakers: The Story of Secret Writing*, Weidenfeld and Nicolson, London, 1967, pp. 588–599.

82. E. Zwicker and R. Feldtkeller, *Das Ohr Als Nachrichtenempfaenger*, 2nd ed., S. Hirzel Verlag, Stuttgart, West Germany, 1967.

83. S. Lin, *An Introduction to Error-Correcting Codes*, Prentice-Hall, Englewood Cliffs, N.J., 1970.

84. R. Hamming, "Error Detecting and Error Correcting Codes," *Bell System Tech. J.*, vol. 29, April 1950, pp. 147–160.

85. G. Forney, "Burst-Correcting Codes for the Classical Bursty Channel," *IEEE Trans. Commun. Technol.*, vol. COM-19, no. 5, 1971, pp. 772–781.

86. J. Massey, "Implementation of Burst-Correcting Convolutional Codes," *IEEE Trans. Inform. Theory*, vol. IT-11, July 1965, pp. 416–422.

87. B. Scharf, "Critical Bands," in *Foundations of Modern Audiotry Theory*, J. Tobias, ed., vol. 1, Academic Press, New York, 1970.

88. R. Orban, "A Program Controlled Noise Filter," *J. Audio Eng. Soc.*, vol. 22, no. 1, 1974, pp. 2–9.

89. H. Kuttruff, *Room Acoustics*, Chapter 4, Wiley (Halstead Press), New York, 1973.

90. B. ATAL, M. SCHROEDER, and G. SESSLER, "Subjective Reverberation Time and Its Relation to Sound Decay," preprint G32 from *Proc. 5th Intern. Congr. Acoust.*, Liege, 1965.

91. *Quadraphony: An Anthology*, preprints from the *J. Audio Eng. Soc.*, the Society, New York, 1975.

92. R. DEFREITES, "A System for Home Reverberation," presented to the 54th meeting of the Audio Engineering Society, Los Angeles, 1976.

93. N. JAYANT, "Adaptive Quantization with One-word Memory," *Bell System Tech. J.*, vol. 52, no. 7, 1973, pp. 1119–1144.

94. K. BAEDER, "Anwendung der Nachhallplatte," *Kino Tech.*, no. 6, West Germany, 1960, pp. 183–188.

95. W. KUHL, "Eine Kleine Nachhallplatte ohne Klangfarbung," presented to the 7th International Congress on Acoustics, Budapest, *Proceedings*, 1971, pp. 461–464.

96. W. KUHL, "The Acoustical and Technological Properties of the Reverberation Plate," *European Broadcast Union Rev.*, part A, no. 49, 1958, pp. 8–14.

97. AKG Division of Phillips, Vienna, Austria, manufacturer of reverberation spring.

98. M. RETTINGER, "Reverberation Chambers," *J. Audio Eng. Soc.*, vol. 20, no. 9, 1972, pp. 734–737.

99. A. KROKSTAD, S. STROM, and S. SORSDAL, "Calculating the Acoustical Room Response by the Use of Ray Tracing Techniques," *J. Sound Vibration*, vol. 8, 1968, pp. 118.

100. K. BAEDER, "Bestimmung der Eigenshaften von Nachhallerzeugern," *Radio Mentor*, vol. 5, 1970, pp. 346–346.

101. E. MEYER, "Definition and Diffusion in Rooms," *J. Acoust. Soc. Amer.*, vol. 26, 1954, pp. 630–636.

102. M. SCHROEDER, B. ATAL, G. SESSELER, and J. WEST, "Acoustical Measurements in Philharmonic Hall, New York" *J. Acoust. Soc. Amer.*, vol. 40, 1966, pp. 434–440.

103. W. KUHL, "Notwendige Eigenfrequenzdicte zur Vermeidung der Klangfaerbung von Nachhall," presented to the 6th International Congress on Acoustics, *Proceedings*, Tokyo, 1968, pp. E69–72.

104. B. ATAL and M. SCHROEDER, "Perception of Coloration in Filtered Gaussian Noise—Short Time Spectral Analysis by the Ear," presented to the International Congress on Acoustics, *Proceedings*, Copenhagen, 1962, Sess. H31.

105. H. KUTTRUFF and R. THIELE, "Ueber die Frequenzabhangigkeit des Schalldrucks in Raumen," *Acustica*, vol. 4, 1954, pp. 614–617.

106. M. SCHROEDER, "Die Statistichen Parameter des Frequenzkurven von grossen Raumen," *Acustica*, vol. 4, 1954, pp. 594–600.

107. W. SABINE, *Collected Papers on Accoustics*, Harvard University Press, Cambridge, Mass., 1927.

108. B. BLESSER, unpublished memorandum.

109. M. SCHROEDER and B. LOGAN, "Colorless Artificial Reverberation," *J. Audio Eng. Soc.*, vol. 9, 1961, pp. 192–197.

110. M. SCHROEDER, personal communications, 1971.

111. M. SCHROEDER, "Natural Sounding Artificial Reverberation," *J. Audio Eng. Soc.*, vol. 10, 1962, pp. 219–223.

112. H. KUTTRUFF, "Kuenstlicher Nachhall," *Frequenz*, vol. 16, 1962, pp. 91–96.

113. H. DATE and Y. TOZUKA, "An Artificial Reverberator Whose Amplitude and Reverberation Time Characteristics Can Be Controlled Independently," *Acustica*, vol. 17, 1966, pp. 42–47.

114. B. ATAL, M. SCHROEDER, and G. SESSLER, "Subjective Reverberation Time and Its Relation to Sound Decay," presented to the 5th International Congress on Acoustics, Liege, Sept. 1965.

115. B. BLESSER, K. BAEDER, and R. ZAORSKI, "A Real-Time Digital Computer for Simulating Audio Systems," *J. Audio Eng. Soc.*, vol. 23, no. 9, 1975, pp. 698–707.

116. A. OPPENHEIM, R. SCHAEFER, and T. STOCKHAM, "Non-linear Filtering of Multiplied and Convolved Signals," *Proc. IEEE*, vol. 56, Aug. 1968, pp. 1264–1291.

117. R. SCHAEFER, "Echo Removal of Discrete Generalized Linear Filtering," Technical Report 466, Research Laboratory of Electronics, Massachusetts Institute of Technology, Cambridge, Mass., 1969.

118. O. READ and W. WELCH, *From Tin Foil to Stereo*, Howard W. Sams Co., Indianapolis, Ind., 1959.

119. T. STOCKHAM, T. CANNON, and R. INGEBRETSEN, "Blind Deconvolution Through Digital Signal Processing," *Proc. IEEE*, vol. 63, Apr. 1975, pp. 678–692.

120. T. STOCKHAM, "The Application of Generalized Linearity to Automatic Gain Control," *IEEE Trans. Audio Electroacoustics*, vol. AU-16, June 1968, pp. 267–270.

121. J. BERMAN, "Loudspeaker Evaluation Using Digital Techniques," presented to 50th Convention of the Audio Engineering Society, preprint in the proceedings, London, Mar. 4, 1975.

122. L. FINCHAM, "Loudspeaker System Simulation Using Digital Technique," presented to 50th Convention of the Audio Engineering Society, preprint in the proceedings, London, Mar. 4, 1975.

123. L. RABINER and B. GOLD, *Theory and Application of Digital Signal Processing*, Prentice-Hall, Englewood Cliffs, N.J., 1975, pp. 565–571.

124. D. ROBINSON and R. DADSON, "A Re-determination of the Equal Loudness Relations for Pure Tones," *Brit. J. Appl. Phys.*, vol. 7, 1956, pp. 166–181.

125. C. Robinson and I. Pollack, "Interaction Between Forward and Backward Masking: A Measure of Integrating Period of the Auditory System," *J. Acoust. Soc. Amer.*, vol. 53, 1973, pp. 1313–1316.

126. T. Stockham, personal communication, 1976.

127. R. Easton, "Soundstream—The First Digital Studio," *recording engineer/producer.*

128. M. F. E. Barron, "The Effects of Early Reflections on Subjective Acoustical Quality in Concert Halls," thesis submitted for the Ph.D. degree, University of Southampton, Southampton, England, 1974.

129. R. Berkovitz and D. McIntosh, "A 16-Channel Programmed Delay Network," presented at the 54th Convention of the Audio Engineering Society, Los Angeles, May 1976.

130. I. Nomoto, M. Iwahara, and H. Onoye, "A Technique for Observing Loudspeaker Wave Front Propagation," *J. Audio Eng. Soc.*, vol. 24, no. 1, Jan.–Feb. 1976, pp. 9–13.

131. A. Levy, "Preliminary High Linear Density Recording Study: Phase II—Pushing the Limits," Internal Memorandum, Bell and Howell, 1975.

3

DIGITAL PROCESSING OF SPEECH

A. V. Oppenheim

Massachusetts Institute of Technology
Cambridge, Mass. 02139

3.1 INTRODUCTION

Speech processing has been an active area for several decades with a wide variety of applications ranging from communications to automatic reading machines.[4] Prior to the mid-1960s, essentially all speech-processing systems were based on analog hardware implementations, although, during that period, a number of speech-processing systems were implemented on general-purpose digital computers. However, these were generally viewed as non-real-time simulations of analog systems, and were based on algorithms that were matched to available analog hardware.[41]

With the inherent flexibility of digital computers there was a natural tendency to experiment with more sophisticated algorithms, even when it appeared that such algorithms might have no practical analog implementations. As the field of digital signal processing developed, both in terms of digital hardware capabilities and the development of new signal-processing algorithms, it became increasingly clear that the associated set of techniques and implementations would have a dramatic impact on the area of speech processing. Many of the developments in the area of digital signal processing were, in fact, carried out within the context of speech processing, in part because the bandwidths associated with speech were well matched to the processing speeds available. Recently we have seen a major thrust toward the incorporation of digital signal processing into speech-processing systems;

almost all modern speech-processing systems now rely, at least in part, on digital signal-processing algorithms.

Speech-processing problems can generally be divided into three classes. In one set of problems we are interested only in speech analysis. For example, for an automatic speech-recognition system, we begin from the speech waveform and the desired result is an action based on recognition of the speech. Other examples, involving analysis only, are speaker identification and speaker verification. A second class of problems involves synthesis only, as in an automatic reading machine for which the input is written text and the output is speech. Another example lies in data-retrieval systems. For example, it is sometimes of interest to obtain data from a computer verbally, as when a data base is to be interrogated from an ordinary telephone. Such a system could, for example, permit a doctor in a remote location to access medical records stored in a central computer.

A third class of problems involves speech analysis followed by speech synthesis. Secure voice transmission and data rate compression of speech represent two common examples. If speech is transmitted by simply sampling and digitizing, the data rate required is in the order of 90,000 bits per second of speech. Through the use of speech analysis followed by appropriate coding, transmission, and resynthesis at the receiver, this can be reduced by a factor of between 10 and 50, depending on the type of system used and the desired quality of the synthesized speech. Another common example of analysis-synthesis lies in the implementation of speech storage and retrieval systems, such as the use of an automatic intercept operator in the telephone system. In speech storage and retrieval systems, speech analysis can be carried out in nonreal time and the results of this analysis stored in the computer memory. To generate the required verbal response, these analysis parameters would then be used to drive a speech synthesizer. Other examples in which speech analysis followed by synthesis is of interest are time expansion and compression of speech and enhancement of degraded speech.

The techniques for digital speech processing can generally be divided into two broad classes.[10] One class utilizes the waveform coding methods as they apply to general audio signals. These methods include pulse code modulation (PCM), delta modulation (DM), differential pulse code modulation (DPCM), and others. All these methods assume that the signal is band limited, but make no other assumptions about the signal. The second class of speech-processing techniques capitalizes more specifically on the structure of the speech waveform, as represented by a model consisting of a slowly varying linear system excited by an appropriate excitation signal.

In Chapter 2, where the general class of audio signals was considered, some of the general techniques not specific to speech were discussed. In this chapter our focus will be entirely on processing techniques that apply specifically to speech. Toward this end, we shall discuss in Sec. 3.2 a simplified model of speech production. Although this model is somewhat of an oversimplification, it has been the basis for many speech-processing systems, and is generally useful for focusing on many of the important aspects of the speech waveform.

3.2 MODEL OF THE SPEECH WAVEFORM

The basic techniques for speech analysis and synthesis can be viewed in terms of a model of the speech waveform as the response of a slowly time varying system to either a periodic or a noiselike excitation.[13,4] More specifically, the speech-production mechanism consists essentially of an acoustic tube, the vocal tract, excited by an appropriate source to generate the desired sound. In the case of *voiced* speech sounds, the excitation corresponds to a quasi-periodic pulse train representing the air flow through the vocal cords as they vibrate. The fricative sounds are generated by forcing air through a constriction in the vocal tract, thereby creating turbulence, which produces a source of noise to excite the vocal tract. A cross-sectional view of the vocal mechanism is shown in Fig. 3-1. Figure 3-2 is an example of a speech waveform in which these two classes of sounds are illustrated.

As suggested by the above discussion, the speech waveform can be modeled as the response of a linear time-varying system, the vocal tract, with the appropriate excitation. If the vocal-tract shape is fixed, the output of the system is the convolution of the excitation and vocal-tract impulse response. Different sounds, however, are produced by changing the shape of the vocal tract. If the vocal-tract shape changes slowly, it is reasonable to still approximate the output on a short-time basis as a convolution of the excitation and vocal-tract impulse response. Figure 3-3 depicts this model, together with an illustration of the time-domain response and the corresponding spectra for a segment of voiced speech. For voiced speech we observe that, if the input on a short-time basis is periodic, corresponding to a fixed fundamental frequency, the output is also periodic. Alternatively, viewing the system in the frequency domain, the Fourier transform of the speech waveform is the product of the Fourier transforms of the excitation function and vocal-tract impulse response. This is depicted in Fig. 3-3(b).

In particular, for a periodic excitation the corresponding spectrum is a line spectrum with harmonics equally spaced by $2\pi/T$ and with an envelope reflecting the shape of the glottal pulse. The frequency response of the vocal tract is a relatively smooth function of frequency; since the vocal tract represents an acoustic cavity, it is characterized primarily by resonances corresponding to the resonant frequencies of the acoustic cavity, which are commonly referred to as *formant frequencies*. The composite spectrum consists of the product of the line spectrum due to the excitation and the spectrum due to the vocal tract and, consequently, is a line spectrum with an envelope that is characteristic of the vocal-tract transfer function. As the vocal-tract

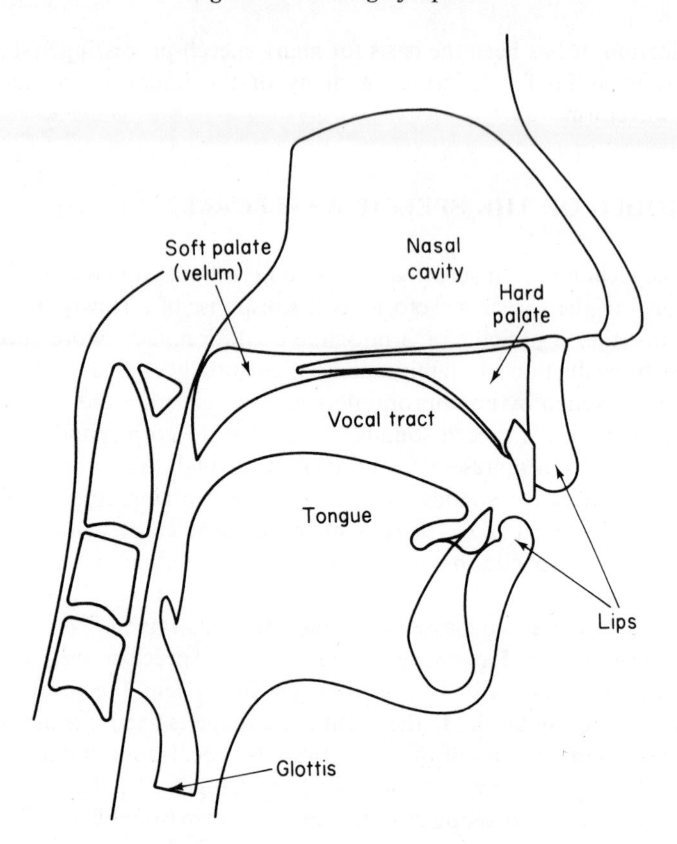

FIGURE 3-1. Cross-sectional view of the vocal mechanism, (after Markel and Gray).

shape changes to generate different sounds, the envelope of the speech spectrum will, of course, also change with time. Similarly, as the excitation period changes for voiced sounds, the spacing of the pitch harmonics will change. Consequently, it is generally desirable to view the spectrum of speech on a short-time basis, as it changes with time.

A time–frequency–intensity display of the short-time spectrum of speech is referred to as a *speech spectrogram*.†[29] It is common practice to consider both wideband and narrowband spectrograms. In a wideband spectrogram, time resolution is relatively high and, in fact, individual periods of the time waveform are evident. The frequency resolution, however, is not sufficient to resolve the fine structure due to the excitation. In a narrowband spectrogram, individual harmonics of the excitation are resolved in frequency,

† It is sometimes also referred to as a *voiceprint*, the trade name of a commonly used commercial speech spectrogram machine.

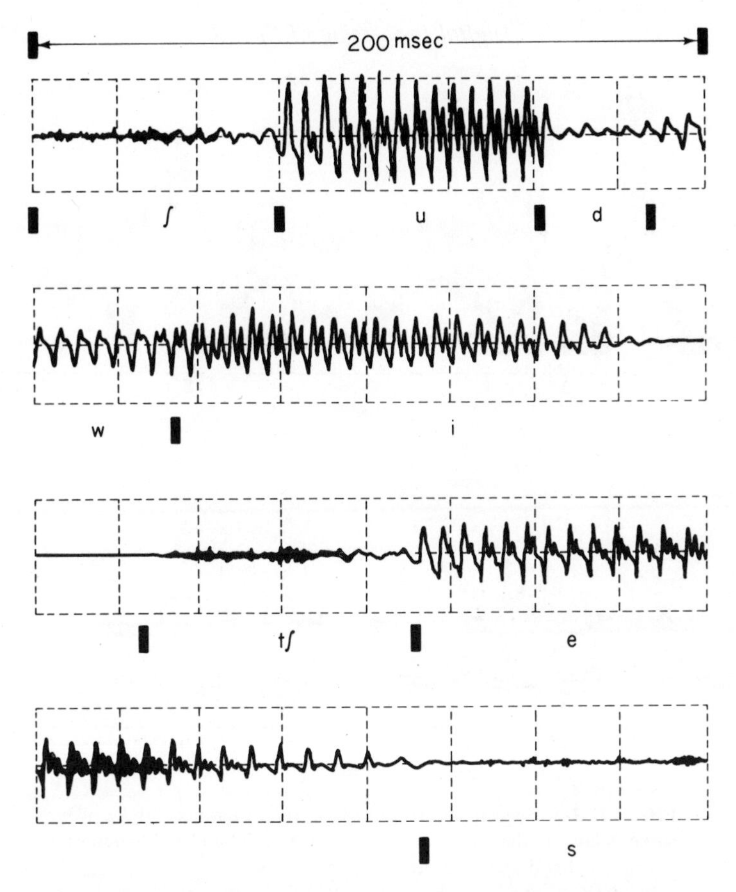

FIGURE 3-2. Example of a speech waveform illustrating different classes of sounds. The utterance is "should we chase . . .".

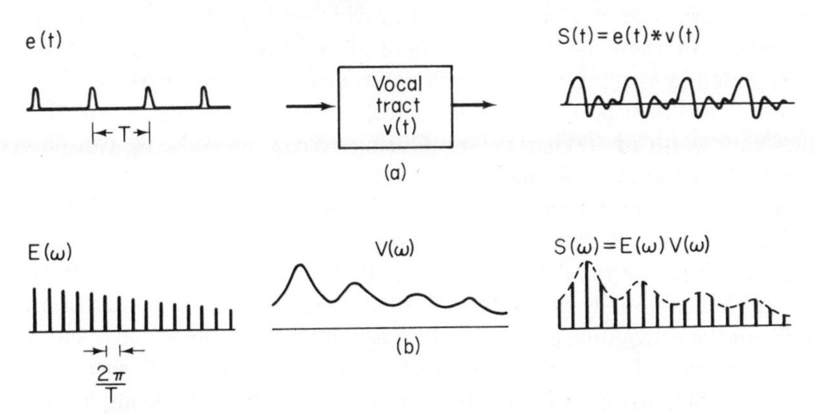

FIGURE 3-3. Model of speech production as the response of a quasi-stationary linear system; (a) time-domain characterization, (b) frequency-domain characterization.

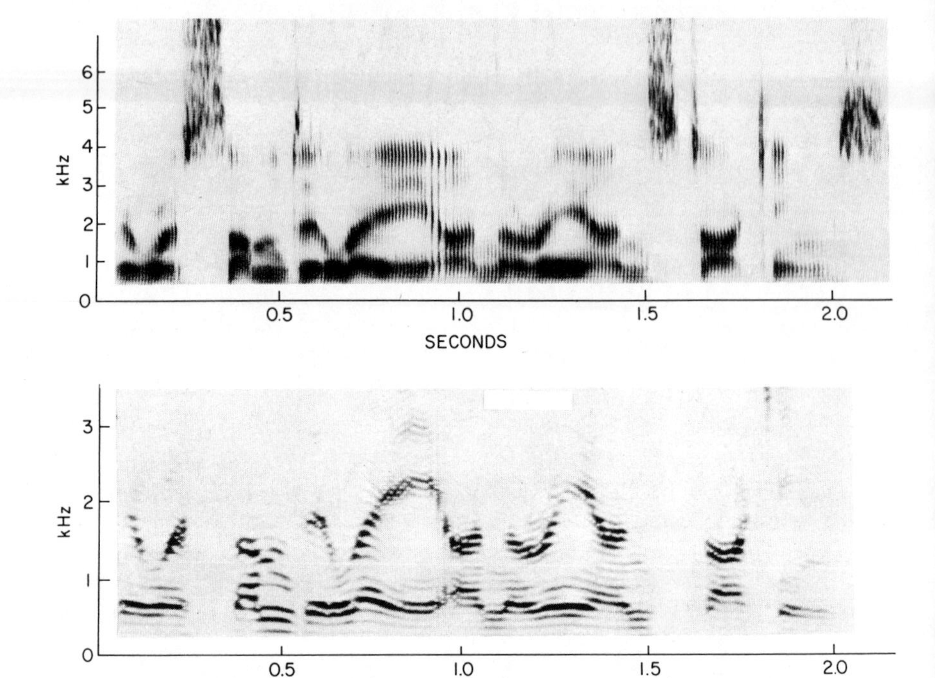

FIGURE 3-4. Speech spectrograms of the utterance "there was some delay on the rayon stockings"; (a) wideband spectrogram; (b) narrowband spectrogram.

but the time resolution is not as high as in a wideband spectrogram. Figure 3-4 shows an example of a narrowband and wideband spectrogram of the same utterance. In these figures, the magnitude of the short-time spectrum is represented by shading, with larger magnitude represented by darker shading. Figure 3-4(a) depicts a wideband spectrogram corresponding to an effective filter bandwidth of 300 hertz (Hz). During voiced intervals, we can observe clearly the vocal-tract resonances, which appear as dark bands in the spectrogram. Also in voiced intervals we can see the individual pitch periods resolved in time. We can also observe unvoiced regions in which the excitation is noise-like rather than periodic. Note that the individual harmonics of the excitation during voiced sounds are not resolved in frequency. Thus, in essence, the wideband spectrogram provides an approximation to the spectral envelope as a function of time. Figure 3-4(b) depicts a narrowband spectrogram, corresponding to an effective filter bandwidth of 45 Hz. Comparing this with the spectrogram in Fig. 3-4(a), we see that the individual harmonics of the excitation are now resolved in frequency, although the time resolution is decreased. In Sec. 3.4 we shall return to a more detailed discussion of the short-time spectral analysis of speech.

Many basic techniques for speech analysis and synthesis are conveniently described in terms of the model that we have presented. Speech analysis systems generally attempt to separate the excitation function from the vocal-tract characteristics. Then, depending on the specific analysis strategy, parameters descriptive of each component are obtained. For example, the excitation function may be characterized in terms of whether it is periodic or noiselike and, for periodic excitation, its fundamental frequency. Vocal-tract characteristics might be represented by samples of its spectrum or in terms of a parametric model. From the model of Fig. 3-3, speech synthesis can be viewed in terms of the response of a slowly time varying linear system to a periodic or noiselike excitation.

For speech synthesis carried out within the context of a speech analysis–synthesis system, the parameters of the time-varying linear system and the excitation are provided by the analysis, and the detailed configuration of the synthesizer is dictated to a large extent by that of the analyzer. In Sec. 3.4–3.6 we shall discuss a number of such analysis–synthesis systems. For speech synthesis alone, there are a variety of configurations that can be used, two specific classes of which we shall discuss in the next section. When discussing speech analysis–synthesis systems, we shall also consider several other synthesis systems.

3.3 TERMINAL ANALOG AND ACOUSTIC TUBE ANALOG SPEECH SYNTHESIZERS

For speech synthesis alone, two classes of synthesizers are commonly used, (1) *terminal analog* and (2) *acoustic tube analog* synthesizers. The class of terminal analog synthesizers is directed at implementing a system whose transfer function approximates the vocal-tract transfer function, but whose implementation bears no direct relation to the details of the vocal-tract structure. Thus they attempt to represent the vocal tract from a terminal point of view only. Acoustic tube analog synthesizers simulate the air pressure or flow as a function of time and position in an acoustic tube for which the cross-sectional area is a function of position. Thus, in some sense, an acoustic tube analog synthesizer attempts to more directly represent physical variables in the vocal tract.

The general structure of a terminal analog synthesizer, also referred to as a *formant* synthesizer, is shown in Fig. 3-5. The basis for this class of synthesizers is the observation that, since the vocal tract is an acoustic cavity, it is characterized by a set of modes or resonant frequencies. Thus its transfer function can be approximated by a cascade combination of resonant circuits, each one representing one of the modes or resonances of the vocal tract.[17,19,20] As the shape of the vocal tract changes, the resonances change. Consequently, the resonant circuits are provided with a set of time-varying

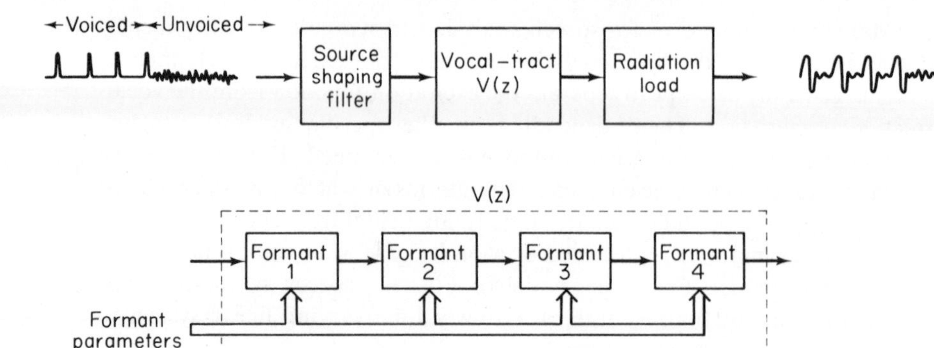

FIGURE 3-5. General structure of a terminal-analog speech synthesizer.

parameters that control the center frequencies and bandwidths of the resonators. If the synthesizer is driven by an impulse train for voiced speech and white noise for unvoiced speech, a source-shaping filter is required to provide the appropriate spectral coloration. This filter is generally not time varying. In addition, a filter that accounts for the effect of the coupling of the acoustic tube into space (i.e., a tube with infinite cross-sectional area) is required. This is also a fixed filter with a characteristic approximately corresponding to that of a differentiator.

For a formant synthesizer implemented in continuous time with lumped analog circuits, an additional filter, referred to as a higher-pole correction filter, is required. This filter accounts for the fact that a distributed system such as an acoustic tube has an infinite number of resonances. Although only a finite number (four or five) fall within the bandwidth of the synthesizer, the higher-frequency poles influence the overall shape of the spectrum within the bandwidth of the synthesizer. Thus, for a cascade of lumped analog poles, an additional frequency boost is required. As was first noted by Gold and Rabiner,[23] this requirement is not present in a digital speech synthesizer because the frequency response is periodic and, consequently, the higher poles are, in effect, present. An alternative way of viewing this is suggested by considering a simple acoustic tube closed at one end and open at the other. Its impulse response is an impulse train, which is more easily represented by a discrete-time filter than by a continuous-time filter with a rational transfer function. Figure 3-6 compares the frequency response of a simple acoustic tube, a five-pole digital filter, and a five-pole continuous-time filter, and illustrates the need for higher pole correction in the continuous-time case.

The formant synthesizer depicted in Fig. 3-5 assumes that the vocal-tract transfer function can be represented by all poles (i.e., does not have antiresonances associated with it). In fact, for a variety of speech sounds, spectral zeros are present owing to the effect of the nasal cavity when the vellum is

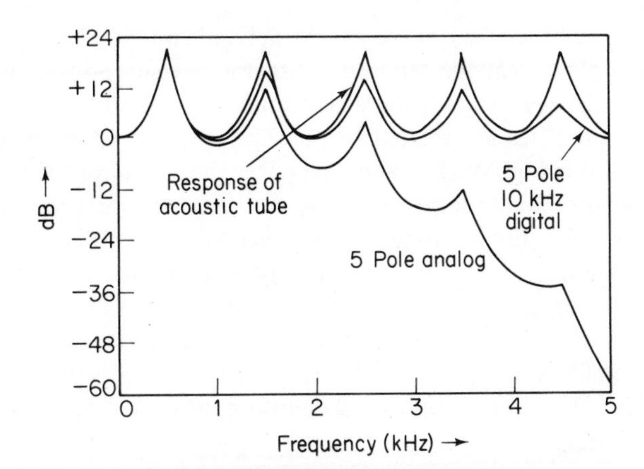

FIGURE 3-6. Comparison of the frequency response of a five pole analog and digital filter with that of a simple acoustic tube; (after Gold and Rabiner).

opened for nasal sounds such as /*m*/, or when the tongue is used to divide the vocal cavity into two loosely coupled cavities as with /ℓ/.[13,16] Some formant synthesizers include a network for spectral zeros, but their effect can often be approximated by adjustment of the bandwidth of the first formant.

An acoustic tube analog speech synthesizer is based on an approximation of the vocal tract as a set of interconnected acoustic tube sections of equal length, as depicted in Fig. 3-7.[4,21,83] It is generally assumed that sound propagation through each section can be treated as a plane wave, and that

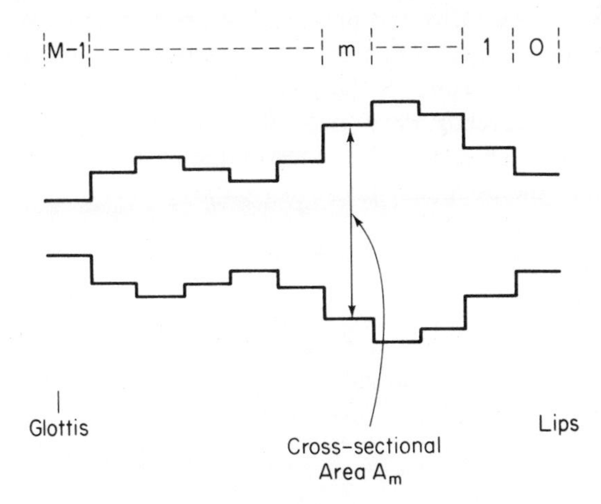

FIGURE 3-7. Representation of the vocal-tract as a set of interconnected sections of equal length and varying cross-sectional area.

internal losses and the effect of the nasal tract and coupling between the vocal tract and the glottis can be ignored. With these assumptions, the acoustic tube model is relatively straightforward to analyze and leads to a filter structure in which the variables are related to physical variables in an acoustic tube. Specifically, following Markel and Gray,[83] the behavior of the acoustic cavity can be described in terms of pressure or in terms of the volume velocity of air flow as a function of time and distance along the tube. Furthermore, within each section, they can each be represented as a combination of a forward- and reverse-traveling wave; the forward-traveling wave moves in the direction from the glottis to the lips, and the reverse-traveling wave moves in the direction from the lips to the glottis. If $u_m(x, t)$ denotes the volume velocity within the mth section, with $x = 0$ denoting the center of the section, then $u_m(x, t)$ is expressed as

$$u_m(x, t) = u_m^+\left(t - \frac{x}{c}\right) - u_m^-\left(t + \frac{x}{c}\right) \tag{3.1}$$

where u_m^+ and u_m^- are the forward- and reverse-traveling waves, respectively, and c is the velocity of sound in air.

The pressure $p_m(x, t)$ in the mth section is also related to the forward- and reverse-traveling volume velocity waves by

$$p_m(x, t) = \frac{\rho c}{A_m}\left[u_m^+\left(t - \frac{x}{c}\right) + u_m^-\left(t + \frac{x}{c}\right)\right] \tag{3.2}$$

where ρ is air density.

Furthermore, the forward- and reverse-traveling waves in each section can be related to each other by virtue of the fact that at the boundary between sections the volume velocity and pressure must be continuous. As a result, at the boundary between each section some fraction of the forward-traveling wave gets transmitted through to the next section, and some fraction is reflected back as a reverse-traveling wave. A similar statement applies to the reverse-traveling wave in each section. For the volume velocity waves, the reflection coefficient μ_m at the boundary between sections m and $m - 1$ is given by

$$\mu_m = \frac{A_{m-1} - A_m}{A_{m-1} + A_m} \tag{3.3}$$

where A_m and A_{m-1} are the cross-sectional areas of the mth and $(m - 1)$st sections, respectively. With ℓ denoting the section length and imposing the condition that the volume velocity and pressure are continuous across the interfaces between sections, it follows that

$$u_{m-1}^+(t + \tau) = \mu_m u_{m-1}^-(t - \tau) + (1 + \mu_m)u_m^+(t - \tau) \tag{3.4a}$$

$$u_m^-(t + \tau) = (1 - \mu_m)u_{m-1}^-(t - \tau) - \mu_m u_m^+(t - \tau) \tag{3.4b}$$

where τ, defined as $\ell/2c$ corresponds to half the time required for a wave to propagate from one end of a section to the other.

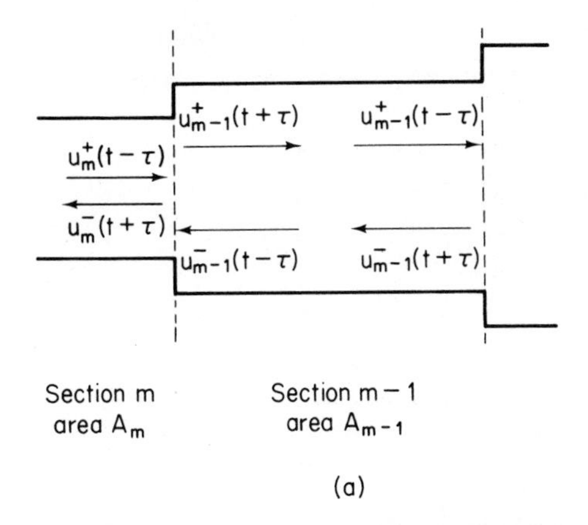

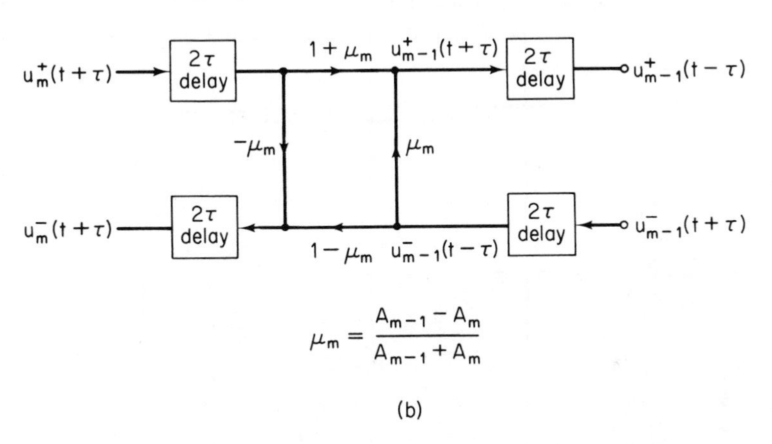

FIGURE 3-8. (a) Two sections of the acoustic tube model of Fig. 7 indicating the forward and reverse travelling volume velocity waves. (b) Signal flow graph representing the relationships between the volume velocity waves in (a), (after Markel and Gray).

Based on Eqs. (3.4a) and (3.4b), the reverse- and forward-traveling volume velocity waves can be related through a linear signal-flow graph. Figure 3-8(a) shows two sections of the acoustic tube model, and Fig. 3-8(b) is the corresponding linear signal-flow graph relating the forward- and reverse-traveling volume velocity waves. A linear signal-flow graph depicting the relationships between the forward- and reverse-traveling volume velocity waves throughout the acoustic tube model is shown in Fig. 3-9. For the boundary condition at the lips, the tube is assumed open so that the pressure is zero. At the glottal end, the $(M - 1)$st section is assumed connected to a volume velocity source with a specified impedance that can be represented by a reflection coefficient μ_M at that end. This flow graph can be interpreted

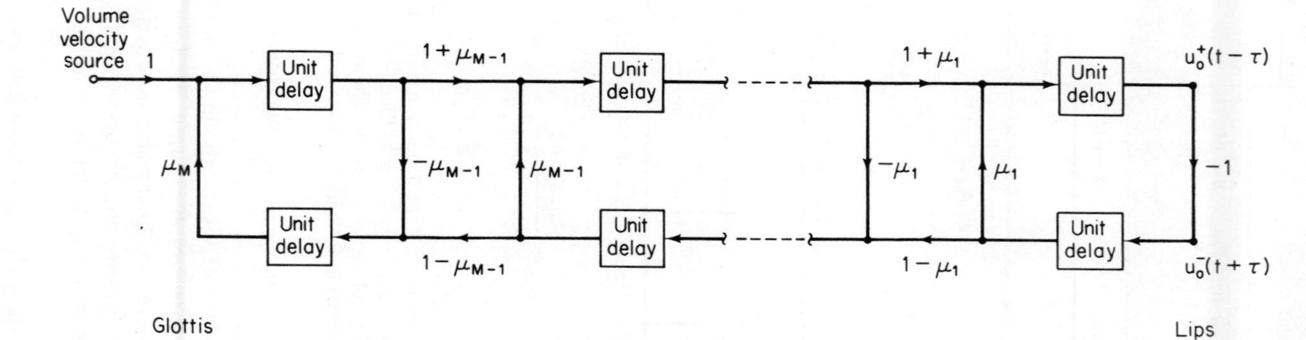

FIGURE 3-9. Linear signal-flow graph depicting the relationships between the forward and reverse traveling volume velocity waves throughout the acoustic tube model of Fig. 7, (after Markel and Gray).

directly as a digital filter structure by taking the delay time $2\tau = \ell/c$ of each section to correspond to a unit delay. As the configuration of the vocal tract changes for different sounds, the cross-sectional area of each section, or equivalently the reflection coefficients μ_m, are modified.

A model of the type depicted in Fig. 3-9 for speech synthesis was first considered by Kelly and Lochbaum.(21) A number of variations of this structure, which offer some potential advantages in the number of multiplications, word lengths, and the like, required, have since been proposed by Markel and Gray.(83)

The preceding discussion completes a preliminary look at speech synthesis. In the next several sections we shall discuss a number of additional synthesizer configurations within the context of analysis–synthesis systems. As we shall see in Sec. 3.6, the acoustic tube analog synthesizer is well matched to a particular analysis–synthesis procedure referred to as linear prediction.

3.4 SHORT-TIME FOURIER ANALYSIS AND SYNTHESIS OF SPEECH

In Sec. 3.2 we saw that, in the frequency domain, speech can be represented on a short-time basis in terms of the product of a spectral envelope, characterizing the vocal tract, and a fine structure, characterizing the excitation. Since the excitation parameters for voiced speech are reflected in the frequency spacing of the pitch harmonics, and the characteristics of the vocal tract are represented to a large extent by the formant frequencies, it is particularly convenient to base the analysis of speech on a frequency-domain representation. As the excitation and shape of the vocal tract change to produce different sounds, the spectrum will change. Consequently, a spectral representation of speech must be based on a short-time Fourier transform, which, of course, varies with time.

If we consider the speech signal to be sampled so that it is represented by a sequence $s(n)$, the short-time Fourier transform $S(\omega, n)$ is defined as

$$S(\omega, n) = \sum_{k=-\infty}^{+\infty} s(k)h(n-k)e^{-j\omega k} \tag{3.5}$$

Thus it represents the Fourier transform of a windowed segment of the speech waveform as the window $h(n)$ slides in time, as depicted in Fig. 3-10. There are two common ways of implementing a short-time spectral analysis as specified by Eq. (3.5). The first is by means of a filter bank and represents the typical implementation when the spectral analysis is to be carried out with an analog system. For digital implementation of the short-time Fourier transform, either a filter bank may be implemented directly or the fast Fourier transform (FFT) algorithm may be used. To see how Eq. (3.5) corresponds

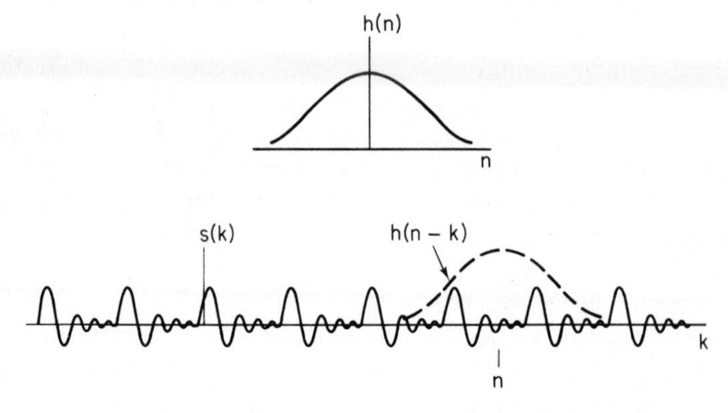

FIGURE 3-10. Representation of short-time Fourier analysis.

to a filter bank, we note that it corresponds to the convolution of the window $h(n)$, with $s(n)$ modulated by $e^{-j\omega n}$; that is,

$$S(\omega, n) = [s(n)e^{-j\omega n}]*h(n) \tag{3.6}$$

where $*$ denotes convolution. Thus $S(\omega, n)$ can be obtained from the system in Fig. 3-11. An alternative implementation of the filter bank follows by rewriting Eq. (3.5) as

$$S(\omega, n) = e^{-j\omega n} \sum_{k=-\infty}^{+\infty} s(k)h(n-k)e^{j\omega(n-k)} \tag{3.7}$$

or

$$S(\omega, n) = e^{-j\omega n}\{s(n)*[h(n)e^{j\omega n}]\} \tag{3.8}$$

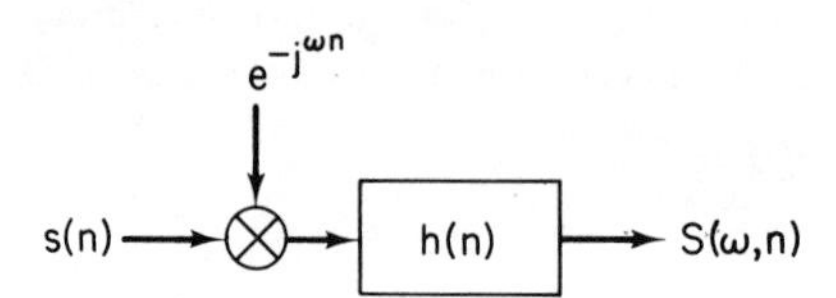

FIGURE 3-11. System for determining the short-time Fourier transform.

The filter with impulse response $h(n)e^{j\omega n}$ has a bandpass characteristic centered around the frequency ω. The system represented by Eq. (3.8) is indicated in Fig. 3-12. The choice of the system of Fig. 3-11 or Fig. 3-12 is largely a matter of convenience. In the case of Fig. 3-11, a lowpass prototype filter with impulse response $h(n)$ is used, and the input signal is modulated with the modulation depending on the value of ω at which $S(\omega, n)$ is to be measured. The system of Fig. 3-12 is more suitable for parallel measurement of $S(\omega, n)$ at a set of values of ω.

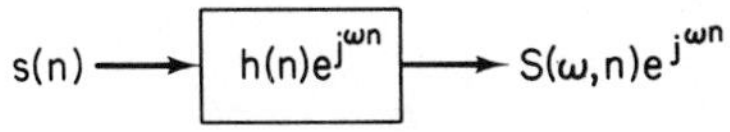

$s(n) \longrightarrow$ $h(n)e^{j\omega n}$ $\longrightarrow S(\omega, n)e^{j\omega n}$

FIGURE 3-12. Alternative system for determining the short-time Fourier transform.

Evaluation of the short-time Fourier transform at a set of equally spaced frequencies $\omega_r = 2\pi r/N$, $r = 0, 1, \ldots, N-1$, can be accomplished using the FFT algorithm. In particular, denoting the samples of $S(\omega, n)$ by $S_r(n)$, so that

$$S_r(n) = S(\omega_r, n) \tag{3.9}$$

we have

$$S_r(n) = \sum_{k=-\infty}^{+\infty} s(k)h(n-k)e^{-j\frac{2\pi}{N}rk} \tag{3.10}$$

By means of the substitution of variables $k = \ell + n$, Eq. (3.10) can be rewritten as

$$\begin{aligned} S_r(n) &= \sum_{\ell=-\infty}^{+\infty} s(\ell + n)h(-\ell)e^{-j\frac{2\pi}{N}r(\ell+n)} \\ &= e^{-j\frac{2\pi}{N}rn} \sum_{\ell=-\infty}^{+\infty} s(\ell + n)h(-\ell)e^{-j\frac{2\pi}{N}\ell r} \end{aligned} \tag{3.11}$$

The summation can now be divided into intervals of length N and the contributions from each interval summed, so that

$$S_r(n) = e^{-j\frac{2\pi}{N}rn} \sum_{m=-\infty}^{+\infty} \sum_{\ell=mN}^{mN+N-1} s(\ell + n)h(-\ell)e^{-j\frac{2\pi}{N}\ell r} \tag{3.12}$$

With a change of variables on the inner sum, and taking advantage of the periodicity of the complex exponential factor $e^{-j(\frac{2\pi}{N})\ell r}$, Eq. (3.12) can be rewritten as

$$S_r(n) = e^{-j\frac{2\pi}{N}rn} \sum_{k=0}^{N-1} \tilde{s}(k, n)e^{-j\frac{2\pi}{N}rk} \tag{3.13a}$$

where

$$\tilde{s}(k, n) = \sum_{m=-\infty}^{+\infty} s(n + k + mN)h(-k - mN)$$

$$k = 0, 1, \ldots, N-1 \tag{3.13b}$$

The summation in Eq. (3.13a) for any fixed n is recognized as the N point (in k) discrete Fourier transform (DFT) of the sequence $\tilde{s}_k(n)$ and, consequently, can be computed using the FFT algorithm. The process of obtaining $S_r(n)$ from $s(n)$ according to Eqs. (3.13) is depicted in Fig. 3-13.

Another important approach to computing the short-time Fourier transform at a set of equally spaced frequencies is obtained by substituting into Eq. (3.10) the identity

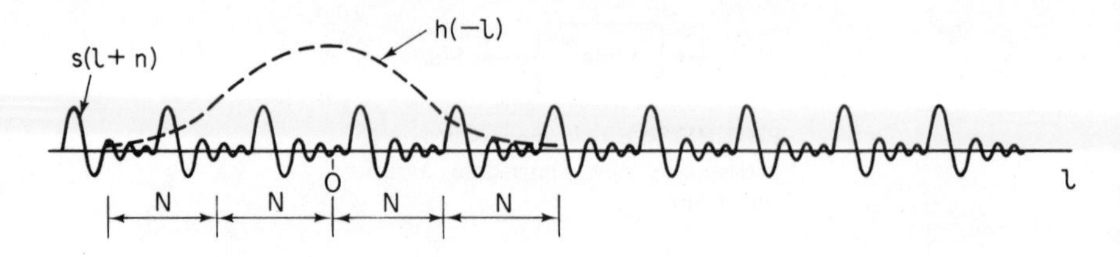

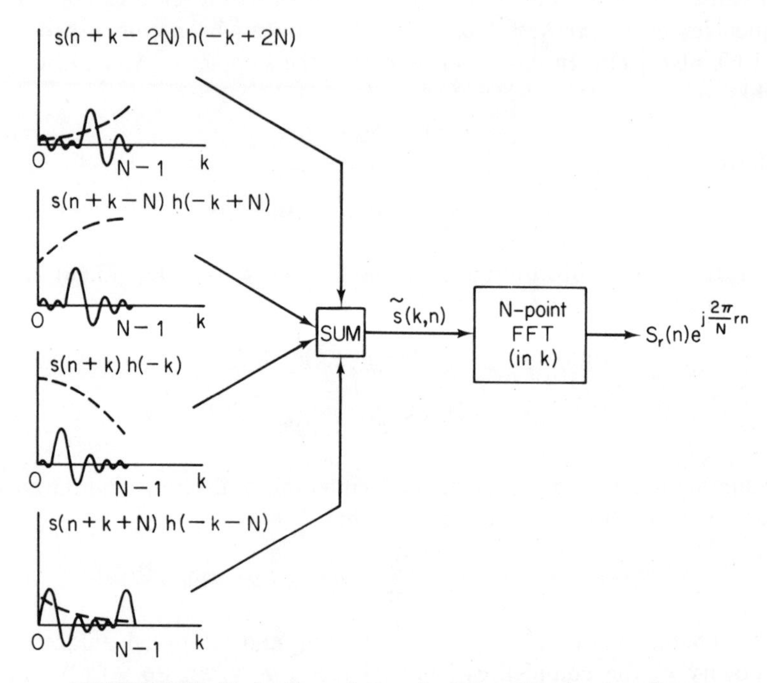

FIGURE 3-13. Evaluation of the short-time Fourier transform using the FFT algorithm.

$$rk = \frac{r^2}{2} + \frac{k^2}{2} - \frac{(r-k)^2}{2} \tag{3.14}$$

to obtain

$$S(\omega_r, n) = e^{-j\frac{\pi}{N}r^2} \sum_{k=-\infty}^{+\infty} s(k)h(n-k)e^{-j\frac{\pi}{N}k^2} e^{j\frac{\pi}{N}(r-k)^2} \tag{3.15}$$

For any fixed n, this can be viewed as

$$S(\omega_r, n) = e^{-j\frac{\pi}{N}r^2}[g_n(k) * e^{j\frac{\pi k^2}{N}}] \tag{3.16a}$$

where

$$g_n(k) = s(k)h(n-k)e^{-j\frac{\pi}{N}k^2} \tag{3.16b}$$

Thus the evaluation of the short-time transform for each n is converted to a convolution. It is important to note that although Eqs. (3.6) and (3.8) also express the short-time transform as a convolution, there is a distinct difference; as opposed to Eqs. (3.6) and (3.8), the convolution of Eqs. (3.16) produces a set of spectral samples for a fixed value of n.

The formulation of Fourier analysis by means of Eqs. (3.16) was first proposed by Bluestein[31] and subsequently developed by Rabiner, Schafer, and Rader[30] into a form referred to as the *chirp z-transform* (CZT) algorithm. Although this algorithm appears to be more complex computationally, it in fact offers an important advantage in some contexts. Specifically, if the window $h(n)$ is of finite length, the convolution in Eq. (3.16a) can be implemented by processing the sequence $g_n(k)$ with a finite impulse response filter, the impulse response of which is a segment of the complex sequence $e^{j\pi k^2/2}$. Since finite impulse response filters have a particularly convenient implementation with charge-coupled devices (CCD) and other charge-transfer technology, spectral analysis based on the use of Eqs. (3.16) can be conveniently implemented with this technology.[36,38,39]

A time-frequency intensity display of the magnitude of the short-time Fourier transform corresponds to the speech spectrogram as discussed in Sec. 3.2. The examples presented in Fig. 3-4 were obtained from an analog speech spectrograph machine. Equivalently, as we have seen, speech spectrograms can be obtained digitally using the FFT.[32,33,35] Figure 3-14 shows

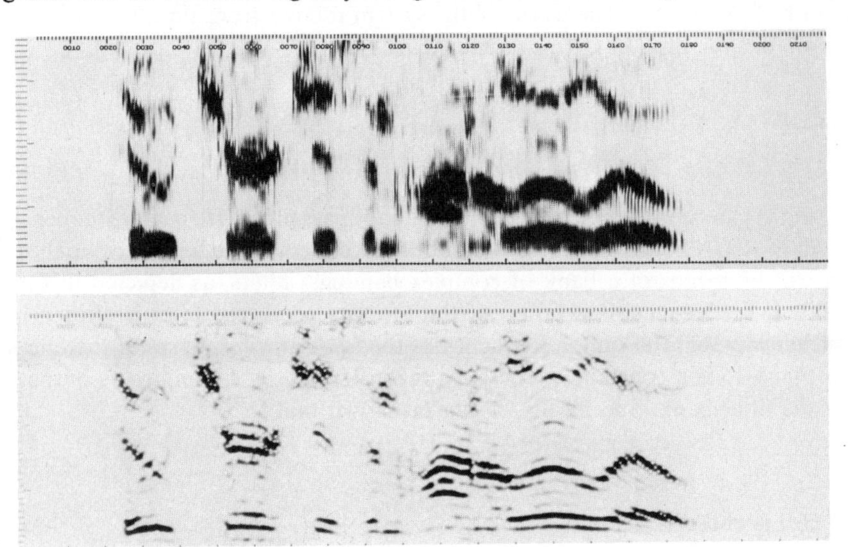

FIGURE 3-14. Speech spectrograms of the utterance "Two churches in Honolulu . . ." (a) wideband analysis, (b) narrowband analysis. The time axis full scale is 2.2 secs. and the frequency axis full scale is 3.5 kHz.

examples of a wideband and narrowband spectrogram computed by the procedure corresponding to Eqs. (3.13) and Fig. 3-13. As is evident in the difference between the narrowband and wideband spectrogram, in both Figs. 3-4 and 3-14, in carrying out a short-time spectral analysis of speech, there is a trade off between time resolution and spectral resolution. In particular, in order for the short-time Fourier transform of speech to be responsive to changes in the excitation and vocal tract, it is desirable for the window $h(n)$ to be short or, equivalently, for the memory of the analysis filters to be short. Thus, in a wideband spectrogram, the duration of the window $h(n)$ is chosen to be on the order of a pitch period to provide good time resolution. The resulting frequency resolution is not sufficient to resolve the individual pitch harmonics, but will in general resolve the formants. For a narrowband spectral analysis, the equivalent filter of Fig. 3-11 or 3-12 is chosen to be narrow in frequency, which in turn requires that the duration of $h(n)$ be on the order of several pitch periods; in this case the individual harmonics of the excitation are resolved at the expense of time resolution.

The graphical display of speech through the use of spectrograms represents an important application of the short-time Fourier transform. Other applications occur in analysis–synthesis. One such system has its origins in the phase vocoder developed by Flanagan and Golden,[42] and the subsequent digital formulation and refinements introduced by Schafer and Rabiner[47] and by Portnoff.[48] The basis for the system follows from Eq. (3.10), which, as we did for Eq. (3.5), we can rewrite as

$$S_r(n)e^{+j\frac{2\pi}{N}nr} = [s(n)*h_r(n)] \tag{3.17a}$$

where

$$h_r(n) = h(n)e^{j\frac{2\pi}{N}nr} \tag{3.17b}$$

$h_r(n)$ can be interpreted as a complex bandpass filter with center frequency at $\omega = (2\pi/N)r$; thus Eq. (3.17) can be interpreted as the frequency analysis of $s(n)$ by means of a bank of complex bandpass filters, as depicted in Fig. 3-15. If the frequency response of the lowpass prototype filter is chosen in such a way that the sum in frequency of the frequency responses of the filters in Fig. 3-15 is a constant, $s(n)$ can be reconstructed by summing the outputs of the filter bank. Specifically, it can be shown that[47,48]

$$s(n) = \frac{1}{N}\sum_{r=0}^{N-1} S_r(n)e^{j\frac{2\pi}{N}nk} \qquad \text{for all } n \tag{3.18}$$

if $h(n)$ is chosen such that

$$\begin{aligned} h(0) &= 1 \\ h(n) &= 0 \qquad \text{for } n = \pm N, \pm 2N, \ldots \end{aligned} \tag{3.19}$$

Equation (3.17) represents the basic equation for a short-time Fourier analyzer, and Eq. (3.18) represents the basic equation for the corresponding

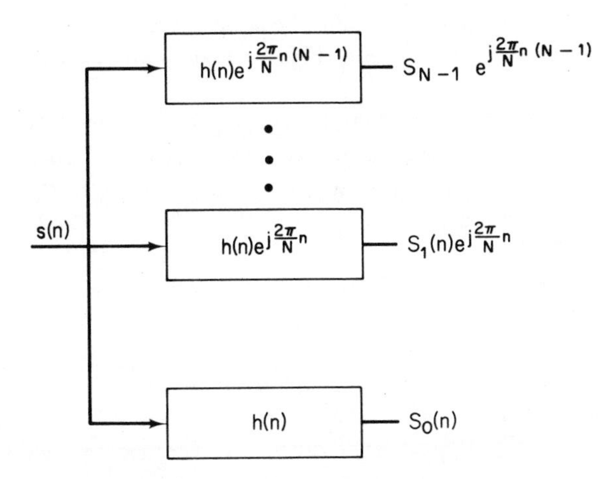

FIGURE 3-15. Filter bank implementation of short-time Fourier analysis.

synthesizer. We have seen through Eq. (3.13) how the analysis can be implemented using the FFT. As expressed in that equation, a new FFT is computed for each n. However, as suggested by Fig. 3-11, $S_r(n)$ for any r is the output of a lowpass filter; consequently, $S_r(n)$ can be sampled in n and can then be reconstructed by interpolation, using any of a variety of interpolating filter designs.

In carrying out the synthesis, it is important to note that Eq. (3.18) does not correspond to a discrete Fourier transform and that, consequently, the FFT cannot be applied directly in its evaluation. However, Portnoff[48] has developed a particularly efficient procedure for implementation of the synthesis equation (3.18) by taking advantage of the FFT algorithm.

The analysis and synthesis of speech using the short-time Fourier transform are not strongly dependent on the basic speech model of Fig. 3-3. The output of the overall system is of very high quality, and does not suffer from some of the distortions characteristic of many of the other analysis–synthesis systems that we shall discuss. On the other hand, since it does not take strong advantage of the speech model, its use for bandwidth compression is limited. One class of applications in which it has proved to be particularly useful is in the speed transformations of speech, where it is of interest to modify a speech signal such that the transformed speech corresponds to a faster or slower speaking rate.

To capitalize more specifically on the basic speech model of Fig. 3-3, most speech analysis–synthesis systems attempt in some way to deconvolve the speech signal (i.e., to separate the excitation function from the vocal tract characteristics). Such systems generally lead to rather significant data-rate compression for speech storage or transmission, although they generally introduce some degradation in quality because the deconvolution cannot be implemented precisely; the model of Fig. 3-3 is only an approximate description.

One classical approach to obtaining an approximation to the vocal-tract transfer function is based on the use of the short-time Fourier transform. In discussing speech spectrograms, we observed that the frequency resolution in a wideband spectrogram is too low to resolve the individual harmonics of the excitation. Thus, in fact, the short-time spectrum with sufficiently low frequency resolution can provide an approximation to the spectral envelope or vocal-tract transfer function. Viewed in equivalent terms, the speech is analyzed by means of a wideband filter bank. The envelope of the output of each bandpass filter then represents an estimate of the amplitude of the spectral envelope of the speech at the center frequency of that bandpass filter. The envelope of the bandpass filter output is obtained by rectification and lowpass filtering. The resulting envelope signals then correspond to the analyzer output. In addition, as will be discussed at the end of this section, a separate analysis is carried out to obtain the excitation parameters. In the corresponding synthesizer, an approximation to the original speech spectrum is obtained by exciting a bank of bandpass filters with an excitation signal obtained from the excitation parameters. A gain factor is applied corresponding to the channel signals obtained from the analyzer. The outputs of the synthesizer bandpass filters are then added to form the speech output. The basic structure of the overall system, referred to as a *channel vocoder*, is depicted in Fig. 3-16.(40,43,44)

The channel vocoder has been one of the most commonly used structures for analog speech bandwidth compression systems. A variety of issues must be considered in the choice of the specifications for the bandpass and lowpass filters utilized in the analyzer and synthesizer(45) and elaborations on the basic structure of Fig. 3-16. A digital implementation of the channel vocoder can be carried out by implementing the bandpass and lowpass filters digitally.(41) Alternatively, the basic spectral analysis represented by the filter bank can be carried out by utilizing the discrete Fourier transform. One such channel vocoder system developed by Bially and Anderson(46) replaces the filter bank in the analyzer by a computation of the short-time Fourier transform; the rectifier and lowpass filter used to obtain the envelope of the channel signals in the analog channel vocoder is replaced by a computation of the magnitude of the short-time Fourier transform. The Fourier transform can be computed either directly or through the use of the FFT algorithm. In their implementation a direct computation was used. The realization of an individual channel is illustrated in Fig. 3-17.

In the synthesizer, the channel signals obtained from the analyzer are used as amplitudes applied to sinusoidal segments at the frequencies corresponding to the analyzer channels, and the resulting weighted sinusoids are summed.

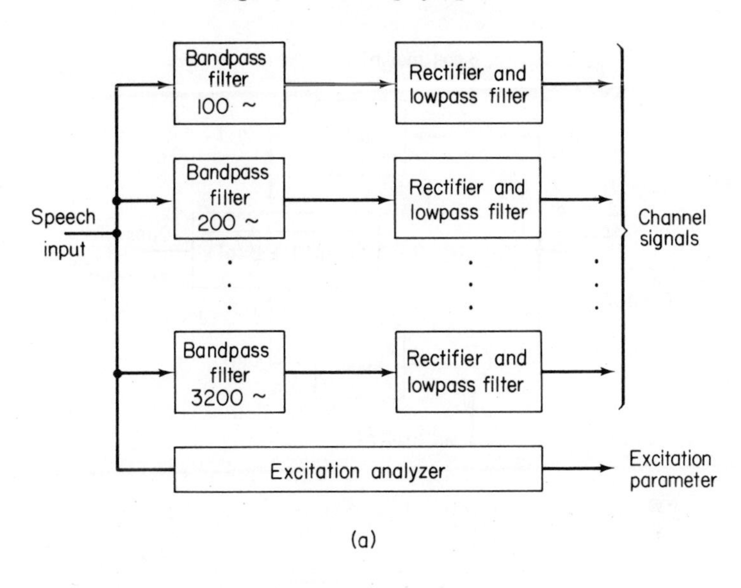

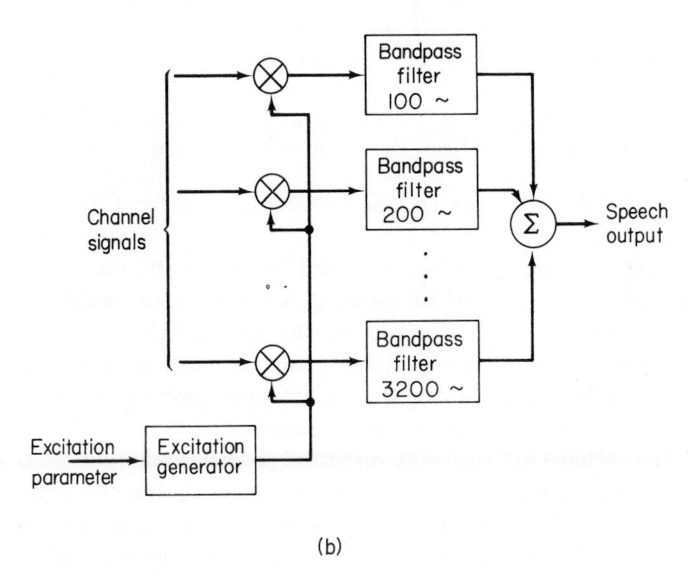

FIGURE 3-16. Basic structure of a channel vocoder, (a) analyzer, (b) synthesizer.

With each sinusoidal segment aligned in phase, the resulting impulse response function is symmetric (i.e., has zero or linear phase). The excitation parameters obtained from the analyzer are used to generate an excitation consisting of pulses spaced by the pitch period for voiced speech and a noiselike sequence for unvoiced speech. The synthesizer structure is shown in Fig. 3-18(a).

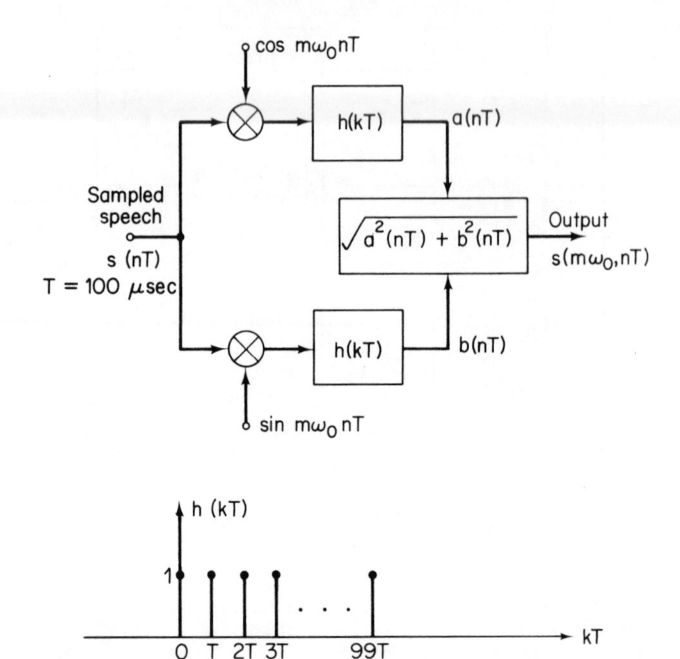

FIGURE 3-17. Structure of an individual channel for a digital channel vocoder analyzer; (after Bially and Anderson).

Figure 3-18(b) illustrates the specific procedure used for formation of the impulse response.

Although there are a number of variations on the structure of Figs. 3-17 and 3-18, including the use of a data window on the input speech to improve the effective frequency characteristics of each channel filter, it represents the digital counterpart of the conventional analog channel vocoder. In Sec. 3.5 and 3.6 we shall consider other digital analysis–synthesis systems that do not have corresponding analog implementations.

Thus far we have not explicitly discussed the determination of the excitation parameters. As we shall see in following sections, some speech analysis–synthesis systems are oriented quite naturally toward a particular means for obtaining the excitation parameters. For channel vocoder systems, that is not the case. Consequently, these systems have tended to utilize a class of algorithms referred to as *time-domain excitation analyzers* that are not specific to any speech analysis–synthesis system.

Voiced speech is approximately periodic on a short-time basis, in which case the objective of the excitation analyzer is to measure the fundamental frequency or, equivalently, the period. For an exactly periodic signal, a variety of relatively straightforward time-domain measurements can be made to

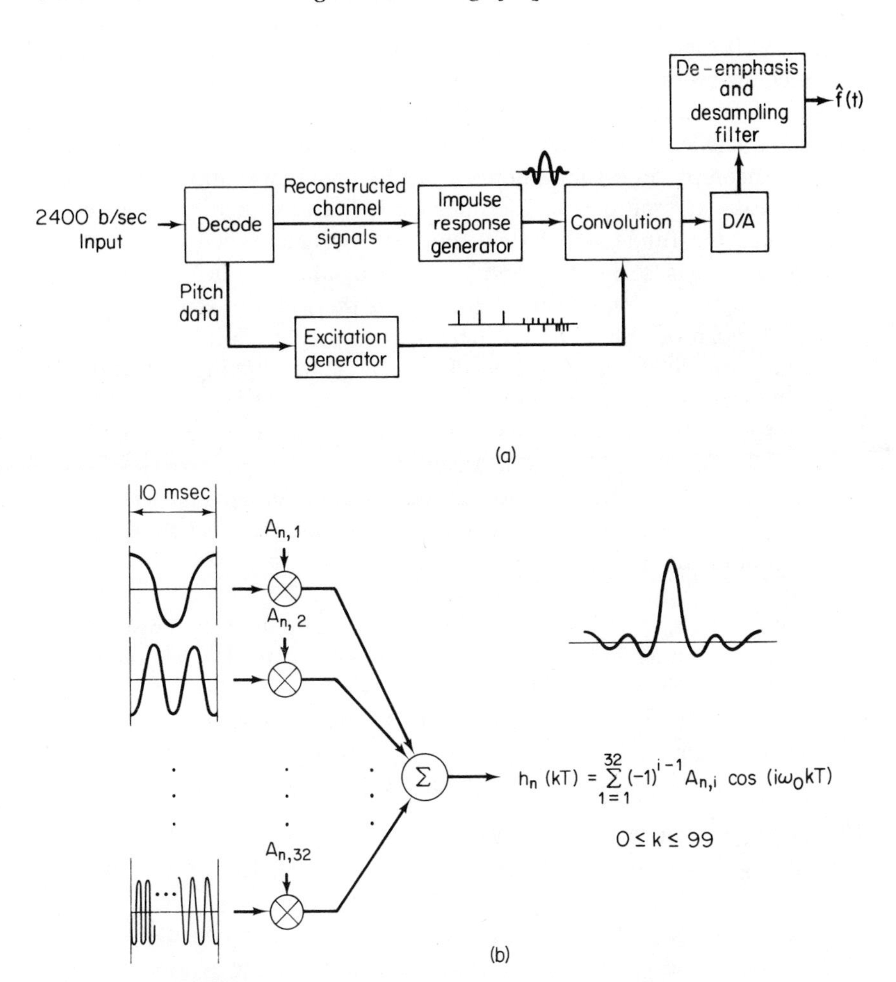

FIGURE 3-18. (a) Synthesizer structure for a digital channel vocoder, (b) algorithm for generation of the impulse response; (after Bially and Anderson).

determine the fundamental frequency. For example, if the periodic waveform is lowpass filtered to retain only a few harmonics, a simple algorithm, such as detection of time between peaks or zero crossings, will suffice. When the waveform is not precisely periodic, however, such simple procedures will often give ambiguous or incorrect results. Consequently, it is generally necessary to use somewhat more elaborate methods. One approach originally proposed by Gold[49] and refined by Gold and Rabiner,[54] which has been particularly successful, is the use of a number of elementary measurements in parallel, followed by a scoring algorithm to determine the fundamental frequency. If there is sufficient ambiguity among the various measurements, the speech is presumed to be unvoiced.

Another successful analysis procedure is the use of the *short-time autocorrelation function* of the speech waveform. If the length of the autocorrelation window encompasses several pitch periods, the autocorrelation function will exhibit peaks at multiples of the pitch period. Consequently, by detecting these peaks the fundamental frequency can be determined. Likewise, the absence of a peak serves to identify unvoiced intervals. In autocorrelation excitation analyzers, one inherent difficulty is the width of the main peak in the autocorrelation function. To sharpen the autocorrelation function near the origin, Sondhi(53) has proposed the use of center clipping and peak clipping followed by autocorrelation. The introduction of these nonlinearities generally results in a sharp, easily detectable peak in the autocorrelation function at multiples of the pitch period.

Although both classes of algorithms provide a decision on whether the speech is voiced or unvoiced, it is common to combine these measurements with a preliminary voiced–unvoiced decision based on a short-time energy measurement. During unvoiced intervals the short-time energy is generally significantly lower than during voiced intervals. By comparing the short-time energy against a threshold, a preliminary decision as to whether the speech is voiced or unvoiced can be made. This measurement can then be used in combination with the measurements made from the pitch detection to reach a final voiced–unvoiced decision.

3.5 HOMOMORPHIC ANALYSIS AND SYNTHESIS OF SPEECH

As observed in Sec. 3.2, the speech waveform consists of a convolution of the excitation function and vocal tract impulse response. The general nonlinear filtering technique referred to as *homomorphic filtering*(60,59,8) is particularly well suited to deconvolution of speech.

The general class of homomorphic systems, as applied to deconvolution of sequences, is depicted in Fig. 3-19. The system D_* is defined by the property that

$$\hat{X}(z) = \log X(z) \tag{3.20}$$

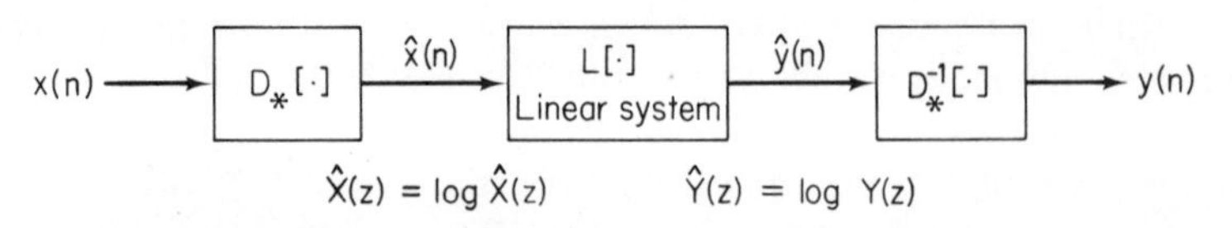

FIGURE 3-19. Canonic representation of the class of homomorphic systems for deconvolution.

where $X(z)$ and $\hat{X}(z)$ are the z transforms of $x(n)$ and $\hat{x}(n)$, respectively. Since, in general, $X(z)$ is a complex-valued function, its logarithm must be appropriately defined.[60] The system L is a linear system, and the system denoted D_*^{-1} is the inverse of the system D_*. Based on the definition of the system D_*, if $x(n)$ is the convolution of two components $x_1(n)$ and $x_2(n)$ so that

$$X(z) = X_1(z)X_2(z) \tag{3.21}$$

then

$$\hat{X}(z) = \hat{X}_1(z) + \hat{X}_2(z) \tag{3.22}$$

Consequently,

$$\hat{x}(n) = \hat{x}_1(n) + \hat{x}_2(n) \tag{3.23}$$

Therefore, the system transforms a convolution of components to a sum, thus permitting the separation of these additive components by linear filtering. The output of the system D_* is generally referred to as the *complex cepstrum.*[60]

A number of properties of the complex cepstrum make this form of analysis particularly well suited to speech analysis and synthesis. These properties are summarized below:

1. Consider a sequence $v(n)$ with a rational z transform of the form

$$V(z) = A\frac{\prod_{k=1}^{m_i}(1 - a_k z^{-1})\prod_{k=1}^{m_o}(1 - b_k z)}{\prod_{k=1}^{p_i}(1 - c_k z^{-1})\prod_{k=1}^{p_o}(1 - d_k z)} \tag{3.24}$$

where $|a_k|$, $|b_k|$, $|c_k|$, and $|d_k|$ are all less than unity so that factors of the form $1 - a_k z^{-1}$ and $1 - c_k z^{-1}$ correspond to zeros and poles inside the unit circle, and the factors $1 - b_k z$ and $1 - d_k z$ correspond to zeros and poles outside the unit circle. Then the complex cepstrum of $v(n)$ has the general form

$$\hat{v}(n) = \begin{cases} \log|A| & n = 0 \\ -\sum_{k=1}^{m_i}\frac{a_k^n}{n} + \sum_{k=1}^{p_i}\frac{c_k^n}{n} & n > 0 \\ \sum_{k=1}^{m_o}\frac{b_k^{-n}}{n} - \sum_{k=1}^{p_o}\frac{d_k^{-n}}{n} & n < 0 \end{cases} \tag{3.25}$$

From this we observe that the complex cepstrum decays at least as fast as $1/|n|$. Thus, for sequences with z transforms of the form of Eq. (3.24), corresponding to sequences representable as sums (or convolutions) of complex exponentials, the complex cepstrum tends to be concentrated around $n = 0$.

2. For a sequence $p(n)$ of the form

$$p(n) = \sum_{k=-\infty}^{+\infty} \alpha_k \delta(n - kN) \tag{3.26}$$

the complex cepstrum is of the form

$$\hat{p}(n) = \sum_{k=-\infty}^{+\infty} \beta_k \delta(n - kN) \tag{3.27}$$

In other words, a sequence of equally spaced (but not necessarily equal amplitude) pulses results in a complex cepstrum of the same form.

3. Let $v_m(n)$ denote a *minimum phase** sequence with Fourier transform $V_m(e^{j\omega})$. Then $\hat{v}_m(n) = 0$ for $n < 0$. Furthermore, let us consider $\hat{v}_e(n)$ defined such that

$$\hat{V}_e(e^{j\omega}) = \text{Re}[\hat{V}_m(e^{j\omega})] = \log|V_m(e^{j\omega})| \tag{3.28}$$

On the basis of Eq. (3.28), $\hat{v}_e(n)$ is the even part of $\hat{v}_m(n)$ and can be obtained through the log magnitude of $V_m(e^{j\omega})$, rather than the complex logarithm as required by Eq. (3.20). It can be shown that

$$\hat{v}_m(n) = \begin{cases} v_e(0) & n = 0 \\ 2v_e(n) & n > 0 \\ 0 & n < 0 \end{cases} \tag{3.29}$$

The basic consequence of this property is that for a minimum phase sequence the complex cepstrum can be obtained through the computation of the log magnitude, rather than requiring the computation of the complex logarithm.

4. Let $v(n)$ denote a *nonminimum* phase sequence with Fourier transform $V(e^{j\omega})$, and let $\hat{v}_e(n)$ denote the inverse Fourier transform of $\log|V(e^{j\omega})|$ [i.e., $\hat{v}_e(n)$ corresponds to the even part of the complex cepstrum of $v(n)$]. Finally, consider $\hat{v}_m(n)$ as defined as

$$\hat{v}_m(n) = \begin{cases} \hat{v}_e(0) & n = 0 \\ 2\hat{v}_e(n) & n > 0 \\ 0 & n < 0 \end{cases} \tag{3.30}$$

Then $\hat{v}_m(n)$ is the complex cepstrum of a minimum phase sequence $v_m(n)$, the Fourier transform of which has the same magnitude as the Fourier transform of $v(n)$; that is,

$$|V(e^{j\omega})| = |V_m(e^{j\omega})| \tag{3.31}$$

These four properties make the notions of homomorphic processing particularly well suited to speech analysis and synthesis. As discussed in Sec. 3.2, the speech waveform can be modeled on a short-time basis as the response of a linear system to an excitation consisting of a pulse train for voiced speech and noise for unvoiced speech. The transfer function of the linear system

* There are a variety of equivalent definitions of a minimum phase sequence. It is generally convenient to think of a minimum phase sequence as one for which the poles and zeros of the z transform all lie within the unit circle.

representing the vocal tract is generally considered to be a rational function of z, that is, of the form of Eq. (3.24). Consequently, on the basis of property 1, the complex cepstrum of the vocal-tract impulse response is of the form of Eq. (3.25), and in particular, then, tends to be concentrated around $n = 0$. For voiced speech, the vocal-tract excitation is of the form of Eq. (3.26); consequently, its complex cepstrum is of the form of Eq. (3.27). Thus the complex cepstrum of the excitation consists of pulses occurring at intervals corresponding to the pitch period. Since the complex cepstrum of the vocal-tract impulse response is concentrated around $n = 0$, the complex cepstra of the vocal-tract impulse response and the excitation for voiced speech tend to occupy somewhat disjoint time intervals. Thus the cepstral values representing the vocal tract can be extracted from the total cepstrum by means of a linear system that multiplies the low-time values by unity and the remainder by zero.

Such a deconvolution is illustrated in Fig. 3-20. Figure 3-20(a) shows a portion of a vowel, and Fig. 3-20(b) its complex cepstrum obtained by first applying a Hamming window to the input in Fig. 3-20(a). A peak in the complex cepstrum at a time corresponding to the pitch period is clearly evident. By choosing the linear filter in Fig. 3-19 to retain only the low-time portion of the complex cepstrum, the excitation component shown in Fig. 3-20(c) is obtained. By choosing the linear filter in Fig. 3-19 to retain only the low-time portion of the complex cepstrum, the deconvolved vocal-tract impulse response shown in Fig. 3-20(d) is obtained. The effect of the window applied to the data before computing the complex cepstrum is clearly evident in Fig. 3-20(c). To verify that the pulse obtained in Fig. 3-20(d) is indeed a good representation of the vocal-tract impulse response, it was convolved with an ideal excitation function consisting of an impulse train with a period corresponding to the pitch period of the original speech as measured from Fig. 3-20(b). The resulting resynthesized waveform is shown in Fig. 3-20(e). The close agreement between the resynthesized waveform and the original is clearly evident.

Thus far our discussion has assumed that the cepstrum is computed according to Eq. (3.20), retaining both spectral magnitude and phase information. It is commonly known that the ear tends to be insensitive to phase. On the basis of properties 3 and 4, if we assume the input speech to be minimum phase, we can compute the inverse transform of the log magnitude of its transform. If, in fact, the input speech is nonminimum phase (as it almost certainly would be), the resulting cepstral values retain information only about the spectral magnitude and not about the phase.

We can understand somewhat less formally how a deconvolution of speech is carried out through the use of the cepstrum. As discussed previously, the spectrum for voiced speech consists of the product of the spectral envelope, representing the vocal tract, and a fine structure representing the

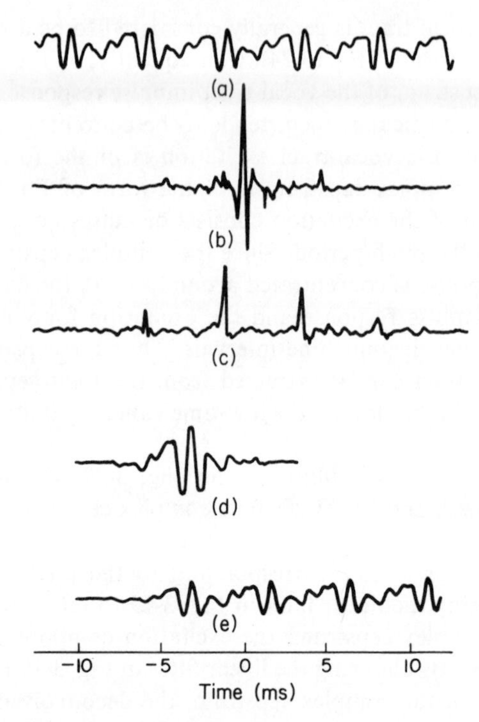

FIGURE 3-20. Illustration of deconvolution of speech by homomorphic filtering; (a) portion of a vowel, (b) complex cepstrum of (a), (c) recovered train of windowed pitch pulses, (d) recovered vocal-tract impulse response, (e) resynthesized speech using impulse-response function of (d) and pitch as measured from (b); (after Oppenheim and Schafer).

excitation. The log spectrum is thus the *sum* of the log of the spectral envelope and the log of the excitation spectrum. The log of the spectral envelope tends to be slowly varying in frequency; the log of the excitation is more rapidly varying in frequency and, in fact, periodic. The inverse Fourier transform of the log of the spectral envelope is thus concentrated around low-time values, whereas the inverse transform of the log of the excitation function consists of a set of lines, reflecting its periodicity in the frequency domain. To obtain the log of the spectral envelope from the composite log spectrum, we in effect want to *smooth* the log spectrum. This is accomplished by applying a low-time window to the cepstrum.[59] This procedure for smoothing the composite log spectrum to obtain the log spectral envelope will be referred to as *cepstral smoothing*.

The effect of cepstral smoothing is illustrated in Fig. 3-21. In Fig. 3-21(a) are cepstra for consecutive segments of a speech waveform. Figure 3-21(b)

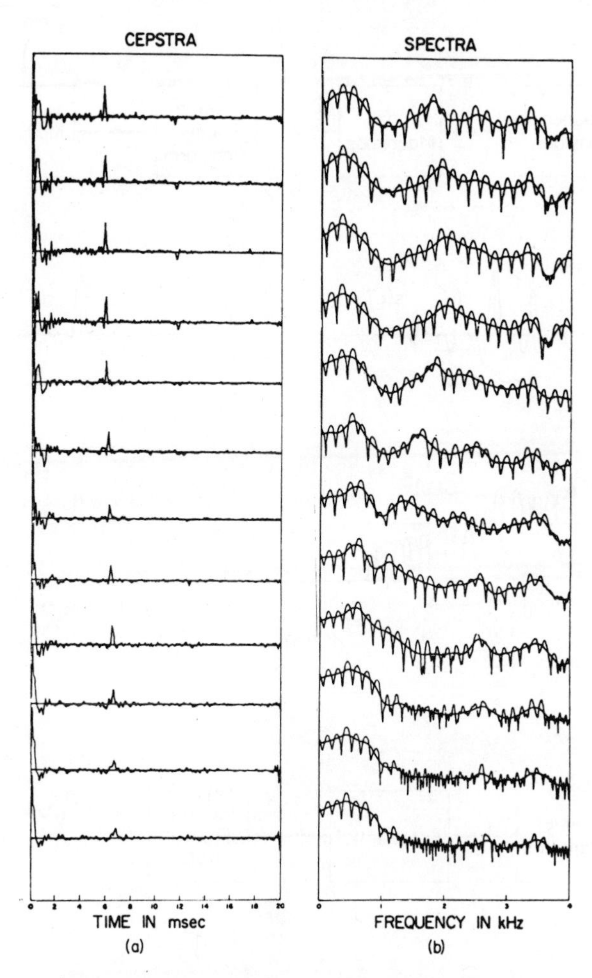

FIGURE 3-21. Illustration of estimation of the spectral envelope by cepstral smoothing; (a) cepstra for consecutive segments of a speech waveform, (b) log spectra and cepstrally smoothed log spectra corresponding to (a); (after Schafer and Rabiner).

shows the corresponding log spectra superimposed, on which are the log spectral envelopes obtained by cepstral smoothing.

The measurement of the spectral envelope by cepstral smoothing has been useful in a variety of contexts. It has been applied by Oppenheim(61) as the basis for a speech analysis–synthesis system and by Schafer and Rabiner(62) for automatic measurement of formants.

For a speech analysis–synthesis system based on homomorphic filtering, the low-time cepstral values are used as parameters to represent the vocal tract or, equivalently, the spectral envelope of the speech. The long-time

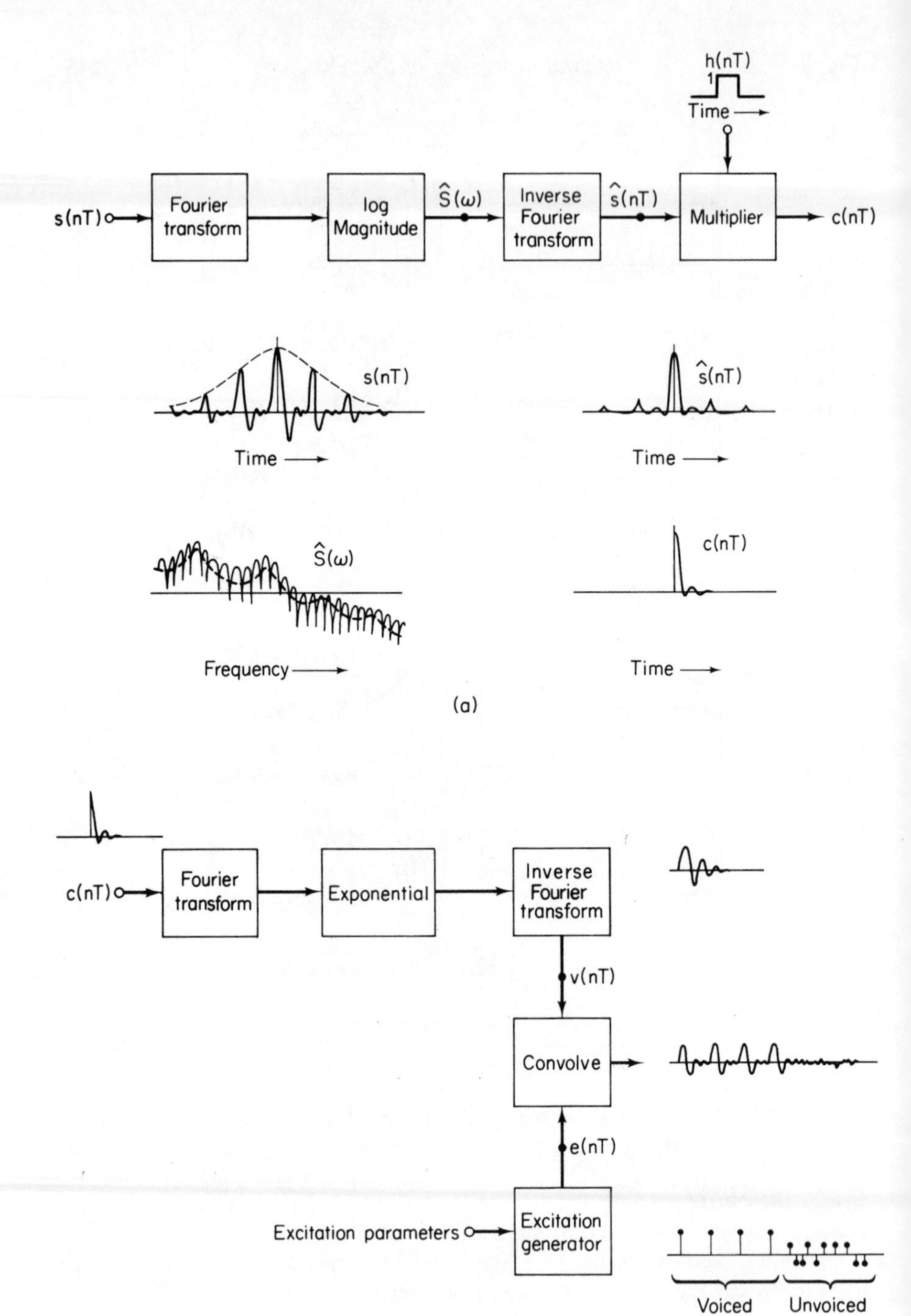

FIGURE 3-22. Block diagram of homomorphic analysis-synthesis system; (a) analyzer configuration, (b) synthesizer configuration.

values are used to extract the excitation parameters. A block diagram of the homomorphic analysis is shown in Fig. 3-22(a).

The excitation parameters are taken from the long-time portion of the cepstrum. Specifically, for voiced speech, peaks occur in the cepstrum at multiples of the pitch period, and, in fact, can be clearly seen in the examples in Fig. 3-21. For unvoiced speech, no such peaks occur. Consequently, a basic structure for the excitation analyzer consists of the determination of the presence or absence of a peak in the long-time portion of the cepstrum as an indication of the presence or absence of voicing, and, if voiced, a measurement of the location of the peak to obtain the pitch period.[51]

The input to the homomorphic synthesizer consists of the low-time cepstral values and the excitation parameters. In the synthesizer the low-time cepstral values are converted to an approximation of the vocal-tract impulse response. The excitation parameters are used to generate an excitation function, which is convolved with the impulse response obtained from the low-time cepstral values, resulting in the output speech.

The cepstrum in the analyzer was obtained from the log magnitude, thus resulting in an even function. If in the synthesizer the vocal-tract impulse response is obtained from a symmetric set of cepstral values, the resulting impulse response function will likewise be symmetrical; that is, it will be zero (or linear) phase. Alternatively, by using Eq. (3.30), the minimum phase impulse response can be obtained. A block diagram of the synthesizer is shown in Fig. 3-22(b). Informal listening tests have verified that the minimum phase and zero phase resynthesis are essentially indistinguishable. In general, the zero phase synthesis is easier to implement because the Fourier transform of an even sequence is real.

Clearly, the implementation of the homomorphic analysis–synthesis system relies heavily on the computation of the short-time Fourier transform. Currently, the most promising hardware implementation of this system relies on the use of charge-coupled devices to implement the spectral analysis.[39] Within the context of such an implementation, a variety of modifications of the analyzer and synthesizer offers potential advantages. In one such modification, rather than smoothing the log spectrum by applying a window to the cepstrum, a finite impulse response (FIR) lowpass filter can be applied directly to the log spectrum. In this case, only one Fourier transform is required in the analysis and only one in the synthesis.

3.6 ANALYSIS–SYNTHESIS BY LINEAR PREDICTION

As an alternative to the representation of vocal-tract information by spectral smoothing, as exemplified by the analysis procedures used in either filter bank or homomorphic analysis, it is possible to consider an approach based on estimating parameters of a vocal-tract model. One such model might consist of representing the vocal tract in terms of a general rational transfer function of the form

$$H(z) = G\frac{1 + \sum_{\ell=1}^{q} b_\ell z^{-\ell}}{1 - \sum_{k=1}^{p} a_k z^{-k}} \tag{3.32}$$

In this case, the parameters used to describe the vocal tract are the numerator and denominator coefficients and the gain factor G. In general, the impulse response (or frequency response) associated with the transfer function (3.32) is a nonlinear function of the numerator and denominator coefficients; thus, estimating these parameters for a segment of speech generally requires the solution of a set of nonlinear equations. For the special case in which the order of the denominator polynomial is zero, the determination of the parameters based on a mean-square-error criterion reduces to the solution of a set of linear equations. For the case in which the order of the numerator polynomial is zero, so that Eq. (3.32) corresponds to an all-pole representation, the determination of the parameters likewise reduces to the solution of a set of linear equations, if the determination is based on minimizing the mean-square error associated with the *inverse* filter.

A speech segment is, of course, sufficiently complex so that we could not expect it to match exactly a model such as that of Eq. (3.32), much less a more simplified model, such as an all-zero or all-pole model. However, as stressed in Sec. 3.2, one important attribute of the vocal-tract transfer function is that it is characterized primarily by resonances, which, of course, are well represented by poles. Thus it is not unreasonable to expect that many of the important characteristics of the vocal-tract transfer function will be preserved in an all-pole model. This section will focus entirely on an all-pole modeling technique commonly referred to as *linear prediction.*[68,70,82,83] In Sec. 3.7 we shall consider the extension of these ideas together with those of Sec. 3.5 to pole-zero modeling of speech.

Let us first consider the problem of estimating parameters in an all-pole model, given an impulse response function. Let $H(z)$ denote an all-pole transfer function of the form

$$H(z) = G\frac{1}{1 - \sum_{k=1}^{p} a_k z^{-k}} \tag{3.33}$$

The impulse response $v(n)$ associated with $H(z)$ satisfies the difference equation

$$v(n) = G\delta(n) + \sum_{k=1}^{p} a_k v(n-k) \tag{3.34}$$

or, for $n > 0$,

$$v(n) = \sum_{k=1}^{p} a_k v(n-k) \qquad n > 0 \tag{3.35}$$

Thus, for $n > 0$, the response $v(n)$ is a linear combination of (i.e., is linearly predictable from) its previous p values. If the data to be modeled corresponds *exactly* to the impulse response of an all-pole filter, Eq. (3.35) will, of course, be satisfied exactly. If not, a linear combination of past values will only correspond to an approximation to $v(n)$. Denoting this approximation by $\tilde{v}(n)$,

$$\tilde{v}(n) = \sum_{k=1}^{p} a_k v(n-k) \qquad n > 0 \tag{3.36}$$

and the associated error $e(n)$, alternatively referred to as the residual, is

$$\begin{aligned} e(n) &= v(n) - \tilde{v}(n) \\ &= v(n) - \sum_{k=1}^{p} a_k v(n-k) \qquad n > 0 \end{aligned} \tag{3.37}$$

If the predictor coefficients a_k are chosen to minimize the mean-square value of the error, they are specified by a set of linear equations. In particular, consider the total mean-square error given by

$$E_T = \sum_{n=1}^{N-1} e^2(n) = \sum_{n=1}^{N-1} \left[v(n) - \sum_{k=1}^{p} a_k v(n-k) \right]^2 \tag{3.38}$$

where the upper limit $N - 1$ is specified by the length of data available. The parameters a_k can be determined by setting

$$\frac{\partial E_T}{\partial a_i} = 0 \qquad i = 1, 2, \ldots, p \tag{3.39}$$

with the result that

$$\sum_{k=1}^{p} a_k \phi_{ik} = \phi_{i0} \qquad i = 1, 2, \ldots, p \tag{3.40a}$$

where

$$\phi_{ik} = \sum_{n=1}^{N-1} v(n-i) v(n-k) \tag{3.40b}$$

By expanding Eq. (3.38) and substituting in Eq. (3.40), we can also obtain an expression for the mean-square error E_T, specifically

$$E_T = \phi_{00} - \sum_{k=1}^{p} a_k \phi_{k0} \tag{3.41}$$

In addition to the predictor coefficients, the gain parameter G needs to be specified in Eq. (3.34). One strategy for this will be discussed shortly.

When $v(n)$ is available for all n, the upper limit $N - 1$ in Eq. (3.38) can be taken as infinite, and Eq. (3.40b) becomes

$$\phi_{ik} = \sum_{n=1}^{\infty} v(n-i) v(n-k) \qquad \begin{array}{l} k = 0, 1, 2, \ldots, p \\ i = 1, 2, \ldots, p \end{array} \tag{3.42}$$

Since $v(n) = 0$, $n < 0$, the coefficients ϕ_{ik} in Eq. (3.42) are then the autocorrelation coefficients of $v(n)$.† If only a finite-length segment of $v(n)$ is available, the upper limit in Eq. (3.40b) must be chosen such that the summation only includes the available values of $v(n)$. In this case, the coefficients ϕ_{ik} are not autocorrelation coefficients.

In our discussion we have assumed that a finite-length segment of the impulse response $v(n)$ is available. In applying all-pole modeling to speech, we begin from the speech waveform, corresponding to the vocal-tract impulse response convolved with the excitation function, from which, we must obtain the coefficients ϕ_{ik} in Eqs. (3.40). Two procedures are commonly used. The first, referred to as the *autocovariance method*, uses a single pitch period (or other finite-length segment) of $s(n)$ as an approximation to a finite-length segment of $v(n)$, and applies the analysis specified by Eqs. (3.40) to obtain the parameters a_k.[70] In the second method, a smooth window is applied over several pitch periods.[73] For the small lags required, the autocorrelation function of this windowed segment will approximate the autocorrelation function of the vocal-tract impulse response $v(n)$. Thus the autocorrelation coefficients of the windowed speech segment are used directly as the coefficients ϕ_{ik} in Eqs. (3.40). For obvious reasons, this method is generally referred to as the *autocorrelation method.* With either of the two methods, the analysis procedure is applied to successive segments of the speech signal so that the coefficients of the model are continually updated to reflect the fact that the vocal-tract characteristics change with time.

Both the autocorrelation and autocovariance methods have been used for speech analysis, but the autocorrelation method has been the more generally accepted. One of its principal advantages is that the coefficient matrix is Toeplitz; that is, the elements along each diagonal are equal.[65] Thus, in the computation of the entries in the matrix, only one row need be generated. Furthermore, as will be discussed later, efficient techniques are available for the solution of the resulting equations (i.e., for the inversion of a Toeplitz matrix). As an additional consideration, a Toeplitz matrix is guaranteed to be nonsingular, and, in the absence of computational inaccuracies, the resulting all-pole filter is guaranteed to be stable. No such assurances exist for the autocovariance method. The principal advantage of the autocovariance method is that in some sense it is more exact. Specifically, if a sequence can in fact be modeled *exactly* as the impulse response of an all-pole filter, and only a finite-length segment is available, then with the autocovariance method the output of the all-pole filter will match the sequence exactly, whereas with the autocorrelation method this will no longer be true. However, in applying linear prediction analysis to speech, this difference is perhaps of secondary

† It should be noted that the coefficients ϕ_{ik} are autocorrelation coefficients only for $i, k \geq 0$ and i and k not zero simultaneously.

consequence, because in practice no data segment will correspond exactly to the output of an all-pole filter. On the other hand, the practical advantages of avoiding singular and unstable solutions to the model are clear. For the remainder of the discussion in this section, we shall focus entirely on the autocorrelation method.

Letting $R(k)$ denote the autocorrelation coefficients of $v(n)$, we rewrite Eqs. (3.40) and (3.41) in terms of $R(k)$ as

$$\sum_{k=1}^{p} a_k R(i-k) = R(i) \qquad i = 1, 2, \ldots, p \tag{3.43}$$

and

$$E_T = R(0) - v^2(0) - \sum_{k=1}^{p} a_k R(k) \tag{3.44}$$

To determine the gain parameter G in Eq. (3.34), the generally accepted and reasonable approach is to choose it such that the total energy in $v(n)$ equals that in the impulse response of the all-pole filter. Let $h(n)$ denote the impulse response of the filter corresponding to Eq. (3.34), and $R_h(k)$ its correlation function, where the coefficients a_k in Eq. (3.34) are determined from the solution of the set of equations (3.43). Then $h(n)$ satisfies the difference equation

$$h(n) = \sum_{k=1}^{p} a_k h(n-k) + G\delta(n) \tag{3.45}$$

Multiplying both sides of Eq. (3.45) by $h(n-i)$, for $i > 0$, and summing over n, we obtain

$$\sum_{k=1}^{p} a_k R_h(i-k) = R_h(i) \tag{3.46}$$

Since G is to be chosen such that

$$\sum_{n=0}^{\infty} v^2(n) = \sum_{n=0}^{\infty} h^2(n)$$

[i.e., so that $R_h(0) = R(0)$], it follows by comparing Eq. (3.46) with Eq. (3.43) that

$$R(k) = R_h(k) \qquad k = 0, 1, \ldots, p \tag{3.47}$$

Thus the method of all-pole modeling using linear prediction matches the first $p + 1$ autocorrelation coefficients of the all-pole filter impulse response with those of the data.

Multiplying both sides of Eq. (3.45) by $h(n)$, summing over n, and recognizing that $h(0) = G$, we obtain

$$R_h(0) = \sum_{k=1}^{p} a_k R_h(k) + G^2 \tag{3.48}$$

Finally, using Eqs. (3.47) and rearranging Eq. (3.48), we obtain

$$G^2 = R(0) - \sum_{k=1}^{p} a_k R(k) \tag{3.49}$$

Thus, once we have solved for the predictor coefficients a_k, the overall gain factor G is obtained in a straightforward manner from Eq. (3.49). Finally, by comparing Eqs. (3.44) and (3.49), we observe that the total mean-square error E_T and the gain factor G are related by

$$E_T = G^2 - v^2(0) \tag{3.50}$$

One important advantage of the method of linear prediction is that efficient methods are available for the solution of the set of equations (3.43). In the autocovariance method, the matrix of coefficients ϕ_{ik} is not Toeplitz. However, it can still be inverted by several relatively efficient methods.[81,82] In the case of the autocorrelation method, a class of highly efficient iterative algorithms are available, originally proposed by Levinson[64] and modified by Durbin.[66,67] In Durbin's method, the solution for the coefficients of the difference equation (3.43) of order i is expressed recursively in terms of the coefficients of the difference equation (3.43) of order $i - 1$ by means of the following set of equations:

$$k_i = \frac{\sum_{j=1}^{i-1} a_j^{(i-1)} R(i-j) - R(i)}{E_{i-1}} \tag{3.51a}$$

$$a_i^{(i)} = -k_i \tag{3.51b}$$

$$a_j^{(i)} = a_j^{(i-1)} + k_i a_{i-j}^{(i-1)} \qquad 1 \leq j \leq i-1 \tag{3.51c}$$

$$E_i = (1 - k_i^2)E_{i-1} \tag{3.51d}$$

where $a_j^{(i)}, j = 1, 2, \ldots, i$, represent the coefficients of the ith-order predictor. The coefficient E_i is related to the total error $E_T^{(i)}$ in the ith-order predictor by $E_T^{(i)} = E_i - v^2(0)$. The set of equations is solved recursively for $i = 1, 2, \ldots, p$, initialized by means of the relation $E_0 = R(0)$. The final solution is given by

$$a_j = a_j^{(p)} \qquad j = 1, 2, \ldots, p \tag{3.52}$$

Since, at each iteration, we obtain E_i, it is possible to easily examine the error as the predictor order increases. The other set of intermediate parameters obtained, k_i, is referred to as the *reflection coefficients* and, in fact, corresponds to the reflection coefficients at the boundaries between successive sections of an acoustic tube with sections of fixed length and varying cross-sectional area.

These are two primary ways in which linear prediction analysis of speech is commonly utilized. One involves the short-time spectral analysis of speech for spectrographic analysis and display. The second is the implementation of an analysis–synthesis system. Spectral analysis of speech using the linear prediction parameters is relatively efficient. When the linear prediction parameters have been obtained, Eq. (3.33) can be evaluated to obtain the spectral values $H(e^{j\omega})$ at a discrete set of frequencies. More specifically, we observe that equally spaced samples on the unit circle of the denominator polynomial can be obtained by applying the FFT to the finite-length sequence $h^{-1}(n)$ given by

$$
\begin{aligned}
h^{-1}(n) &= 1 && n = 0 \\
&= -a_n && n = 1, 2, \ldots, p \\
&= 0 && \text{otherwise}
\end{aligned}
$$

This sequence is, in fact, the impulse response of the inverse of the all-pole filter; that is,

$$h(n)*h^{-1}(n) = G\delta(n)$$

where $\delta(n)$ is a unit impulse. If N samples of $H(z)$ equally spaced on the unit circle are desired, the $p + 1$ point sequence $h^{-1}(n)$ is augmented with $N - p - 1$ zeros.

Because of the spectral matching properties of linear prediction,[74,83] it tends to provide a good approximation to the spectral envelope. An illustration of this is shown in Fig. 3-23, where the spectrum of the impulse response of the all-pole filter obtained through linear prediction is shown superimposed on the Fourier transform of the windowed speech data.

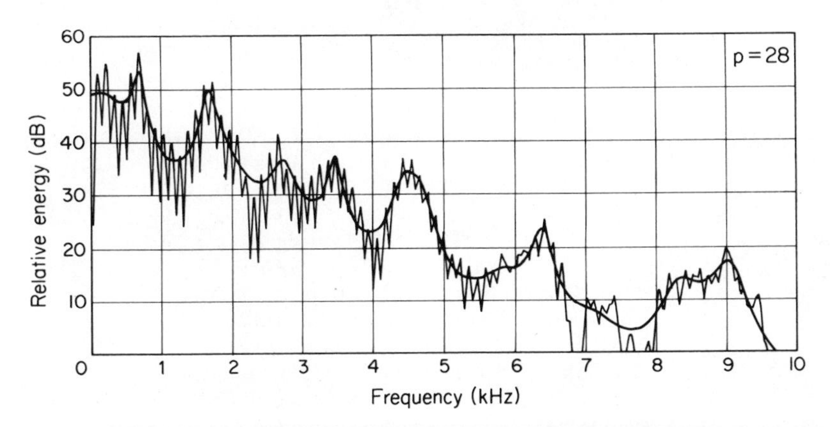

FIGURE 3-23. Spectral analysis of speech using linear prediction with 28 poles. The linear prediction spectrum is shown superimposed on the Fourier transform of the windowed speech; (after Makhoul).

Figure 3-24 compares spectral analysis using short-time Fourier analysis, homomorphic analysis, and analysis using linear prediction. The segment was a synthetic vowel with known formant frequencies. The first two spectra correspond to the Fourier transform of the segment weighted with a Hamming window of length 51.2 and 12.8 milliseconds (ms), respectively. As discussed in Sec. 3.4, this corresponds in essence to a filter bank analysis. Figure 3-24(a) represents a high-resolution analysis, as evidenced by the clearly apparent

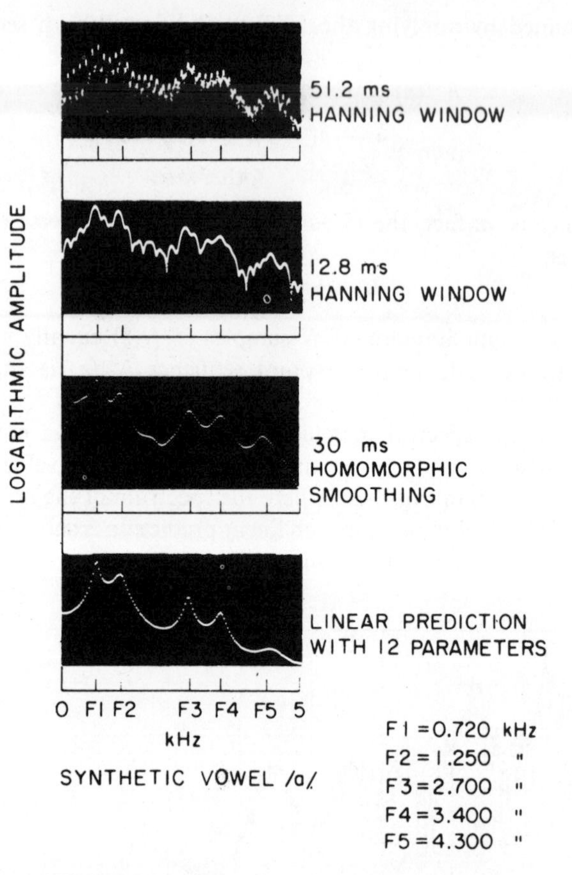

FIGURE 3-24. Spectral analysis of a synthetic vowel; (a) Fourier transform of data with a 51.2 msec. Hamming window, (b) Fourier transform of data with a 12.8 msec. Hamming window, (c) spectral envelope obtained by homomorphic filtering, (d) spectral envelope obtained from linear prediction analysis.

spectral fine structure associated with the excitation. In Fig. 3-24(b) the effective filters are wider band, as evidenced by the observation that the spectral fine structure is not well resolved. The peaks in the spectral envelope corresponding to the formants are generally in evidence, although somewhat broad owing to the effect of the spectral window.

Figure 3-24(c) represents the spectral envelope obtained by homomorphic filtering. The formants and absence of spectral fine structure are clearly in evidence. Finally, Fig. 3-24(d) shows the spectrum obtained using linear prediction analysis with 12 coefficients. This spectrum has the cleanest overall appearance, with the formants clearly in evidence. The generally clean appearance of the linear prediction spectrum is due to the fact that since the spectrum is modeled by a twelfth-order predictor polynomial, it is constrained to have no more than six resonances. This argument does not, of course, ensure that the spectral peaks will occur at the formant frequencies. In this example, that clearly is the case, however, and the experience has been that it is generally, but not always, true.

In addition to their use for the spectral analysis of speech, linear prediction techniques play an important role in speech analysis–synthesis systems. Through Eq. (3.33), the predictor parameters specify the parameters of the vocal-tract transfer function. With the specification of these parameters the speech can be resynthesized along the lines discussed in Sec. 3.3. In implementing the resynthesis, several choices are available for the structure used to implement the vocal-tract transfer function. A direct-form structure can be generated directly from the predictor coefficients. Alternatively, Eq. (3.33) can be factored to obtain a cascade-form structure. Both of these can be viewed as corresponding to a terminal analog synthesizer.

A common difficulty tends to arise, however, in either of these implementations. The synthesizer parameters are, of course, continually updated, corresponding to successive analysis time frames. To avoid the effects of large changes in parameters, it is generally preferable to provide a smooth transition in the parameters by employing interpolation between time frames. In the direct form, however, although successive sets of parameters may correspond to a stable filter, interpolated sets of values may not. For the cascade form, stability is more easily controlled. Difficulty arises in this case, however, because in a cascade configuration the order in which sections are cascaded must be maintained between successive analysis frames owing to the effect of the initial conditions just prior to updating coefficients.

The synthesis structure that avoids both of these difficulties, and that for other reasons seems to be particularly well matched to linear prediction analysis of speech, is the set of structures that correspond to direct acoustic tube modeling. There are a number of such structures, all of which are closely related. Somewhat surprisingly, the parameters of these structures are generated in the process of solving the equations obtained in the autocorrelation method using the Levinson recursion. Specifically, the coefficients k_i in Eq. (3.51), referred to as reflection coefficients, correspond to the reflection coefficients at the boundaries between successive sections of an acoustic tube with sections of fixed length and varying cross section, and correspond directly to the reflection coefficients μ_m in the structures of Figs. 3-8 and 3-9.

The determination of the excitation parameters for a linear prediction analysis–synthesis system is generally based on examination of the error signal, obtained by filtering the original speech with the *inverse* of the approximation of the vocal-tract transfer function. This error signal is an approximation to the excitation component of the speech waveform. Any of a variety of time-domain pitch-detection and voiced–unvoiced algorithms, such as autocorrelation analysis, can then be utilized to extract the excitation parameters.[56,83]

In summary, linear prediction represents an extremely powerful technique for the digital analysis of speech. Its spectral matching properties seem to be particularly well suited to the characteristics of speech. Furthermore, the algorithms associated with this set of techniques are both elegant and efficient, and are well matched to the current state of the art in microprocessors and other digital technology. In this chapter we have only been able to touch briefly on a few of the many aspects of this set of techniques. For a more detailed treatment, the reader is referred to the excellent text by Markel and Gray.[83]

3.7 POLE-ZERO MODELING OF SPEECH

In discussing the basic model for speech production and the characteristics of the vocal tract, we commented that the vocal tract could be reasonably approximated in terms of a rational transfer function represented by poles and zeros, the poles corresponding to vocal-tract resonances and the zeros introduced due to such effects as coarticulation and coupling between the vocal tract and the nasal cavity. It has been customary in many speech analysis and synthesis systems to represent the vocal-tract characteristics in terms of an all-pole model, approximating the effects of the zeros by adjustment of the bandwidth of the first formant. At present, it is not totally clear whether or not accurate representation of the zeros is important in speech analysis–synthesis systems. The uncertainty stems in part from the fact that there has not generally been available a reliable technique for measurement of the zeros. For more general speech analysis, as carried out, for example, within the context of automatic speech recognition, or for studies directed toward linguistic aspects of speech and physiological studies of speech production, a variety of important cues is provided by information about spectral zeros.

In the all-pole analysis of speech using linear prediction, the analysis procedure was applied to windowed segments of the speech waveform. Because of the spectral matching properties of linear prediction, the resulting spectrum tends to depend primarily on the spectral envelope, and, until the number of poles in the model becomes excessive, is not sensitive to the spectral fine structure. In contrast, any of the currently proposed techniques for determination of spectral zeros is sensitive to the presence of fine structure in the spectrum. Reference to Fig. 3-3 suggests the reason for this: any technique for measuring spectral zeros will tend to be sensitive to the fact that between pitch harmonics the spectral amplitude approaches zero, an effect that may be misinterpreted as being due to the presence of spectral zeros. Thus, with any of the potential techniques for the measurement of spectral zeros, it appears to be important to precede such analysis with a deconvolution of the speech.

Three methods are generally considered for such a deconvolution. One is by simply extracting individual pitch periods of the speech waveform, and considering a single pitch period to correspond to the impulse response of the vocal tract. This, of course, would only be exactly correct if the vocal-tract impulse response were shorter in duration than a pitch period. Techniques that rely on this method of deconvolution are generally referred to as *pitch synchronous analysis.* A second technique for implementing a deconvolution is through the use of an *all-pole linear prediction analysis.* If the speech spectrum contains both poles and zeros, a very high order all-pole analysis is required, since the zeros must be approximated by poles. This high-order all-pole analysis then provides an approximation to the vocal-tract impulse response or, equivalently, the spectral envelope to which a lower-order pole-zero analysis can be applied. A third technique for implementing the deconvolution is through the use of *homomorphic deconvolution* to obtain an approximation to the vocal-tract impulse response, to which a pole-zero analysis can then be applied.[94]

A variety of pole-zero modeling techniques has been formulated theoretically; methods involving large matrix operations and iterative optimization techniques are generally unsatisfactory in the environment of speech analysis, in which memory and speed requirements are important and the potential for real-time processing is often desirable. Thus the most suitable techniques tend to be those based on a least-squares error criterion, formulated in such a way as to involve the solution of linear equations. This generally leads to techniques in which the poles are estimated first, followed by an estimation of the zeros, rather than simultaneous estimation of both. Poles can be estimated independently of zeros using the covariance formulation of linear prediction analysis. Specifically, for pole-zero modeling of the vocal-tract impulse response $v(n)$, we consider $v(n)$ represented by a difference equation of the form

$$v(n) = \sum_{k=0}^{q} b_k \delta(n-k) + \sum_{k=1}^{p} a_k v(n-k) \tag{3.53}$$

For $n > q$, where q is the total number of zeros, Eq. (3.53) becomes

$$v(n) = \sum_{k=1}^{p} a_k v(n-k) \qquad n > q \tag{3.54}$$

so that $v(n)$ for any $n > q$ is linearly predictable from its previous p values. Thus, even when zeros are included in the model, the poles can be estimated by determining the a_k's in Eq. (3.54) to minimize the prediction error. In this case, however, by virtue of the fact that Eq. (3.54) is only valid for $n > q$, we must be careful to use only values of $v(n)$ in that range, even though, after the deconvolution, $v(n)$ is available for all n. Specifically, with $e(n)$ denoting the prediction error, so that

$$e(n) = v(n) - \sum_{k=1}^{p} a_k v(n-k) \qquad n > q \tag{3.55}$$

we choose the coefficients a_k to minimize the total mean-square prediction error given by

$$E_T = \sum_{n=q+1}^{\infty} e^2(n) \tag{3.56}$$

Following a procedure identical to that used to obtain Eq. (3.40), the resulting equations for the predictor coefficients a_k are

$$\sum_{k=1}^{p} a_k \phi_{ik} = \phi_{i0} \qquad i = 1, 2, \ldots, p \tag{3.57a}$$

where

$$\phi_{ik} = \sum_{n=q+1}^{\infty} v(n-i)v(n-k) \tag{3.57b}$$

The coefficient matrix ϕ_{ik} is symmetrical, but, as was the case with the auto-covariance method, is not Toeplitz. After determining the coefficients a_k, one method, proposed by Shanks,[86] is to determine the coefficients b_k in Eq. (3.53) by minimizing the mean-square error between $v(n)$ and the impulse response of the pole-zero filter. This leads to the set of linear equations

$$\sum_{k=0}^{q} b_k R_{ik} = V_i \qquad i = 0, 1, \ldots, q$$

where

$$R_{ik} = \sum_{n=0}^{\infty} w(n-i)w(n-k) \tag{3.58a}$$

$w(n)$ is the impulse response of the all-pole filter; that is,

$$w(n) = \sum_{k=1}^{p} a_k w(n-k) + \delta(n) \tag{3.58b}$$

and

$$V_i = \sum_{n=0}^{\infty} v(n)w(n-i) \tag{3.58c}$$

A second possible procedure for estimating the zeros is first to process the data $v(n)$ by an inverse filter to remove the poles, invert the resulting spectrum to transform the zeros to poles, and then estimate these poles on the inverted signal using linear prediction. This method is similar to a technique originally proposed by Durbin.[66,67]

Figure 3-25 shows the results obtained using these techniques for pole-zero spectral analysis. Figure 3-25(a) depicts the spectrum of the natural nasalized consonant /m/. Two strong antiresonances are clearly evident, one at 650 Hz and the other at 3.2 kilohertz (kHz). Figure 3-25(b) shows the spectrum resulting from homomorphic deconvolution. Figure 3-25(c) shows the spectrum resulting from a 12-pole linear prediction analysis of the homomorphically smoothed spectrum in Fig. 3-25(b). It is evident that this spectrum accurately reflects the presence of the resonances but not the antiresonances. Figures 3-25(d) and 3-25(e) represent the spectrum obtained from a 12-pole, 10-zero model. In Fig. 3-25(d) the zeros are obtained using Shanks's method,

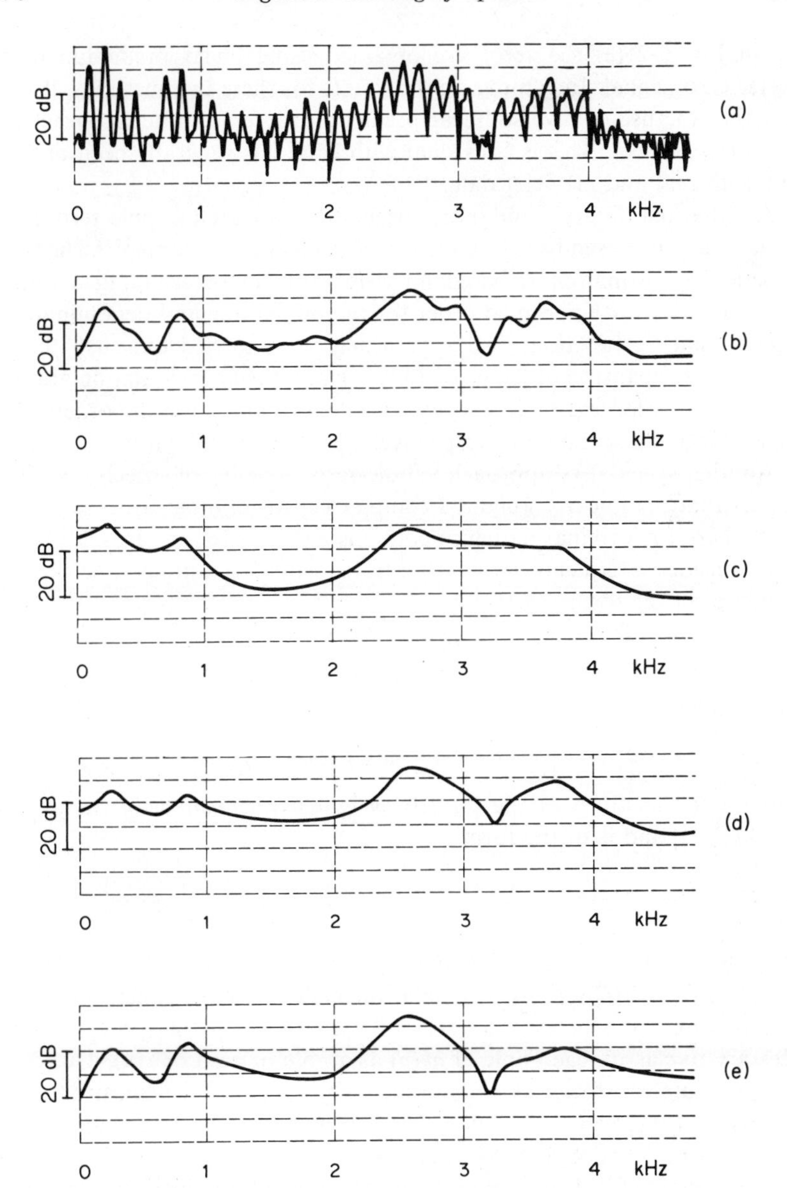

FIGURE 3-25. Illustration of pole-zero analysis of speech; (a) spectrum of the nasalized consonant /m/, (b) spectral envelope obtained by cepstral smoothing, (c) spectrum corresponding to a 12-pole linear prediction analysis of (b), (d) spectrum corresponding to a 12-pole 10-zero model in which the zeroes are obtained using Shanks' method, (e) spectrum corresponding to a 12-pole 10-zero model in which the zeroes are obtained using linear prediction on the inverse error signal. (after Kopec).

and in Fig. 3-25(e) the zeros are obtained using linear prediction on the inverse error signal. In comparing the results for these two methods, it seems clear that the use of inverse linear prediction more accurately characterizes the antiresonances. This is consistent with what has generally been observed with both real and artificial data.[94]

A somewhat different and more speculative approach to pole-zero modeling has been proposed based on the use of Padé approximants.[91] The theory of Padé approximation is based on the notion of determining a rational approximation, with specified order for the numerator and denominator, to a given analytic function. The approximation is carried out such that the power series expansion of the given function and the approximation are equal for terms of order less than $\mu + \nu + 1$, where μ and ν are the orders of the numerator and denominator, respectively, in the approximation.

Another speculative approach to pole-zero modeling of speech stems from the fact that the linearly weighted complex cepstrum of a sequence contains poles where the original sequence contains either poles or zeros.[93] Specifically, consider a sequence $v(n)$ with z transform $V(z)$, and assume that $V(z)$ is rational of the form

$$V(z) = \frac{N(z)}{D(z)}$$

The complex cepstrum $\hat{v}(n)$ is defined such that its z transform $\hat{V}(z)$ is

$$\hat{V}(z) = \log V(z)$$

Thus the z transform of the linearly weighted cepstrum $n\hat{v}(n)$ is given by $-z[d\hat{V}(z)/dz]$ and is of the form

$$-z\frac{d\hat{V}(z)}{dz} = -z\frac{D(z)N'(z) - N(z)D'(z)}{N(z)D(z)} \tag{3.59}$$

where the prime denotes differentiation with respect to z. Thus, if the poles of $n\hat{v}(n)$ are determined, for example, through the use of linear prediction, these represent the poles and zeros of $v(n)$. In using this procedure, it is necessary to classify each pole of $n\hat{v}(n)$ as a pole or zero of $v(n)$. This can be done in a number of ways. One approach is to do a separate linear prediction analysis of $v(n)$ to estimate its poles, and then to identify the remaining poles of $n\hat{v}(n)$ as the zeros of $v(n)$. An alternative strategy is to examine the residues of the poles of $n\hat{v}(n)$. It is easily shown that zeros of $v(n)$ inside the unit circle will generate negative residues, and poles inside the unit circle will generate positive residues. There is a corresponding test to separate the poles and zeros of $v(n)$ outside the unit circle. It is interesting to note, incidentally, that the deconvolution of speech generally required in pole-zero modeling is in essence carried out automatically in this process, in that the low-time cepstral values represent only vocal-tract impulse response information.

Another possibility for modeling the vocal-tract impulse response is to apply a pole-zero model to the linearly weighted cepstrum. This corresponds to modeling the vocal-tract impulse response in terms of a *fractional* pole-zero model (i.e., one in which the poles and zeros may possibly be of noninteger order). At the present time, these ideas have not been carefully explored and consequently must be considered as highly speculative.

3.8 SUMMARY

In this chapter we have reviewed the basic techniques for digital processing of speech. We have concentrated on those techniques and systems that tend to rely on the basic model of speech production. Much of our discussion has focused on the point of view that the basic problem in speech processing is a deconvolution of the speech signal into components representing the excitation and the vocal tract. The use of short-time Fourier analysis and homomorphic filtering relies very heavily on the computation of the Fourier transform and, consequently, the use of the fast Fourier transform algorithm. Speech analysis by linear prediction, which is based specifically on a parametric model, relies on the use of efficient algorithms for inverting a correlation or covariance matrix. Linear prediction tends to offer a number of important advantages related to the fact that it provides a parametric model. This also, of course, restricts flexibility, as compared with the other techniques.

All these techniques are well matched to the current state of digital technology and are able to exploit the use of large-scale integration, the availability of microprocessors, and so on. The systems discussed in Secs. 3.4 and 3.5 are also well matched to the technology of charge-transfer devices. Since these devices are particularly well suited for high-resolution spectral analysis, they potentially can form the basis for low-cost speech systems requiring such analysis. Consequently, we are likely to see continued emphasis on and development of all these techniques and systems.

Many problems remain to be solved in speech processing. Although linear prediction analysis provides an efficient technique for all-pole modeling of speech, there is not yet available an equally reliable and efficient method for pole-zero modeling or the use of a more general parametric model. The potential for some of these techniques lies perhaps in the work carried out on system identification and modeling. Some of these techniques have been explored for speech processing, but they often are not acceptable because they rely on different assumptions than would be reasonable for speech problems. Consequently, in exploring the use of such techniques it is important to carefully consider the assumptions and objectives on which the techniques are based. It is likely, however, that future development of algorithms for speech processing will rely more heavily on a recognition of the time-varying characteristics of the system and will make use more heavily of analysis procedures for such systems, which have been successful in other areas such as optimal control and modeling of time-varying systems.

REFERENCES

General References

1. B. Gold and C. M. Rader, "Systems for Compressing the Bandwidth of Speech," *IEEE Trans. Audio Electroacoustics*, vol. AU-15, no. 3, Sept. 1967, pp. 131–135.
2. B. Gold and C. M. Rader, *Digital Processing of Signals*, McGraw-Hill, New York, 1969.
3. J. L. Flanagan, C. H. Coker, L. R. Rabiner, R. W. Schafer, and N. Umeda, "Synthetic Voices for Computers," *IEEE Spectrum*, vol. 7, no. 10, 1970, pp. 22–45.
4. J. L. Flanagan, *Speech Analysis Synthesis and Perception*, 2nd ed., Springer-Verlag, New York, 1972.
5. L. R. Rabiner and C. M. Rader, eds., *Digital Signal Processing*, IEEE Press, New York, 1972.
6. R. W. Schafer, "A Survey of Digital Speech Processing Techniques," *IEEE Trans. Audio Electroacoustics*, vol. AU-20, no. 4, Mar. 1972, pp. 28–35.
7. E. O. Brigham, *The Fast Fourier Transform*, Prentice-Hall, Englewood Cliffs, N.J., 1974.
8. A. V. Oppenheim and R. W. Schafer, *Digital Signal Processing*, Prentice-Hall, Englewood Cliffs, N.J., 1975.
9. L. R. Rabiner and B. Gold, *Theory and Application of Digital Signal Processing*, Prentice-Hall, Englewood Cliffs, N.J., 1975.
10. R. W. Schafer and L. R. Rabiner, "Digital Representations of Speech Signals," *Proc. IEEE*, vol. 63, no. 4, Apr. 1975, pp. 662–677.
11. Digital Signal Processing Committee of the IEEE Acoustics, Speech and Signal Processing Society, eds., *Selected Papers in Digital Signal Processing II*, IEEE Press, New York, 1976.

Model of Speech Production

12. R. L. Miller, "Nature of the Vocal Cord Wave," *J. Acoust. Soc. Amer.*, vol. 31, 1959, pp. 667–677.
13. G. C. M. Fant, *Acoustic Theory of Speech Production*, Mouton and Co., The Hague, The Netherlands, 1960.

14. J. L. Flanagan, "Source-System Interactions in the Vocal Tract," *Ann. N.Y. Acad. Sci.*, vol. 155, 1968, pp. 9–15.

15. A. E. Rosenberg, "Effect of Glottal Pulse Shape on the Quality of Natural Vowels," *J. Acoust. Soc. Amer.*, vol. 49, 1971, pp. 583–590.

16. G. C. M. Fant, *Speech Sounds and Features*, The MIT Press, Cambridge, Mass., 1973.

Speech Synthesizers

17. J. Q. Steward, "An Electrical Analogue of the Vocal Organs," *Nature*, vol. 110, 1922, pp. 311–312.

18. H. R. R. Dudley and S. S. A. Watkins, "A Synthetic Speaker," *J. Franklin Inst.*, vol. 227, 1939, pp. 739–764.

19. H. K. Dunn, "The Calculation of Vowel Resonances and an Electrical Vocal Tract," *J. Acoust. Soc. Amer.*, vol. 22, 1950, pp. 740–753.

20. K. N. Stevens and G. C. M. Fant, "An Electrical Analog of the Vocal Tract," *J. Acoust. Soc. Amer.*, vol. 25, 1953, pp. 734–742.

21. J. L. Kelly, Jr., and C. Lochbaum, "Speech Synthesis," *Proc. 4th Intern. Congr. Acoust.*, vol. G42, 1962, pp. 1–4 (also see ref. 27).

22. J. L. Flanagan, C. H. Coker, and C. M. Bird, "Digital Computer Simulation of a Formant-Vocoder Speech Synthesizer," *15th Ann. Meeting Audio Engr. Soc.*, Preprint 307, 1963.

23. B. Gold and L. R. Rabiner, "Analysis of Digital and Analog Formant Synthesizers," *IEEE Trans. Audio Electroacoustics*, vol. AU-16, Mar. 1968, pp. 81–94.

24. L. R. Rabiner, "Digital-Formant Synthesizer for Speech Synthesis," *J. Acoust. Soc. Amer.*, vol. 43, 1968, pp. 822–828.

25. L. R. Rabiner, L. B. Jackson, R. W. Schafer, and C. H. Coker, "Digital Hardware for Speech Synthesis," *IEEE Trans. Commun. Tech.*, vol. COM-19, 1971, pp. 1016–1020.

26. J. L. Flanagan, "Voices of Men and Machines," *J. Acoust. Soc. Amer.*, vol. 51, 1972, pp. 1375–1387.

27. J. L. Flanagan, L. R. Rabiner, eds., *Speech Synthesis,* Dowden, Hutchinson & Ross, Stroudsburg, Pa., 1973.

28. A. H. Gray, Jr., and J. D. Markel, "Digital Lattice and Ladder Filter Synthesis," *IEEE Trans. Audio Electroacoustics*, vol. AU-21, 1973, pp. 491–500.

Spectral Analysis of Speech

29. R. K. Potter, G. A. Kopp, and H. G. Kopp, *Visible Speech*, Dover, New York, 1966.

30. L. R. Rabiner, R. W. Schafer, and C. M. Rader, "The Chirp Z-Transform

Algorithm and Its Applications," *Bell System Tech. J.*, vol. 48, May 1969, pp. 1249–1292.

31. L. I. Bluestein, "A Linear Filtering Approach to the Computation of the Discrete Fourier Transform," *IEEE Trans. Audio Electroacoustics*, vol. AU-18, Dec. 1970, pp. 451–455.

32. A. V. Oppenheim, "Speech Spectrograms Using the Fast Fourier Transform," *IEEE Spectrum*, vol. 7, Aug. 1970, pp. 57–62.

33. P. Mermelstein, "Computer Generated Spectrogram Displays for On-Line Speech Research," *IEEE Trans. Audio Electroacoustics*, vol. AU-19, March 1971, pp. 44–47.

34. R. W. Schafer and L. R. Rabiner, "Design of Digital Filter Banks for Speech Analysis," *Bell System Tech. J.*, vol. 50, no. 10, Dec. 1971, pp. 3097–3115.

35. J. F. Murphy, "An Improved Sound Spectrograph," *1972 Conf. Speech Commun. and Processing*, pp. 420–422.

36. R. W. Means, D. D. Buss, and J. H. Whitehouse, "Real Time Discrete Fourier Transforms Using Charge Transfer Devices," *Proc. CCD Applications Conf.* Proceedings, San Diego, Calif., Sept. 1973, pp. 127–139.

37. H. R. Silverman and N. R. Dixon, "A Parametrically Controlled Spectral Analysis System for Speech, *IEEE Trans. Acoust. Speech Signal Processing*, vol. ASSP-22, Oct. 1974, pp. 362–381.

38. R. W. Brodersen, C. R. Hewes, and D. D. Buss, "Spectral Filtering and Fourier Analysis Using CCD's," IEEE Advanced Solid-State Components for Signal Processing, *IEEE Intern. Symp. Circuits and Systems*, Newton, Mass., Apr. 1975, pp. 43–68.

39. D. D. Buss, R. W. Brodersen, C. R. Hewes, and A. F. Tasch, Jr., "Communication Applications of CCD Transversal Filters," *Natl. Telecommunications Conf.*, New Orleans, La., Dec. 1975, pp. 1–1.

Speech Analysis–Synthesis Using the Short-time Fourier Transform

40. H. Dudley, "Remaking Speech," *J. Acoust. Soc. Amer.*, vol. 11, 1939, pp. 169–177.

41. R. M. Golden, "Digital Computer Simulation of a Sampled-Data Voice Excited Vocoder," *J. Acoust. Soc. Amer.*, vol. 35, 1963, pp. 1358–1366.

42. J. L. Flanagan and R. M. Golden, "Phase Vocoder," *Bell System Tech. J.*, vol. 45, 1966, pp. 1493–1509.

43. M. R. Schroeder, "Vocoders: Analysis and Synthesis of Speech," *Proc. IEEE*, vol. 54, May 1966, pp. 720–734.

44. B. Gold and C. M. Rader, "The Channel Vocoder," *IEEE Trans. Audio Electroacoustics*, vol. AU-15, no. 4, Dec. 1967, pp. 148–160.

45. R. Golden, "Vocoder Filter Design: Practical Considerations," *J. Acoust. Soc. Amer.*, vol. 43, Apr. 1968, pp. 803–810.

46. T. Bially and W. Anderson, "A Digital Channel Vocoder," *IEEE Trans. Commun. Tech.*, vol. COM-18, no. 4, Aug. 1970, pp. 435–442.

47. R. W. Schafer and L. R. Rabiner, "Design and Simulation of a Speech Analysis–Synthesis System Based on Short-Time Fourier Analysis," *IEEE Trans. Audio Electroacoustics*, vol. AU-21, June 1973, pp. 165–174.

48. M. R. Portnoff, "Implementation of the Digital Phase Vocoder Using the Fast Fourier Transform," *IEEE Trans. Acoust. Speech Signal Processing*, vol. ASSP-24, no. 3, June 1976, pp. 243–248.

Excitation Analyzers

49. B. Gold, "Computer Program for Pitch Extraction," *J. Acoust. Soc. Amer.*, vol. 34, 1962, pp. 916–921.

50. B. Gold, "Note on Buzz-Hiss Detection," *J. Acoust. Soc. Amer.*, vol. 36, 1964, pp. 1659–1661.

51. A. M. Noll, "Cepstrum Pitch Determination," *J. Acoust. Soc. Amer.*, vol. 41, Feb. 1967, pp. 293–309.

52. M. R. Schroeder, "Period Histogram and Product Spectrum: New Methods for Fundamental-Frequency Measurement," *J. Acoust. Soc. Amer.*, vol. 43, no. 4, Apr. 1968, pp. 829–834.

53. M. M. Sondhi, "New Methods of Pitch Detection," *IEEE Trans. Audio Electroacoustics*, vol. AU-16, June 1968, pp. 262–266.

54. B. Gold and L. R. Rabiner, "Parallel Processing Techniques for Estimating Pitch Periods of Speech in the Time Domain," *J. Acoust. Soc. Amer.*, vol. 46, no. 2, Aug. 1969, pp. 442–449.

55. A. M. Noll, "Pitch Determination of Human Speech by the Harmonic Product Spectrum, the Harmonic Sum Spectrum, and a Maximum Likelihood Estimate," *Computer Processing in Communications Proceedings*, J. Fox, ed., Polytechnic Press, New York, 1969.

56. J. D. Markel, "The Sift Algorithm for Fundamental Frequency Estimation," *IEEE Trans. Audio Electroacoustics*, vol. AU-20, Dec. 1972, pp. 367–377.

57. J. N. Maksym, "Real-Time Pitch Extraction by Adaptive Prediction of the Speech Waveform," *IEEE Trans. Audio Electroacoustics*, vol. AU-21, June 1973, pp. 149–153.

58. M. J. Ross, H. L. Shaffer, A. Cohen, R. Freudberg, and H. J. Manley, "Average Magnitude Difference Function Pitch Extractor," *IEEE Trans. Acoust. Speech Signal Processing*, vol. ASSP-22, 1974, pp. 353–362.

Homomorphic Speech Processing

59. A. V. Oppenheim and R. W. Schafer, "Homomorphic Analysis of Speech," *IEEE Trans. Audio Electroacoustics*, vol. AU-16, 1968, pp. 221–226.

60. A. V. Oppenheim, R. W. Schafer, and T. G. Stockham, "Nonlinear Filtering

of Multiplied and Convolved Signals," *Proc. IEEE*, vol. 56, 1968, pp. 1264–1291.

61. A. V. Oppenheim, "Speech Analysis–Synthesis System Based on Homomorphic Filtering," *J. Acoust. Soc. Amer.*, vol. 45, 1969, pp. 459–462.

62. R. W. Schafer and L. R. Rabiner, "System for Automatic Analysis of Voiced Speech," *J. Acoust. Soc. Amer.*, vol. 47, part 2, 1970, pp. 634–648.

63. C. Weinstein and A. Oppenheim, "Predictive Coding in a Homomorphic Vocoder," *IEEE Trans. Audio Electroacoustics*, vol. AU-19, no. 3, Sept. 1971, pp. 243–249.

Linear Prediction

64. N. Levinson, "The Wiener RMS (Root Mean Square) Error Criterion in Filter Design and Prediction," *J. Math. Phys.*, vol. 25, 1947, pp. 261–278.

65. U. Grenander and G. Szegö, *Toeplitz Forms and Their Applications*, University of California Press, Berkeley, Calif., 1958.

66. J. Durbin, "Efficient Estimation of Parameters in Moving-Average Models," *Biometrika*, vol. 46, parts 1 and 2, 1959, pp. 306–316.

67. J. Durbin, "The Fitting of Time-Series Models," *Rev. Inst. Intern. Statist.*, vol. 28, no. 3, 1960, pp. 233–243.

68. B. S. Atal and M. R. Schroeder, "Predictive Coding of Speech Signals," *Proc. 1967 Conf. Speech Commun. Processing*, 1967, pp. 360–361.

69. B. S. Atal and M. R. Schroeder, "Adaptive Predictive Coding of Speech Signals," *Bell System Tech. J.*, vol. 49, no. 6, Oct. 1970, pp. 1973–1986.

70. B. S. Atal and S. L. Hanauer, "Speech Analysis and Synthesis by Linear Prediction of the Speech Wave," *J. Acoust. Soc. Amer.*, vol. 50, 1971, pp. 637–655.

71. F. Itakura and S. Saito, "Speech Information Compression Based on the Maximum Likelihood Spectral Estimation, *J. Acoust. Soc. Japan*, vol. 27, 1971, pp. 463–472.

72. J. P. Burg, "The Relationship Between Maximum Entropy Spectra and Maximum Likelihood Spectra," *Geophysics*, vol. 37, no. 2, Apr. 1972, pp. 375–376.

73. J. D. Markel, "Digital Inverse Filtering—A New Tool for Formant Trajectory Estimation," *IEEE Trans. Audio Electroacoustics*, vol. AU-20, June 1972, pp. 129–137.

74. F. Itakura, "Speech Analysis and Synthesis Systems Based on Statistical Method," Doctor of Engineering Dissertation, Department of Engineering, Nagoya University, Japan, 1972 (in Japanese).

75. J. Makhoul and J. Wolf, "Linear Prediction and the Spectral Analysis of Speech," NTIS No. AD-749066, *BBN Report No. 2304*, Bolt Beranek and Newman, Inc., Cambridge, Mass., 1972.

76. S. F. BOLL, "A Priori Digital Speech Analysis," Computer Science Division, University of Utah, Salt Lake City, Utah, UTEC-CSC-73-123, 1973.

77. J. D. MARKEL and A. H. GRAY, "On Autocorrelation Equations as Applied to Speech Analysis," *IEEE Trans. Audio Electroacoustics*, vol. AU-20, Apr. 1973, pp. 69–79.

78. J. MAKHOUL, "Spectral Analysis of Speech by Linear Prediction," *IEEE Trans. Audio Electroacoustics*, vol. AU-21, June 1973, pp. 140–148.

79. P. EYKHOFF, *System Identification: Parameter and State Estimation*, Wiley, New York, 1974.

80. E. M. HOFSTETTER, "An Introduction to the Mathematics of Linear Predictive Filtering as Applied to Speech Analysis and Synthesis," *MIT Lincoln Laboratory Technical Note 1973–36*, Rev. 1, Apr. 12, 1974.

81. M. MORF, "Fast Algorithms for Multivariable Systems," Ph.D. Dissertation, Stanford University, Stanford, Calif., 1974.

82. J. MAKHOUL, "Linear Prediction: A Tutorial Review," *Proc. IEEE*, vol. 63, Apr. 1975, pp. 561–580.

83. J. D. MARKEL and A. H. GRAY, JR., *Linear Prediction of Speech*, Springer-Verlag, New York, 1976.

Pole-zero Modeling of Speech

84. C. G. BELL, H. FUJISAKA, J. M. HEINZ, K. N. STEVENS, and A. S. HOUSE, "Reduction of Speech Spectra by Analysis-by-Synthesis Techniques, *J. Acoust. Soc. Amer.*, vol. 33, 1961, pp. 1725–1736.

85. O. FUJIMURA, "Analysis of Nasal Consonants," *J. Acoust. Soc. Amer.*, vol. 34, 1962, pp. 1866–1875.

86. J. L. SHANKS, "Recursion Filters for Digital Processing," *Geophysics*, vol. 32, no. 1, 1967, pp. 33–51.

87. S. A. TRETTER and K. STEIGLITZ, "Power-Spectrum Identification in Terms of Rational Models," *IEEE Trans. Automat. Control*, vol. AC-12, Apr. 1967, pp. 185–188.

88. A. G. EVANS and R. FISCHL, "Optimal Least Squares Time-Domain Synthesis of Recursive Digital Filters," *IEEE Trans. Audio Electroacoustics*, vol. AU-21, Feb. 1973, pp. 61–65.

89. A. V. OPPENHEIM and J. M. TRIBOLET, "Pole-Zero Modeling Using Cepstral Prediction," *QPR No. 111*, Research Laboratory of Electronics, Massachusetts Institute of Technology, Cambridge, Mass., 1973, pp. 157–159.

90. J. M. TRIBOLET, "Identification of Linear Discrete Systems with Applications to Speech Processing," Master's Thesis, Department of Electrical Engineering, Massachusetts Institute of Technology, Cambridge, Mass., Jan. 1974.

91. M. MORF, T. KAILATH, and B. DICKINSON, "General Speech Models and

Linear Estimation Theory," *IEEE Speech Proc. Conf.*, Carnegie-Mellon University, Pittsburgh, Pa., Apr. 15–19, 1974.

92. G. E. KOPEC, "Speech Analysis by Homomorphic Prediction," S.M. Thesis, Department of Electrical Engineering, Massachusetts Institute of Technology, Cambridge, Mass., 1975.

93. A. V. OPPENHEIM, G. E. KOPEC, and J. M. TRIBOLET, "Signal Analysis by Homomorphic Prediction," *IEEE Trans. Acoust. Speech Signal Processing*, vol. ASSP-24, no. 4, August 1976, pp. 327–332.

94. G. E. KOPEC, A. V. OPPENHEIM, and J. M. TRIBOLET, "Speech Analysis by Homomorphic Prediction," *IEEE Trans. Acoust. Speech and Signal Processing*, vol. ASSP-25, no. 1, Feb. 1977, pp. 40–49.

Speech Recognition

95. D. R. REDDY, "Computer Recognition of Connected Speech," *J. Acoust. Soc. Amer.*, vol. 42, no. 2, Aug. 1967, pp. 329–347.

96. B. S. ATAL, "Automatic Speaker Recognition Based on Pitch Contours," Ph.D. Thesis, Polytechnic Institute of Brooklyn, Brooklyn, N.Y., 1968.

97. D. J. BROAD, "Formants in Automatic Speech Recognition," *Intern. J. Man–Machine Studies*, vol. 4, 1972, pp. 411–424.

98. B. S. ATAL, "Effectiveness of Linear Prediction Characteristics of the Speech Wave for Automatic Speaker Identification and Verification," *J. Acoust. Soc. Amer.*, vol. 55, 1974, pp. 1304–1312.

99. F. ITAKURA, "Minimum Prediction Residual Principle Applied to Speech Recognition," *IEEE Trans. Acoust. Speech Signal Processing*, vol. ASSP-23, 1975, pp. 67–72.

100. J. MAKHOUL, "Linear Prediction in Automatic Speech Recognition," *Speech Recognition: Invited Papers Presented at the 1974 IEEE Symposium*, D. R. Reddy, ed., Academic Press, New York, 1975, pp. 183–220.

101. D. R. REDDY, ed., *Speech Recognition*, Academic Press, New York, 1975.

4

DIGITAL IMAGE PROCESSING

B. R. Hunt

University of Arizona
Tucson, Ariz. 85721

4.1 INTRODUCTION

The purpose of this chapter is to give the reader an introduction to a field of applications of digital signal processing that has perhaps been more greatly affected by recent developments in signal-processing technology than many other applications in this book. Processing images by digital techniques has benefited from the discovery of algorithms, such as the fast Fourier transform (FFT), and from the achievement of low cost semiconductor integrated circuits, such as used for refresh memory in the newer image display systems now available. The reason that digital image processing has benefited so much by these technology advancements is cost: a digital image may contain from 10^5 to 10^6 data values; processing and storage of these values would be constrained if the technology for doing so were still tied to the technology of 1965.

The burst of activity in digital image processing in the past 5 to 10 years has outrun the ability to give a truly comprehensive survey in the space of a chapter. Indeed, several different books in this area have been published or are in the process of being published, and the sum of these books is still not completely comprehensive or current. Thus, in this chapter our goal is not to be comprehensive, but to outline some areas of digital image processing that are either "classic" (i.e., have been of research and applications interest for 10 years or more) or have made a great impact in a short period of time by virtue of importance (e.g., reconstruction from projections and its impact on diagnostic medical radiology). A second point is that our discussion is oriented toward signal *processing*, as opposed to signal analysis, classification, and so on. A recent activity in digital image processing has been the infusion of various techniques from the field of artificial intelligence to automatically extract various kinds of information from an image. Although the theory and practice of these techniques is of great interest, it is a matter of preference as to whether they constitute signal processing (as the term is classically used) or signal analysis and classification. The author tends toward the former viewpoint, and image analysis and pattern recognition have been excluded from this chapter.

The chapter is organized into five major sections. In Sec. 4.2 we review the basic concepts associated with image formation and recording, and introduce the basic processes, notation, and terminology. For the reader who has not ventured into image processing, but who has an understanding of linear systems theory, we shall describe the basic processes of image formation by using linear systems and convolution integrals. In contrast to the basic one-dimensional linear systems associated with time functions, however, linear systems associated with images are two-dimensional functions of two spatial variables. We shall also look at the processes by which an image is sensed and recorded. To the engineer familiar with linear systems, image sensing and recording will introduce a disturbing problem: sensor and recording nonlinearity. Finally, we shall examine the effects of sampling and redisplaying images, which are directly affected by recording system nonlinearity, and then conclude with a discussion of the basic properties of human vision, the end user of digital images.

In Sec. 4.3, we turn to our first application, digital image data compression. Intuition tells us that there is much redundant information in the average image. But how do we measure that redundancy, and having measured it, how do we remove it? We shall see that the redundancy can be described in a statistical way, and that removal of the redundancy can be achieved either in the original spatial domain of the image or by transforming the image to a new domain. Spatial domain compression schemes will be seen to have a natural relationship to data prediction techniques that have been used in compressing one-dimensional (nonimage) signals; the transform domain compression schemes will be seen to benefit directly from the discovery of fast transform algorithms, such as the FFT, which have been so important in other applications of digital signal processing.

In Sec. 4.4, we consider digital image restoration. Image formation involves a convolution of energies from the object with the impulse response of the image formation system. This has the effect of altering the image. Sometimes the degree of alteration is slight or acceptable, as in the case of a high-quality image; sometimes the distortion is gross, as when an image is taken with the camera out of focus or in motion. In either case, image restoration is concerned with the removal of the degradations in the image caused by the impulse response of the image formation system. It can be shown that this is a problem without a unique solution and made worse by the presence of noise. Hence there is not one method of image restoration, but many, and within a given method alternatives exist as to how one decides to treat the sensor and recording system nonlinearities. A further complication is that the image formation system that creates the degraded image need not be stationary and can be subject to random variations, as when imaging through turbulent atmosphere.

In Sec. 4.5, we discuss reconstruction from projections. Some of the image formation processes associated with medical diagnosis result in projecting a three-dimensional volume onto a two-dimensional plane: a conventional x ray is a typical example. However, three-dimensional relationships can be very critical in many medical procedures, such as surgery, and there is a need to remove the ambiguity associated with projection images. The Fourier transform domain description of an image gives insight into the means by which this ambiguity can be removed. Furthermore, the use of the FFT algorithm makes it possible to carry out the reconstruction from the transform domain for large volumes of data.

In Sec. 4.6, we discuss image enhancement. Image enhancement is a subjective process; one man's enhancement is another man's noise. Consequently, there is less of the formal mathematics usually associated with signal processing. However, the hardware technology in digital signal processing has had a more visible impact on image enhancement than the other topics discussed. High-quality, interactive displays are now available that do on-line, real-time enhancement processes such as contrast alteration and pseudocolor substitution. These displays are a direct result of digital technology in refresh memories, read-only memories, and high-speed digital-to-analog conversion.

The basic theme that underlies the discussion in this chapter is digital signal processing (as, indeed, it underlies all chapters in this book). The relation of digital signal-processing technology to image processing may seem obscure to the reader who first encounters this topic; this is probably due to the familiar aspect of images (we "see" them everyday) and to the psychological disposition of "signals" to be quantities found in circuits or wires. But digital signal processing is inherent in the applications of this chapter, as a careful reading will reveal. Most interesting of all, since human vision begins at the retina of the eye, the discrete structure of visual receptors in the retina (rods and cones) justifies an argument that human vision itself may be a process involving digital image processing.

The background assumed of the reader is that of basic maturity in digital signal processing and the associated mathematical concepts of same. No reader background in either the physics or optics of images is assumed. The purpose of including Sec. 4.2 on image formation and recording is to give the reader the sufficient background and terminology for understanding the basic concepts associated with images; the reader is advised to refer to this first section as he reads the others in order to have these fundamental concepts fixed in mind as the material in the other sections is encountered. For the remaining concepts, it is assumed that the reader has a maturity in the material at the level of Oppenheim and Schafer's book[11] on digital signal processing. Thus we use nomenclature and terms that can be found there.

4.2 FUNDAMENTAL CONCEPTS

Digital image processing refers to that part of digital signal processing in which the signal is an image. A dictionary definition of image is "a reproduction or representation of the form of a person or thing." The physical mechanism by which this reproduction or representation is created is of importance. Our inherent association with the visual senses predisposes us to conceive of an image as a stimulus on the retina of the eye, in which case the mechanisms of optics govern the image formation. Technology has given us many ways to capture images by other than the retina, however. Recently, there has been a great expansion of images generated by sensors operating on energies with no relation to the human visual system (e.g., synthetic aperture radar, acoustic holography, and penetrating radiation). Fortunately, the great diversity of cases where images are created and recorded can still be encompassed within a compact body of mathematics. It is the purpose of this section to describe the elements of image formation and recording processes.

4.2.1 Image Formation and Recording

The principal elements in imaging are represented in a schematic form in Fig. 4-1; the "box" is a device that is capable of acting upon a radiant energy component of the object (this does not rule out the possibility that the "box" itself emits some type of energy and then acts upon the interaction of this energy with the object; such is the case in synthetic aperture radar, for example). The actions of the "box" are culminated in the *image plane*, where an *image* of the object is created; a sensor of some kind is located in the image plane, and the image formed by the "box" is sensed and recorded. Thus we see that images are inherently involved with the sensing by indirect means (energy transport) of a remote region, and the physics of the various energy-transport mechanisms that govern the sensing process are basic to the image formation process. The physical form of the devices in the "box" is less important than the mathematics that govern all such processes; fortunately, the associated mathematics is well developed and easily related to the linear systems theory, which is familiar to workers in other branches of signal processing.

Realistic image formation systems possess *neighborhood* properties in the

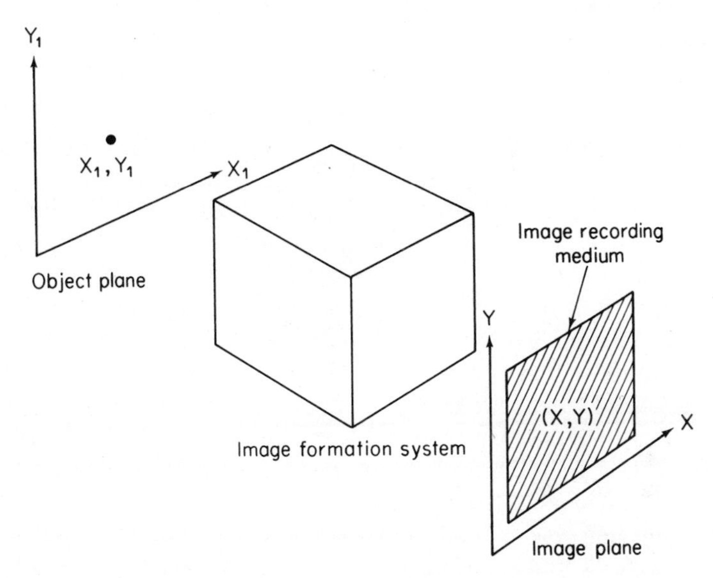

FIGURE 4-1. Schematic diagram of image formation geometry.

processes by which images are formed. By this, we mean that a point (x, y) in the image plane is the image not only of a corresponding point (x_1, y_1) in the object plane, but also may be a function of radiant energy contributions in a (possibly infinite) neighborhood of x_1, y_1. The image, $g(x, y)$, corresponding to an object radiant energy distribution, $f(x_1, y_1)$ is the accumulation of those infinitesimal contributions. Assuming linearity in the process by which the "box" acts upon radiant energies, and assuming linearity in the process by which radiant energies accumulate in the image plane, the formation of the image can be described by the equation

$$g(x, y) = \int_{-\infty}^{\infty}\int_{-\infty}^{\infty} h(x, y, x_1, y_1) f(x_1, y_1)\, dx_1\, dy_1 \tag{4.1}$$

In Eq. (4.1) the function h governs how radiant energy components in the neighborhood of a given object point contribute to the radiant energy distribution in the neighborhood of the image of the object point. The function h depends upon all four coordinate variables in general to allow a change in the neighborhood effects with changes in spatial position.

In practice, image formation equations can be handled (with reasonable effort) only if the function h is invariant in the coordinate variables of the object plane. In such a case the contribution to an image point from the neighborhood of an object point depends only on the relative displacement between the two. The function h can be simplified so that

$$h(x, y, x_1, y_1,) = h(x - x_1, y - y_1) \tag{4.2}$$

and the image formation equation becomes

$$g(x, y) = \int_{-\infty}^{\infty} \int_{-\infty}^{\infty} h(x - x_1, y - y_1) f(x_1, y_1) \, dx_1 \, dy_1 \tag{4.3}$$

This equation governs a *two-dimensional shift-invariant* linear system, and is the natural extension to two dimensions of equivalent one-dimensional systems behavior. In image processing the terminology *space invariant* is often used instead of shift invariant. The function h, which is the system impulse response function, is commonly referred to as the *point-spread function*.

The term "point-spread function" is usually associated with optical images, and Eq. (4.3) naturally arises from an analysis of optical image formation.(1) Many image formation systems other than optical (visible light and lenses) can be described by equations such as Eq. (4.3), however. For example, the formation of images from penetrating radiations can be described by an equation such as Eq. (4.3); the book by Andrews and Hunt(2) contains discussion of image formation processes for both active and passive penetrating radiation imaging systems, and should be consulted for further details. Equations such as (4.1) or (4.3) govern so many image formation processes because the sensing of imagery rests upon radiation transport; diverse imaging processes are unified by the nature of the radiation transport process, which is usually a linear, additive, and space-invariant process.(1,2)

The equations governing image formation have received much treatment, particularly in optics. Less well analyzed in the same framework, however, are the equations governing image sensing and recording. A system for forming images is of no utility without the means to sense and record the image. When discussing digital image processing, current computer technology is usually not capable of performing many operations of interest (linear filtering, for example) on real-time images, where a real-time image might be considered to be an image of the spatial resolution and frame rate (temporal bandwidth) of commercial broadcast television. Digital image processing does not usually take place in real time, as do many one-dimensional digital signal-processing operations.

Image sensing and recording is divided into two major technologies: photochemical recording and photoelectronic recording. Each technology is exemplified by a readily available product: photographic film and television,† respectively. The general usage of both is widespread, and both also have been used as media for input to digital image processing.

Photographic film utilizes the fact that halogen salts of silver are changed by exposure to light; the resulting changes cause the deposition of free silver when the salts are subjected to chemical reduction (development). Unexposed

† We are using "television" in the generic sense rather than implying commercial broadcast television.

silver halide grains are then removed by a *fixer* compound [see Mees[3] for a complete discussion of photochemistry]. The resulting pattern of deposited silver carries the image. Quantitative analysis of the equations of image recording by silver halide films rests upon the work of Hurter and Driffield, who found experimentally that the mass of silver deposited is logarithmically related to the *exposure* E, defined as

$$E = \int_{t_1}^{t_2} I_0(t)\, dt \tag{4.4}$$

where I_0 is the incident light intensity during the exposure interval. Hurter and Driffield showed also that the mass of silver deposited could be related to an optically measured quantity, *optical density*:

$$D = \log_{10} \frac{I_1}{I_2} \tag{4.5}$$

In Eq. (4.5), I_1 is the intensity of a reference source of light shining on a piece of developed film, and I_2 is the intensity of the light transmitted through (or reflected from) the film; $I_1 \geq I_2$ always. Given the definition of optical density as in Eq. (4.5), Hurter and Driffield demonstrated that

$$D = k_1 m_{\mathrm{Ag}} \tag{4.6}$$

where m_{Ag} is the mass of silver deposited and k_1 is a normalizing constant. The relation between the quantities in Eq. (4.4) to (4.6) was summarized by Hurter and Driffield in a *characteristic curve* (now usually referred to as the *H–D* curve) of D versus the logarithm of E. A typical characteristic curve for film is seen in Fig. 4-2. There is a linear region and two nonlinear regions at high and low exposure, referred to as *saturation* and *fog*.

The implications of Hurter and Driffield's work can be seen by taking a piece of film and shining light through it. Defining I_2 and I_1 as associated with Eq. (4.5), the intensity transmitted through (or reflected from) the film is governed by Bouger's law[4]:

$$I_2 = I_1 \exp(-k_1 m_{\mathrm{Ag}}) \tag{4.7}$$

which becomes, using Eq. (4.6),

$$I_2 = I_1 \exp(-D) \tag{4.8}$$

If the film was originally exposed in the linear region of the *H–D* curve in Fig. 4-2, then

$$D = \gamma \log E - D_0 \tag{4.9}$$

where γ is the slope of the linear region and D_0 is a constant, resulting from the linear region not passing through the origin. We also assume that the intensity $I_0(t)$ was constant during exposure, and set $t_2 - t_1 = 1$ (without loss of generality). Substituting into Eq. (4.8), we have

$$I_2 = I_1 \exp(-\gamma \log I_0 + D_0) = k(I_0)^{-\gamma} \tag{4.10}$$

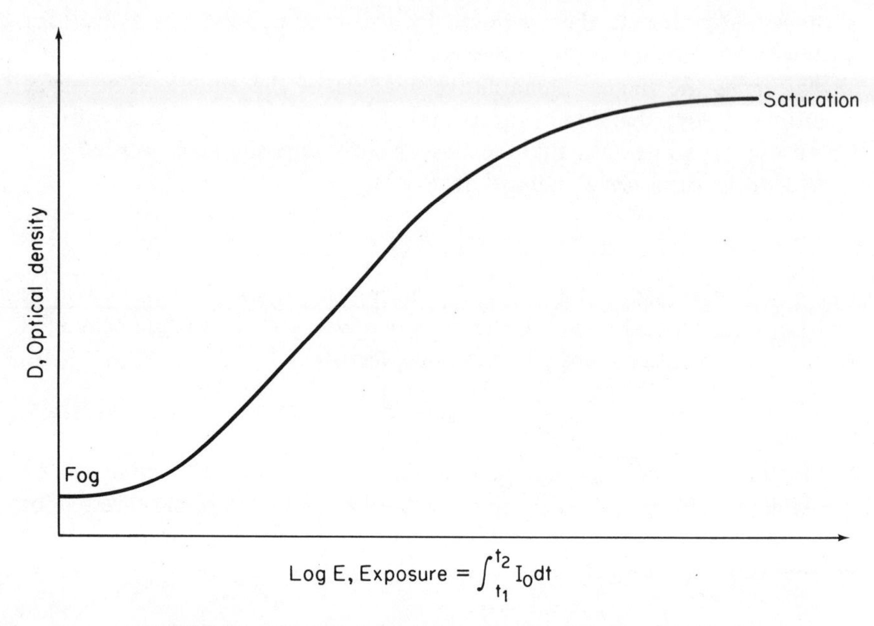

FIGURE 4-2. Characteristic (H-D, D-Log E) curve of photographic film.

Thus the intensity observed, I_2, is a nonlinear mapping of the incident intensity, I_0. For $\gamma = -1$, the intensity representation is exact. Films for which γ is negative are called *positive* films, films for which γ is positive are called *negative* films. This confusion in terminology is unfortunate, but probably too ingrained to change.

Equation (4.10) shows that photochemical sensors make a power-law, nonlinear recording of the intensity of radiant energy incident upon the sensor surface. This nonlinearity is unavoidable (except for $\gamma = -1$ recording), and is something that must often be dealt with in many digital image operations (often it is dealt with by assuming that it can be neglected, which may be pragmatic but not precise).

The nonlinear properties of photochemical image sensing are not alleviated by the use of a photoelectronic sensor. Such devices sense light intensity by means of photoemission of electrons; a considerable amount of circuit and device complexity is usually required to collect and record the emitted electrons, and actual device characteristics are usually empirically measured rather than modeled in detail. Surprisingly, the input–output characteristics of photoelectronic sensors are usually specified by a power-law relationship, such as

$$e = k_e(I_0)^{\gamma_e} \tag{4.11}$$

where k_e is a constant, γ_e the gamma of the device (directly analogous to the gamma of film), and e the electron current. The form of Eq. (4.11) is identical to that of Eq. (4.10), except for the sign of the power. This sign difference is important, because it makes the nonlinear relation between stimulus and response less severe. Furthermore, it is not uncommon to design devices with γ_e close to unity, so that the overall system response is only weakly nonlinear. For example, $\gamma_e = 0.8$ is typical in commercial systems of broadcast television.[5,6]

The sensing of image intensity information cannot be divorced from the introduction of sensor noise, however. Any sensor corrupts the data it senses by noise, and image sensors are no different. Noise arising in photoelectronic sensors can be described in a fairly straightforward fashion. First, there are statistical fluctuations in the emission of photoelectrons. At low light levels, these fluctuations have a Poisson probability density, converging to a Gaussian probability density function as the light level increases. The variance of the fluctuations increases as the light level increases, which means that the noise is dependent upon the signal.[7] A second source of noise is in thermal electron fluctuations in the circuits that sense and process the photoelectron current; this noise is independent of the signal. Both photoelectron and thermal noise are uncorrelated (white noise) processes.

Film sensors introduce noise also, but it is a more complex type of noise. In film, the image is formed by the grains of silver deposited after exposure and development. The grains are not uniform in shape or size. They are also randomly located in the film emulsion rather than being distributed uniformly in space. Consequently, the net effect is of a very complex noise process, which is signal dependent, as in the case of photoelectron noise; however, the dependence between signal and noise is more complex [and not entirely agreed upon; see Huang[8] or Falconer,[9] or Walkup and Choens,[10] for example]. The amplitude distribution of film-grain noise fluctuations is usually assumed to be Gaussian.[3] Film-grain noise is an uncorrelated process only if the film is examined in regions that are spaced farther apart than the grain size of the film.

The details of image formation, sensing, and recording can be idealized in a single block diagram, as shown in Fig. 4-3. Object radiant energies are transformed by a linear system, with system point-spread function h, into image radiant energies. The radiant energies in the image are transformed by the sensor response, s. This is a nonlinear point-mapping operation (i.e., it has "no memory," in contrast to h, which is extended over a spatial neighborhood). The sensed intensities cannot be recorded without noise, however, and an additive noise term is included. This noise term is, in general, not simple; it depends on the sensed intensities g_i in the case of the signal-dependent analysis. The resulting recorded image is g.

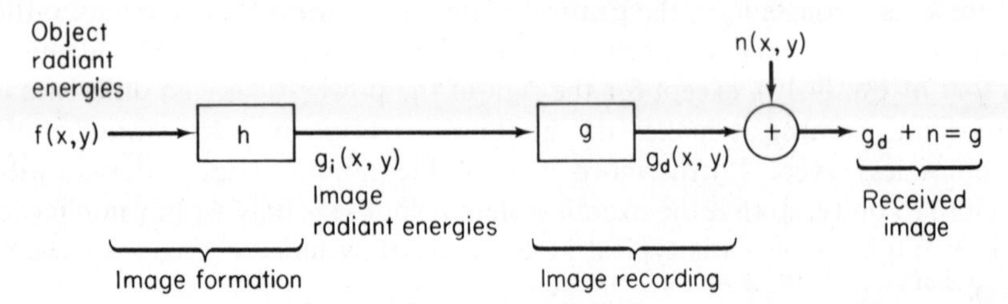

FIGURE 4-3. Block diagram representation of image formation and recording processes.

The model of Fig. 4-3 is applicable to both film and photoelectronic sensing. In the case of film, the nonlinear mapping s is the characteristic or *H–D* curve of the film, which relates incident light to optical density (and, hence, mass of silver). The noise is a fluctuation of silver mass in film, and this is additive in the domain in which the signal is recorded; that is, film records incident intensity by local masses of silver, and the function s and the noise are representative of the physical processes that actually occur. In photoelectronic systems, the similarity of power-law relations between Eqs. (4.10) for film and (4.11) for photoelectronics indicates that a similar model is valid, except electron current is the response from the incident light stimulus. In this latter case, Eq. (4.11) shows clearly that a logarithmic relation between response and stimulus can be derived for photoelectronic systems, exactly as in the form of Eq. (4.9).

The preceding discussion makes it possible to point out an important distinction: the image radiant energies, g_i in Fig. 4-3, are transformed by the recording process into a response variable g_d. In film, the variable g_d is local density of silver. Hence the recorded image is a *density image* (optical density, proportional to silver density), that is, the logarithm of the original *intensity image.* (A similar discussion can be made for photoelectron image sensing; it is not conventional to do so, however.) The presence of noise in the recorded density domain is critical. Since

$$g = s(g_i) + n \tag{4.12}$$

operating on both sides with s^{-1} gives

$$s^{-1}(g) = s^{-1}[s(g_i) + n] \tag{4.13}$$

Since s is a nonlinear function, Eq. (4.13) means that transforming the noisy recorded image back to the original intensity domain (operating with s^{-1}) does *not* yield an intensity image that has additive combination of signal and noise. On the other hand, in the domain wherein noise and signal are additive, the incident intensities are nonlinearly transformed, as we see in Eq. (4.12). These relations between signal and noise are unique to image processing. Dealing with these relations in an optimum fashion is a problem that is still at the edge of current research. Usual practice is to ignore this problem with an approximation such as

$$s^{-1}(g) \simeq g_i + s^{-1}(n) \tag{4.14}$$

or

$$\begin{aligned} g &= s(g_i) + n = s(h^{**}f) + n \\ &\cong h^{**}f + n \end{aligned} \tag{4.15}$$

Fortunately, good results can be obtained with these approximations, justifying their usage.

4.2.2 Image Sampling and Quantization

After an image has been formed and recorded, it still must be converted into a form suitable for digital processing. In the case of images recorded by photoelectronic means, this is usually not difficult, since the electric current sensed by the photoelectronic scanner is in a form amenable to sampling and quantization; it thus can be analyzed as extensions into two dimensions of similar techniques in one-dimensional digital signal processing. Thus quantization errors are equivalent to the addition of another noise source to the data.[11] Sample spacings must satisfy the Nyquist theorem, which has a natural generalization to two dimensions.[1]

Devices that sample and quantize images are developed on the technology of microdensitometry. Such systems project a spot of light with intensity I_1 onto a film. The amount of light I_2 transmitted through (or reflected from) the film is collected by photomultiplier. The *transmittance* is defined as

$$T = \frac{I_2}{I_1} \tag{4.16}$$

and can be used to compute the optical density, as in Eq. (4.5). The spot of light projected on film is then moved in a raster scanning sequence to sample the film. The mathematical model that describes this is

$$g_1(x, y) = \int_{-\infty}^{\infty}\int_{-\infty}^{\infty} h_a(x - x_1, y - y_1)g(x_1, y_1)\, dx_1\, dy_1 \tag{4.17}$$

where g is the image on film, h_a the intensity profile of the spot of light projected on film, and g_1 the image that is actually sampled (i.e., g_1 is evaluated by the scanner at fixed increments $x = j\,\Delta x$, $y = k\,\Delta y$). The matrix of samples $g_1(j\,\Delta x, k\,\Delta y)$ is the *sampled* or *digital* image.

Equation (4.17) (which is also a valid description of photoelectronic image sampling) shows that the recorded image is further modified in the sampling process. By proper choice of the profile h_a and sample spacings, it is possible to prefilter the image as it is being sampled. Since an image is usually not band limited (owing to grain noise and other high-frequency components), the prefiltering inherent in the sampling process of Eq. (4.17) can be used to suppress the effects of aliasing that would otherwise occur.(12)

It should be obvious that quantization of the transmittance is equivalent to sampling an intensity image, and quantization of density is equivalent to sampling a density image. One often hears statements that density sampling is preferable because of the compaction inherent in the logarithm. Such rules of thumb can be deceptive, however.(13)

4.2.3 Reconstruction and Display of Digital Images

In one-dimensional digital signal processing, the reconstruction of an analog signal from digital data is achieved by a lowpass filter, and is theoretically justified by the band-limited interpolation theorem.(11) Ideal band-limited interpolation requires the use of the sin x/x (sinc) function. There is no possible extension of this function to analog image reconstruction, since the sinc function impulse response of the ideal lowpass filter has negative values; these negative values imply negative light, which is an impossibility in the context of image reconstruction.

Analog reconstruction of an image can be made by a device similar to that which samples the image. A spot of light is projected onto unexposed film, and the intensity of this *display spot* is modulated according to the digital data values. (Alternatively, cathode-ray tube (CRT) phosphors may be the source of light and of direct-view display). The spot is moved across the film in a raster scan fashion. It is direct to see that the reconstruction process can be described as

$$g_2(x, y) = \int_{-\infty}^{\infty} \int_{-\infty}^{\infty} h_d(x - x_1, y - y_1) g_1(x_1, y_1)\, dx_1\, dy_1 \tag{4.18}$$

where h_d is the light intensity profile of the display spot, and g_1 the matrix of samples from Eq. (4.17), represented here as a set of weighted Dirac impulses spaced x, y apart; g_2 is the reconstructed image. The display spot profile is the impulse response of an interpolating filter, analogous to the similar case in one-dimensional reconstruction of analog signals. Display spots have a simple shape in virtually all reconstruction systems (e.g., a Gaussian profile). Consequently, there will not be a perfect reconstruction, since the simple display spots do not attenuate completely the higher-ordered replications of the sampled image spectrum. Fortunately, this is usually not a serious problem, and good displays with simple systems are readily available.

The preceding discussion indicates the spectral distortions that result from sampling and display of images. Such effects can be corrected for by digital filtering of the sampled data.(12)

Image reconstruction encounters another problem, that of display fidelity. If a number in the computer represents a particular image optical density value, then a *true fidelity display* would result in the creation of a display image with the same measurable optical density on film as contained in the computer. (Similar requirements can be posed for transmittance or for photoelectronic data.) Such a display would have an input–output transfer characteristic as seen in Fig. 4-4(a). Such ideal characteristics are seldom seen, however. An actual display characteristic is likely to be as in Fig. 4-4(b), where there

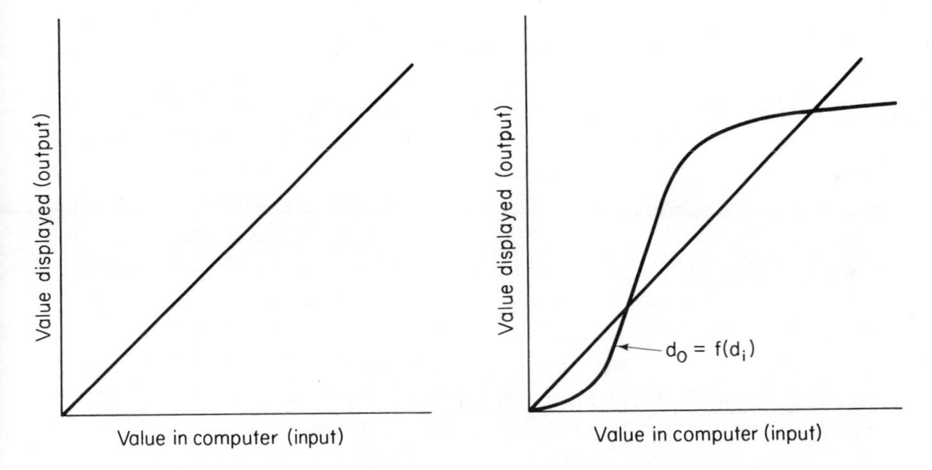

FIGURE 4-4. (a) Input-output transfer characteristics of an ideal display system; (b) Input-output transfer characteristics of a typical display system.

is substantial deviation from the 45 degree line of the ideal. The ideal behavior can be closely approximated by *linearization* of the display. The following steps are necessary:

1. Fixed steps of transmittance or density are generated and sent to the display, and the actual response of the display to a given computer-generated step is measured.
2. The measurements from step 1 define a *display characteristic*, $d_0 = f(d_i)$. The linearized characteristic is given by $d_i = f^{-1}(d_0)$. This inverse transformation can be constructed empirically and implemented either in a table look-up or a polynomial fit by least-squares analysis.
3. Before displaying any image, it is first transformed by processing it through the f^{-1} function. The result is to "predistort" the data so that values in the computer are rendered accurately on the display.

Techniques for linearization of displays have been successfully applied at many laboratories. Exact linearization is not possible, of course, since the precise form of the nonlinear characteristic will vary as a function of film-processing control, purity of chemicals, aging of (or damage to) a CRT phosphor, and the like. However, exercising care can lead to linearized displays in which the departure from linearity is in the neighborhood of $\pm 5\%$ of full-scale values. Display linearization, it should be noted, is a calibration step for analog image reconstruction, which is not usually encountered in one-dimensional signal processing with linear electronics and circuits.

4.2.4 Human Visual System Properties

The ultimate judge of any image is most often a human being. If the human visual system were such as to respond with exact linearity and perfect fidelity to stimuli, we could neglect examination of it. However, the human visual system possesses characteristics that make it nonlinear and without absolute fidelity of response. The importance of these characteristics has been recognized for some time,[13] but the complete application of image processing methods in the context of human visual system properties is mostly unexplored.

The first facet of the human visual system is the perception of light intensity. Experiments with human observers in judging the "just-noticeable difference" in the change of a light stimulus from a reference value of intensity have shown that the perception of light intensity is not linear; instead, the variation in the magnitude of the just-noticeable difference, when plotted against the reference intensity, shows logarithmic behavior over several orders of magnitude of intensity.[14] This subjective experimental finding is also consistent with objective evidence, in which it has been shown, for animal eyes, that light-sensitive neurons in the retina and optic nerve fire at rates which are proportional to the logarithm of the light intensity that falls upon them.[15] Similar objective evidence in human beings has not been measured, for the obvious reason. Nonetheless, the combination of objective evidence in animals and subjective evidence in humans seems more than persuasive to support the conclusion that the perception of light intensity is logarithmic. This is a substantial deviation from linearity, of course.

A second facet of human visual perception is in the spatial frequency response of the eye. Viewed as a two-dimensional linear system (i.e., linear after the initial logarithmic transformation of incident intensity), the impulse response of the eye is not a Dirac function. The eye operates on incident visual fields with a point-spread function, which in profile cross section appears as in Fig. 4-5(a).[16] The central Dirac spike and negative side lobes of this impulse response indicate that the eye processes spatial frequencies as though it were a highpass filter. The exact shape of the frequency response of the eye has been studied by a variety of psychovisual experiments and has confirmed the highpass behavior of the eye, as well as attenuation of high frequencies; the combination is a crude bandpass shape to the spatial-frequency response of the eye. Figure 4-5(b), for example, shows the bandpass shape recently measured in a series of experiments by Mannos and Sakrison.[17]

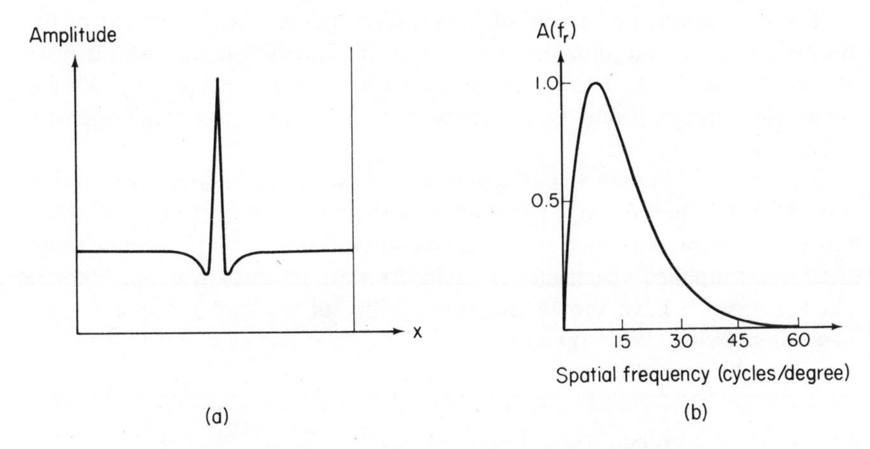

FIGURE 4-5. (a) Point-spread-function (radially-symetric) cross-section of human eye; (b) Frequency response (radially-symetric) cross-section of human eye.

A final property of the human visual system is *saturation*, that is, a limit in response for either very high or low values of incident intensity. The properties of the visual system can be described in a block diagram or schematic model, as in Fig. 4-6. Other known visual properties are not explained at all by this model. This is evidence, for example, that some perceptual processes can be explained only by parallel linear systems, replacing the single system in Fig. 4-6, the *frequency-channel model.*[18] Other visual phenomena, such as the illusion of simultaneous contrast, suggest that the logarithmic preprocessing in Fig. 4-6 is a gross oversimplification. In spite of the known drawbacks to Fig. 4-6, it is a useful model because:

1. It explains some important phenomena, such as light intensity perception and Mach bands.

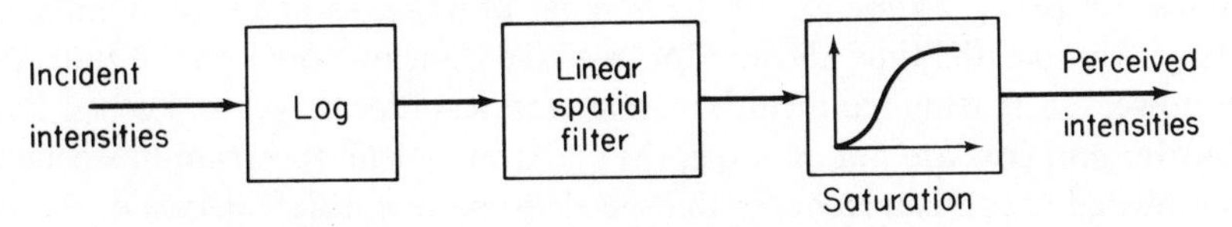

FIGURE 4-6. Block diagram representation of human visual system.

2. It indicates some of the information-processing structure of the visual system; in particular, the human visual system is seen to have some facets of homorphic information processing, as defined by Oppenheim et al.[19]

The logarithmic processing of the eye is of interest in the context of the discussion in the preceding sections concerning intensity images and density images. We see that, because of the logarithm, the eye perceives a density image even though it may be presented, through a display, with an intensity image.

The usage of human visual system models seems to be logical for future applications of digital image processing. Such gains must be made cautiously, however. The human visual system is so complex that an inappropriate usage of an oversimplified visual model could do more harm than good. Mannos and Sakrison[17] have shown the applicability of a visual model in image data compression. Widespread and major applications of visual models still remain to be seen, however.

4.3 APPLICATIONS IN IMAGE DATA COMPRESSION

The first application area that we wish to discuss is data compression of images by digital means. The digital technology revolution has affected all areas in the communication and storage of data, because of the inherent advantages in digital systems with respect to noise protection and error correction, unique message switching capabilities, decreasing cost and increasing reliability of digital systems, and so on. Concurrent with this trend in digital technology has been the increased use of images in a number of scientific and engineering disciplines, such as medicine, remote sensing, physics experimentation, and resource planning. An increased use of images occurring simultaneously with the increased use of digital technology has had the easily anticipated result: a burst of activity in the communication and storage of images by digital means.

A typical image has a great amount of redundant information in it, as a casual glance at most images will show. This redundancy implies an economic waste. Communication of an image in digital form requires a channel bandwidth that is a function of the number of image samples, the number of bits per sample, the time allowed for message transmission, and the transmitter power; increasing bandwidth increases transmitter power and costs. Even if power and cost are not an issue, the electromagnetic spectrum has become so crowded that it is a resource that needs to be zealously conserved. Reduction of image redundancy is vital, therefore, and likewise in the digital storage of images. If only *one* image were to be stored, probably no concern would be voiced. But many current and future systems, such as NASA's Earth Resources Technology Satellite, have the capability to produce great numbers of images for which digital storage and retrieval is of interest. Even though digital storage systems keep getting cheaper in cost, the sheer numbers of images anticipated make the reduction of image redundancy of first importance.

4.3.1 Considerations in Image Data Compression

The redundancy in image data can be described in terms of correlation between data samples. The visible redundancy in an image is exhibited by a high degree of statistical predictability between adjacent samples extracted from an image. The purpose of image data compression is to remove this statistical predictability (i.e., to decorrelate the data as much as possible), and then to properly prepare the data for digital transmission or storage. The diagram in Fig. 4-7 shows the basic elements of an image data compression system. The first step is to process the image data by an operation that attempts to remove as much as possible of the data correlation. This decorrelated data must then be properly quantized, in the second step; then, in the third step, the quantized samples are coded into a form suitable for transmission (coding may include such criteria as error detection or correction, of course).

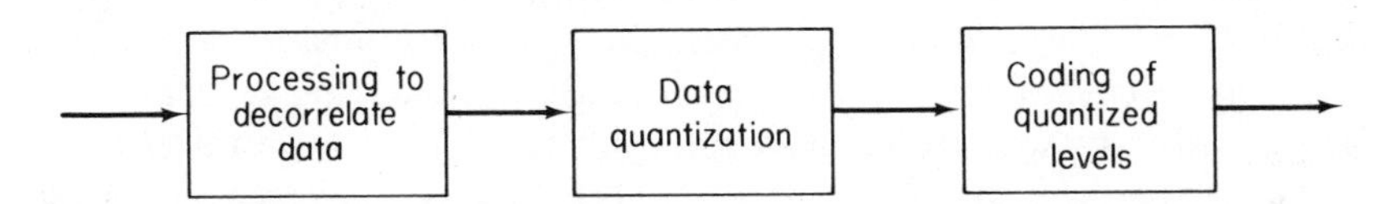

FIGURE 4-7. Block diagram components of image data compression system.

Steps 2 and 3, quantization and coding, are basically governed by the same considerations no matter what particular decorrelation scheme is chosen in step 1. It is in the schemes proposed for step 1, processing the data to remove correlation, that differences in image data compression are to be found. Consequently, we shall spend more effort in discussing the different methods for implementing the first block in Fig. 4-7, than implementations of the second and third blocks. This is also consistent with the objectives of this book for applications of digital signal processing, a topic that primarily directs interest to the first block.

In developing methods to implement the first block in Fig. 4-7, a number of considerations are of importance. First, consider the matter of image statistics. If an image is sampled as an N by N matrix of samples, with each sample quantized to P bits, then a total of N^2P bits is required to represent the image for storage and transmission using a standard pulse-code modulated (PCM) code. However, as observed above, there is a great amount of redundancy in a typical image. One way to measure this redundancy and to compare it to the nominal N^2P bits is the use of the *histogram* statistics and the associated entropy statistics. Given P bits, there are 2^P quantization levels; if each of the N^2 pixels is examined and a count of quantization levels is made, the resulting tabulation of relative frequency of occurrences of quantization levels is the image histogram. When divided by the total pixels, N^2, it is an approximation to the probability density function which can be assumed to underlie the generation of image intensities. If we denote the normalized frequencies by p_i, for $i = 1, 2, \ldots, 2^P$, then the entropy is defined to be

$$h = -\sum_{i=1}^{2^P} p_i \log_2 (p_i) \tag{4.19}$$

and is the average (in bits per pixel) information associated with each picture element. Typically, studies of images have shown h to be much less than the number of quantization bits, P, that a standard PCM representation would require; values of entropy on the order of 1 bit/pixel have been reported.[20] This means that it should be possible, theoretically at least, to compress images to an average of 1 bit/pixel, with no loss of information.

The entropy provides a measure of statistical redundancy without any clue as to the source of this redundancy. The source of redundancy lies, as the eye naturally tells the viewer, in the high degree of spatial similarity in local picture regions. A way to measure this spatial redundancy is by means of the image covariance matrix. First, we take an image that is sampled as an N by N matrix and then lexicographically order it into a vector of length N^2 [i.e., the first row (or column) of $g(j, k)$ becomes vector elements 1 through N, the second row (or column) becomes vector elements $N + 1$ through $2N$, etc.). Then we form the image covariance matrix:

$$[C_g] = E\{(\mathbf{g} - E(\mathbf{g}))(\mathbf{g} - E(\mathbf{g}))^T\} \tag{4.20}$$

where the E denotes ensemble expectation, and $\mathbf{g}$ is the vector formed from ordering the sampled image. In practice, an ensemble expectation is seldom possible, and the covariance matrix is approximated by a spatial correlation operation.[21]

A covariance structure such as $[C_g]$ is a very weak constraint on images. Cole[21] has shown that many images are basically the same in a covariance (or power spectrum) sense. Consequently, there is motivation for replacing a detailed matrix covariance structure, such as $[C_g]$, with a simpler model. The use of an autoregressive Markov process model of order n, where n is typically small (say 3), has been explored; for example, see Habibi.[22] The fact that such models work and can be justified in the context of methods such as differential pulse-code modulation (DPCM) compression, points directly to the high degree of predictability between adjacent pixels in an image.

Besides the statistical considerations in image data compression, the properties of the ultimate user of an image are also an important consideration. The human visual system has some limitations and some (partially) known idiosyncrasies. The use of particular properties of the human visual system in image data compression is known as *psychophysical* processing. For example, the human visual system is known to behave with a logarithmic-type nonlinearity in transformation of the light intensities that enter the eye. Furthermore, the human visual system is insensitive to the very highest and very lowest spatial frequencies, and behaves much like a bandpass filter in midrange spatial frequencies. This latter behavior is a result of the phenomenon of neural inhibition. The meaning of these nonlinear and frequency-sensitive properties is that an optimum image data compression system should process the image data in a fashion analogous to the visual system for greater immunity to coding and channel errors. This procedure was first advocated by Stockham.(23)

The rigorous mathematical basis of much of data compression lies in the concepts of rate-distortion theory.(24) As pointed out by Mannos and Sakrison,(17) the powerful theorems of rate-distortion theory have not been applied to the problem of image data compression, and a principal reason for this failure is the difficulty of specifying an error fidelity criterion that accounts for the behavior of the human visual system. Mannos and Sakrison were able to show that a fidelity criterion based upon the nonlinear and spatial-frequency-sensitive properties of the human visual system is valid, and that the unknown parameters in such a function can be identified. Their work seems to be of great importance for future developments in image data compression. In all the schemes that we shall discuss next, the inclusion of appropriate preprocessing could significantly improve the overall compression system performance.

4.3.2 Spatial-Domain Image Compression Schemes

In terms of the block diagram of Fig. 4-7, one possibility as a process to occupy the first box is the identity operation; that is, the original picture is not changed in any way, whatever data compression is achieved occurs in the quantization and coding blocks. Data compression cannot be achieved without reference to data-dependent and observer-dependent criteria, however. For example, if observer preferences require accuracy of 1 part in 1000, then quantization of 10 bits is necessary; however, if accuracy of 1 part in 8 is allowable, then 3-bit quantization can be used. Consequently, the role that quantization per se can play in data compression is constrained. Data compression can take place at the coding block, however, and one of the chief efforts since the development of Shannon's information theory has been the construction of codes that have optimal properties with respect to data compression. Shannon's theory assures that there is a code which achieves the bit rate of the data sourse; that is, in the case of images with entropies on the order of 1 bit/pixel, there exist coding schemes that would have an average of 1 code bit/pixel. Unfortunately, the existence of such codes is of no comfort without algorithms to construct them. Algorithms for constructing approximations to the optimal codes are known; for example, the Huffman code is an efficient procedure for matching the code to the statistics of the data source, and thereby achieving compression over standard PCM data representations. However, codes such as this are variable length (message code words are of different lengths in bits), and the coding and decoding operations require complicated buffering, storage, and data synchronizing algorithms to function. Furthermore, such codes are very dependent upon the probability of source symbols, and any changes can degrade the code's performance (drastically in some cases). Consequently, the use of quantization coding as a principal means of image data compression is limited, and we must look to other techniques.

The spatial-domain compression technique that has been the most widely explored to occupy the first block of Fig. 4-7 is that of *differential pulse-code modulation* (DPCM). A DPCM scheme is identical in overall structure to the linear predictive coding (LPC) methods used in speech compression, and image DPCM is sometimes referred to as a predictive compression scheme, as a result. An overview of a DPCM scheme is in Fig. 4-8. The method uses the statistical predictability between pixels and forms an estimate of each pixel as a linear combination of previous pixels; *previous pixels* is a term that has direct meaning in the context of top-to-bottom, left-to-right scanning (as in commercial television), which imposes a specific sequence on pixel occurrence. It is also a valid scheme even when the image has already been scanned, of course. The difference between the linear estimate of the pixel and the real value is then computed and quantized. The quantized difference is coded and transmitted. At the receiver, the symbols are decoded and the data are reconstructed by an nth-order linear predictor (identical to the transmitter predictor, of course), which generates a pixel estimate to be added back into the differences.

The predictor structures in Fig. 4-8 are characterized as *feedback* predictors, since they generate a quantization signal in the loop and reconstruct from the feedback of the prediction. It is possible to develop DPCM schemes with feed-forward predictors, and also to generate DPCM with a quantizer that is downstream (outside of) the prediction loop. Such systems produce more accumulated error in the reconstructed picture, however. The feedback predictor *at the receiver* is necessitated by the sequential nature of arriving data values. The use of an identical feedback predictor *at the transmitter*

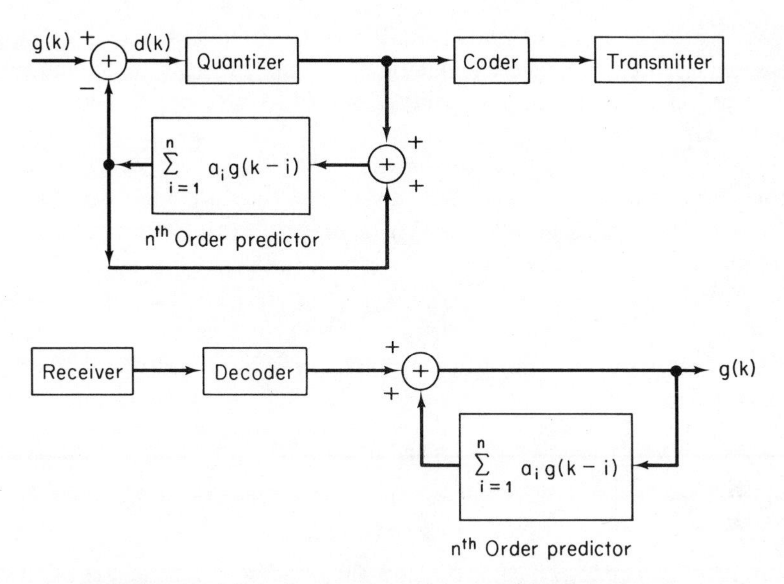

FIGURE 4-8. Block diagram components of DPCM compression, with nth order predictor.

means that, in the absence of errors introduced by quantization, the reconstruction would be exact. Placing the quantizer inside the transmitter prediction loop means that both transmitter and receiver will be operating from the same quantized prediction values, thus minimizing the errors between them in the reconstruction process.

The DPCM scheme achieves compression from the differencing step, since the differences will have a much smaller dynamic range. For example, suppose that the original PCM image required the representation of image intensities with values (on an arbitrary scale) of 0 to 255 units; then 8-bit quantization would be required if a 1-unit error was necessary in the least significant bit (LSB) of the representation. However, the difference values will be much less; if the differences occupied a range of 0 to 7 units (on the same scale), then a 1-unit error in the LSB could be obtained from a 3-bit quantization.

Since the DPCM structure is conceptually simple, as seen in Fig. 4-8, the heart of DPCM image data compression is in the specification of the order of the predictor n, the values of the predictor coefficients a_i, and the quantization thresholds and number of quantization levels.

The order of the predictor is determined by the data. In general, if a data sequence is modeled as an nth-order autoregressive Markov process, then an optimally designed nth-order predictor will cause the resulting difference values to be an uncorrelated sequence.[20] Images are obviously not nth-order Markov processes, but experience with image data has shown that it is possible to model the overall covariance statistics of images by third-order

Markov processes, which leads to a third-order predictor ($n = 3$).[22] Likewise, experience with image simulations has shown little to be gained, in either objective or subjective quality, for DPCM predictors of higher than third order.

The determination of the predictor coefficients, a_i, can be made by the application of mean-square-error analysis. If the data along a scan line are $g(k)$ and the predicted value is $\hat{g}(k)$, then the mean square error is to be minimized:

$$\min_{\text{all } k, a_i} e = E\left\{g(k) - \sum_{i=1}^{n} a_i g(k-i)\right\} \tag{4.21}$$

This is a common problem, and its solution, assuming stationary processes, can be written as[25]

$$\sum_{i=1}^{n} r(j-i)a_i = r(j), \qquad \text{for } j = 1, 2, \ldots, n \tag{4.22}$$

where

$$r(j-i) = E[g(k-j)g(k-i)] \tag{4.23}$$

is usually referred to as the autocorrelation of g. The coefficients a_i are found by solving Eq. (4.22).

The optimum predictor coefficients are seen to depend upon the data's spatial predictability, as embodied in the autocorrelation function. For stationary data, we see from Eq. (4.20) that the autocorrelation differs from the covariance statistics, discussed above, by at most a constant. For nonstationary data, the function r of Eq. (4.23) is spatially dependent, and the optimum predictor coefficients would have to be changed as a function of spatial position. This is certainly the case in images. Fortunately, image statistics, though nonstationary, can be usually approximated by stationary assumptions well enough so that a stationary linear predictor performs quite well. Most DPCM errors in image data compression are usually seen at edges or boundaries in the image, which are regions where the stationarity assumptions are least valid, and are perceived visually as anomalous bright or dark spots in the reconstructed DPCM image.

The selection of the number of quantization levels and the positioning of quantization thresholds for each level is a task that is part qualitative and part quantitative. The positioning of thresholds is a task that can be carried out in a quantitative fashion. A paper by Max[26] first established techniques for nonuniform quantization of data, in a fashion dependent on the data's probability distribution function, so as to minimize the mean-square error induced by a finite number of quantization levels. Max's algorithm yields the optimum threshold positions, given the *number* of quantization levels. However, the number of quantization levels must be selected by subjective and visually qualitative criteria.

The minimum number of quantization levels is two (1 bit) and corresponds to quantizing the image at a fixed positive and negative value for the pixel differences. This is usually referred to as *delta modulation,* and the DPCM structure in Fig. 4-8 can be simplified to a threshold clipper to replace the quantizer and an integrator to replace the *n*th-order predictor. Delta modulation for image compression suffers from the same defects as delta modulation of other signals, such as speech,[27] that is, slope overload and granularity distortion. However, an image that has been initially oversampled by a great degree can be compressed by delta modulation with little subjective error. An image sampled more closely to the Nyquist rate will suffer from more objectionable slope overload (near edges in the image) and granularity (in uniform intensity regions). Adaptive delta modulation, as in speech compression,[27] reduces these errors. In general, however, delta modulation has not been found as satisfactory for image compression as has been demonstrated for speech.

The use of more than two quantization levels produces compressed images with better subjective quality. A DCPM compression system using an eight-level (3-bit) quantizer, with thresholds optimally designed for minimum error, will produce compressed images with the same subjective quality as 6- to 8-bit PCM representations of the images, with the exception of errors in the region of sharp edges in the images.

The output of the quantizer is suited for further coding, of course, since the probability density of quantized pixel differences is not uniform. The use of a proper code, such as the Shannon–Fano or Huffman code, can further reduce the overall data rate. Pratt[28] states that final rates of 2.5 bits/pixel can be achieved if quantizer outputs are Huffman coded. The increased cost and complexity of data buffer, storage, and synchronization unit to process the Huffman code must be balanced against this additional reduction in bit rate.

Figure 4-9 demonstrates DPCM image data compression, and is a coded image using 3.0 bits/pixel and a third-order predictor in DPCM compression; the original image used 8 bits/pixel and appears visually identical to this one.

The preceding discussion was presented in the context of line-scanned DPCM, (i.e., the predictor uses pixels only from the current scan line). The two-dimensional nature of images makes it possible (and reasonable) to develop DPCM methods that utilize not just the previous pixels in a scan line, but pixels in previous scan lines. DPCM compression schemes based on this two-dimensional prediction can be designed by the same methods as those used in the design of one-dimensional predictors. Since images have two-dimensional covariance structure, the use of a two-dimensional predictor should be expected to achieve better results in compression, since the image can be decorrelated in two dimensions by the prediction and differencing operations. Spatial predictors do, indeed, show superior quality in the compressed images; Habibi[22] reported that a two-dimensional third-order pre-

FIGURE 4-9. DPCM compressed image: 3 bits/pixel, 3rd order predictor (8 bits/pixel original image).

dictor with eight-level (3-bit) quantization produced compressed images that could not, by visual subjective criteria, be distinguished from the original images, which were an 11-bit PCM representation.

The prediction and differencing concepts associated with DPCM can be extended into the temporal domain for images, like commercial television, that consist of a number of frames per second. In a typical video image, many pixels do not change or change only slowly from frame to frame. Consequently, a DPCM compression scheme can be derived in which the next pixel estimate is obtained on the basis of a two-dimensional set of pixels in the current frame plus a similar set of pixels in the previous frames. In practice, the temporal contribution to the predictor cannot involve high-order terms, because each temporal term in the predictor requires a buffer for storage of a full image frame. Simulation studies using a third-order predictor, based upon the pixel immediately to the left, the pixel immediately above, and the pixel in the previous frame at the same spatial position, have shown very good compressed images at 1 bit/pixel.[28]

4.3.3 Transform-Domain Image Compression Schemes

The basic operations of a transform-domain compression system can be understood by returning to the covariance matrix defined in Eq. (4.20). The matrix $[C_g]$ describes the correlation of the image data in the natural xy space of the image coordinate system. An important technique in multivariate statistical analysis is to examine a set of data not in its natural coordinate system, but in a coordinate system with more desirable properties. In particular, the coordinate system defined by the eigenvalues and eigenvectors of the covariance matrix are useful:

$$[C_g] = [\Phi][\Lambda][\Phi]^T = \sum_{i=1}^{N^2} \lambda_i \mathbf{\Phi}_i \mathbf{\Phi}_i^T \tag{4.24}$$

where $[\Phi]$ is the matrix of orthogonal column eigenvectors $\mathbf{\Phi}_i$, and $[\Lambda]$ is the (diagonal) matrix of eigenvalues.

The coordinate transformation defined by the eigenvector matrix $[\Phi]$ has the property of decorrelation of the data into components of decreasing statistical significance. Let the eigenvalues of $[C_g]$ be arranged into decreasing order and indexed so that

$$\lambda_1 \geq \lambda_2 \geq \lambda_3 \geq \cdots \geq \lambda_{N^2} \tag{4.25}$$

and let the eigenvectors associated with each eigenvalue be arranged in the same order. Then the matrix $[\Phi]$ of eigenvectors has the property that the product of $[\Phi]$ with a (lexicographically ordered) image vector $\mathbf{g}$,

$$\mathbf{G} = [\Phi]\mathbf{g} \tag{4.26}$$

is such that the vector $\mathbf{G}$ has uncorrelated components, and the components of $\mathbf{G}$ are in order of decreasing statistical variance.[29] This behavior is a property of the discrete Karhunen–Loeve expansion, which is the meaning associated with Eqs. (4.24) to (4.26).

The utility of the covariance or Karhunen–Loeve (K–L) transformation for data compression is evident. By decomposing the data into a set of uncorrelated components of decreasing statistical significance, compression can be achieved by selecting those components of greatest statistical importance, and discarding the rest. For example, $M < N^2$ components of $\mathbf{G}$ are selected and transmitted, along with "bookkeeping" information to indicate the components used, and the bandwidth is reduced by a factor N^2/M. At the receiver, the M data components received are made into a vector of length N^2, again by inserting zero values in place of the $N^2 - M$ components not transmitted. This new vector, which we call $\mathbf{G}'$, is then used to reconstruct the original image by the transformation

$$\mathbf{g}_c = [\Phi]^T\mathbf{G}' \tag{4.27}$$

and the mean-square error associated with the compression is

$$||\mathbf{g} - \mathbf{g}_c|| = \epsilon, \tag{4.28}$$

and the K–L transform has the property of yielding the smallest mean-square error of all possible linear trensformations.

Equations (4.25) and (4.26) indicate that the K–L transformation operations require computation proportional to N^4, since the original data contain N^2 samples. For typical N ($N = 256$ or 512), this is an impossible amount of computation. Even worse is the computation of eigenvalues and eigenvectors for the N^2 by N^2 covariance matrix. Fortunately, experimental studies have shown that a very large number of the entries in the N^2 by N^2 covariance matrix $[C_g]$ are near zero (i.e., the covariance between two pixels rapidly approaches zero as the distance between the pixels increases). The point at which interpixel covariance has become small enough to be assumed zero (say 5 or 10% of the maximum covariance value) defines the *interpixel correlation* distance, which can be expressed in terms of an integer number of samples. This distance is utilized by breaking the image up into blocks of dimension greater than but comparable to the interpixel correlation distance. If each block is of size P, a P^2 by P^2 covariance matrix of all the blocks can be computed:

$$[C_{gp}] = \sum_{i=1}^{Q^2} E[(\mathbf{g}_i - E(\mathbf{g}_i))(\mathbf{g}_i - E(\mathbf{g}_i))^T] \tag{4.29}$$

where $Q = N/P$, and $\mathbf{g}_i$ is the ordered vector from each subblock. Then if $[\Phi_p]$ is the matrix of eigenvectors associated with the P^2 eigenvalues ordered as in Eq. (4.26), the compression operations are carried out for these subblocks exactly as described for the entire image in Eqs. (4.26) and (4.27), with $[\Phi_p]$ replacing $[\Phi]$. Typically, most images have correlation distances such that $P = 16$ is a reasonable compromise between the size of the covariance matrix and the rapidity with which interpixel correlation approaches zero.[30] For subblock compression, the computation time is proportional to Q^2P^4.

Even though subblock decomposition makes compression by K–L transform feasible, it is still not a highly efficient process; the computation requirements certainly are beyond the application of such methods to high data rates, such as video image sources.

The discovery of fast transform algorithms (Fourier, Hadamard, etc.) has had a great impact on many applications of digital signal processing. There has been a similar impact on image data compression. Any linear transformation, such as the Karhunen–Loeve, accomplishes nothing more than a rotation of the image into a new coordinate space. The K–L transform has the property of producing a coordinate space in which the image statistical components are uncorrelated. The pragmatic question to ask is whether or not other transforms, particularly the fast transforms such as Fourier, would possess similar desirable properties. Fortunately, the answer is yes. Although fast transform algorithms do not result in perfect decorrelation of the data, as does the K–L transform, they have been found to work remarkably well; their advantages in computation greatly outweigh the marginal decreases in compression efficiency that they possess.

Fast transform compression schemes can be described in a fashion similar to the K–L. Fast transforms have the additional advantage of being separable, so that the two-dimensional transform domain can be reached with one-dimensional operations. The mathematical description is simplified as well.

If $[W]$ is an orthogonal, unitary, one-dimensional transform operator (i.e., the core matrix of the Fourier transform, Hadamard transform, etc.[31]), then the rotation into transform coordinates is carried out by the computation

$$[G] = [W]^T[g][W] \tag{4.30}$$

where $[g]$ is the original N by N matrix of image samples, and $[G]$ is the transform of $[g]$. The statement of Eq. (4.30) is easily recognized as the implementation of the transform first on the rows of the image, followed by the transformation of the row transforms along the columns. Writing out in full product-sum notation,

$$\begin{aligned} G(m, n) &= \sum_{k=0}^{N-1} \sum_{j=0}^{N-1} g(j, k)w(m, n, j, k) \\ &= \sum_{k=0}^{N-1} w(n, k) \sum_{j=0}^{N-1} g(j, k)w(m, j) \end{aligned} \tag{4.31}$$

where the separability of the transform kernel is used to derive the last step. Separability is well known for the most commonly encountered kernel, the Fourier:

$$\begin{aligned} w(m, n, j, k) &= \exp\left[-\frac{i2\pi}{N}(mj + nk)\right] \\ &= \exp\left[-\frac{i2\pi}{N}(mj)\right]\exp\left[-\frac{i2\pi}{N}(nk)\right] \end{aligned} \tag{4.32}$$

and is also valid for less-well-known kernels, such as Hadamard and Haar. See Andrews[31] for detailed discussion.

In the K–L transform, the eigenvalues λ_i correspond to actual variance statistics projected on the coordinate axes of a space in which the image ensemble data are uncorrelated. In the coordinate space produced by a fast transform, the transform coefficients (i.e., the entries in the matrix $[G]$) are not variances but projections on the axes of the coordinate system established by the transform matrix $[W]$. However, energy compaction occurs in both the K–L and fast transform domains. In the K–L domain, the largest variances (and hence energies) are associated with those columns of $[\Phi]$ or $[\Phi_p]$ that are the preferred or "natural" directions for image data variability. Likewise, in a fast transform domain the largest transform coefficients are those which correspond to the preferred or "natural" directions for the image data in the direction of the transform coordinate system. In this context, transform methods of compression, whether K–L or fast transform, amount to nothing but expanding the image in a set of basis vectors (or basis pictures, since the vector must represent a two-dimensional structure), and truncating the expansion in such a way that little error is made in the expansion, but a large number of transform coefficients are eliminated. Truncation is possible because most image energy is found in a small number of components.

To illustrate, consider a transform compression scheme operating with the Fourier transform. From Eqs. (4.31) and (4.32), we see that the (m, nth) transform coefficient $G(m, n)$ is the projection of the image $g(j, k)$ along the basis vector (or basis picture) formed by the Fourier kernel value:

$$w(m, n) = \exp\left(\frac{i2\pi}{N}mn\right) \tag{4.33}$$

Typical images are such that the low-index terms in the spatial-frequency domain are very large compared to the high-index terms; that is, images are usually dominated by the low-frequency structure. These low frequencies are the overall gross shapes and outlines of features in the image, and the black–white illumination and contrast characteristics of the image. High frequencies represent sharp edges and "crispness" in the image, but contribute little spatial frequency energy. For example, a typical image might contain 95% of its energy in the lowest 5% of the spatial frequencies of the two-dimensional Fourier domain. Retention of these Fourier components, plus sufficiently many higher frequency components to yield an image with enough sharpness to be acceptable to the human eye, can yield appreciable data compression.

Once the concept of transform compression as a selective retention of expansion coefficients is grasped, transform compression systems appear to be deceptively simple. The sophistication in such coding schemes results, therefore, from comparisons of different transform operators, development of procedures to determine the desired transform coefficients to select for retention, quantization of the selected transform coefficients, and coding of the quantized coefficients. The following paragraphs summarize the state of activity in these areas.

A number of fast transform algorithms have been studied and compared for compression efficiency to the Karhunen–Loeve (or optimum) transform. Among these have been the Fourier, Hadamard, and Haar transforms[32]; the slant transform[33]; the cosine transform[34]; and discrete linear basis.[35] Comparison of all possible transforms to determine which is optimum requires that they all be evaluated simultaneously on the same images and with all other parameters (selection, quantization, and coding of samples) held the same. Such has not been done, but the evidence in the referenced literature appears to be sufficient to draw the following conclusions:

1. No fast transform achieves the ideal or optimum compression performance of the Karhunen–Loeve.

2. In performance measures such as mean-square error, the slant transform approaches closest to the Karhunen–Loeve, with Fourier, Hadamard, and Haar following in that order, for small picture subblocks, say of size 16 by 16 or 32 by 32.

3. The subjective and objective differences are not great between the slant-transform best performance and the Haar-transform worst performance.

The selection of which transform coefficients will be retained is done in one of two ways. In *threshold* sampling, a threshold is set (usually determined by total mean-square error); coefficients greater than the threshold are retained, and the remainder are deleted. In *zonal* sampling, a mask is overlaid on the transform domain; elements within the mask are retained, and the remainder are deleted. Since transform operations are usually structured with a general frequency or sequency index, with transform coefficients ordered in terms of increasingly complex (more oscillations per unit length) basis vectors, and since image energy compaction occurs in the lower-frequency or sequency indexes, zonal sampling is equivalent to a generalized lowpass filtering of the image. Conversely, threshold sampling is sensitive to important transform coefficients located anywhere in the transform domain. Consequently, threshold sampling has been found to have superior performance over zonal sampling in reconstructed image quality for the same total number of samples deleted. Unfortunately, a threshold sampling scheme requires that information on the position in the transform domain of each retained sample be transmitted along with the sample itself. This can add an appreciable number of data bits if a simple position code is used; however, run-length coding is capable of achieving a position code with only a modest increase in code bits.[32]

The samples that are selected from the transform domain must be quantized. Unfortunately, the transform-domain samples usually show a much greater variation in magnitude than the original spatial-domain samples, as any experience with Fourier transform data, for example, will verify. This magnitude disparity suggests the use of a variable number of quantization bits, depending upon the magnitude of the sample, but complicates the process greatly. Furthermore, the design of a quantizer to minimize quantization error must utilize the probability density function of the data samples. Various analyses have been made of the probability density function of transform-domain samples.[32,33] The best compromise between simplicity and accuracy appears to be a quantization, with a fixed number of bits per sample, based upon the Gaussian density. High-quality reconstructions can be obtained with Gaussian quantized transform-domain samples using as few as 64 levels (6 bits) per sample.[32] This is probably a reflection of the fact that transform operators are weighted linear sums, and the sum of arbitrary random variables tends toward the Gaussian.

Coding of transform-domain samples is a function of the sample selection algorithm. As discussed above, threshold sample selection requires a code to indicate the sample position in the transform domain, which is implemented with a fixed number of code bits per sample. Zonal sampling, however, makes use of the general lowpass nature of images; this lowpass behavior means that samples in the low-frequency (or low-sequency) regions of the transform domain will be much greater in magnitude than high-frequency values. This can be exploited by using a decreasing number of code bits per sample, as samples in the selected zone vary from low- to high-frequency positions.[33] Position information is not required since zone shape is known, and the order in which samples are extracted from the zone can be fixed.

The complete data compression process in transform compression (data transformation, sample selection, quantization, and coding) results in good performance. Figure 4-10 shows comparisons of several different transform schemes. The images in Figs. 4-9 and 4-10 were 256 by 256 pixels, at 8 bits/pixel. It is evident that transform compression can achieve better performance than DPCM compression.

4.3.4 Other Aspects of Image Data Compression

Since many images are multiframe and sequential in time, like commercial television, and little changes from scene to scene, interframe compression (as opposed to intraframe compression) seems destined to receive more attention in the future. As discussed above, DPCM schemes for interframe compression have already been considered. It appears that the combination of inter- and

(a)

(b)

FIGURE 4-10. (a) Original image; (b to f) all are zonal sampling with transforms in 16 × 16 pixel blocks, and 1.5 bits/pixel final (compressed) pixel rate. Specific transforms are the following: (b) Hadamard transform, (c) Haar transform, (d) Slant transform, (e) Cosine transform, (f) Karhunen-Loeve transform.

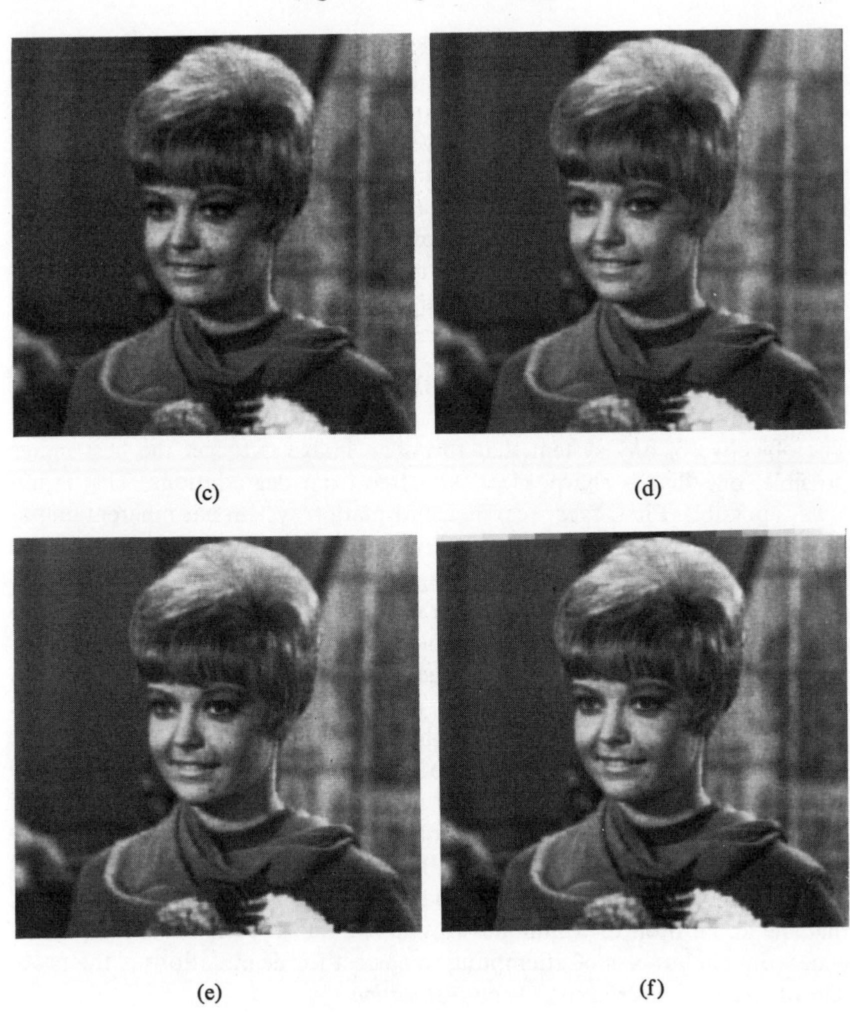

(c) (d)

(e) (f)

FIGUR 4-10. (*Continued*)

intraframe coding has the potential for overall data compression on the order of 30 to 50:1, for those applications where it is suited.

Color imagery is also of increasing interest, and the application of DPCM and transform compression schemes to color imagery has been made.[33,36] The methods are similar to those discussed above, except for the extra attention that must be paid to the color tristimulus description of the image.

The use of hybrid compression systems has been demonstrated recently. These schemes use a transform compression system to decorrelate the data in one image direction (usually the rows of the sampled image, the horizontal direction) and a DPCM system to decorrelate the image in the other direction (columns or vertical direction). The result is a system that is much simpler in structure (two-dimensional transforms are not required), and yet achieves coding efficiency equal to or better than either DPCM or transform.[20]

Finally, the schemes discussed above are not adaptive; that is, they do not vary code bit allocations, quantization levels, and so on, as a function of the image data. Tescher has shown[37] that even greater performance in image coding can be achieved by adaptive compression schemes, and has achieved data compression on images as much as twice as great, for the same quality of reconstructed image, as with other systems.

4.4 APPLICATIONS IN IMAGE DEBLURRING

The objective of any system that forms an image is to get the best image possible, one that is sharp, clear, and free from degradations. This is not always possible. First, every real image formation system has inherent limitations; the impulse response of a real system is of finite width and causes an inevitable loss of resolution in the image. If important details, of a size comparable with the impulse response, are sought in the image, the loss of resolution becomes of concern. As an example, the NASA remote planetary probes send back images of remarkable quality (particularly under the circumstances), but planetary scientists are always seeking surface detail that is marred by the resolution limits of the cameras. A second case of image degradation arises out of unfortunate happenstance. Every precaution may be taken to ensure the best possible image, but a certain fraction emerge degraded anyway by the relative motion between object and camera, by improper focus of the camera, and the like. Of this fraction of degraded images, a certain number will be valuable enough, or of such unique origin, that the application of techniques to undo the degradation is a worthwhile endeavor. The process of attempting to correct for degradations is the problem of *image deblurring* or *image restoration.*

As the reader will observe in the following, a number of different methods have been proposed for the deblurring or restoration problem. This is in distinct contrast to image data compression, discussed in the previous section, where only two basically different methods were presented. Image restoration is a problem without a unique solution, as we shall see; the lack of a unique solution has led to many different attempts at better solutions.

4.4.1 Important Aspects of Image Deblurring

As discussed previously, the basic equation of image formation is given as

$$g(x, y) = \int_{-\infty}^{\infty} \int_{-\infty}^{\infty} h(x - x_1, y - y_1) f(x_1, y_1) \, dx_1 \, dy_1 \qquad (4.34)$$

where g is the formed image, h the point-spread function or impulse response, and f the intensity distribution of the object. Of course, the image g is not known directly; the image is only the intensity modulation of a suitable radiation that propagates from the object. An image is known only from being sensed and recorded by some suitable medium (e.g., film, photosurface, retina). This sensing and recording also induce noise, as there is no way to record a signal without recording-system noise being included. The overall processes of image formation, sensing, and recording were shown in Fig. 4-3 in block-diagram form. It is the recorded, noisy image that must be deblurred.

The deblurring problem is made more difficult by the nature of the recording system and its inherent noise. As discussed in Sec. 4.2, the most common image-recording system, photographic film, has a nonlinear response characteristic and injects noise that is modulated with the signal. A typical photographic response characteristic is seen in Fig. 4-11(a), in which is plotted optical film density, as defined above, as a function of incident intensity, assuming that the intensity is constant during the interval of exposure. The usual plot of such information uses a logarithmic scale on the horizontal axis, as in Fig. 4-11(b), and the resulting *D*–log *E* curve with its linear region sometimes causes misconceptions about the severe nonlinearity associated with photographic film. Since it is the density of silver deposited on the film (which is proportional to optical density) that records the image, we see the inherent nonlinearity in film processes.

Film noise processes are equally complex. The noise that results from random deposition of film grains has a variance which is proportional to the

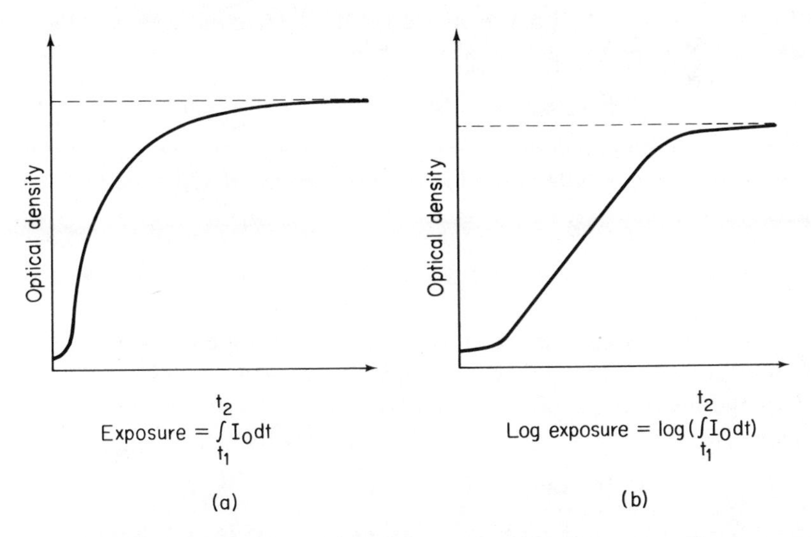

FIGURE 4-11. (a) Optical density as a function of exposure; (b) Optical density as a function of the logarithm of exposure.

local density of developed silver grains, the proportionality being further complicated by a power-law relationship.[8] The noise is thus signal-dependent noise in the optical density that records the image. Dealing with signal-dependent noise, in general, is a problem that digital signal processing has not been able to meet. What few results are available indicate that signal dependency may not be a problem[10] Consequently, it is a common assumption that the noise is a signal-independent process in the density domain. However, the logarithmic relation between film density, which stores the recorded image, and incident radiant intensities means that even this signal-independent density assumption gives a signal-dependent multiplicative noise in the light intensities that exposed the film. Similar effects are found in photoelectronic image sensors (such as television and its generic counterparts), which have a power-law response.

Given the above discussion, the complete model of the image formation and recording process becomes

$$g(x, y) = s\left[\int_{-\infty}^{\infty}\int_{-\infty}^{\infty} h(x - x_1, y - y_1) f(x_1, y_1)\, dx_1\, dy_1\right] + n(x, y) \tag{4.35}$$

where g is the object as actually recorded, s the response characteristic of the recording process, and n a noise process. We assume that n is independent of the recorded image, given the discussion above. The problem of *image deblurring* or *image restoration* is to infer the original distribution of object radiant intensities, $f(x, y)$, from the noisy recorded image, $g(x, y)$.

The difficulty of image deblurring is seen in terms of Eq. (4.35). The original image was deblurred in incident intensities, but a nonlinear version of those intensities is all that is available. If we attempt to remove the nonlinearity by taking the inverse of s, we have

$$s^{-1}\{g(x, y)\} = s^{-1}\{s(h(x, y)**f(x, y)) + n(x, y)\} \tag{4.36}$$

where $**$ represents two-dimensional convolution. Since the inverse nonlinearity does not distribute over addition, the presence of the noise term means that (1) the true inverse cannot be constructed to obtain the original incident intensities, and (2) a nonlinear combination of signal and noise is present in the data obtained by transforming g by the inverse of the nonlinear characteristic.

Present theory cannot encompass the problem posed by Eq. (4.36), and in practice one of two different assumptions is made in deblurring images. First, the distribution of s^{-1} over addition can be *assumed* in Eq. (4.36), which gives

$$s^{-1}\{g(x, y)\} \cong h(x, y)**f(x, y) + s^{-1}\{n(x, y)\} \tag{4.37}$$

which has the assumed effect of replacing the noise by a new noise process $s^{-1}\{n\}$. A second approach is to work directly with the data as recorded, and assume that the nonlinear function s can be neglected, which yields

$$g(x, y) \cong h(x, y)**f(x, y) + n(x, y) \tag{4.38}$$

We see that Eqs. (4.37) and (4.38) can result from either of the following assumptions being true: (1) the function s is a *mild nonlinearity* and can be approximated by a linear relation, or (2) the data g are of small amplitude variation (*low-contrast* image), and so s and s^{-1} are approximately linear for small variations. Figures such as 4.11(a) are conclusive that the nonlinearity associated with s is usually not mild. Conversely, the full range of black to white in a properly exposed image violates the low-contrast assumption. In spite of these difficulties, one can only note, on a pragmatic level, that images can be successfully deblurred in the *intensity domain*, associated with the transformation in Eq. (4.37), or the *density domain*, associated with Eq. (4.38), in spite of the theoretical difficulties presumably entailed by assuming away the nonlinearities.

A cursory observation of either Eq. (4.37) or (4.38) indicates that the image restoration problem is a problem in *deconvolution* (i.e., the processing of the recorded data by a function h^{-1}, which is the inverse, with respect to the convolution operation $**$, of h). The problem of deconvolution is difficult because it is equivalent to solving an ill-conditioned set of linear equations.[38] In image deblurring the problem difficulty is increased because of the fact that many image point-spread functions are singular; that is, the Fourier transforms of the point-spread functions possess zeros; motion blur and out-of-focus blur are two examples. Consequently, much effort in image restoration centers on the necessity to find methods that circumvent the singularity problem.[2]

Digital processing for image deblurring requires that the equations be represented in sampled or discrete form. Thus Eq. (4.35) becomes

$$g(j\,\Delta x, k\,\Delta y) \cong s\left[\sum_{p=0}^{N-1}\sum_{q=0}^{N-1} h((j-p)\,\Delta x, (k-q)\,\Delta y) f(p\,\Delta x, q\,\Delta y)\right] + n(j\,\Delta x, k\,\Delta y) \tag{4.39}$$

where the approximation symbol indicates that the discrete form is not an exact representation of the original integrals. A similar expression can be formulated for either Eq. (4.37) or (4.38). It is of interest that Eq. (4.39) is a system of simultaneous equations in the unknown values of f. In the case of the assumption leading to Eq. (4.38), the corresponding discrete equations (letting $\Delta x = \Delta y = 1$, without loss of generality) yields

$$g(j, k) \cong \sum_{p=0}^{N-1}\sum_{q=0}^{N-1} h(j-p, k-q) f(p, q) + n(j, k) \tag{4.40}$$

in which the unknown values of f are found in a set of simultaneous linear equations.

Equation (4.40) suggests that the image deblurring problem is equivalent to the solution of a set of linear equations. Such is the case, and it is possible to demonstrate the representation of Eq. (4.40) in a matrix–vector product. In this context, the significance of digital signal-processing techniques, such as linear filters and the FFT, is that they provide a rapid computational tool for solving, or approximating the solution to, a very large set (N^2 unknowns) of simultaneous linear equations. This insight is very important for the development of more sophisticated image deblurring methods, but requires too much matrix theory discussion to be examined herein. The reader is urged to consult Hunt[39,40] and Andrews and Hunt[2] for complete details of this duality between matrix representations and deblurring by discrete Fourier transform computations.

4.4.2 Basic Deblurring Methods

The discrete convolution operations represented in Eqs. (4.40) have a counterpart in the discrete Fourier transform (DFT) domain. The DFT of Eq. (4.40) is given by

$$G(u, v) = H(u, v)F(u, v) + N(u, v) \tag{4.41}$$

where $u, v = 0, 1, \ldots, N - 1$, and capital letters represent the DFT of corresponding lowercase quantities. Thus

$$H(u, v) = \sum_{k=0}^{N-1} \sum_{j=0}^{N-1} h(j, k) \exp\left[-\frac{i2\pi}{N}(ju + kv)\right] \tag{4.42}$$

is the DFT of the sampled point-spread function, and likewise for G, F, and N. Usually, the nonzero bounds of the point-spread function are much less than the size of the picture to be deblurred. Consequently, N may be much greater than the dimensions of $h(j, k)$, and an appropriate number of zeros must be inserted (along with a proper shift of position to correct for phase) into $h(j, k)$ before transforming. In theory, one also must pick N large enough to suppress *wrap-around* effects from a circular convolution. In practice, wrap-around effects are not important. Since the original scene is of indefinite extent compared to the point-spread function, the edges of the blurred image have contributions from outside the scene convolved into the scene by the point-spread function. The effects at the edges from deconvolution with incomplete information (scene information beyond the image's edge) is more important than wrap-around. Baxter[18] has shown how to partially correct for such edge effects. The more critical choice in picking N is to minimize wrap-around and truncation effects associated with the inverse point-spread function of $h(j, k)$, rather than $h(j, k)$ itself.

The simplest approach to image deblurring is to process the recorded image in the spatial-frequency domain with the inverse filter; thus

$$\hat{F}(u, v) = H_i(u, v)G(u, v) = \frac{G(u, v)}{H(u, v)} \tag{4.43}$$

is the deblurred image estimate associated with the inverse filter of the point-spread function. Being the simplest approach, this is also the one fraught with the most difficulties:

1. Many blur point-spread functions are such that their DFTs have zeros. For example, the DFT of a one-dimensional motion blur along the horizontal direction is given by

$$H(u, v) = \frac{\sin(\pi a u)}{\pi a u} \tag{4.44}$$

   where a is the blur distance in pixels. If the blur is severe enough (a is large enough) so that zeros of the sinc function lie within the Nyquist frequency, the inverse filter is singular. A similar problem is encountered with the out-of-focus blur, which convolves the aperture shape with the image. For most regular aperture shapes (circular, square, etc.), the corresponding DFT has zeros, and the zeros will result in singularity if within the Nyquist frequency. Unfortunately, this is usually the case.

2. Equally as bad is the fact that even if the point-spread function is not singular, it is usually *ill conditioned.* That is, the magnitude of the transform goes to zero so rapidly for some values of u and v (typically high-frequency indexes, since blurring implies lowpass behavior) that the inverse $1/H$ causes the noise term in Eq. (4.41) to be drastically magnified, obliterating the image.

The inverse filter, despite its limitations, can be used in image restoration. Figure 4-12 is a 512 by 512 digital image after being blurred by computer with a Gaussian lowpass point-spread function; noise has been added to the blurred image, and the signal-to-noise ratio (SNR), measured in terms of signal variance to noise variance, is SNR = 2000:1 (33 decibels, dB). The point-spread function has no zeros and the SNR is high, so that the inverse filter for deblurring is workable. Figure 4-13(a) is the result of processing Fig. 4-12 by the inverse filter using the FFT for implementation. However, if the SNR is lowered by the addition of more noise power, the inverse filter restoration performs poorly. In Fig. 4-13(b) we see the inverse filter restoration in the case of SNR = 200:1 (23 dB); the blurred picture for this higher noise level appears the same as its counterpart in Fig. 4-12, noise being very difficult to observe visually for SNR greater than 20 dB. The noise was added in the density (logarithm of intensity) domain; the image was blurred in the intensity domain, but deblurred in the density domain, as discussed at length in Sec. 4.4.4.

Figures 4-12 and 4-13 illustrate an important point: the inverse filter can work, but it requires a very high signal-to-noise ratio and a mild degree of blur. Unfortunately, no real rule of thumb can be given to guide the experi-

FIGURE 4-12. 512 × 512 digital image, blurred by Gaussian-shaped point-spread-function; noise added, SNR = 33 dB.

menter desiring to deblur an image; fortunately, the inverse filter is so simple to implement that it can be tried, to see if it works, with little real loss if it fails because of noise or singularity.

The sensitivity of the inverse filter to noise and singularity is avoided by another deblurring technique, the Wiener filter. As suggested by the name, this filter is based upon the optimal estimation theory of Norbert Wiener. The filter is derived by finding the linear estimate

$$\hat{f}(j, k) = L[g(j, k)]$$

where L is a linear operator, such that

$$E\{(f(j, k) - \hat{f}(j, k))^2\}$$

(a)

FIGURE 4-13. (a) Inverse filter restoration of Figure 4-12; (b) Inverse filter restoration of Figure 12, in case where SNR = 23 dB.

is minimized. The derivation of the estimator has been done by a number of authors. The first to make explicit use of it in the context of image processing was Hellstrom,[41] who derived spatial- and frequency-domain versions of the estimator. For applications in digital image processing, the frequency-domain version is used, and digital filtering of the image is performed with a filter having the transfer function

$$H_w(u, v) = \frac{H(u, v)^*}{|H(u, v)|^2 + \dfrac{\Phi_n(u, v)}{\Phi_f(u, v)}} \tag{4.45}$$

and

$$\hat{F}(u, v) = H_w(u, v)G(u, v) \tag{4.46}$$

(b)

FIGURE 4-13. (*Continued*)

where the asterisk superscript indicates complex conjugate, and Φ_n and Φ_f are the power spectra of noise and signal, respectively. As in the case of the inverse filter, the filtering is implemented by two-dimensional FFTs, and the inverse transform of Eq. (4.46) gives the deblurred image.

It is possible to observe the following properties of the Wiener filter from a quick inspection of Eq. (4.45):

1. If the noise is very small or zero, so that $\Phi_n \rightarrow 0$, the Wiener filter reduces to the inverse filter. Thus, in low noise spatial-frequency regions (typically the low spatial-frequency regions of the image), the Wiener filter has inverse filter behavior.

(a)

FIGURE 4-14. (a) Weiner filter restoration of Figure 4-12; (b) Wiener filter restoration of Figure 4-12 in case where SNR = 23 dB.

2. If the signal power becomes small, so that $\Phi_f \longrightarrow 0$, the Wiener filter has zero gain. This solves the problem of singular behavior in a point-spread function, and also controls ill-conditioned behavior even in the absence of singularity.

Figures 4-14(a) and 4-14(b) are the images corresponding to Figs. 4-13(a) and 4-13(b) when the deblurring is done by Wiener filter. Figure 4-14(a) resembles 4-13(a), indicating the equivalence of Wiener and inverse filters for low noise. Figure 4-14(b) is a great improvement over 4-13(b), however, indicating the control of ill-conditioned behavior and noise in a Wiener filter.

The border effects in Figs. 4-13 and 4-14 are due to the convolution edge

(b)

FIGURE 4-14. (*Continued*)

effects, as discussed above, and the wrap-around of the inverse filter point-spread function.

Close comparison of Figs. 4-13(b) and 4-14(b) leaves one with the following impression: the Wiener filter reduces the visual noise associated with the inverse filter, but possibly at the expense of some detail and sharpness. This can be explained in two ways: (1) the minimum-mean-square-error constraint is a very powerful requirement and could be relaxed; (2) the nonlinear and adaptive properties of the human visual system may not be "matched" to the minimum-mean-square-error criterion. In Sec. 4.3.3, a different approach is taken to deblurring filter development, which also reduces the a priori information requirements of the Wiener filter.

4.4.3 Blind Deblurring and Power-Spectrum Equalization

A method that produces superior restorations to the inverse or Wiener filter is due to Cannon[42] and is known as the *power-spectrum equalization* or *homomorphic* filter. [The filter was first derived on the basis of homomorphic system theory by Cole,[21] and then an equivalent form was derived by Cannon using the power-spectrum-equalization approach, which we follow in this discussion.] The filter transfer function is derived on the basis of a simple constraint, much less stringent than the minimum-mean-square-error constraint of the Wiener filter. The filter is derived by seeking a linear estimate,

$$\hat{f}(x, y) = L[g(x, y)]$$

where L is a linear operator, such that the power spectrum of the estimate is equal to the power spectrum of the original image. Thus

$$\Phi_{\hat{f}}(u, v) = \Phi_f(u, v) \tag{4.47}$$

is the constraint. Since $g(x, y)$ is defined by Eq. (4.37) or (4.38), the power spectrum of $\hat{f}$ is direct to compute:

$$\Phi_{\hat{f}}(u, v) = |L(u, v)|^2[|H(u, v)|^2\Phi_f(u, v) + \Phi_n(u, v)] \tag{4.48}$$

which can be equated to the right side of Eq. (4.47) and then solved explicitly for $|L(u, v)|$, the magnitude spatial-frequency response of the linear filter that equalizes the power spectrum. The resulting filter response is

$$H_{\text{PSE}}(u, v) = |L(u, v)| = \left[\frac{1}{H(u, v)^2 + \dfrac{\Phi_n(u, v)}{\Phi_f(u, v)}}\right]^{1/2} \tag{4.49}$$

and

$$\hat{F}(u, v) = H_{\text{PSE}}(u, v)G(u, v) \tag{4.50}$$

gives the frequency-domain values of the deblurred image, which are inverse transformed to generate the actual image. Again, processing would be done by using the FFT to implement the computations.

The following properties of the power-spectrum-equalization filter are noted.

1. For low noise, $\Phi_n \longrightarrow 0$, the filter reduces to the magnitude response of the inverse filter.
2. For low signal, $\Phi_f \longrightarrow 0$, the filter gain goes to zero.
3. In between these two extremes, the filter gain is greater than H_w and less that H_i. This is because of the absence of the H^* term seen in the numerator of H_w and the square root operation. It is, in fact, possible to show that the power-spectrum-equalization filter is the geometric mean between the inverse filter and the Wiener filter.[21]

Since the power-spectrum-equalization filter has greater gain than the Wiener filter, but without the ill-conditioned behavior of the inverse filter, the result is a filter that admits into the deblurred image more of the detailed structure associated with high frequencies, which is where the Wiener filter usually cuts off. There is also an increase in visual noise, but the human eye is usually willing to accept the increased noise in return for the additional fine structure that is meaningful.

Figures 4-15(a) and 4-15(b) correspond to the Figs. 4-13(a) and 4-13(b) when the deblurring is done by the power-spectrum-equalization filter. Figure 4-15(a) resembles Figs. 4-13(a) and 4-14(a), indicating again the convergence

(a)

FIGURE 4-15. (a) Power-spectrum-equalization filter restoration of Figure 4-12; (b) Power-spectrum-equalization filter restoration of Figure 4-12, where SNR = 23 dB.

(b)

FIGURE 4-15. *(Continued)*

toward the inverse filter in a low-noise environment. Figure 4-15(b) is the deblurring for the case of high noise; comparison to Figs. 4-13(b), 4-14(b), and the original, Fig. 4-12(a), indicates the superiority of this method of deblurring.

Although the transfer function in Eq. (4.49) is a magnitude-only response, the filter can also be implemented with a phase response. For many known image degradations (e.g., motion blur and out-of-focus blur), the phase response of the blur is 0 and $\pm\pi$. This can be inserted separately into the response of Eq. (4.49) to give a deblurring filter, which corrects for magnitude and phase effects in the image. Cannon[42] demonstrated that such a deblurring is superior to the magnitude-only correction, as would be guessed from basic principles.

Most important is the insight that the power-spectrum equalization gives us into the problem of *blind deblurring*, that is, using the blurred image itself to estimate the parameters necessary to carry out the deblurring operation. The Wiener filter, effectively, requires that values of Φ_n, Φ_f, and H be known a priori, and a similar observation might be made about the power-spectrum-equalization filter. Explicit concentration upon power-spectrum relations in Eq. (4.49) leads to the realization, however, that the relevant information can be estimated from the blurred image.

Consider an image $g(x, y)$ that is subject to an image formation law such as Eqs. (4.37) and (4.38). We break the image up into segments of size M by M, where M is large compared to the dimension of the point-spread function, but small compared to the actual image dimensions N by N; $M = 64$ is typical. The sections need not be disjoint either. Neglecting end effects, each section of the image can be described as the convolution of the point-spread function with an equivalent section from the original distribution of object intensity. Thus

$$g_i(x, y) \cong \int_{-\infty}^{\infty} \int_{-\infty}^{\infty} h(x - x_1, y - y_1) f_i(x_1, y_1)\, dx_1\, dy_1 + n_i(x, y) \tag{4.51}$$

is the approximate description of the image formation law for each section g_i. The power spectrum of each g_i can be computed, after conversion to discrete samples of size M by M, as

$$\Phi_g^i(u, v) \cong |H(u, v)|^2 \Phi_f^i(u, v) + \Phi_n^i(u, v) \tag{4.52}$$

where the superscripts denote the ith section. Assuming that the image and noise can be approximated by stationary random processes, the Φ_f^i and Φ_n^i are samples from the same underlying power-spectra functions. Summing over i sections, therefore, will average out the statistical fluctuations in these power spectra that would be seen in any one section. Thus

$$\begin{aligned} \frac{1}{N} \sum_{i=1}^{Q^2} \Phi_g^i(u, v) &\cong \frac{1}{N} \sum_{i=1}^{Q^2} [|H(u, v)| \Phi_f^i(u, v) + \Phi_n^i(u, v)] \\ &= |H(u, v)|^2 \hat{\Phi}_f(u, v) + \hat{\Phi}_n(u,v) \end{aligned} \tag{4.53}$$

where $Q = N/M$, and $\hat{\Phi}_f$ and $\hat{\Phi}_n$ are estimates of the corresponding signal and noise power spectra.

The importance of Eq. (4.53) can be seen in the following points:

1. The magnitude of the Fourier transform of the blur point-spread function will have a characteristic effect upon the average of the power spectra of the sections. For blurs such as motion or out of focus, the blur will leave a characteristic signature that allows the blur to be identified, and all the pertinent parameters of the blur, such as pixel size and phase reversals, can also be determined.[42]

2. The denominator of the power-spectrum-equalization filter is given in toto by Eq. (4.53), and if an estimate of Φ_f by itself is known, the deblurring can be carried out. This estimate of Φ_f can be obtained in several ways. First, a class of unblurred images with similarity to the blurred image can be used to determine Φ_f, since it has been shown[21] that most images are much alike in the power-spectrum sense. Or a measure of Φ_n can be determined (usually possible in quiescent image regions), the values of H determined from identifying its characteristic signature, and the solution for $\hat{\Phi}_f$ carried out in Eq. (4.53) from the data values.

Figure 4-16(a) is an image taken with a camera in motion during the exposure interval. The motion was horizontal, and the effects are clearly visible in causing the smaller letters of the sign to be blurred beyond legibility. Figure 4-16(b) shows the same image after deblurring by the method described above; the motion blur left a characteristic signature in the power spectrum, which was automatically detected and used to construct a power-spectrum-equalization restoration filter. The text information is clearly legible after this deblurring step. For further information see Cannon.[42]

4.4.4 Comments on Intensity Versus Density Deblurring

All the preceding images were blurred in the intensity domain either by computer simulation of film characteristic curve or by actual photography, as in Fig. 4-16. However, they were deblurred in the density domain, which is the logarithm of intensity, as discussed above. Why this difference? First, it is obviously associated with the linearity assumption posed as Eq. (4.38). Second, density deblurring is usually preferable to intensity-domain deblurring, associated with Eq. (4.37), for pragmatic reasons of image quality. Intensity data span many orders of magnitude (two to three orders is typical), which can accentuate the side lobes of a deblurring filter at points of sharp black-to-white transitions in an image. Density data, being much less than an order of magnitude in dynamic range, are much less sensitive to such effects. Extensive work by both Cannon[42] and Cole[21] proved the desirable visual qualities of images that were deblurred in the density domain. Thus the linearity assumption associated with Eq. (4.38) is to be preferred on pragmatic grounds.

4.4.5 Nonlinear Deblurring Methods

From the viewpoint of digital signal processing, the methods discussed above are implemented by linear filters, using fast Fourier transforms. This is not to imply that efficient implementation is trivial or simple; considerable sophistication can be expended in optimizing the filtering methods and

(a)

(b)

FIGURE 4-16. (a) Image photographed by camera in horizontal motion during exposure interval; (b) Power-spectrum-equalization restoration of Figure 4-16(a), with blind-deburring to detect and estimate blur point-spread-function.

their computer implementation. However, the philosophy is that of linear processing and is readily recognized in the context of classical digital signal processing.

However, realism requires that the shortcomings of linear processing be recognized. First, a real image has properties that are not dealt with in linear processing. For example, an image consists of always positive light intensities, and a linear-processing scheme can generate negative numbers from the side lobes of a deblurring filter. Second, a linear-processing scheme is only an approximation to reality, since image-recording media such as film are inherently nonlinear. Deblurring methods that take into account the nonlinearity are of interest, therefore.

The basic difficulty in nonlinear deblurring (as with almost all nonlinear processes) is in computation. Nonlinear systems lack the computational efficiency that is found in linear processing by FFT. Consequently, few of the nonlinear deblurring methods that have been proposed have ever been implemented on large sampled images; the computational requirements associated with large samples are too excessive. The solutions of these problems seem to rest in numerical and computational analysis more than digital signal processing; hence our discussion in this section will be much shorter.

One nonlinear method that can be implemented is associated with density deblurring in assumptions related to Eq. (4.38). Converting the blurred intensities to density values by taking the logarithm, deblurring by FFT, and then exponentiating the results is a system that is nonlinear in overall response, although implemented by FFT. The resulting intensities are always positive also. The theoretical foundations of such processing are found in the theory of homomorphic signal processing, and are associated with a multiplicative model of image formation.[19] The logarithmic spatial filtering is also seen to be consistent with the human visual system model, which we presented in the first section.

A positive deblurred image is also assured by the method of Frieden,[43] which generates the deblurred image as the solution of the simultaneous nonlinear equations

$$\begin{aligned} g(j, k) &= h(j, k) ** \exp[-1 + h(j, k) ** \lambda(j, k) + \mu] \\ &\quad + \exp[-1 + \lambda(j, k)] \\ P &= \sum_j \exp[-1 + h(j, k) ** \lambda(j, k) + \mu] \end{aligned} \tag{4.54}$$

where the restored image is

$$\hat{f}(j, k) = \exp[-1 + h(j, k) ** \lambda(j, k) + \mu] \tag{4.55}$$

and where $**$ represents two-dimensional discrete convolution and P is the total energy in the incident image. The solution is, thus, positive and bounded. Solution is difficult, however. Doing the convolutions in Eqs. (4.54) and (4.55) by FFT is of no great aid, because the basic difficulty is in solving the equations for the unknowns $\lambda(j, k)$ and μ (Lagrange multipliers from an optimization problem). The method has been implemented for very small images (50 by 50) with separable point-spread functions by using a Newton–Raphson iteration scheme to solve the nonlinear equations. Bigger systems encounter difficulties in implementation of the nonlinear solution. A direct optimization approach has been proposed, but never implemented.[2]

Frieden's method results in a nonlinear restoration from positive and bounded image assumptions. Nonlinear methods also result from the treatment of the recording-medium nonlinearity. A Bayesian approach to restoration with a nonlinear recording-medium response has been demonstrated[40,44]; the actual deblurred image is the solution of a nonlinear matrix equation. The solution of the equation for the large number of variables present in a digital image requires a considerable effort in computation, in which the role of digital signal processing is reduced to convolution implementations.[44] The method has been applied to images as large as 512 by 512.

A full formulation of the deblurring problem with positive and bounded image intensities and nonlinear recording media results in a problem in nonlinear programming.[2] The solution of a general nonlinear programming problem with the number of variables found in a digital image is beyond current capabilities. Special algorithms have been constructed and demonstrated on small images, say 32 by 32, using the simplex method and related mathematical programming concepts.[45] The extension to large digital images has yet to be attempted. In general, nonlinear deblurring is a field in digital image processing in which much remains to be done.

4.4.6 Space-Variant Deblurring

The previous sections have shown the role of convolutions and two-dimensional digital filtering in image deblurring. These methods require space-invariant image formation; in the case of a space-variant process, Eq. (4.34) becomes

$$g(x, y) = \int_{-\infty}^{\infty} \int_{-\infty}^{\infty} h(x, x_1, y, y_1) f(x_1, y_1)\, dx_1\, dy_1 \tag{4.56}$$

which does not result in a discrete convolution when converted into a sampled form. Unfortunately, a number of image formation situations of interest result in space-variant equations; nonuniform motion blur and optical aberrations are two prominent examples.

As an example, consider a one-dimensional horizontal blur in which the blur function is a minimum at the left of the image and increases linearly to a maximum on the right side of the image. If the image is sampled with a uniform raster, then, clearly, one pixel at the right of the image is the sum of contributions over a greater number of adjacent pixels than is the case at the left of the image. One way to rectify this is to increase the interval between samples, in moving from left to right, in such a way that each pixel sample is the convolution sum of a fixed number of adjacent pixels. This is, in effect, a coordinate transformation that yields a space-invariant blur. After deblurring, the original image is recreated by a coordinate transformation that is inverse to the original.

Sawchuk[46] demonstrated that this general method could be extended to a number of space-variant motion and optical aberration blurs. The process has the general structure shown in Fig. 4-17. A space-variant blur is first decomposed in terms of two geometric distortions; one maps from space-variant to space-invariant coordinates; the second maps from space-invariant to space-variant oordinates. The deblurring system is constructed from the inverse of these geometric distortions, and the actual deblurring is implemented in space-invariant coordinates by a linear processor. Thus convolution and FFT computations are possible for rapid implementation on large sampled images, the distortions themselves presenting only a small to moderate computational requirement. Processing of this kind has been successfully applied to space-variant motion blur[46] and to optical aberrations such as coma.[47]

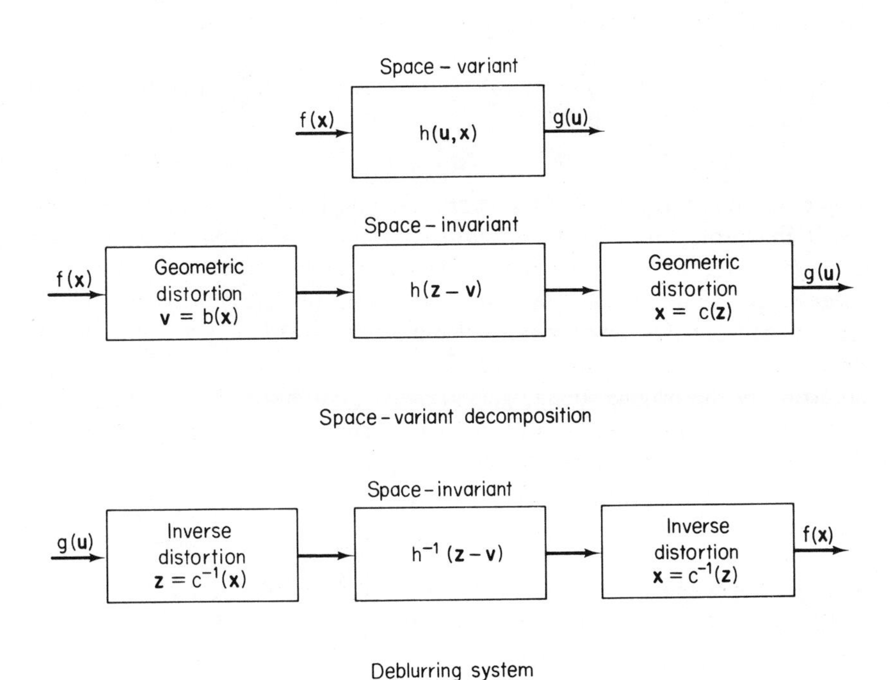

FIGURE 4-17. Block diagram representation of space-variant deburring processes.

Atmospheric turbulence in images produces a blur that changes in time, owing to the thermal gradients in the air that produce random phase delays in the wavefronts of the light carrying the image. Averaged over a long exposure, these perturbations approach a Gaussian-like point-spreading function that is stationary; deblurring of such images has been accomplished in some cases.[48] Generally, the degradation is great enough to benefit little from deblurring. A recently introduced method for atmospheric deblurring is by Knox.[49]

At any instant of time, the image formation in atmospheric turbulence is governed by a random point-spread function. The succession of images observed when viewing the same object is thus

$$\begin{aligned} g_i(x, y) &= \int_{-\infty}^{\infty} \int_{-\infty}^{\infty} h_i(x - x_1, y - y_1) f(x_1 y_1) \, dx_1 \, dy_1 \\ &= h_i(x, y) ** f(x, y) \end{aligned} \tag{4.57}$$

which treats the blur function as space invariant but changing at each instant of time, i. The time average of Eq. (4.57) leads to

$$\langle g_i(x, y) \rangle = \langle h_i(x, y) \rangle ** f(x, y) \tag{4.58}$$

and the random superposition of phases in h_i leads to a time-averaged point-spread function that is so broad as to make unrecoverable much high-frequency information in $f(x, y)$.

A different result is obtained if the images are Fourier transformed and squared before averaging. Then

$$\langle G_i(u, v) G_i(u, v)^* \rangle = \langle H_i(u, v) H_i(u, v)^* \rangle \, F(u, v) F(u, v)^* \tag{4.59}$$

where the superscript * indicates complex conjugate. The squaring operations result in "protecting" the high-frequency information destroyed by phase averages in Eq. (4.58). If a point source is available in the field of view, the mean-square point-spread function $\langle H_i H_i^* \rangle$ can be determined and deblurring carried out. However, phase information is lost in Eq. (4.59), and the deblurred image is the inverse transform of the square root of the power spectrum of the original image $f(x, y)$.

A different, but closely related, processing step is to compute the statistical autocorrelation of the image transforms:

$$\begin{aligned} \langle G_i(u, v) G_i(u + u_1, v + v_1)^* \rangle &= \langle H_i(u, v) H_i(u + u_1, v + v_1)^* \rangle \\ &\quad \times F(u, v) F(u + u_1, v + v_1)^* \end{aligned} \tag{4.60}$$

We see that when $u_1 = v_1 = 0$, then Eq. (4.60) is identical to Eq. (4.59). If the point source is again available, then

$$\frac{\langle G_i(u, v) G_i(u + u_1, v + v_1)^* \rangle}{\langle H_i(u, v) H_i(u + u_1, v + v_1)^* \rangle} = F(u, v) F(u + u_1, v + v_1)^* \tag{4.61}$$

since the point source will allow the calculation of the complex autocorrelation of the time-averaged Fourier transform of the point-spread function.

Consider now what happens if we divide both sides of Eq. (4.61) by the magnitude of these quantities.

$$\frac{\langle G_i(u, v)G_i(u + u_1, v + v_1)^* \rangle}{\langle H_i(u, v)H_i(u + u_1, v + v_1)^* \rangle} \times \frac{|\langle H_i(u, v)H_i(u + u_1, v + v_1)^* \rangle|}{|\langle G_i(u, v)H_i(u + u_1, v + v_1)^* \rangle|}$$
$$= \frac{F(u, v)F(u + u_1, v + v_1)^*}{|F(u, v)F(u + u_1, v + v_1)^*|} \tag{4.62}$$
$$= \exp[i\Phi(u, v) - i\Phi(u + u_1, v + v_1)]$$

where Φ is the phase function of F in the two-dimensional Fourier plane. The right side is a two-dimensional difference equation of the phase-function, with the left side given from the measured quantities. Integration of this difference equation over the Fourier plane gives the phase function. Combined with the magnitude of F from the deblurring of Eq. (4.61) the result is the deblurred image.

Simulations of the Knox–Thompson method show that it offers much promise for increasing the resolution of atmospherically degraded images. Figure 4-18 illustrates the process on a simulated "asteriod" image; Fig.

(a)

FIGURE 4-18. (a) Original image; (b) Four frames of (simulated) atmospheric turbulence degradations of Figure 4-18(a); (c) Restoration from 100 frames such as in Figure 4-18(b).

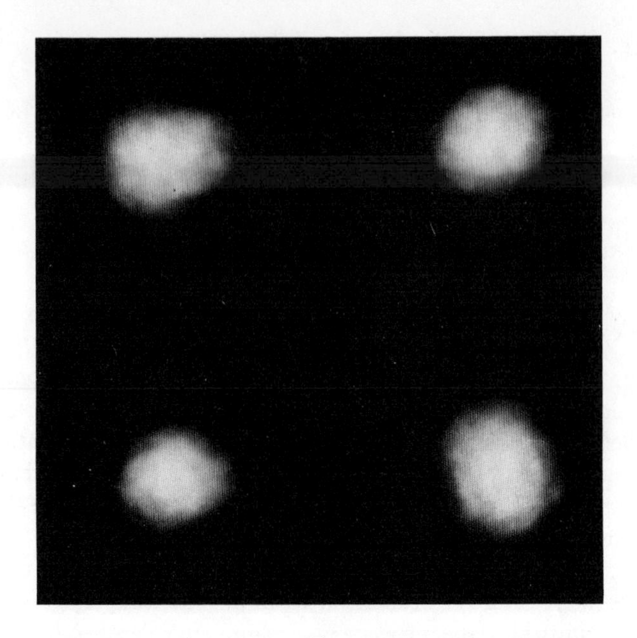

(b)

(c)

FIGURE 4-18. (*Continued*)

4-18(a) is the original, Fig. 4-18(b) shows four individual atmospherically degraded images, and Fig. 4-18(c) the reconstruction when phase information from Eq. (4.62) is included from processing 100 images such as in Fig. 4-18(b).

4.5 APPLICATIONS IN RECONSTRUCTION FROM PROJECTIONS

The discovery of penetrating radiations (x rays, neutrons, etc.) made possible the formation of images from regions that were originally inaccessible, or accessible only by drastic and possibly undesirable means; the interior of the brain is an example of such a region. The resulting imagery has been of great significance to the progress of medicine; the emphasis on quality control in a large system has also made such imagery very important to nondestructive testing methods. Yet penetrating radiation images suffer from a drawback: they are two-dimensional shadow projections from three-dimensional objects. The important spatial relationships between the internal components of the object being projected are obscured, at best, and lost, at worst. In the case of surgical procedures, for brain tumors, for example, this loss of internal structure is obviously critical.

The problem of reconstructing internal structure from projections has received much attention recently, and digital signal-processing techniques have had a great impact in the solution of this problem. A number of methods have been proposed; the references for the paper by Mersereau and Oppenheim[50] should be consulted for a bibliographic survey. In the following we shall discuss the nature of the reconstruction problem and its solution by a method that is so integral to signal processing: the Fourier transform.

4.5.1 Projection Problem

Penetrating radiation images are formed by the attenuation of radiation by material in the path of the radiation beam; the more dense the material, the less intense is the emergent beam after penetration. Thus the *integral* of material along the beam direction determines the image observed in the emergent beam. Let $f(x_1, x_2, x_3)$ be an object distribution of matter in space coordinates x_1, x_2, x_3. Let the illuminating beam of penetrating radiation be parallel to the x_1 axis, as in Fig. 4-19. Then the distribution of penetrating radiation observed in the x_2, x_3 plane is proportional to the function g, defined as

$$g(x_2, x_3) = \int_{-\infty}^{\infty} f(x_1, x_2, x_3)\, dx_1 \tag{4.63}$$

An important property of *projections*, such as defined in Eq. (4.63), can be seen by examining the Fourier transform of $g(x_1, x_2)$:

$$G(w_2, w_3) = \int_{-\infty}^{\infty}\int_{-\infty}^{\infty} g(x_2, x_3) \exp[-i(w_2 x_2 + w_3 x_3)]\, dx_2\, dx_3 \tag{4.64}$$

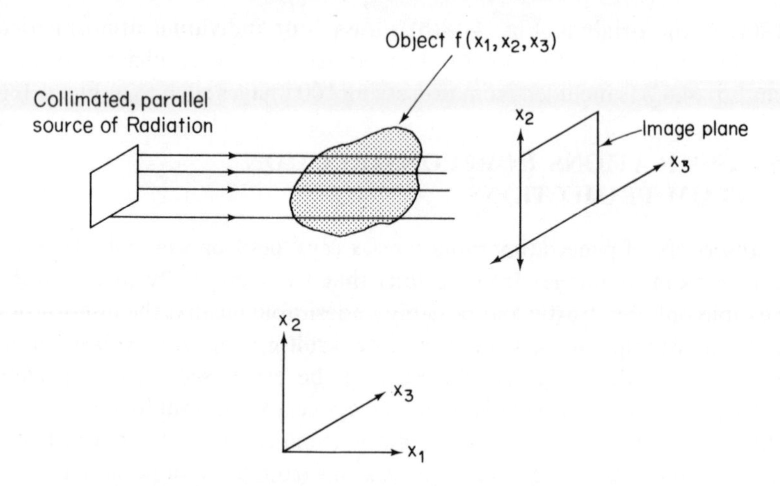

FIGURE 4-19. Geometry for reconstruction from projections.

But consider the three-dimensional transform of the original object f:

$$F(w_1, w_2, w_3) = \int_{-\infty}^{\infty}\int_{-\infty}^{\infty}\int_{-\infty}^{\infty} f(x_1, x_2, x_3) \exp[-i(w_1x_1 + w_2x_2 + w_3x_3)] \times dx_1\, dx_2\, dx_3 \tag{4.65}$$

Comparing G and F, it is clear that

$$G(w_2, w_3) = F(w_1, w_2, w_3)|_{w_1=0} \tag{4.66}$$

That is, the transform of the projection is the transform of the original object evaluated at $w_1 = 0$. This is called a *slice* of the Fourier transform, inasmuch as it is a slice along a two-dimensional plane of the original three-dimensional transform.

It should be obvious that a similar property holds for lower-dimension projections. Assume that the illuminating three-dimensional beam of radiation in Fig. 4-19 is now collimated to a *planar* beam (i.e., a beam which is infinitesimally thin in coordinate x_3 and perpendicular to x_3, but uniform and of extent greater than the object in the other coordinate, x_2). The beam will project the cross section of the object features along the beam location in coordinate x_3. Calling the object features at beam location x_3 and the distribution $f_3(x_1, x_2)$, the one-dimensional projection of the two-dimensional cross section is

$$g(x_2) = \int_{-\infty}^{\infty} f_3(x_1, x_2)\, dx_1 \tag{4.67}$$

and as before

$$G(w_2) = F_3(w_1, w_2)|_{w_1} = 0 \tag{4.68}$$

is the relation between the projection and original transforms.

Now assume that the two-dimensional illuminating beam remain perpendicular to x_3, but that the source is rotated about a fixed center in the object so that the angle between the source and x_1 axis is not zero but a value θ (see Fig. 4-20). Obviously, a change of coordinates can be made so the projection is parallel to coordinate u_1 in a coordinate system:

$$\begin{bmatrix} u_1 \\ u_2 \end{bmatrix} = \begin{bmatrix} \cos\theta & \sin\theta \\ -\sin\theta & \cos\theta \end{bmatrix} \begin{bmatrix} x_1 \\ x_2 \end{bmatrix}$$

The theorems of Eqs. (4.67) and (4.68) apply in this coordinate system u_1, u_2, leading to the following *projection-slice theorem:* the one-dimensional Fourier transform of a projection at an angle θ is equal to the Fourier transform of the original two-dimensional data evaluated along the angle θ in the two-dimensional Fourier plane; that is, it is a slice at an angle θ.

The reconstruction of a three-dimensional body from projections can be made by utilizing the above theorems. If the two-dimensional source is held at a fixed coordinate x_3, as in Fig. 4-20, but the angle θ is varied over $0 < \theta < \pi$, then the collection of one-dimensional projections can be used to reconstruct the cross section at the fixed coordinate x_3, discussed in the next section. Then the value of coordinate x_3 is incremented and a new cross section reconstructed as the source rotates. The process is continued until the complete assemblage of cross sections traces out the internal three-dimensional structure of the object.

Of course, a set of two-dimensional projections can (in theory) be used to reconstruct the three-dimensional object without the intervening reduction

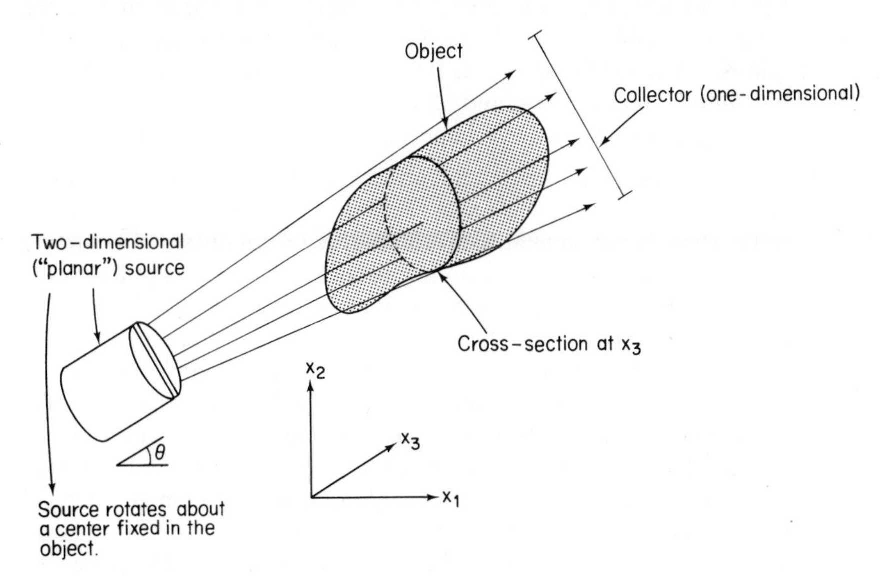

FIGURE 4-20. Two-dimentional projections and rotation geometry.

to one-dimensional projections and cross sections, as discussed in the previous paragraph. In practice, however, the computational requirements of direct three-dimensional reconstruction are so great as to make the cross-sectional two-dimensional method preferable, even though repetitious. Thus we consider the reconstruction problem to be basically one of recovering two-dimensional information from one-dimensional projections; next we shall discuss digital signal-processing methods to implement such a reconstruction.

4.5.2 Reconstruction Methods

A basic procedure for reconstructing from the projection is seen in the projection-slice theorem. The Fourier transform of the projection gives the values of the Fourier transform of the original unprojected data along a slice through a transform domain. Therefore, if the slices are made dense enough in the Fourier domain, the Fourier domain is *filled in* by the slices; an inverse Fourier transform is then used to complete the reconstruction process.

This procedure is simple enough to describe, but requires considerable detailed investigation in actual implementations. The following points are important for reconstruction from Fourier domain slices.

1. As the planar source rotates, it makes a projection for each angle θ. The number of projections required to adequately fill in the Fourier domain is obviously a function of the Fourier transform of the function being reconstructed. In general, however, the smaller the increment between each projection angle, the more dense is the slicing of the Fourier domain (see Fig. 4-21). Two practical constraints exist on the number of projections: (1) the data-processing requirements imposed by additional projections, and (2) the increasing radiation exposure for additional projections (very important for medical applications).
2. The slices of the Fourier domain are equispaced samples in the coordinate system defined by the angle of rotation θ. The spacing of samples in the slices is not compatible with the usual form of two-dimensional transforms, however. A two-dimensional DFT is computed on a rectangular array of rows and columns, and the values of the DFT from the slices are situated on a polar raster, as we see in Fig. 4-21. Therefore, it is necessary to interpolate from the polar form into the rectangular raster. This is simple in principle, but a nontrivial process for computational practice, since each sample in the concentric circles in the polar raster must be referenced to the proper rectangular coordinates before interpolation. This two-dimensional coordinate referencing and interpolation can be simplified somewhat by a method known as a concentric-squares raster, but the simplification is obtained at the price of a variable sampling interval.[50]

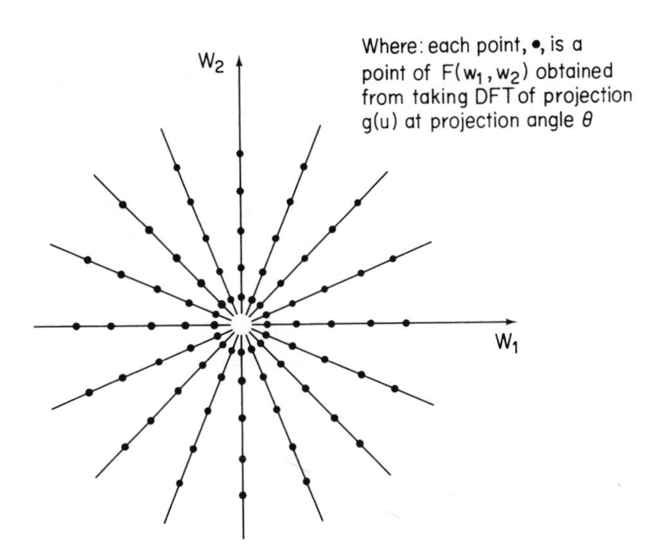

FIGURE 4-21. Fourier domain samples obtained from a set of projections.

3. The nature of projections, as can be seen from Fig. 4-21, is to fill in the two-dimensional Fourier domain quite heavily at low frequencies and to more sparsely represent the high frequencies. Therefore, to get a representation of gross, low-frequency structure requires only a small number of projections, say 25 to 30 projections. To adequately represent the high frequencies requires a large number of projections, say 100 or more. One possibility, if the number of projections is limited, is to use the projections that are known and interpolate between those known projections to more densely fill in the Fourier space. This yields slightly better results, but is ultimately limited by the fact that interpolated projections do not necessarily reflect the reality of the Fourier domain. It is the elements in the Fourier domain that cannot be predicted from interpolation that create the highest quality in an image.

Another method of reconstruction that utilizes digital signal processing is the convolution or smearing method. The basis of this method can be seen by taking two projections at right angles. If each projection is smeared (i.e., a two-dimensional function is created which is constant in one dimension and has a cross section in the orthogonal dimension that is everywhere equal to the original unsmeared projection), the summation at right angles of these two smeared functions will very crudely approximate the original object. The reader should try this with a square and projections at right angles to the square sides as an example. In general, it seems possible to conceive of smearing each projection (which can be done by convolution with the proper function), then summing the smears after suitably weighting them. Such a method can be implemented by digital signal-processing techniques and also is computable by optical means. The method is highly sensitive to noise, however, because the convolution and weighting operations lead to differentiation-like actions on the data.

As an example of the results obtainable from digital methods of reconstruction, Fig. 4-22 is a cross section image of a human brain, obtained from a commercially marketed system for x-ray reconstruction of the interior of the head.

The topic of reconstruction from projections is not exhausted by this brief discussion. The bibliography in the paper by Mersereau and Oppenheim[50] can be consulted for further methods.

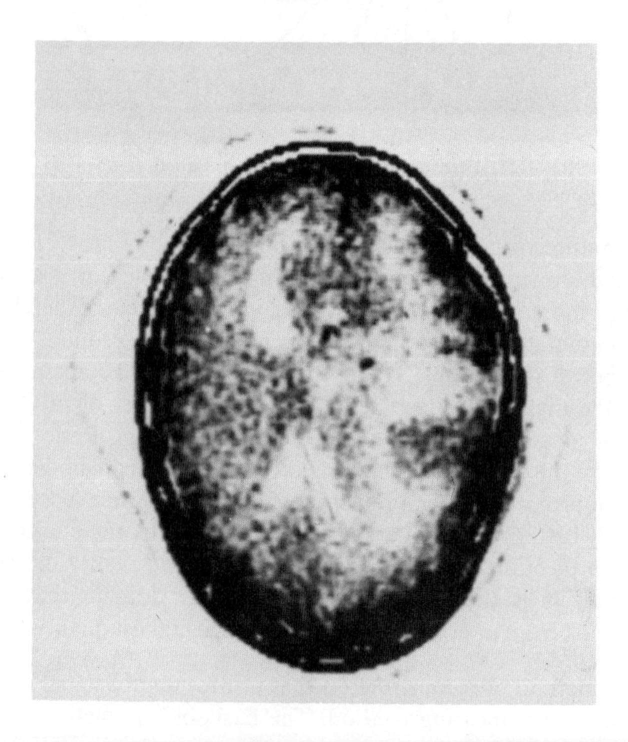

FIGURE 4-22. Actual brain cross-section reconstructed from X-ray projection scanner.

4.6 IMAGE ENHANCEMENT

In image enhancement the goal is, quite literally, to make a picture "look better." With the goal being so loosely structured, it is not surprising that the variety of methods used in image enhancement is equally loosely structured. The subjective judgment of making the image "look better" is also associated with a criteria of usage: make the picture look better for what purpose? If precise analysis or numerical measurement is associated with the use of the image, drastic operations that result in appreciable distortion in space and/or intensity relations may be unacceptable. On the other hand, if the image is used in a purely subjective fashion, operations that result in considerable distortion of space and/or intensity relations may be acceptable if the net effect is better performance in the subjective task. Consequently, image enhancement is composed of a variety of methods whose suitability depends upon the goals at hand when enhancement is originally applied.

4.6.1 Enhancement by Spatial Filtering

Spatial filtering is frequently employed in image enhancement. Even though an image may not suffer from any known defect, a mild high-boost spatial filter can be employed to give an overall appearance of increased sharpness. Equally useful is dc notch filtering, in which a portion (or all) of the frequencies in the neighborhood of the dc term are removed or attenuated. The result is an image in which the overall saturation of large black-and-white objects is removed, and a rescaling of image intensities leaves an image in which small details with small intensity fluctuations are readily seen. Figures 4-23(a) and 4-23(b) are examples of spatial enhancement filtering on a radiograph of a nuclear reactor fuel element. The perceived detail is obviously much greater in the enhanced image, and fuel and sleeve are indicated.

A particularly attractive enhancement technique is based upon homomorphic processing[19] and a multiplicative law of image formation. The physics of surface reflectivity dictates that an image is formed of two components:

$$i(x, y) = i'(x, y)r(x, y) \tag{4.69}$$

where i' is the illumination component and r is the reflectivity of illuminated objects. The illumination component is usually a process composed of spatial low frequencies, where the reflection component is specular and rich in detail. If the logarithm of i is computed,

$$\log i(x, y) = \log i'(x, y) + \log r(x, y) \tag{4.70}$$

then the multiplicative relations between object reflectivity and illumination become additive. A dc notch filter on the logarithm of the image suppresses the illumination component; a simultaneous high boost enhances the fine detail in the reflectivity component. Performing an exponentiation to convert the processed image back into intensity results in an image without negative numbers also. Note further that this filtering takes place in the film density domain—by virtue of the logarithm—and is another aspect of the preferable nature of density processing discussed previously in the section on image restoration.

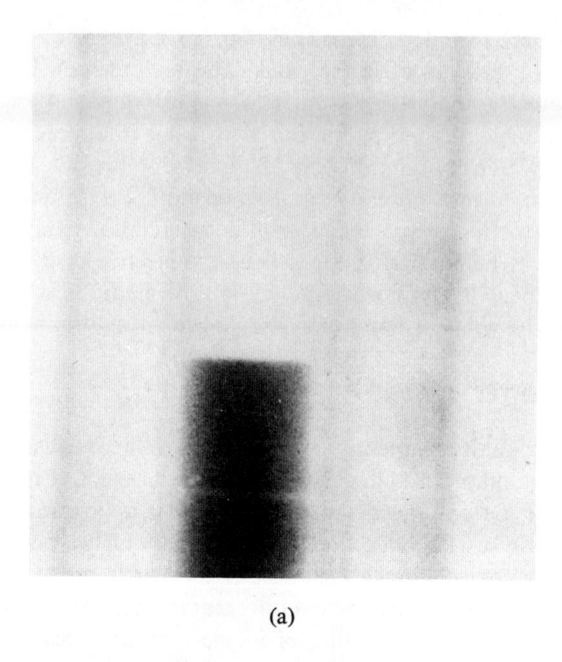

(a)

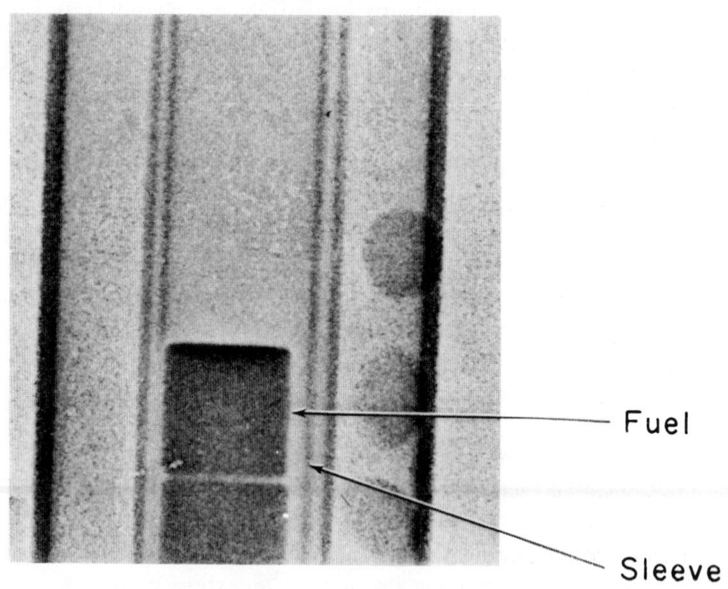

FIGURE 4-23. (a) Original image; (b) Enhancement of Figure 4-23(a) by high-pass filtering.

Figures 4-24(a) and 4-24(b) are examples of an image enhanced by the homomorphic process. Note that simultaneously the image has been sharpened while details in the dark interior of the building have been made more visible.

(a)

(b)

FIGURE 4-24

4.6.2 Point Operations for Enhancement

Enhancement based upon spatial filtering can be contrasted with enhancement in which the operations are not localized in a neighborhood (as in convolution), but in which the operations are point mappings. That is, one image point is transformed into a new image point in a way that is not dependent upon its neighbors. We can group and summarize point operations as follows:

CONTRAST MANIPULATIONS: The enhanced image is generated by altering the image contrast by mapping image intensities through a nonlinear function. For example, if an image is to be enhanced which contains regions that are underexposed, a function that "stretches" out low-intensity regions and translates them into a more suitable intensity interval for viewing could be used. One obvious situation for employing such mappings is in D–log E correction of film images and in the calibration of image displays, as previously discussed in the first section.

STATISTICAL ENHANCEMENT: The development of functions to carry out contrast manipulation can be partially automated by the use of summary statistics from an image (e.g., intensity mean or variance) to adjust the parameters of a contrast transformation. The ultimate in this is the technique of histogram equalization. In information theory, a uniform histogram corresponds to a maximum information message. Therefore, if the histogram of a digital image (a count of the frequency of occurrence of the quantized image intensities) is transformed to the uniform (equal values in all bins) histogram, the resulting image should be a maximum information display. The technique usually works best for images quantized in intensity, since they usually show a greater skew in the histogram.[13] The result can be a dramatic enhancement for little effort.[51]

ARTIFACT GENERATION: The above methods seek to enhance the image, not drastically alter it. If the purpose of an enhancement scheme is to make certain information readily and unambiguously perceptible, it is often common to employ artifact generation methods, in which appreciable distortion of intensity and/or space relations may occur. The most common method is pseudocolor, the arbitrary assignment of a range of (usually) monochrome intensities to a set of colors. The displayed image will be contoured in colors at the boundaries between color assignment to intensities. The result can be vivid highlighting of important detail, or a destructive and possibly misleading confusion of features, or an amusing pattern of color that neither reveals or conceals. Other artifact generation schemes are contouring in black and white by the reduction of quantization levels to a small (usually 10 or less) number. The deletion of one to three of the most significant bits also leads to contours, and the creation of artifact detail by such a method can be quite vivid.

Of inherent interest in the employment of point operations for enhancement is the recent development of digital image displays with instantaneous point-mapping capabilities. Figure 4-25 is the schematic diagram of a commercially available digital display.[52] The use of high-speed read-only memory (ROM) makes it possible to carry out any point mapping as the image is digitally transmitted from refresh memory to CRT display. The original image remains unchanged on the digital refresh disc. Therefore, a reloading of the contents of the ROMs makes it possible to instantaneously switch from one point-mapping operation to another. The resulting flexibility in contrast enhancement, *D*–log *E* correction, pseudocolor assignment, and the like, must be seen to be appreciated. Digital displays such as these make image enhancement by point operations a powerful, convenient, and user-interactive tool.

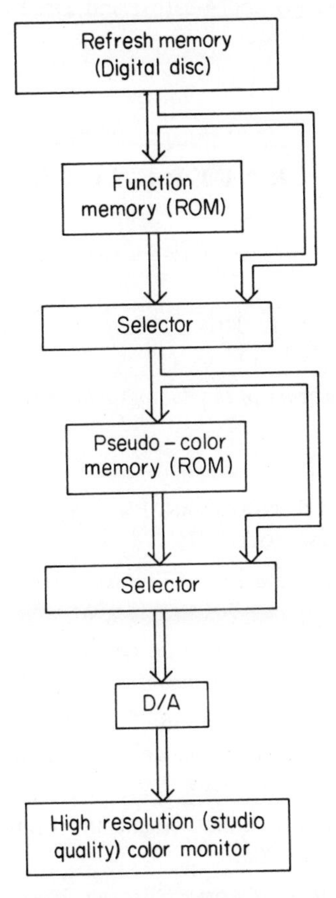

FIGURE 4-25. Block diagram of interactive, point-mapping-enhancement, display system.

ACKNOWLEDGMENTS

The author would like to acknowledge the perceptive eye of Dr. Thomas G. Stockham, Jr., in reading and commenting on portions of this manuscript as it evolved. Grateful thanks are also extended to those individuals who gave images to illustrate various facets of the material. In particular, the following individuals supplied the images indicated, and their generosity was absolutely crucial in allowing the production of the final manuscript: Dr. William K. Pratt, University of Southern California, for the images in Figs. 4-9 and 4-10 in the section on image data compression; Dr. T. M. Cannon, Los Alamos Scientific Laboratory, for Figs. 4-16(a), 4-16(b), and 4-22; Drs. T. G. Stockham, Jr., and B. Baxter of the University of Utah for Figs. 4-23(a) and 4-23(b); and Dr. E. Barrett of ESL, Inc., for Figs. 4-18(a), 4-18(b), and 4-18(c).

REFERENCES

1. J. W. Goodman, *An Introduction to Fourier Optics*, McGraw-Hill, New York, 1968.
2. H. C. Andrews and B. R. Hunt, *Digital Image Restoration*, Prentice-Hall, Englewood Cliffs, N.J., 1977.
3. C. E. K. Mees, *The Theory of the Photographic Process*, Macmillan, New York, 1954.
4. *American Institute of Physics Handbook*, McGraw-Hill, New York, 1972.
5. L. M. Biberman and S. Nudelman, *Photoelectronic Imaging Devices*, vols. 1 and 2, Plenum, New York, 1971.
6. G. M. Glasford, *Fundamentals of Television Engineering*, McGraw-Hill, New York, 1955.
7. L. Mandel, "Fluctuations of Photon Beams: The Distribution of the Photoelectrons," *Proc. Phys. Soc., London*, vol. 74, 1959, pp. 233–243.
8. T. S. Huang, "Some Notes on Film-Grain-Noise," Appendix 14, *Restoration of Atmospherically Degraded Images*, NSF Summer Study Report, Woods Hole, Mass., 1966, pp. 105–109.
9. D. G. Falconer, "Image Enhancement and Film-Grain-Noise," *Opt. Acta*, vol. 17, 1970, pp. 693–705.
10. J. F. Walkup and R. C. Choens, "Image Processing in Signal Dependent Noise," *Opt. Eng.*, vol. 13, 1974, pp. 258–266.
11. A. V. Oppenheim and R. W. Schafer, *Digital Signal Processing*, Prentice-Hall, Englewood Cliffs, N.J., 1975.

12. B. R. HUNT and J. R. BREEDLOVE, "Sample and Display Considerations in Processing Images by Digital Computer," *IEEE Trans. Computers*, vol. 24, 1975, pp. 848–853.

13. T. G. STOCKHAM, JR., "Image Processing in the Context of a Visual Model," *Proc. IEEE*, vol. 60, 1972, pp. 828–842.

14. L. M. HURVITCH and D. JAMESON, *The Perception of Brightness and Darkness*, Allyn and Bacon, Boston, 1966.

15. T. N. CORNSWEET, *Visual Perception*, Academic Press, New York, 1970.

16. P. C. BAUDELAINE, "Digital Picture Processing and Psychophysics: A Study of Brightness Perception," Ph.D. Dissertation, University of Utah, Department of Computer Science, Salt Lake City Utah, 1973.

17. J. MANNOS and D. L. SAKRISON, "The Effects of a Visual Fidelity Criterion on the Encoding of Images," *IEEE Trans. Inform. Theory*, vol. 20, 1974, pp. 525–536.

18. B. BAXTER, "Image Processing in the Human Visual System," Ph.D. Dissertation, University of Utah, Department of Computer Science, Salt Lake City, Utah, 1976.

19. A. V. OPPENHEIM, R. W. SCHAFER, and T. G. STOCKHAM, JR., "Nonlinear Filtering of Multiplied and Convolved Signals," *Proc. IEEE*, vol. 56, 1968, pp. 1264–1291.

20. A. HABIBI and G. ROBINSON, "A Survey of Digital Picture Coding," *Computer*, vol. 7, May 1974, pp. 22–35.

21. E. R. Cole, "The Removal of Unknown Image Blurs by Homomorphic Filtering," Ph.D. Dissertation, University of Utah, Department of Computer Science, Salt Lake City, Utah, 1973.

22. A. HABIBI, "Comparison of *n*th Order DPCM Encoder with Linear Transformations and Block Quantization Techniques," *IEEE Trans. Commun. Technol.*, vol. 19, 1971, pp. 948–956.

23. T. G. STOCKHAM, JR., "Intra-frame Encoding for Monochrome Images by Means of a Psychophysical Model Based on Nonlinear Filtering of Signals," *Proc. 1969 Symp. on Picture Bandwidth Reduction*, Gordon and Breach, New York, 1972.

24. T. BERGER, *Rate Distortion Theory*, Prentice-Hall, Englewood Cliffs, N.J., 1971.

25. J. MAKHOUL, "Linear Prediction: A Review," *Proc. IEEE*, vol. 63, 1975, pp. 561–580.

26. T. MAX, "Quantizing for Minimum Distortion," *IRE Trans. Inform. Theory*, vol. 16, 1970, pp. 7–12.

27. R. W. SCHAFER and L. R. RABINER, "Digital Representations of Speech Signals," *Proc. IEEE*, vol. 63, 1975, pp. 662–677.

28. W. K. PRATT, *Digital Image Processing*, Wiley, New York, 1977 (forthcoming).

29. K. FUKUNAGA, *Introduction to Statistical Pattern Recognition*, Academic Press, New York, 1972.

30. P. A. WINTZ, "Transform Picture Coding," *Proc. IEEE*, vol. 60, 1972, pp. 809–820.

31. H. C. ANDREWS, *Computer Techniques in Image Processing*, Academic Press, New York, 1968.

32. W. K. PRATT and H. C. ANDREWS, "Transform Image Coding," USC Report 387, University of Southern California, Department of Electrical Engineering, Los Angeles, 1970.

33. W. CHEN, "Slant Transform Image Coding," USC Report 441, University of Southern California, Department of Electrical Engineering, Los Angeles, 1973.

34. N. AHMED, T. NATARAYAN, and K. R. RAO, "Discrete Cosine Transform," *IEEE Trans. Computers*, vol. 23, 1974, pp. 90–93.

35. R. HARALICK and K. SHANMUGAM, "Comparative Study of a Discrete Linear Basis for Image Data Compression," *IEEE Trans. Systems Man Cybernetics*, vol. 4, 1974, pp. 16–28.

36. W. K. PRATT, "Spatial Transform Coding of Color Images," *IEEE Trans. Commun. Technol.*, vol. 19, 1971, pp. 980–992.

37. A. G. TESCHER, "The Role of Phase in Adaptive Image Coding," USC Report 510, University of Southern California, Department of Electrical Engineering, Los Angeles, 1973.

38. B. R. HUNT, "A Theorem on the Difficulty of Numerical Deconvolution," *IEEE Trans. Audio Speech Signal Processing*, vol. 20, 1972, pp. 94–95.

39. B. R. HUNT, "The Application of Constrained Least-Squares Estimation to Image Restoration by Digital Computer," *IEEE Trans. Computers*, vol. C-22, 1973, pp. 805–812.

40. B. R. HUNT, "Digital Image Processing," *Proc. IEEE*, vol. 63, 1975, pp. 693–708.

41. C. W. HELLSTROM, "Image Restoration by the Method of Least Squares," *J. Opt. Soc. Amer.*, vol. 57, 1967, pp. 297–303.

42. T. M. CANNON, "Digital Image Deblurring by Nonlinear Homomorphic Filtering," Ph.D. Dissertation, University of Utah, Department of Computer Science, Salt Lake City, Utah, 1974.

43. B. R. FRIEDEN, "Restoring with Maximum Likelihood," University of Arizona, Optical Sciences Center Tech. Rept. No. 67, Tucson, Ariz., 1971.

44. B. R. HUNT, "Bayesian Methods of Nonlinear Digital Image Restoration," *IEEE Trans. Computers*, 1977, vol. C-26, pp. 219–229.

45. D. P. MACADAM, "Digital Image Restoration by Constrained Deconvolution," *J. Opt. Soc. Amer.*, vol. 59, 1969, pp. 748–752.

46. A. A. SAWCHUK, "Space-variant Image Motion Degradation and Restoration," *Proc. IEEE,* vol. 60, 1972, pp. 854–961.

47. G. M. ROBBINS and T. S. HUANG, "Inverse Filtering for Linear Shift-variant Imaging Systems," *Proc. IEEE,* vol. 60, 1972, pp. 862–871.

48. J. L. HORNER, "Optical Restoration of Images Blurred by Atmospheric Turbulence Using Optimal Filter Theory," *Appl. Opt.,* vol. 9, 1970, pp. 167–171.

49. K. KNOX, "Diffraction-limited Imaging with Astronomical Telescopes," *Proc. Intern. Opt. Computing Conf.,* IEEE/SPIE, Washington, D.C., April 1975, pp. 94–97.

50. R. M. MERSEREAU and A. V. OPPENHEIM, "Digital Reconstruction of Multidimensional Signals from Their Projections," *Proc. IEEE,* vol. 62, 1974, pp. 1319–1338.

51. H. C. ANDREWS, A. G. Tescher, and R. P. KRUGER, "Image Processing by Digital Computer," *IEEE Spectrum,* vol. 9, July 1972, pp. 20–32.

52. Comtal Corp., Product Information Literature, Series 5000 Display System, Pasadena, Calif., 1974.

5

APPLICATIONS OF DIGITAL SIGNAL PROCESSING TO RADAR

J. H. McClellan

Massachusetts Institute of Technology
Cambridge, Mass. 02139

and

R. J. Purdy

M. I. T. Lincoln Laboratory
Lexington, Mass. 02173

5.1 INTRODUCTION

Radar technology has evolved over the past several decades to become a mature and quite specialized field. It is therefore perhaps somewhat surprising that the last several years have seen substantial activity in the design of new systems and in the modification of old equipment. This is due in no small part to the recent widespread application of high-speed digital logic to these systems. Digital technology has aroused this interest for a variety of reasons: (1) it is remarkably flexible in allowing a vast number of functions to be performed with a finite catalog of components; (2) very high precisions can be obtained simply by increasing word lengths; (3) high-speed digital components can now acheive processing times compatible with the stressing requirements of radar systems; and (4) digital systems are becoming increasingly more economic. In this chapter, we shall first discuss some general principles of radar signal processors and then examine some specific recent digital systems.

Large radar systems have the general structure shown in Fig. 5-1. These systems consist of a data processing system (DPS) which provides overall system control and decision making, a control unit which decodes commands and provides the appropriate system timing, a signal generator, an array for transmitting and receiving radio frequency (RF) energy, and a signal processor that provides matched filtering for the received waveforms, generates the necessary target metrics, and in general, reduces the data volume and rate to values that are appropriate for the DPS.

The earliest radars contained analog equipment throughout all stages of their systems. (The "DPS" was a man in front of an A-scope.) However, with the growth in size and complexity of radar systems, the inclusion of a large digital computer for system command and control became essential. Thus radar systems inevitably began to require the conversion from analog to digital information at some point in the system. At first, the flexibility and versatility of digital processing was most needed in the final signal processing areas where speed limitations were the least stressing, since most of the data reduction had already been accomplished. However, as radar signal designs and processing requirements continued to become more sophisticated, the need for the versatility of digital techniques in front-end signal processing grew. At the same time, higher-bandwidth signals were being used and this drove signal processors in the very difficult direction of high speed and high flexibility. However, with the advent of fast, increasingly economic integrated digital circuitry, it is becoming more and more feasible to achieve this speed and flexibility throughout the entire signal processor. Thus the analog-to-digital interface in the overall radar system now lies at the receiver-signal processor boundary.

5.1.1 Brief History

The physical principle underlying the operation of a radar system was first demonstrated by Hertz in the 1880s. His experiments proved that electromagnetic waves are reflected by metallic objects just as light waves are reflected by mirrors. Although a practical device was demonstrated and patented by a German engineer, Christian Hulsmeyer, in 1904 for the prevention of ship collisions, the development of operational radar sets did not begin in earnest until the 1930s. During this period, the military implications of radar for aircraft location under conditions of limited visibility became important. Work was undertaken in Germany, France, Great Britain, and the United States; the most renowned of these efforts was that of the British. The chain home stations, based on a system proposed by Robert Watson-Watt in 1935, were instrumental in the defense of Britain during the blitz of

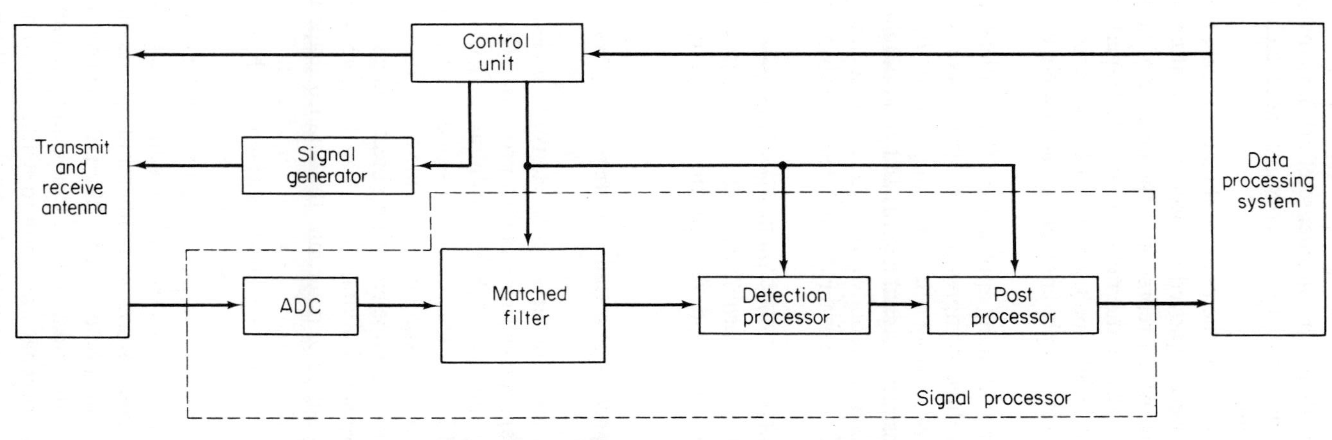

FIGURE 5-1. Block diagram of a large modern radar system.

1940. During the war, there was an intense effort in many countries which resulted in the development of new microwave radar technology and new signal processing techniques.

In the 1950s the nature of the military threat changed, and more sophisticated radar systems were deployed to meet the new situation. An example was the SAGE System, which represented the first use of large computers to replace man as the brains of the radar system. High-performance aircraft and missiles had made such sophistication necessary, because the human operator could not react quickly enough nor accurately enough in such an environment.

The 1950s also saw continuing application of sophisticated signal processing to radar systems. Pulse compression processing was developed at Bell Laboratory[(22)]. This processing allowed signals to simultaneously provide high energy (long transmitted signal length) and good ranging characteristics (short compressed length). Woodward's study of the ambiguity function provided great insight into the general signal design problem and led to much research in this area.

In recent years, radar has found application in innumerable military systems, as well as in such civilian problems as air traffic control, ship navigation, weather mapping, and, of course, the detection of speeding automobiles by police.

5.1.2 Basic Components of a Radar System

All radar systems measure target characteristics, including position, by transmitting electromagnetic energy and then processing the reflected energy. In general, obtaining knowledge about an illuminated object requires that the transmitted signal have "structure". Consequently, the waveform design and the resultant signal processing may become quite complex. For example, it is straightforward to determine either an object's range or its velocity, but it is difficult to measure both simultaneously. Furthermore, very sophisticated processing might be required to determine the detailed shape of a target.

A large modern radar system, shown in Fig. 5-1, has many subsystems. In particular, it contains a signal generator (which may be digital), a modulator to convert the signal to the proper radio frequency (RF) and amplify it, an antenna to transmit the RF energy toward the target, an antenna to receive the RF energy reflected from the target, a demodulator to convert the received signal to baseband, a signal processor to filter and operate on the signal, and, finally, a large general-purpose computer (the DPS) to coordinate the activities of all parts of the entire radar system. Commands emanate from the DPS to tell the signal generator what waveform to synthesize, to position the radar antenna, and to specify the signal processing algorithm to be used. To close the loop, the DPS receives the processed data from the signal processor and performs calculations of its own (e.g., Kalman filtering for tracking, system resource allocation, and radar scheduling). This processing by the DPS is vital to the system's primary mission of making decisions about the objects under scrutiny. However, the scope of this activity is sufficiently large and disjoint from that of the signal processor that it will not be considered further in this chapter.

Signal generation is an important companion to signal processing, since this subsystem provides both the transmitted waveforms and test signals used by the processor. In this chapter, we shall discuss a structure for an all-digital signal generator. A digital approach again offers flexibility, since the radar system may require a wide variety of waveforms, depending upon the target environment.

5.1.3 Signal-processing Functions

In general, there are three major signal-processing functions in modern radar: (1) signal compression (matched filtering), (2) data-rate reduction (thresholding), and (3) generation of target metrics (estimation of target position in range, angle, and velocity). When the processing is done digitally, it is necessary to have an analog prefilter and an analog-to-digital converter (ADC) at the radar-signal processor interface. The functional breakdown of the digital signal processor is shown within Fig. 5-1. It is important to emphasize that the signal processor is ultimately a bandwidth compression device. It receives data at a high rate (e.g., the bandwidth of radar signals, perhaps as high as 10 to 100 megahertz, (MHz)), and processes the signal in such a manner that a relatively low data rate can be maintained between the signal processor and the DPS. The data at this point, however, are highly concentrated since they contain only the essential characteristics of the target.

5.2 PARAMETERS MEASURED BY RADAR

We shall now turn our attention to a more detailed discussion of some of the parameters that can be measured by a radar. These include the target's angular direction (azimuth and elevation), range, velocity, and reflectivity (cross section). To perform any of the measurements, it is necessary to illuminate the target with sufficient energy so that there is some minimum reflected signal energy to process. The famous radar equation (of which there are many forms) summarizes the relationship among the transmitted power (P_T), received power (P_R), range (R), antenna gains, and system losses of the radar. The basic form of the equation is as follows:

$$P_R = \frac{KP_T}{R^4} \tag{5.1}$$

The constant of proportionality accounts for antenna gains, target cross section, and the like. Since the received power is inversely proportional to the fourth power of the range, a sixteen-fold increase in transmitted power is necessary to double the range capability of the system. It is this dramatic dependence of transmitted power on range that motivated the development of pulse compression by matched filtering.

5.2.1 Angle

The most straightforward method for determining an object's angular position is the radiation of an extremely narrow antenna beam. The beam width of an electromagnetic wave is proportional to the wavelength λ and inversely proportional to the physical antenna length L.

$$\text{Beam width} \propto \frac{\lambda}{L} \tag{5.2}$$

Thus a high-frequency radar or a large antenna is required for good angular resolution. In more sophisticated schemes, however, signal processing is employed to increase the angular resolution as, for example, in focused synthetic aperture radars which are developed later in this chapter.

5.2.2 Range

The measurement of range in a radar is equivalent to the measurement of time. Consider the transmission of an extremely narrow pulse at time $t = 0$. If the object lies at a range R, then an echo from the target will be received at $t = T$, corresponding to the two-way travel time to the target:

$$T = \frac{2R}{c} \tag{5.3}$$

where c is the velocity of light. The range resolution R_r is proportional to the time resolution, which in the case of a simple pulse is the pulse width δ.

$$R_r = \frac{c\delta}{2} \tag{5.4}$$

The closer two targets are in range, the narrower the transmitted pulse must be in order to distinguish between the two objects. This presents a problem, however, if the radar transmitter has a peak instantaneous power limit, since the energy in the transmitted pulse is proportional to the pulse width. Thus finer range resolution is accompanied by a decrease in the maximum effective range of the radar when a simple pulse is used for range measurements. One of the triumphs of signal-processing theory is the concept of *pulse compression*, wherein a linearly frequency modulated (LFM) pulse is employed to obtain range resolution with a wideband signal of long time duration. The signal duration provides the energy necessary for long range coverage, while the bandwidth of the waveform is exploited through matched filtering to obtain range resolution inversely proportional to the bandwidth of the signal. Digital implementation of this matched filter by employing the fast Fourier transform (FFT) algorithm for high-speed convolution is a major topic to be discussed at length later in this chapter.

5.2.3 Velocity

The determination of the target's velocity involves the measurement of frequency shift. When a narrowband electromagnetic wave of center frequency f_0 is reflected from an object moving with constant velocity v, the received signal is shifted in frequency by an amount that is proportional to f_0. This frequency shift f_D is known as the Doppler shift and is given by the expression

$$f_D = \frac{2vf_0}{c} = \frac{2v}{\lambda} \tag{5.5}$$

The measurement of the Doppler shift can be accomplished by spectrum analysis and the velocity resolution is determined by the degree to which two signals at different frequencies can be distinguished. Since a very narrow bandwidth is achieved by long-duration signals, the velocity resolution is proportional to the time duration of the transmitted signal. More details of this result are presented in the discussion of the ambiguity function. Finally, notice that the velocity resolution can be improved by increasing the frequency of the RF carrier and thus decreasing the wavelength λ. In the theory developed in Sec. 5.4 and in the systems described later, the usefulness of digital signal processing for computing angle, range, and velocity will become obvious.

5.3 PULSED RADARS

The modern high-performance radar transmits a sequence of individual, perhaps different, waveforms. The timing of events in such a radar is keyed on the interpulse period (IPP) as shown in Fig. 5-2. At time $t = 0$, a signal of

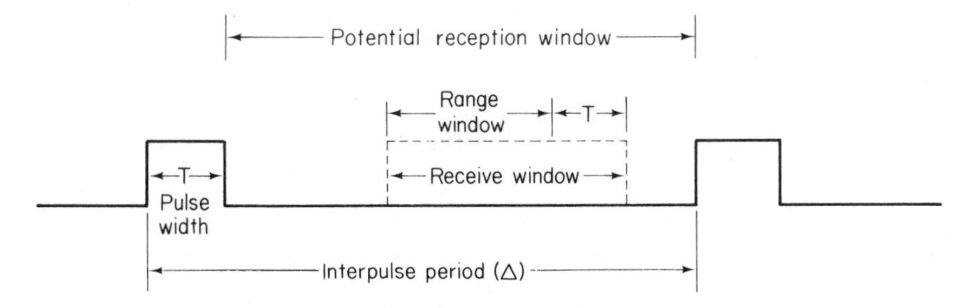

FIGURE 5-2. Timing of events in a pulsed radar.

time duration T is transmitted by the radar. The same signal or perhaps a different one is transmitted Δ seconds later where Δ is the IPP and $1/\Delta$ is the pulse repetition frequency (PRF). In high-performance radars, the IPP may vary with each transmission, since each transmission will have different measurement objectives and/or target range. If the same radar antenna is used for both transmission and reception of signals, it is not possible to start receiving the target echoes until time $t = T$. The potential reception interval extends from $t = T$ to $t = \Delta$ and contains the receive window and range window.* The targets of interest lie in a range interval bounded by R_{min} and R_{max}. The *range window* is the corresponding time interval from $t_1 = 2R_{\text{min}}/c$ to $t_2 = 2R_{\text{max}}/c$, during which echoes will return from the targets. Since pulse compression filtering may be employed in the signal processor, it is necessary to receive data for a time duration equal to the range window plus the signal length (T). This time interval is referred to as the *receive window*. Figure 5.2 summarizes the relationships between these time windows. The receive window is an important signal-processing parameter, especially in digital systems. Basically, the receive window tells us how much valid data are to be processed during one IPP and this dictates both the storage capacity and the computational speed of the digital signal processor.

5.3.1 Ambiguities

It is important to note that multiple transmissions with a constant IPP give rise to ambiguities in the measurement of range by the radar. Echoes from targets at ranges greater than $R_1 = c\Delta/2$ will not return during the same IPP as the transmitted signal and may, therefore, lead to ambiguous measurement of range. For example, a target at the range $R = (c/2)(2T + \Delta)$ will appear at time $t = 2T + \Delta$ in response to a pulse transmission at $t = 0$, whereas a target at $R = cT$ will also show up at $t = 2T + \Delta$ owing to transmission at $t = \Delta$. One can define an unambiguous range interval, located between any two pulses, for the measurement of range. It is also possible to work around this ambiguity problem by varying the PRF to change the relative arrival times of the ambiguous responses. Simple algorithms can then be employed to estimate the true response.

In our previous discussion of velocity measurement by Doppler frequency shift, we tacitly assumed that the signal was a continuous sinusoidal wave. Now suppose that a pulse train modulates the RF carrier of frequency f_0. Furthermore, assume that the phase of the transmitted signal is zero at the beginning of each subpulse, and that the range to the target remains constant (relative to the radar's range resolution) over many IPPs. What we shall show is that the pulsed radar samples a sinusoid whose frequency is the

* Sophisticated scheduling is sometimes employed by interleaving the transmission and reception of multiple pulses.

Doppler frequency and, hence, the measurement of velocity is just discrete-time spectrum analysis of a continuous time signal. Note that the discrete-time nature of the spectrum analysis is independent of the type of signal processor, whether analog or digital. The signal $s(t) = p(t)\, e^{j2\pi f_0 t}$ is transmitted starting at $t = 0$ and a response from the moving target located at range R_0 is received at time $t = T_0$ and has the form

$$s(t - T_0) = p(t - T_0)e^{j2\pi f_0(t-T_0)}.$$

When the pulse is transmitted again at $t = \Delta$, the target will have moved in range by an amount $v\Delta$; if we examine the phase at $t = T_0 + \Delta$ we will find it has changed from the phase at $t = T_0$. During this second IPP the received signal is $r(t) = s(t - \Delta - (T_0 - 2v\Delta/c))$. At $t = T_0 + \Delta$, we obtain

$$r(T_0 + \Delta) = p(2v\Delta/c)e^{j2\pi f_0(2v\Delta/c)} \tag{5.6}$$

while the received signal at $t = T_0$ is

$$r(T_0) = s(T_0 - T_0) = s(0) = 1 \tag{5.7}$$

In general, the response at $t = T_0 + n\Delta$ will be

$$r(T_0 + n\Delta) = e^{j4\pi nv\Delta/\lambda} \tag{5.8}$$

Thus $r(T_0 + n\Delta)$ represents samples at $t = n\Delta$ of a sinusoid at the Doppler frequency $f_D = 2v/\lambda$. The measurement of f_D can be accomplished by saving up N of these samples and then calculating the discrete Fourier transform (perhaps by employing the FFT). Since the samples form a discrete-time signal, the frequency spectrum will exhibit a periodic structure. The spectrum repeats when $f_D = 1/\Delta$. Thus the measurement of velocity by pulsed radar is also subject to ambiguities. Again one might attempt to resolve these velocity ambiguities by varying the PRF of the pulse train.

5.3.2 Pipelined Computation

A radar that transmits a rapid fire sequence of signals lends itself quite naturally to a pipelined architecture for signal processing. Recall from Fig. 5-1 that our canonic radar system has a digital signal processor with three major subsystems: (1) a digital matched filter, (2) a detection processor, and (3) a postprocessor. The data flow sequentially through these three units. One possible way of organizing the computation is to do the entire processing job on all the data from one receive window before data from the next receive window begin to arrive. However, each subsystem of the signal processor would be idle one third of the time on the average. The solution to this problem is to pipeline the computations. If we assume for the moment that all three subsystems take an equal amount of time to process one data set (i.e., the data from one receive window), then the signal processing system is pipelined by inserting storage buffers between the subsystems so that each subsystem can process a different data set in one IPP.* All three subsystems can then be made to work in parallel. The schedule of computations is shown in Fig. 5-3.

The radar transmits its first signal P0 and subsequently good data arrive at the receiver during the time period RW0 (*receive window* 0). During RW0, the received signal is stored in a high-speed buffer memory BM1. The input port of BM1 must be capable of handling a data rate that, according to the Nyquist criterion, is at least twice the highest frequency present in the baseband radar signal. A medium-bandwidth radar signal may have a signal bandwidth of 10 to 60 MHz. With present technology, the implementation of BM1 might require the multiplexing of slower memories and a complicated memory control to achieve the proper rate. However, this performance can be achieved.

At the end of RW0, the matched filter will begin processing the first data set. The signal is read out of the memory BM1 at a rate consistent with the processing speed of the matched filter. Note that the input sampling rate and the matched-filter clock rate are independent. The matched filter, of course, must run fast enough to process all the data from RW0 within one IPP. Interestingly, depending on the clock rate, the matched filter may process the data in a time shorter than the signal duration (i.e., faster than real time). Note BM1 (and all the buffer memories) must be double buffered, since the matched filter will be taking data from the memory during the time interval marked MF0, and, at the same time, BM1 will be receiving new data for RW1. The outputs from the matched filter are stored in BM2 for subsequent use by the detection processor. At the end of receive window RW1, only one element in the processing chain is in operation, because the pipeline has yet to fill up with data. At this instant of time, we are ready to activate the detection processor to work on the data set RW0. The matched filter will process data set RW1 during the epoch MF1; the input will assemble the data set RW2. Finally, at the end of RW2, all three processing units are working in parallel on three different data sets and the pipeline is full.

Thus three IPPs elapse during the processing of one data set, but sufficient throughput is maintained for real-time operation. That is, one unprocessed data set is presented to the signal processor during one IPP and one processed data set is transferred to the DPS. This segmentation of the computational load of the signal processor allows the design of the subprocessors to be done individually. For example, one can minimize the hardware complexity of the matched filter subject to the constraint that the filtering operation be completed in less than one IPP.

The interaction of the three subsystems is embodied in the buffer memories

* In some systems the detection processor is not a separate unit in the pipeline, but rather is lumped together with the matched filter.

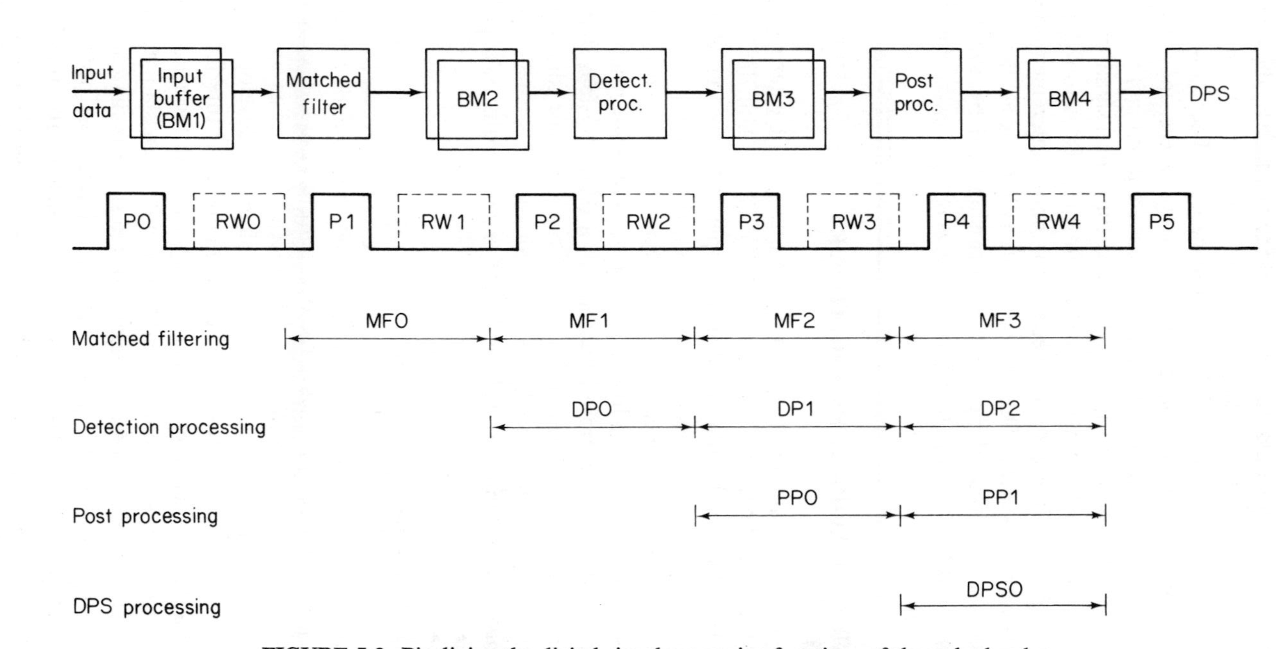

FIGURE 5-3. Pipelining the digital signal processing functions of the pulsed radar.

that provide the data interface at each stage. The implication of Fig. 5-3 is that each buffer memory would be sized to hold two data sets, the one being fed into the memory by the processor preceding the memory, and the one being worked on by the processor following the memory. There are situations (e.g., synthetic aperture radar and burst processing) where one must store data over many IPPs prior to integration or coherent summing. This will increase the amount of buffer memory required in the digital signal processor. It will become increasingly obvious in the following discussion that large amounts of high-speed memory are needed for both buffer memories and internal processor memories. This is a characteristic of radar digital processors, but the gains in smoothing out the processing load by allowing the use of the entire IPP for computation often outweigh the extra complexity of these memories.

5.4 RADAR SIGNAL-PROCESSING THEORY

An important issue in any radar system is the question of what signals the radar will generate and process. The study of radar signals using the ambiguity function of Woodward[45] is an excellent example of the power of Fourier transform theory. In the discussion that follows, the ambiguity function is developed for the discrete-time case, since the continuous-time case has been treated elsewhere.[7,9] We would like to point out, as an aside, that the problem of signal design may be more important in a digital system than in an analog system, because the flexibility of a digital processor allows one to implement virtually any waveform, whereas analog systems usually* synthesize and process fixed, simple waveforms such as continuous-wave (CW) or chirp pulses.

5.4.1 Matched-Filter Theory

The simple notion of transmitting a signal and measuring the round trip delay to the target underlies all pulsed radars. However, the $1/R^4$ dependence of received power on range means that the received signal is usually very faint. Early radar designers recognized this problem and developed ways of improving the received signal-to-noise ratio by filtering out interfering noise. The matched filter[27] is the filter which optimizes the peak received signal-to-noise power ratio in the presence of additive white Gaussian noise. Subsequently, other workers in the field demonstrated that the matched filter is the optimum processor in the sense that it maximizes the probability of detecting the target.[46] This latter criterion is much more appealing and fundamental, since it deals directly with one objective of the radar system; detection. In this section, we explore the application of digital filtering to matched filters. As one might imagine, when filtering sampled data, there are some interesting differences.

The classical tool for studying the behaviour of a matched filter when the received signal is delayed and Doppler shifted is the well-known radar ambiguity function [7,9,36]. The derivation presented here follows closely that of Blankenship and Hofstetter[3].

If the transmitted signal has a complex envelope denoted by $s(t)$, then the received signal has a complex envelope

$$r(t) = s(t - \tau)e^{-j2\pi f(t-\tau)} \tag{5.9}$$

where τ is the relative delay and f the relative Doppler shift. For the moment we asume that the received signal is processed by a filter that is exactly matched to the transmitted signal, in which case the impulse response of the matched filter is the time-reversed complex conjugate, $s^*(-t)$, of the transmitted signal (we ignore the finite delay required to make the filter realizable). The output $y(t)$ of the matched filter is then the convolution of $r(t)$ with the matched filter impulse response $s^*(-t)$.

$$y(t) = \int_{-\infty}^{\infty} s^*(-u)s(t - u - \tau)e^{-j2\pi f(t-u-\tau)}\, du \tag{5.10}$$

A change of variables yields

$$y(t) = e^{-j2\pi f(t-\tau)} \int_{-\infty}^{\infty} s^*(v)s(v + (t - \tau))e^{-j2\pi fv}\, dv \tag{5.11}$$

Since the output response is simply delayed by τ, a simpler expression, referenced to the origin is obtained

$$A_a(t, f) \triangleq \tilde{y}(t, f) = \int_{-\infty}^{\infty} s^*(v)s(v + t)e^{-j2\pi fv}\, dv \tag{5.12}$$

where the phase factor has been arbitrarily dropped, since it is the magnitude of the output (the envelope) which is usually studied. The above function is the well-known ambiguity function.† It is interpreted simply as the output time function of a signal filtered by its matched filter.

This function has many interesting and important properties that allow a radar signal designer to predict the performance of a signal in complicated target environments. Many excellent available radar books treat the subject in great detail.[7,36,39]

When the matched filter is implemented digitally, the received signal is first sampled every T_s seconds resulting in the sampled values

$$s(nT_s - \tau)e^{-j2\pi f(nT_s-\tau)},$$

and the unit sample response of the digital matched filter is given by $s^*(-nT_s)$.

* A notable exception is surface acoustic wave (SAW) correlation filters.

† Alternative forms of the ambiguity function exist and differ only in the location of the conjugate and in the sign of f.

Thus the *digital ambiguity function* is easily seen to be

$$A_d(t, f) = \sum_{n=-\infty}^{\infty} s(nT_s + t)s^*(nT_s)e^{-j2\pi f nT_s} \tag{5.13}$$

A very simple relationship exists between the analog and digital ambiguity functions for any given waveform. This relationship is precisely the relationship between the Fourier transforms of an analog signal and its sampled version; that is, if

$$G_a(f) = \int_{-\infty}^{\infty} G_a(t)e^{-j2\pi ft}\, dt \tag{5.14}$$

and if

$$G_d(f) = \sum_{n=-\infty}^{\infty} G_d(nT_s)e^{-j2\pi f nT_s} \tag{5.15}$$

then

$$G_d(f) = \frac{1}{T_s}\sum_{n=-\infty}^{\infty} G_a\left(f + \frac{n}{T_s}\right) \tag{5.16}$$

Comparison of (5.14) and (5.15) with (5.12) and (5.13) reveals immediately the desired relationship between the analog and digital ambiguity functions:

$$A_d(t, f) = \frac{1}{T_s}\sum_{n=-\infty}^{\infty} A_a\left(t, f + \frac{n}{T_s}\right) \tag{5.17}$$

Equation (5.17) is particuarly useful for visualizing the effect of sampling rate on the structure of the digital ambiguity function in terms of known analog ambiguity functions. This point can be illustrated by considering the LFM waveform given by

$$s(t) = p(t)e^{j\pi\frac{Wt^2}{T}} \tag{5.18}$$

where

$$p(t) = \begin{cases} 1 & |t| \leq T \\ 0 & \text{elsewhere} \end{cases} \tag{5.19}$$

and where T denotes the signal duration and W the swept bandwidth. The analog ambiguity function of this waveform is

$$A_a(t, f) = \begin{cases} \dfrac{T\sin\left[\pi(Wt - fT)\left(1 - \frac{|t|}{T}\right)\right]}{\pi(Wt - fT)} & |t| \leq T \\ 0 & |t| > T \end{cases} \tag{5.20}$$

where, as usual, a phase term has been dropped from the expression. The structure of this function can be most readily appreciated by means of an ambiguity contour diagram such as depicted in Fig. 5-4. The shaded area indicates those regions of the t-f plane where $A_a(t, f)$ is substantially greater than zero.

Equation (5.18) can now be used to construct ambiguity diagrams for the digital LFM signal. Two of these are shown in Figs. 5-5(a) and 5-5(b) for

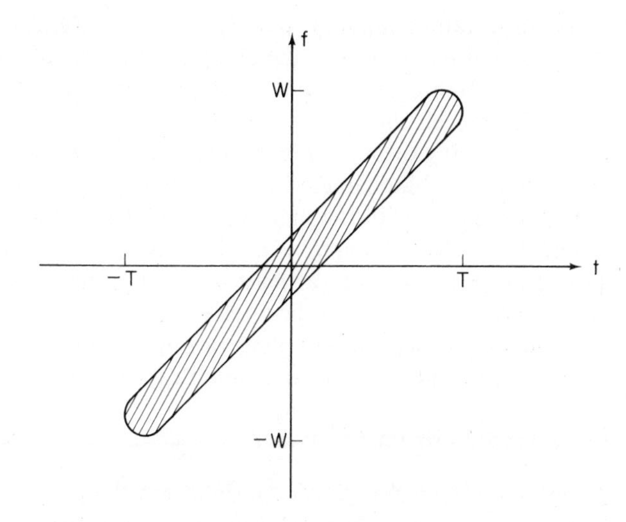

FIGURE 5-4. Contour plot of the LFM ambiguity function (shaded region indicates the region where the ambiguity function is significantly greater than zero).

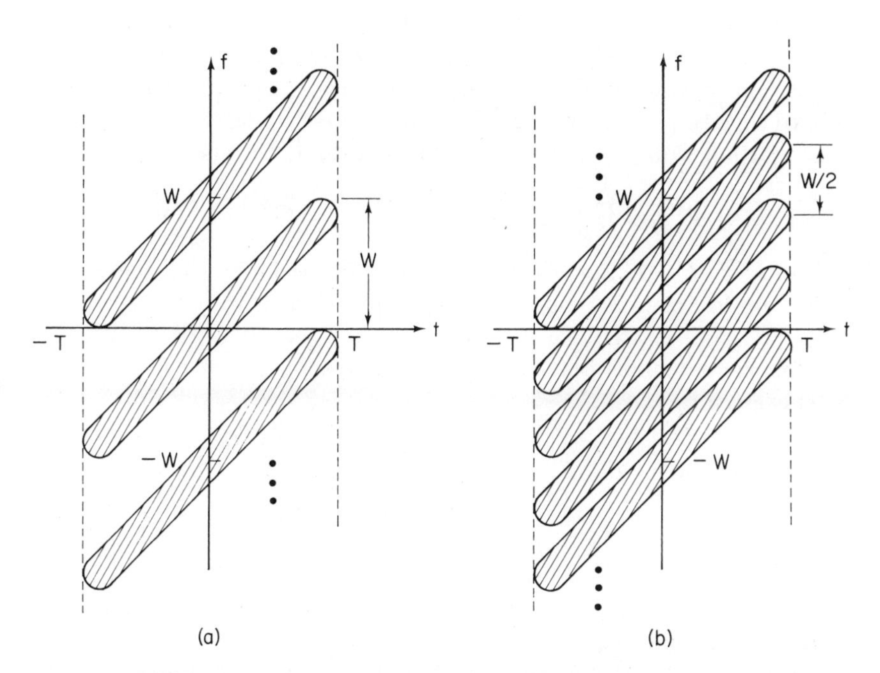

FIGURE 5-5. Contour diagram of the digital LFM ambiguity function: (a) Nyquist rate sampling; (b) Sampling at one-half the Nyquist rate.

$T_s = (W)^{-1}$ (Nyquist rate sampling) and $T_s = (W/2)^{-1}$ (undersampling). Recall that the essential bandwidth of the complex envelope of LFM is W, the swept bandwidth. Hence sampling the complex envelope at baseband, in-phase and quadrature channels, at a rate equal to W for each channel statisfies Nyquist's sampling theorem. Figure 5-5(b) shows clearly that undersampling leads to unacceptable characteristics in that large spurious responses at $t = \pm(T/2)$ appear along with the main response at $t = 0$. On the other hand, Fig. 5-5(a) shows that sampling at the Nyquist rate yields acceptable behavior of the ambiguity function, at least along the time axis. Spurious responses start to appear at $t = \pm T$ if the received signal has any appreciable Doppler shift, and in practice the sampling rate would most likely be set equal to the signal bandwidth plus the maximum expected Doppler shift.

5.4.2. Compressed Pulse for LFM

Although ambiguity diagrams provide insight into the gross behavior of the digital ambiguity function of the LFM waveform, it is instructive to derive an exact mathematical expression for this function. This is done using Eq. (5.12), rather than making use of Eqs. (5.17) and (5.20) because the mathematics involved is somewhat simpler in the former case.

Assume that the LFM signal given by Eqs. (5.18) and (5.19) is transmitted and that the received signal is sampled every T/M seconds where M is an integer. This guarantees that there will be exactly M nonzero samples in the received signals. Let $t = (k + \rho)(T/M)$ where k is an integer and $0 < \rho < 1$, is the location within a sample cell. Equation (5.12) becomes

$$A_d\left(t = (k + \rho)\frac{T}{M}, f\right) = \sum_{n=0}^{M-1-k} e^{j\pi\frac{W}{T}(n+k+\rho)^2\left(\frac{T}{M}\right)^2} e^{-j\pi\frac{W}{T}\left(\frac{nT}{M}\right)^2} e^{-j2\pi f\left(\frac{nT}{M}\right)} \tag{5.21}$$

for $0 < k < M - 1$. For $-(M - 1) < k < 0$, the lower and upper limits of the sum in Eq. (5.21) must be changed to $-k$ and $M - 1$, respectively. For $|k| > M - 1$, $A_d(t, f) = 0$.

After some tedious but straightforward algebra, the magnitude of Eq. (5.21) is given by

$$\left|A_d\left(t = (k + \rho)\frac{T}{M}, f = \frac{\nu}{T}\right)\right| = \left|\frac{\sin \pi\frac{N}{M}\left(k + \rho - \frac{M}{N}\nu\right)\left(1 - \frac{|k + \rho|}{M}\right)}{\sin \pi\frac{N}{M^2}\left(k + \rho - \frac{M}{N}\nu\right)}\right|, \quad |k| < M \tag{5.22}$$

where $N = TW$ and $\nu = fT$. The phase can also be computed, but is of no interest in the following.

Equation (5.22) is studied conveniently by specific examples. Figures 5-6(a) and 5-6(b) show plots of $|A_d(t, 0)|$ for a LFM signal whose time-bandwidth product is 512. The signal is sampled at its Nyquist rate ($M = N = 512$)

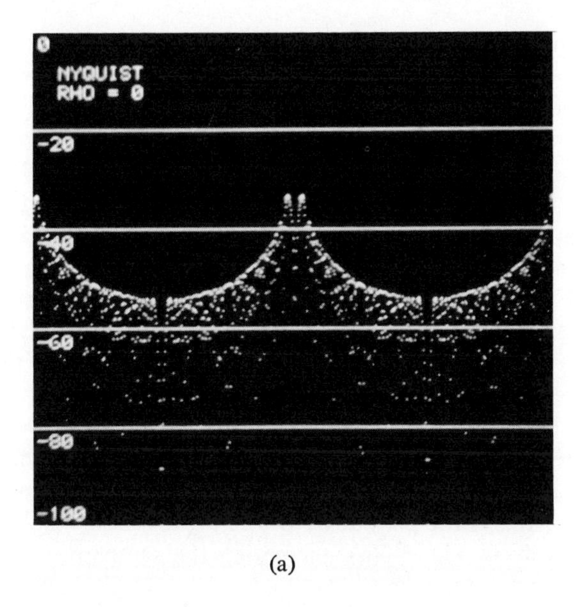

(a)

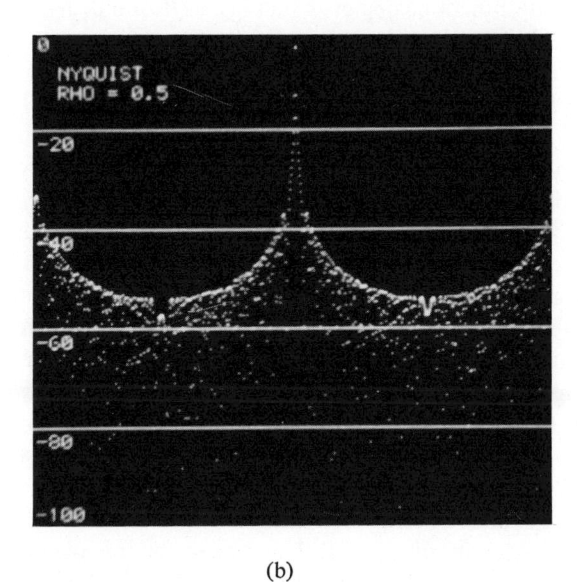

(b)

FIGURE 5-6. Simulation results showing the output of the unweighted LFM matched filter, sampled at the Nyquist rate with a time bandwidth product of 512: (a) Zero delay offset ($\rho = 0$), (b) Half-sample delay offset ($\rho = 0.5$).

and the relative delay between the signal and the matched filter, ρ, is set equal to 0 and 0.5, respectively. These plots were obtained by direct evaluation of Eq. (5.22) using a simulation program that passes the received LFM signal through its matched filter. The central region of Fig. 5-6(a) shows a large response consisting of a single sample value at 0 decibels (dB) surrounded by very low sidelobes near -40 dB. This may seem surprising to those familiar with analog LFM processing, since it is well known that an unweighted matched filter results in a -13 dB sidelobe level. The absence of -13 dB sidelobes in Fig. 5-6(a) is due to the fact that, when $\rho = 0$, the sampling points miss the -13 dB sidelobes and fall close to the zeros of the ambiguity function. When $\rho = 0.5$, as in Fig. 5-6(b), the central region of the ambiguity function does indeed display samples at the familiar -13 dB sidelobe level.

Figures 5-6(a) and 5-6(b) also show a major difference between digital and analog LFM: in the digital case the sidelobes do not fall off uniformly with increasing $|t|$, but increase as $|t|$ nears the ends of the response interval. The reason for this is the periodicity of the digital LFM ambiguity diagram shown previously in Fig. 5-5(a). Increasing the sampling rate reduces this effect as evidenced by Figs. 5-7(a) and 5-7(b) which were computed for $M = 2N = 1024$ (i.e., sampling at twice the Nyquist rate). The behavior of the ambiguity functions depicted in Figs. 5-7(a) and 5-7(b) is essentially indistinguishable from that of analog LFM.

To complete this study of the effect of the sampling rate on the ambiguity function, examine Figs. 5-8(a) and 5-8(b) which were derived for $M = 0.5N = 256$ (i.e., undersampling at one half the Nyquist rate). The spurious responses predicted in Fig. 5-7(b) are now clearly evident.

The sidelobe level of the LFM ambiguity function can be reduced by using a mismatched filter with a frequency response obtained by multiplying the frequency response of the matched filter by a weighting function, such as the Hamming weighting, $W(f) = 0.54 + 0.46 \cos(2\pi f)$. Figures 5-9(a) and 5-9(b) show the time axis of the ambiguity function for Nyquist rate sampling when Hamming frequency domain weighting is used at the receiver. The broadened main response lobe caused by the weighting is clearly evidenced for $\rho = 0$ by the two large responses on either side of the main response. The near-in sidelobe level is reduced from the -13 dB level of Fig. 5-8(b) to the -42 dB level in Fig. 5-9(b). What is somewhat surprising at first glance, however, is that the Hamming weighting has also succeeded in reducing the sidelobes in the vicinity of $|t| = T$. The reason for this phenomenon is that the sidelobe structure in the vicinity of $|t| = T$ is governed by the aliased versions of the main ambiguity function response and thus is very similar to the sidelobes in the vicinity of $|t| = 0$. Since Hamming weighting is specifically designed to reduce the near-in sidelobes, it is not surprising that it also reduces the sidelobes of similar structure appearing near $|t| = T$.

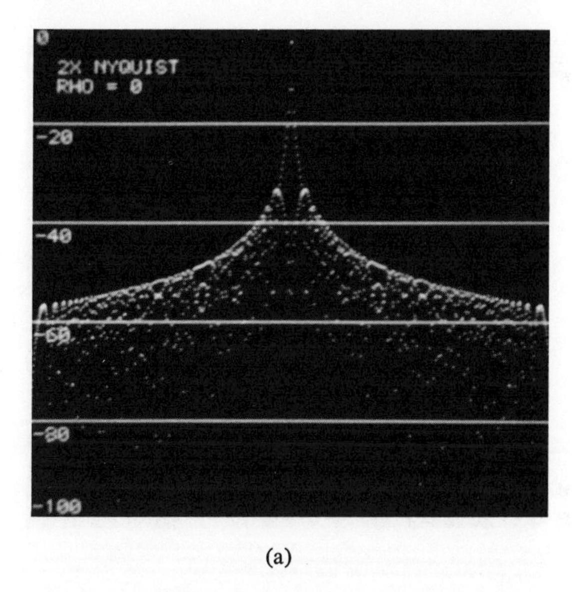

(a)

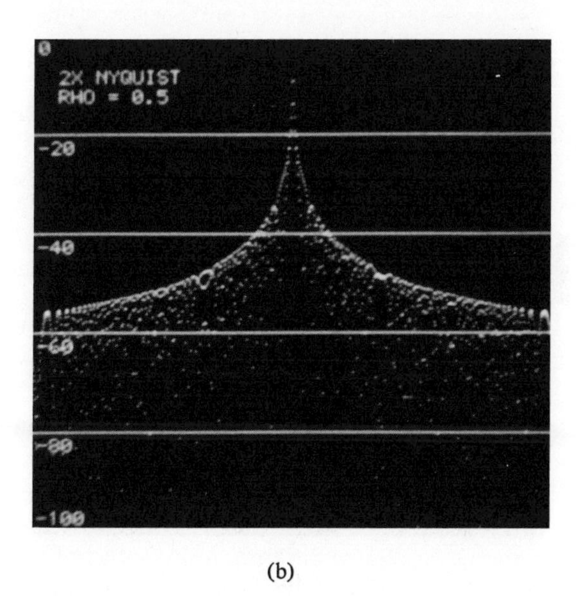

(b)

FIGURE 5-7. Simulation results showing the output of the unweighted LFM matched filter, sampled at two times the Nyquist rate with a time-bandwidth product of 512: (a) Zero delay offset ($\rho = 0$); (b) Half-sample delay offset ($\rho = 0.5$).

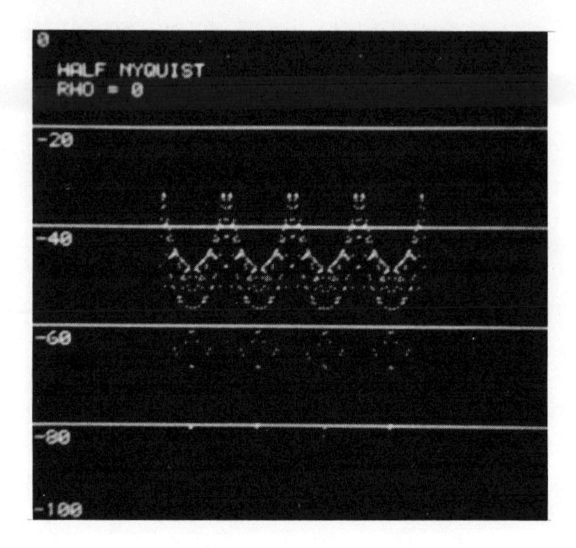

(a)

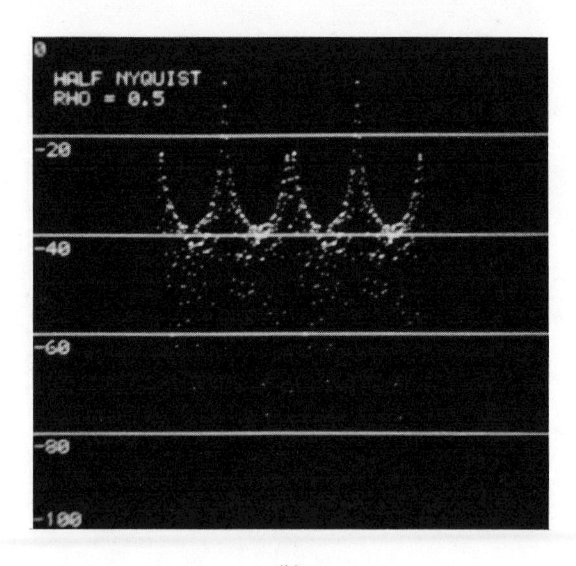

(b)

FIGURE 5-8. Simulation results showing the output of the unweighted LFM matched filter, sampled at one-half the Nyquist rate with a time-bandwidth product of 512: (a) Zero delay offset ($\rho = 0$); (b) Half-sample delay offset ($\rho = 0.5$).

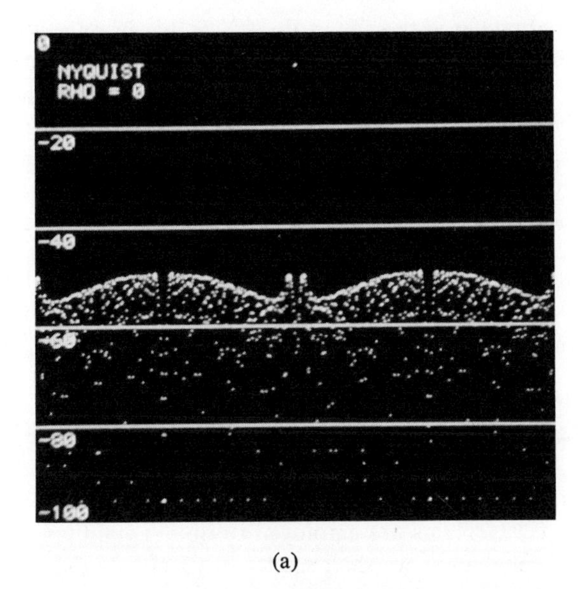

(a)

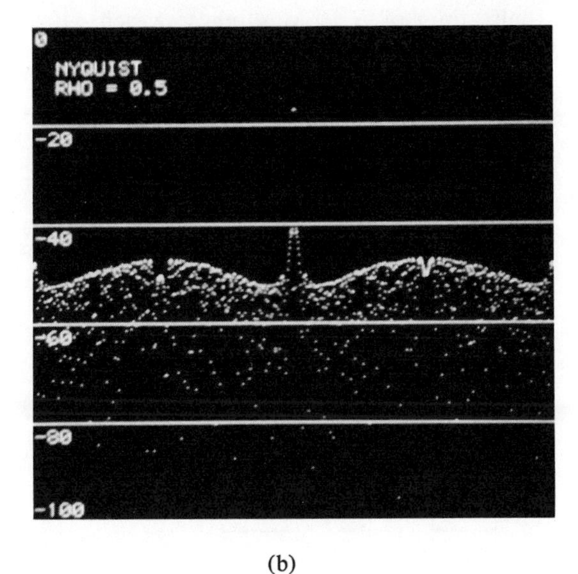

(b)

FIGURE 5-9. Simulation results showing the output of a Hamming weighted LFM matched filter, sampled at the Nyquist rate with a time-bandwidth product of 512: (a) Zero delay offset ($\rho = 0$); (b) Half-sample delay offset ($\rho = 0.5$).

The only parameter appearing in Eq. (5.22) that has not yet been studied is the Doppler shift, $\nu = fT$. It is evident from Eq. (5.22) that the effect of ν should be the same as the effect of ρ, because they appear in the form $\nu + \rho$. This would be true except for the fact that ν is not restricted to lie between 0 and 1, as ρ is. The effect of larger values of ν can easily be predicted from Fig. 5-5(a) which shows that a large positive value of ν should lead to a large spurious response in the vicinity of $t = T$. This effect is not important in most practical LFM systems for which the swept bandwidth is usually many times the expected Doppler shift.

5.4.3 Processing Burst Waveforms

The ambiguity function for an LFM pulse is sometimes described as a "knife-edge" ambiguity function owing to the large response along the line $f = (W/T)t$. As noted above, the effect of a Doppler shift is the same as a time delay (range offset); that is, the LFM pulse exhibits range-Doppler coupling. Burst waveforms are employed to obtain simultaneous range and Doppler measurements as we shall see from a study of the burst ambiguity function.

A uniform pulse train of N LFM subpulses, each with bandwidth W, pulse length T, and spacing Δ is shown in Fig. 5-10. It may be written as

$$s(t) = \sum_{n=0}^{N-1} a_T(t - n\,\Delta) \tag{5.23}$$

where $a_T(t) = Be^{j\pi(W/T)t^2}$ for $0 < t < T$. The analog ambiguity function* is given by

$$|A(p\,\Delta + t', f)| = (N - |p|)E_a \left|\frac{A_a(t', f)}{E_a}\right| \cdot \left|\frac{\sin \pi f(N - |p|)\,\Delta}{(N - |p|)\sin \pi f\,\Delta}\right| \tag{5.24}$$

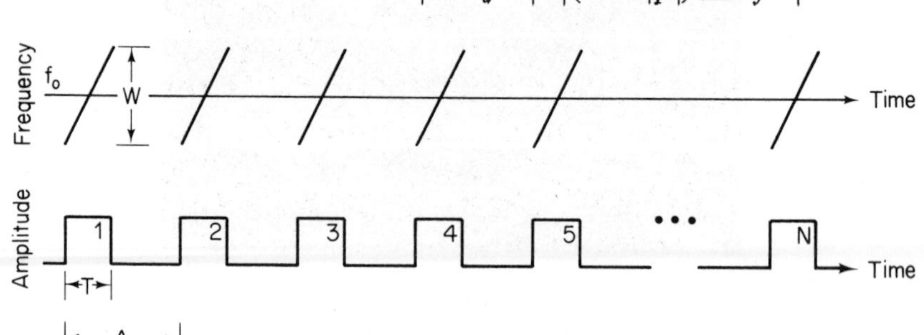

FIGURE 5-10. Uniform burst containing identical LFM subpulses.

* Assuming that the internal burst duty cycle is less than 50%.

where p is an integer denoting the pth range strip, $E_a = B^2T =$ subpulse energy, and the subpulse ambiguity function $A_a(t, f)$ is given by

$$|A_a(t', f)| = E_a\left(1 - \frac{|t'|}{T}\right)\left|\frac{\sin \pi WT\left(\frac{t'}{T} - \frac{f}{W}\right)\left(1 - \frac{|t'|}{T}\right)}{\pi WT\left(\frac{t'}{T} - \frac{f}{W}\right)\left(1 - \frac{|t'|}{T}\right)}\right| \tag{5.25}$$

A contour plot of this function, illustrating the peaks is shown in Fig. 5-11. The uniform burst ambiguity function is the well-known bed of nails with a tilt due to the LFM subpulse. Note that the total extent in frequency is less than $\pm W$, since the signal decorrelates in frequency for Doppler shifts larger than W. Such shifts far exceed physical reality. The burst waveform is typically designed so that the Doppler shift is less than one ambiguous Doppler interval. In addition, there are TW Doppler ambiguities on the tilted ridge, and from Eq. (5.25) the amplitude falls off linearly over the major portion of the ambiguity ridge.

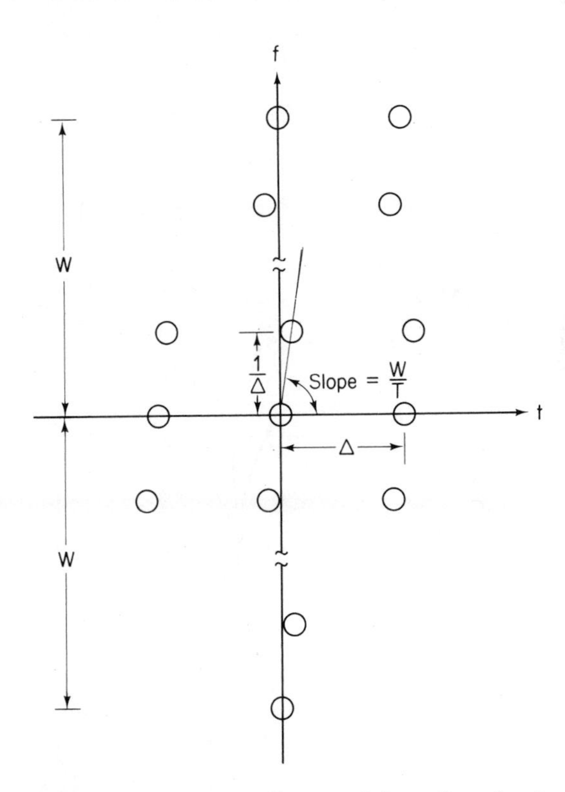

FIGURE 5-11. Contour diagram of the uniform burst ambiguity function.

The effect of sampling the waveform and processing it digitally is analyzed using Eq. (5.17). Sampling at the Nyquist rate ($T_s = 1/W$) requires that the ambiguity diagram of Fig. 5-11 be repeated and added, as shown in Fig. 5-12. Thus, for a filled burst ($\Delta = T$), the extreme Doppler ambiguities (at $f = \pm W$) fold onto the main response. However, since they are virtually zero, the mainlobe is unchanged. Similarly, the second extreme ambiguities (at $f = \pm(W - 1/\Delta)$) fold onto the first Doppler ambiguity. Although the second extreme ambiguities are nonzero, the first Doppler ambiguity is less than the main ambiguity, and the sum is approximately constant. Note that for a nonfilled burst the folded Doppler ambiguities are not aligned and additional range peaks should be observable. Thus Eq. (5.17) predicts that the process of sampling the waveform and providing digital matched filtering leaves the range axis and the first ambiguous Doppler interval relatively unaffected. However, succeeding Doppler intervals should begin to show some small degradation when the burst is not filled.

The *bed of nails*[9] ambiguity function shows the utility of burst wave-

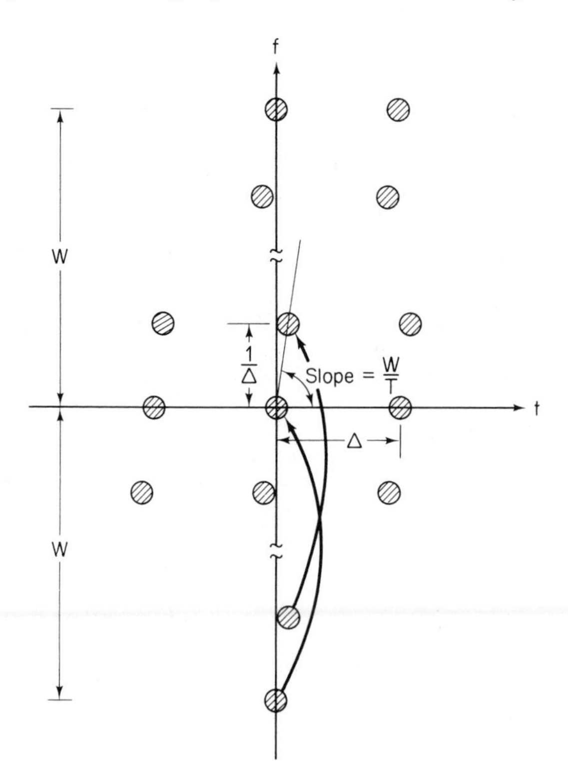

FIGURE 5-12. Contour diagram of the ambiguity function of a filled uniform burst to illustrate periodicity along the Doppler axis.

forms for simultaneous resolution in both range and velocity. The price for this capability, however, is the existence of ambiguous responses in both velocity and range. The ambiguous velocity responses many times can be placed (by suitably selecting waveform parameters) outside the region of anticipated velocities. However, the ambiguous range responses often cannot be placed outside the anticipated range of targets. This is a drawback of uniform bursts which causes problems and requires subsequent processing. However, the ability to simultaneously measure range and velocity, and to suppress returns from targets at uninteresting velocities, is often important enough to warrant use of uniform bursts in spite of the ambiguity problem.

It is possible to design bursts in which the subpulses are nonuniformly spaced. Such bursts yield a third type of ambiguity function, usually called a *thumbtack*[9] ambiguity function. Figure 5-13 shows the general structure

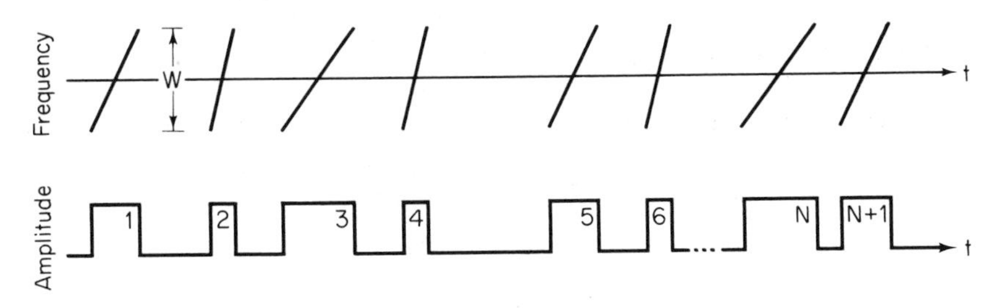

FIGURE 5-13. Non-uniform burst containing LFM subpulses with equal swept bandwidths.

for an N pulse burst. The purpose of the nonuniform spacing is to prevent the subpulses from correlating with one another in range. This eliminates all range ambiguities and yields a thumbtack ambiguity function. Such an ambiguity function would seem to be the solution of the radar signal design problem, since it yields simultaneous range and velocity measurement capability with no range ambiguities (as before, the velocity ambiguity generally can be placed outside the region of expected velocities.) However, compared to a uniform burst with the same TW product, this performance is achieved at the expense of strongly degraded performance in heavy distributed clutter.[9] It is important to note that increasing the time-bandwidth product combats this effect.

5.5 MATCHED-FILTER IMPLEMENTATION

We are now ready to begin a detailed discussion of an implementation of a digital matched filter. We shall concentrate on an implementation that employs the fast Fourier transform (FFT) for the realization of high-speed convolution. This is by no means the only way to implement a matched filter (it could be hard wired), but it is a technique that allows vast flexibility in implementing any filter. Since the signals transmitted by the radars considered here are necessarily time limited, the operation of matched filtering is equivalent to filtering with a nonrecursive or finite impulse response (FIR) filter. For example, a LFM signal with time duration T, bandwidth W, and time-bandwidth product TW, when sampled at ϵ times its Nyquist rate, forms a discrete-time signal of length $L = \epsilon TW$. The time-reversed complex conjugate of this signal can be used as the impulse response of the FIR matched filter. A prime motivation for using a digital realization of the matched filter is the flexibility derived from storing the filter's impulse response in a read-write memory. With such a configuration, the filter characteristics can be changed quickly and easily to implement new signals or to weight the frequency response for sidelobe reduction. Finally, one can envision adapting the filter in real time to meet a changing target environment.

5.5.1 High Speed Convolution

A very efficient implementation exists for discrete-time FIR filters, namely high-speed convolution.[41] This method implements the convolution in the frequency domain as shown in Fig. 5-14. The sequences $x(n)$ and $h(n)$ of length

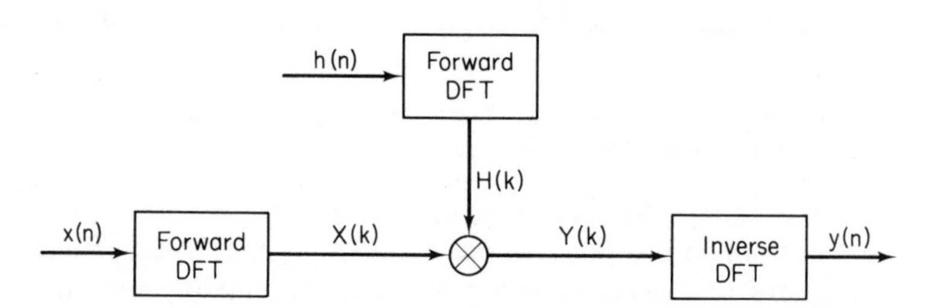

FIGURE 5-14. Block diagram illustrating convolution using the discrete Fourier transform. High-speed implementation results from the use of the FFT algorithm to compute the DFTs.

N are transformed using the discrete Fourier transform (DFT), the transforms $X(k)$ and $H(k)$ are combined by multiplication, and the product $Y(k)$ is inverse transformed to yield the filter's output. The efficiency of the method derives from the use of the FFT algorithm to calculate the DFTs involved. For large N, the savings in computation over direct time domain convolution are dramatic.

Since the convolution implemented using DFTs is circular,[30] each output batch of N points contains only $N - L + 1$ valid output points. (L is the number of samples in the filter.) Interestingly, in a radar with multiple transmissions, we are not usually concerned with continuous matched filtering, but require only one set of filtered data (one range window) per interpulse period (IPP). Thus all the data from one receive window are processed with one pass through the high-speed convolver of Fig. 5-14 and only the good data are saved. If we let R be the size of the range interval, then $M = 2R\epsilon W/c$ samples occur during the time of the range window. Hence the size of the transform required is

$$N = \epsilon TW - 1 + \epsilon W(2R/c) = L + M - 1 \tag{5.26}$$

If we fix the size of the range window and plot curves to show the trade off among time duration, signal bandwidth, and transform size, we obtain Fig. 5-15 for a range window of 30 kilometers (km). As an example, a waveform with a time-bandwidth product of 2048 and a bandwidth of 10 MHz (sampled at its Nyquist rate) requires a length 4096 transform for a 30 km (approximately) range window.

5.5.2 Computational Requirements

The computational savings of the FFT form of high-speed convolution are great when compared to the direct form implementation of an FIR filter for large time-bandwidth product signals. However, the bandwidth of typical radar signals is often 10 to 100 MHz, and this still places a heavy burden on the processing capability of a digital convolver system, even when using the FFT. Since many samples are generated for each IPP, a large amount of processing is required. To examine this point further, consider a radix-2 FFT, where the length of the transform is $N = 2^n$. The basic computational element (CE) of the radix-2 FFT is shown in Fig. 5-16 for the decimation-in-time (DIT) version of the algorithm. The CE (or *butterfly*, as it is often called) consists of one complex multiplication and two complex additions. As a result, the computational complexity of the FFT can be measured by the number of CEs required for a complete transform. For an N point, radix-2 FFT, this number is $(N/2) \log_2 N$, since the algorithm consists of $\log_2 N$ stages and each stage requires $N/2$ butterflies. A glance at Fig. 5-17 should serve to verify these numbers for the case $N = 8$.

The implementation of high-speed convolution with the FFT requires the calculation of two FFTs and the multiplication of two N-point sequences (assuming that the DFT of the filter has been precomputed and stored for use in the reference spectrum multiplication.) Thus the total computation time for one convolution is

$$T_C = (N \log_2 N)T_B + NT_M \tag{5.27}$$

where T_B is the time required for the calculation of one butterfly and T_M is the time required for one complex multiplication. Assuming that these two times are comparable (i.e., T_B is dominated by the complex multiply time), the total time is approximately

$$T_C \cong N(1 + \log_2 N)T_M \tag{5.28}$$

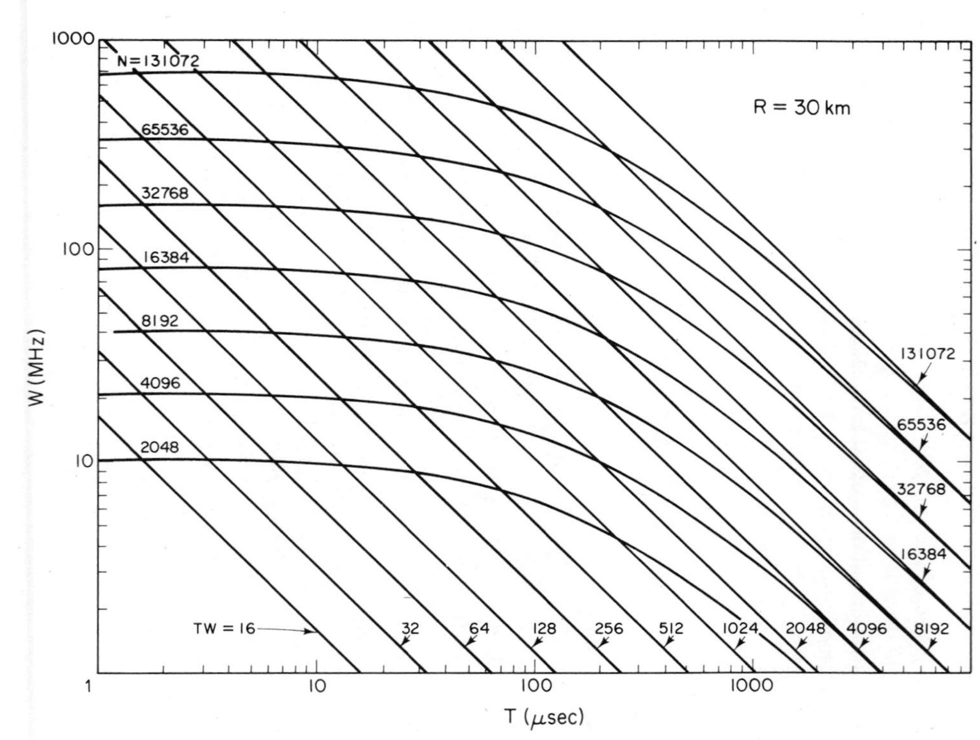

FIGURE 5-15. Transform size ($N = 2^n$) requirement for a 30-km range window and Nyquist rate sampling.

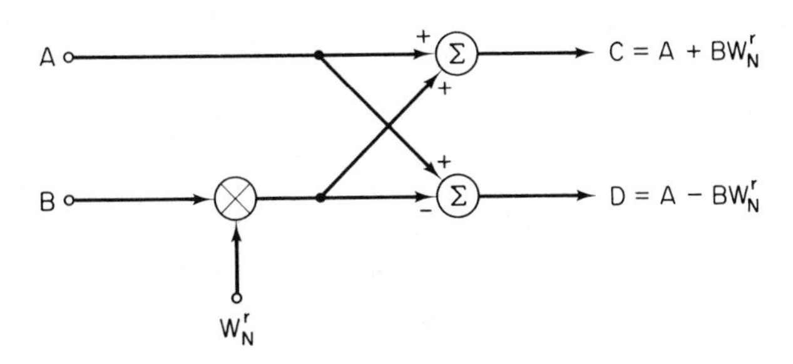

FIGURE 5-16. Computational element (butterfly) of the radix-2, decimation-in-time, FFT algorithm.

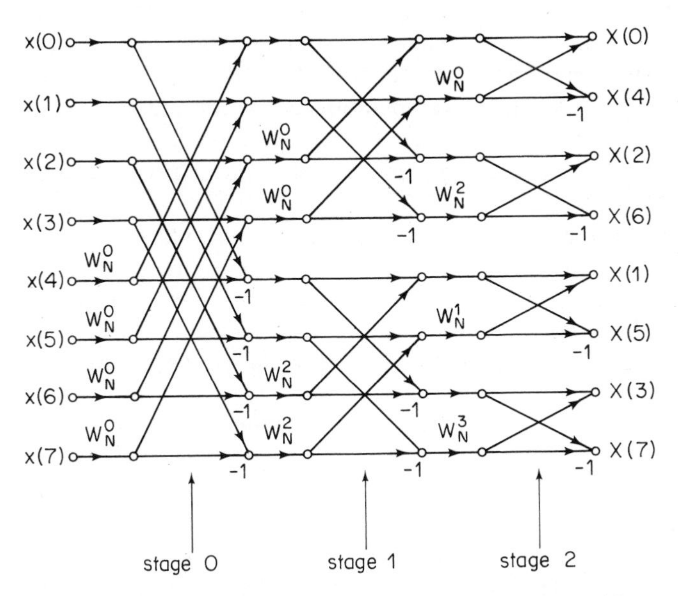

FIGURE 5-17. Eight-point, decimation-in-time, FFT algorithm with normally ordered input data.

The constraint that the convolution be performed entirely within one IPP fixes the minimum multiplication rate. For example, suppose that the PRF of the radar system is 1 kilohertz (kHz) (IPP = 1 millisecond, (ms)) and the transform size is $4096 = 2^{12}$. The requirement of the real-time operation leads to a multiplier speed of approximately 20 nanoseconds (ns), which is difficult to achieve with presently available digital multiplier hardware. Actually, the requirement is even more severe since we have neglected all overhead for data handling. Even though the above calculation suggests that a hardware limit exists to performing high-speed convolution, this is not the case.

The problem comes about because we have implicitly assumed serial computation. Such a computational flow would be typical in a general-purpose computer with an architecture that processes a single stream of instructions. It is natural to ask whether the use of parallel processing might not improve the situation, and indeed this is the case. The pipeline architecture for the FFT[18,29] can substantially improve computation speed using a hardware structure with virtually no overhead for control functions. We next examine this structure in more detail.

5.5.3 Pipeline Architecture

The derivation of the pipeline FFT rests on the observation that at any stage of the FFT it is not necessary to finish the calculation of all the butterflies in that stage before beginning the computation of the butterflies in the succeeding stage(s). In fact, as Fig. 5-17 shows for the case of a radix-2, length 8, FFT one can begin the computation of the top butterfly in stage 1 after the top three butterflies in stage 0 have been completed. Similarly, the top butterfly in stage 2 could be computed after the top two butterflies in stage 1 have been done. Thus a structure with $\log_2 N$ (for the radix-2 case) CEs working in parallel could be arranged to take advantage of this observation.

Note, however, that it is necessary to insert reordering memories in between the separate CEs in order to present the data in the proper sequence to each CE. A more detailed discussion of a pipeline FFT implementation will be undertaken later, but first let us examine the impact of this parallelism on the computation time of a radix-2 FFT.

Consider the calculation of one DFT using a pipeline FFT. At the very best, one could expect a reduction in computation time by a factor of $\log_2 N$. However, since it is not possible to start the butterflies simultaneously, we shall show the total computation time will be reduced by only a factor of $\frac{1}{2}\log_2 N$. When computing multiple DFTs, this start-up time will only be incurred once, and the computation time will be reduced by a factor approaching $\log_2 N$ (see section 5.6.6). Thus the total time for high-speed convolution using a length N transform is:

$$T_C = (N \log_2 N/(1/2 \log_2 N))T_B + NT_M \tag{5.29}$$

Again making the assumption that $T_B \cong T_M$, we obtain

$$T_C \cong 3NT_M \tag{5.30}$$

Using our previous example, the required multiplier speed would be 83 ns, a figure which is within the state of the art with high-speed digital circuitry.

The general arrangement of a radix-r pipeline convolver (DIT algorithm) is indicated in Fig. 5-18. Data is presented to the first stage of the pipeline on each of its r inputs, N/r samples per input port. The forward pipeline is arranged to perform an N-point, radix-r DFT on normally ordered input data. After transforming into the frequency domain, each r-tuple of data is

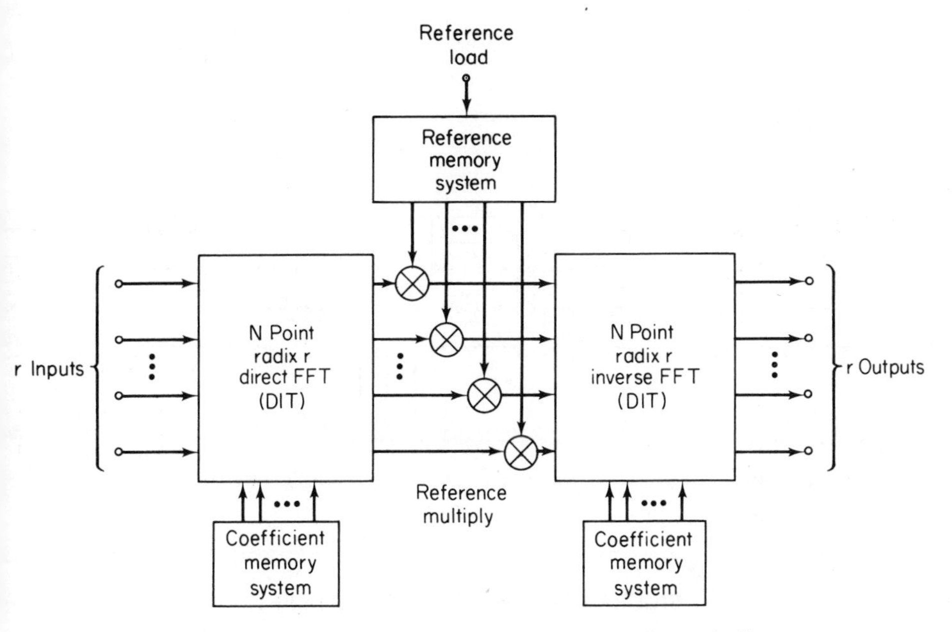

FIGURE 5-18. Digital filtering using cascaded radix *r* pipeline FFTs.

multiplied by the proper sample of the filter frequency response. Subsequently, the product is further processed by a second pipeline arranged to perform an inverse FFT. The data into the inverse transform are digit reversed (in base *r*), but the output will be in correct order. Thus the inverse transform undoes the digit reversing created by the forward transform.

5.6 EXAMPLE OF A DIGITAL RADAR PROCESSING SYSTEM

The design of a radar processing system involves many tradeoffs and the specification of many subsystems. We shall discuss in this section the design of a processor system which was developed at Lincoln Laboratory over the past several years. The major subsystem in this processor is a radix-4 pipelined fast Fourier transform matched filter. We shall first discuss this system.

Detailed block diagrams* for radix-4, 4096-point pipeline transforms are shown in Figs. 5-19 and 5-20. Each transform is comprised of five topologi-

* A system based on these designs is being manufactured by the General Electric Company to a Lincoln Laboratory specification. This system is a radix-4, 16,384-point transform, which is clocked at 30 MHz. Thus a complete 16,384-point circular convolution can be performed each 136 microseconds.

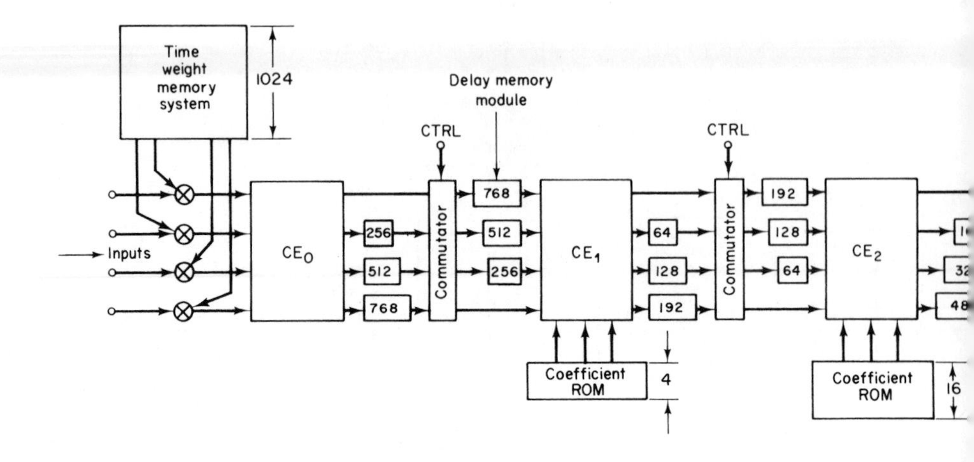

FIGURE 5-19. Forward radix-4 pipeline FFT.

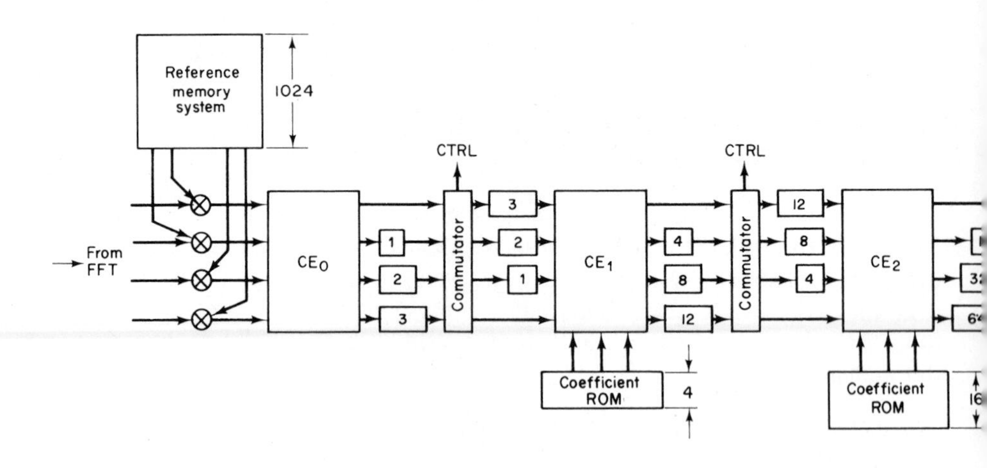

FIGURE 5-20. Inverse radix-4 pipeline FFT.

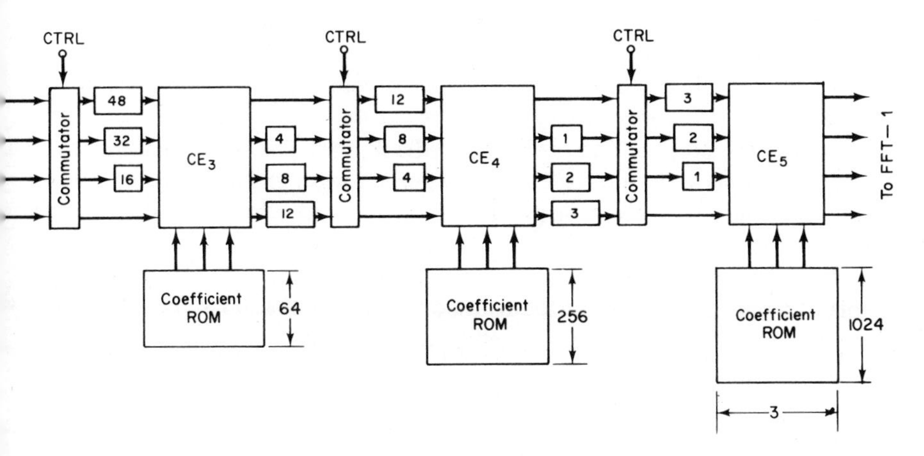

FIGURE 5-19. (*Continued*)

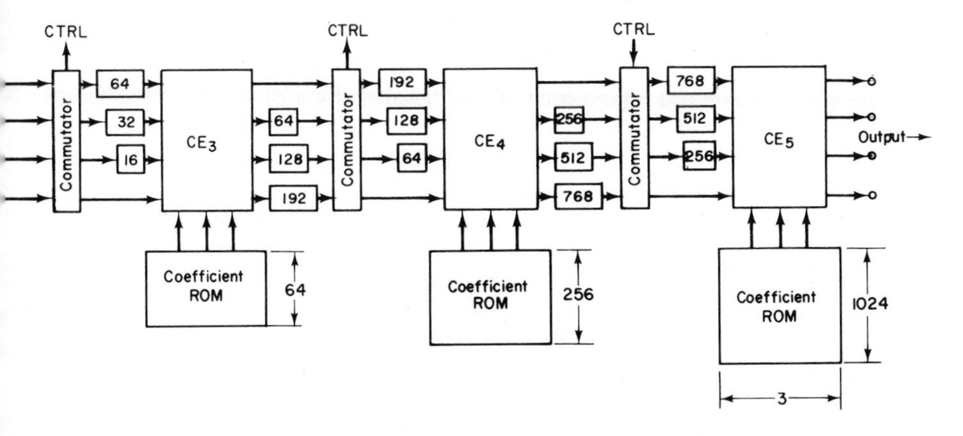

FIGURE 5-20. (*Continued*)

cally similar stages consisting of a temporal skew (delay line) memory system, a computational element (CE), and a rotational coefficient store. Each CE in turn consists of a set of three vector rotational devices and a 4-point DFT matrix. Except for the storage order of rotational coefficients and a minor difference in the DFT matrix wiring, the forward pipeline FFT component stages are identical to those of the inverse pipeline.

The temporal skew memories consist of a commutating switch assembly and a set of shift register delays. The sizes of the requisite delays vary with position in the pipeline and are noted in the figures. The skew memories are necessary to permute the data as they proceed through the transformation so as to ensure that the proper set of samples is seen by each CE at each point in time.

The reference function memory stores the different filters to be implemented. Since the data are digit reversed as they are delivered by the first FFT, the reference memory must provide such addressing.

5.6.1 Computational Conventions and Data Format

The choice of data representation in the digital matched filter is a crucial decision that must be made early in the system design, since it impacts both the hardware complexity and the system accuracy. Some alternatives are fixed point, full floating point, and hybrid floating point; and 2s complement, 1s complement or sign magnitude. The use of fixed point in a pulse compression matched filter requires a large number of bits, since the signal grows by a factor of TW through the filter. Floating point is a more natural choice. In fact, only a limited type of floating point, the hybrid floating point, is required because the exponent will tend to grow only in the positive direction. This data format has been successful because it is easily implemented in hardware and has demonstrated good performance in simulation studies.

In this hybrid-floating-point data format, all arithmetic computations are carried out in full fractional, signed 2s complement representation. Operands are complex and are represented as two fractional mantissas with a common exponent. The specific form is

$$(m_r + jm_i)2^b \tag{5.31}$$

where m_r and m_i are the real and imaginary component mantissas, respectively, and are represented with an equal number of bits. The common exponent, b, is a positive integer whose initial value is defined to be zero for the input signal. By calculating the signal gain, it can be shown that b requires only a few bits (e.g., 5 bits for a 16 K transform). The choice of word lengths for the mantissas is a painstaking process that is best done by simulation. In the following, some processing results for systems using a 9- and 11-bit mantissa are described.

5.6.2 Computational Element Architecture

The architecture of a radix-4 DIT CE is illustrated in Figure 5.21. Information is presented simultaneously on four parallel input ports (or rails), each containing one complex word. At the input to the CE, the mantissa and exponent fields of each datum are separated. The mantissas are fed to the rotational elements for phase modification by the twiddle factors W^k. Note that the upper data path never requires rotation. The rotation process itself may be performed by either complex multiplication or the CORDIC iteration, as discussed later[11,44].

The data exponents are used to perform a format conversion from floating point to fixed point. The exponent information controls a system of four pairs of alignment shifters. One pair of shifters per complex datum is necessary to scale both the real and imaginary mantissas. In this way, the relative weights of the four complex data are adjusted properly for further computations.

Next the aligned complex mantissas are coupled to a 4-point DFT matrix, which consists of two levels of addition (Fig. 5-21). An extra guard bit is added to each mantissa at each level of addition to protect against overflow. Therefore, the data grow by 2 bits in length at the matrix output.

The final process within a CE is simply a conversion back to vector floating-point form. Since there exist three opportunities for overflow in the CE (one in rotation and two in the 4-point DFT), it is necessary to count, starting from the sign bit and working to the right, the number of bits (up to 3) that agree with the sign bit. Three minus this number yields the number of right shifts necessary to scale a given mantissa back into range. The exponent for a given complex datum is then computed as the sum of the largest exponent of the four (as determined earlier in the alignment) and the scale count.

5.6.3 Vector Rotation

The implementation of vector rotation is a large part of the hardware cost in a special purpose FFT. Two techniques will be discussed: complex multiplication and the CORDIC iteration. The latter is of interest because there is a net reduction in the hardware cost of the butterfly.

Since the input data are in Cartesian form, the desired rotations can be implemented simply by performing a complex multiplication with the complex exponential

$$W^k = \cos\left(\frac{2\pi k}{N}\right) + j\sin\left(\frac{2\pi k}{N}\right) \tag{5.32}$$

Multiplication of a complex datum by W^k requires four real multiplications, an addition, and a subtraction. In the high-speed regime of radar signal

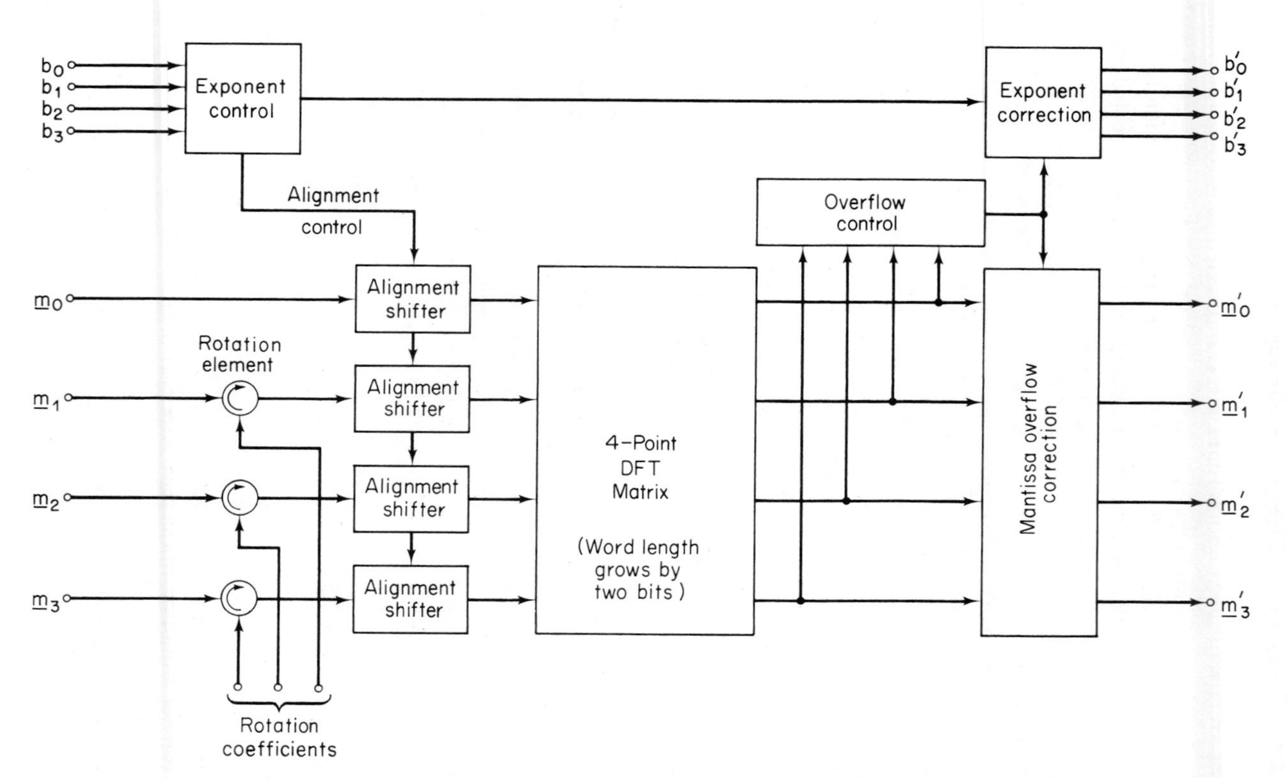

FIGURE 5-21. Typical radix-4 CE with hybrid floating point arithmetic.

processing, the real multipliers would quite likely be implemented using array multipliers[32].

The complex multiplier approach provides more capability than is really necessary, since both the phase and the amplitude of the data can be altered. The FFT algorithm requires only phase modification, and techniques exist which are strictly suited to phase alteration. One such scheme, the CORDIC rotation algorithm, offers some potential hardware savings over the full complex multiplier approach[11,44].

Given a vector of the form $x + jy$ to be rotated by an angle θ, the CORDIC iteration is of the form

$$x_{i+1} = x_i - y_i(\alpha_i 2^{-i}) \tag{5.33a}$$

$$y_{i+1} = y_i + x_i(\alpha_i 2^{-i}) \tag{5.33b}$$

where $x_0 = x$, $y_0 = y$, and $i = 0, \ldots, M - 1$.
The α_i are deduced from a concurrent iteration of the form

$$\xi_{i+1} = \xi_i - \alpha_i \Delta\theta_i \tag{5.34a}$$

where

$$\xi_0 = \theta \tag{5.34b}$$

$$\Delta\theta_i = \tan^{-1}(2^{-i}) \qquad (\textit{arctangent radix}) \tag{5.34c}$$

and

$$\alpha_i = \begin{cases} +1 & \text{if } \xi_i \geq 0 \\ -1 & \text{if } \xi_i < 0 \end{cases} \tag{5.34d}$$

The number of iterations, M, controls the precision to which θ is resolved and the error is bounded by

$$\epsilon_M < \tan^{-1}(2^{-M+1}) \tag{5.35}$$

Since this scheme converges only within a half-plane, a preliminary stage, capable of mapping the input vector into the first and fourth quadrants, is necessary if θ ranges over all four quadrants. This preliminary mapping can be accomplished trivially by changing the sign of x.

A byproduct of the CORDIC algorithm is that the process introduces an unwanted gain. Fortunately, the gain is independent of θ, being solely a function of M. It is given by the formula

$$K_M = \prod_{i=0}^{M-1} \sqrt{1 + 4^{-i}} \tag{5.36}$$

For $M > 4$, K_M approaches an asymptotic value of approximately 1.65. Since only three of the four CE inputs are subject to rotation (Fig. 5-21), the one input that is not rotated is not subject to gain K_M and must be compensated by passing the real and imaginary mantissas on this rail through a hard-wired multiplier pair. In other words, K_M need not be explictly removed at each stage, but can be allowed to propagate through the pipeline FFT.

This presents no problem if the increased overall system gain does not saturate the dynamic range capacity of the hybrid-floating-point data representation. Note also that the multiplier is quite simple and compact, since it multiplies the data by a fixed known number, $K_M \cong 1.65$.

5.6.4 Storage Requirements

There are several different types of memory required in the realization of a pipeline FFT convolution processor, as follows:

1. Interstage or delay line memories
2. Filter frequency response memory
3. Rotation coefficient memory

The memory systems vary in architecture and performance, and are best discussed separately.

Each pair of CEs is interconnected by a system of *first-in, first-out* (FIFO) memory blocks and a commutating switch assembly. A typical configuration is given in Fig. 5-22. The system consists of memory modules of length L_k, $2L_k$, and $3L_k$. The width of a memory stack is the same as the number of bits per complex datum. The stacks behave logically as if they were shift-register delays, but they are often realized with random-access memory devices with appropriate address control.

The fundamental block length L_k is a function of the position in the pipeline and the order of the input data (i.e., normal order or digit-reversed order). For example, in a radix-4 pipeline FFT with digit-reversed input data, the size of the memory increases by a factor of four at each stage.

The commutating switch assembly is necessary to permute the data between the FFT stages, as prescribed by the radix-4 DIT algorithm. The arrangement can be thought of as a four-wafer rotary switch with contacts arranged as shown in Fig. 5-22. Given this interpretation, the control rule becomes simply, rotate the switch assembly counterclockwise once every L_k system clocks, where L_k is the fundamental block length of the associated FIFO memory.

The reference memory must be a large memory capable of storing the various desired filter frequency responses. Four output ports are required to service the four data streams of the pipeline. If the contents of each memory block are arranged to coincide with the digit-reversed ordering of the data as they are delivered by the direct FFT, the addressing control reduces to little more than a counter.

Each CE has associated with it a memory block that stores the relevant rotation information (either the exponent of W or the rotation angle). For a given FFT algorithm and size, the sequence of rotations that must be

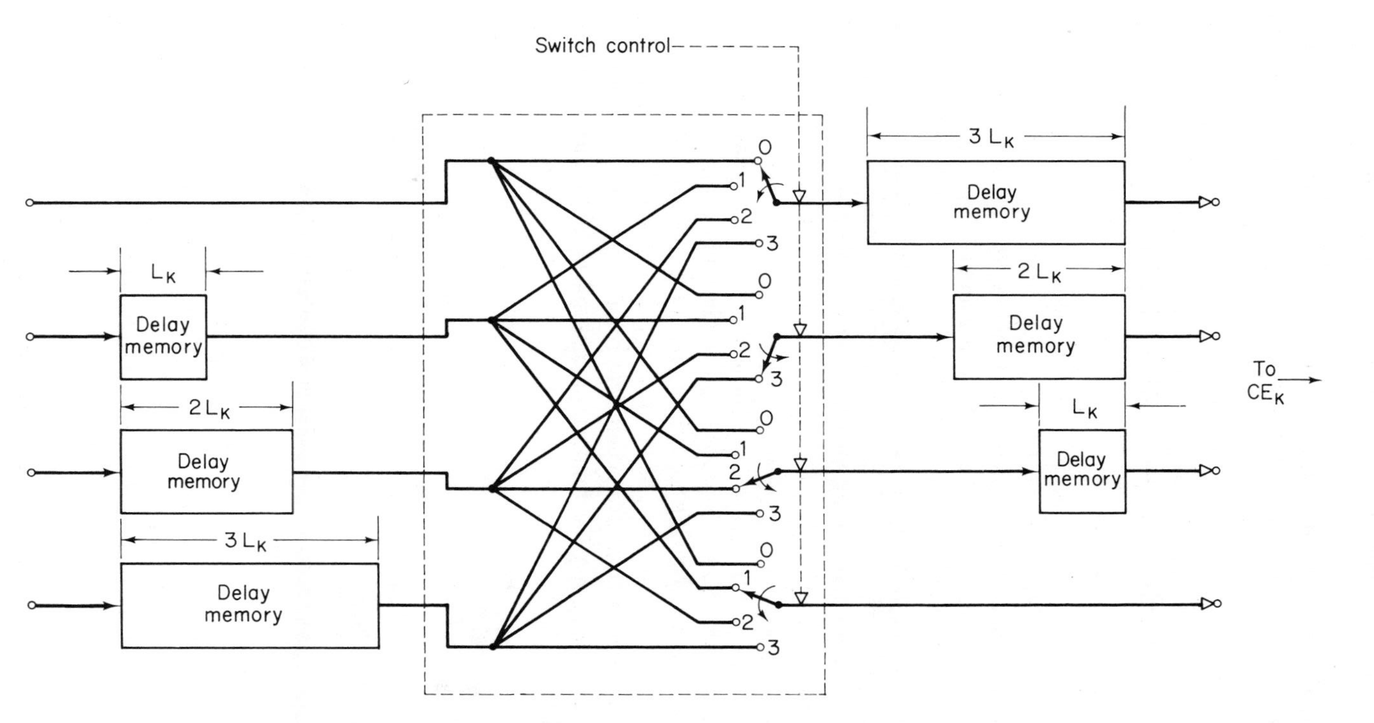

FIGURE 5-22. Typical delay-line memory system with commutator switch.

applied to the data at each stage is uniquely defined. The memory blocks can, therefore, be realized with nonvolatile or read-only devices. If complex multipliers are used as the rotational element, each memory cell would contain $\cos\theta$ and $\sin\theta$. If, on the other hand, the CORDIC algorithm is employed, each cell would contain the angle θ encoded in the arctangent radix (Eq. 5.34) so as to directly control the CORDIC iteration.

5.6.5 Radix Trade Offs

In a design of the pipeline FFT convolver, many parameters influence the choice of radix. One accepted way of making this decision is to compare the hardware complexity of the FFT as measured by the number of multipliers (or rotators) required or by the size of the interstage delay memories. In addition, other parameters, such as the integrated circuit (IC) technology, the desired performance level, and the interfacing with other system components at the input and output, will enter into the final choice, but these are system dependent and harder to determine. First, let us consider the issue of multipliers.

Figure 5-23 shows schematically the structure of the radix-2, radix-4, and radix-8 CEs. The radix-2 CE consists of an adder, subtractor, and one rotator. The radix-4 and radix-8 CEs are built up from the radix-2 block with varying interconnection paths. Note, however, that the radix-8 CE (and higher radix designs) requires internal fixed coefficient rotators that are not needed in the radix-4 case. Counting the number of multipliers and adders per CE, we obtain Table 5.1.

TABLE 5.1

	No. multipliers	*No. adders*
radix-2	1	2
radix-4	3	8
radix-8	9 (including internal)	24

Although, the radix-4 CE has three times the number of multipliers as the radix-2 CE, only half as many radix-4 CEs are required for a given length FFT. Thus the radix-4 FFT requires 50% more multiplier hardware. On a per transform basis, this is actually an effective reduction in multiplier hardware, because the radix-4 FFT can accept data at twice the rate of a radix-2 FFT. To see this, recall that for an N-point, radix-2 pipeline FFT, the data are divided into two sets, each consisting of $N/2$ points, and these are used at the two input ports of the first CE of the pipeline. Thus $N/2$ clock periods are required to get all the data into the radix-2 pipeline. On the other hand, for

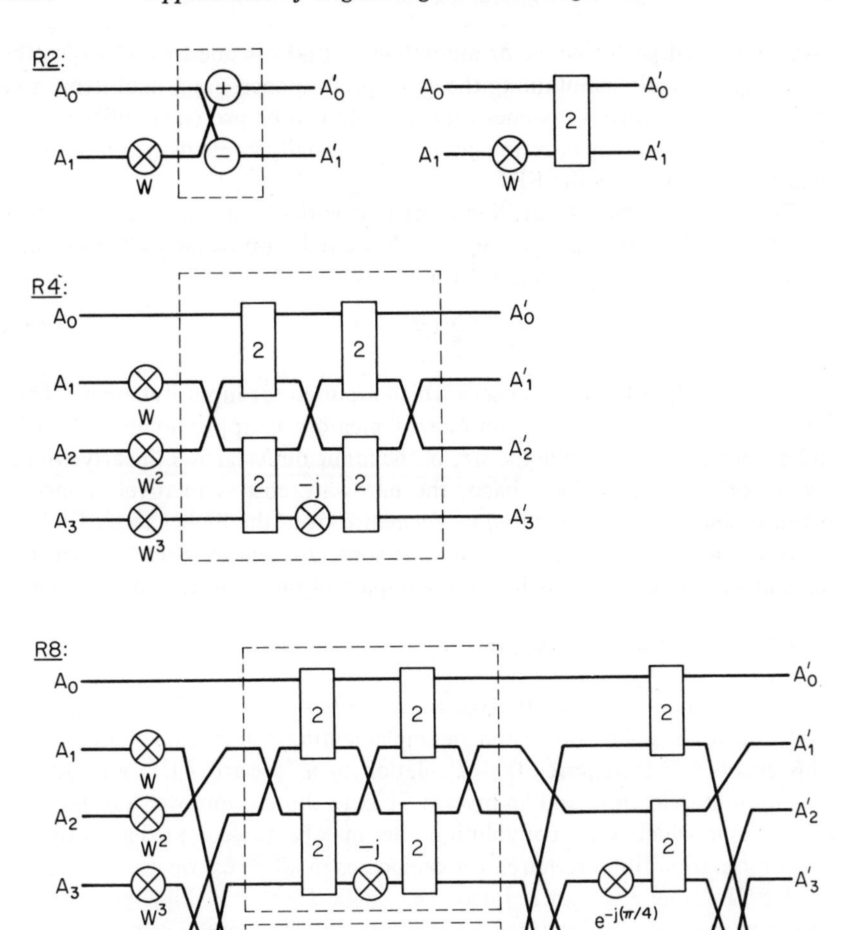

FIGURE 5-23. Radix-2 radix-4, and radix-8 CEs built up from radix-2 modules.

a radix-4 pipeline FFT, the input data are divided into four sets of length $N/4$, and $N/4$ clock periods will suffice. Obviously, for a radix-8 system, $N/8$ clocks are needed. Hence, the added hardware of a radix-4 or radix-8 system gives improved performance or aternatively, it allows one to slow down the system clock while maintaining the same performance. As a final item, note that the above analysis assumes that the data can be presented efficiently to the pipeline FFT. This ordering operation is usually implemented in a buffer memory that precedes the FFT.

The other significant hardware cost that varies as a function of radix is the amount of interstage delay memory. For a radix-r pipeline FFT, Gold and Bially[16] give the approximate formula as

$$\left(\frac{N}{2}\right)(r + 1) \tag{5.37}$$

This includes $(N/2)(r - 1)$ cells which are used for input bufferring. Thus internal to the pipeline, the amount of memory is approximately N cells, independent of radix, while the size of the input buffer grows linearly with r. Again, on a per transform basis, the hardware cost is relatively constant because one is getting increased performance from the higher radix FFT.

In the next section we shall examine some alternative filter implementations including some discussion of the impact of radix on filter performance.

5.6.6 Fast Convolver Rates

The previous sections have discussed the basic components and parameters that need to be considered in implementing a fast convolution filter. This class of filter requires the calculation of a forward FFT, a frequency domain multiplication, and an inverse FFT. In this section, we shall examine four ways in which a fast convolution filter may be implemented and indicate the computation times required for each case. In all cases, we assume a pipelined FFT rather than an iterated (i.e., single CE) FFT. Pipelined systems have higher speeds than iterated systems but at the expense of more hardware.

The most direct way to implement a fast convolution filter is shown in Fig. 5-24(a). In this system, both the forward and inverse radix-r FFTs are implemented in hardware and the entire system, as well as the FFTs, is pipelined. If we assume, as will be done throughout this section, that the initial reordering of the data is performed in an input buffer (IB) memory, then it takes N/r clock pulses to read in the N data points on r input rails. Also the delay through the pipelined FFT of any data point is $(N - r)/r \approx N/r$. This ignores all computation reclocking which is small compared to N/r.

This amount of time, N/r clock pulses, is called the *epoch*. Figure 5-24(b) shows the number of epochs required to deliver a stream of filtered data sets from the system in Fig. 5-24(a). The first data set, labeled number 1, is read into the forward FFT in one epoch. This fills the forward FFT. During the next epoch the first data set is emptied out of the forward FFT

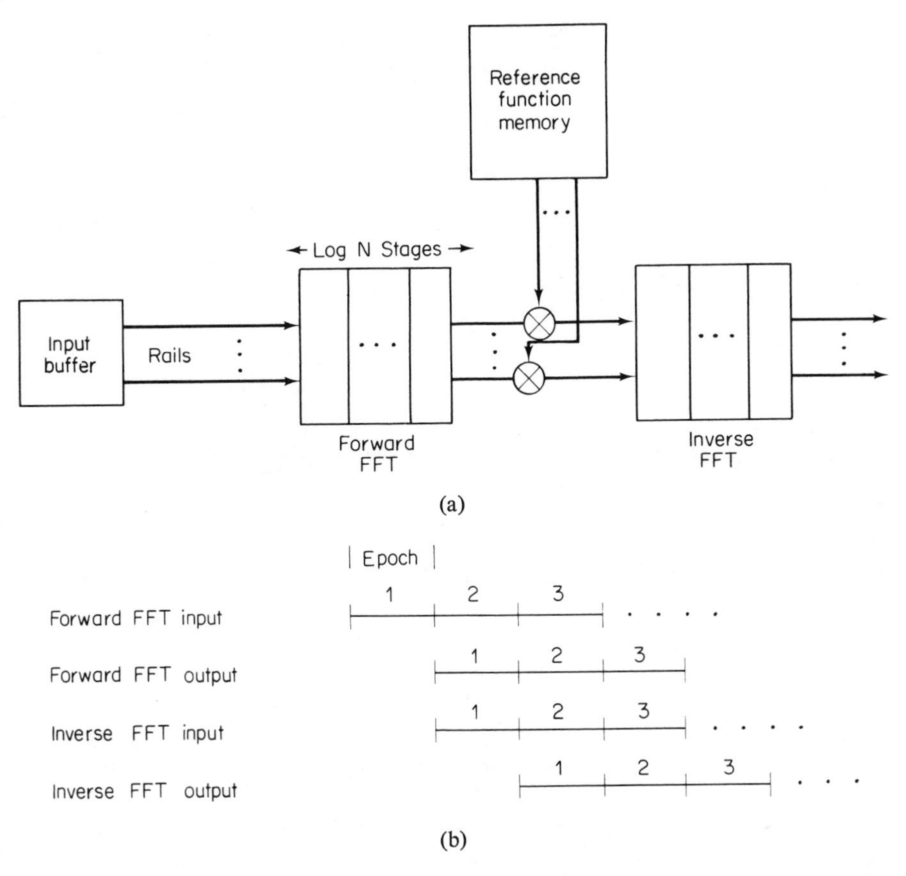

FIGURE 5-24. Dual pipeline FFT matched filter: (a) Block diagram; (b) Timing diagram.

through the multiplier. If the small delays through the multiplier are neglected, the inverse FFT can begin to fill while the forward FFT is emptying. One more epoch is then required to empty the inverse FFT. The forward FFT can begin to fill with a second data set (labeled 2) immediately after the first is loaded, and this second data set can follow the first through the entire system. Thus three epochs are required for the first data set to be completely filtered, and one filtered output is delivered each epoch thereafter. This computation time, in epochs, is plotted in Fig. 5-25.

The dual FFT system allows arbitrary data sets to be filtered sequentially with arbitrary reference functions (selected from the reference function memory). However, many radar applications require that the *same* data set be filtered with several different filters. Since, in this case, only one forward transform needs to be performed, followed by several inverse transforms, it

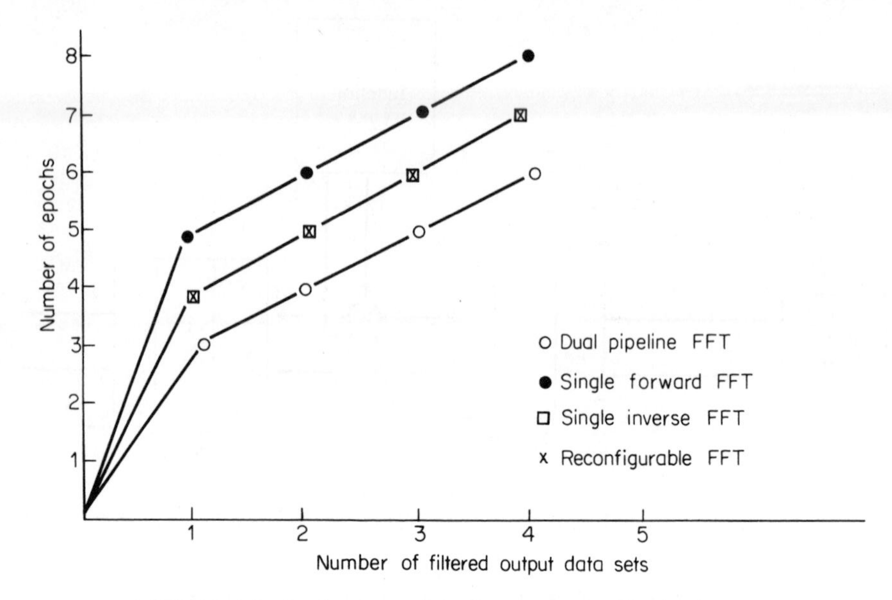

FIGURE 5-25. Performance of the four systems used to implement the matched filter: dual pipeline FFT, single forward FFT, single inverse FFT, and reconfigurable FFT.

should be possible to eliminate one of the pipeline FFTs of Fig. 5-24(a). This is quite desirable since it would save a large amount of hardware. The next three systems exploit this possibility.

The second system is shown in Fig. 5-26(a). This diagram has been simplified by showing only single input and output paths (each path still contains r rails). In this system the data are first forward transformed and the results stored in a temporary storage memory (TSM). The data are then multiplied by the filter function and inverse transformed. This allows multiple readouts of the forward transformed data from the TSM and multiple filtering of the same data set; the output for each filter will appear sequentially.

However, complications develop. First, the data at the output of the forward FFT are in digit reversed order. In the dual pipeline system this is not a problem since the inverse FFT is configured to accept, as input, digit reversed data and produce, as output, normally ordered data (see Figs. 5-19 and 5-20). However, the system of Fig. 5-26(a) has only one FFT for which the input data must be in normal order.* The digit reversal can be corrected by reading the data out of the TSM in digit reversed order. Then the second FFT

* Although we refer to this system as a single forward FFT system by analogy with Fig. 5-19, the important feature is that the input to the pipeline FFT must be in normal order. In particular, it is possible to have an inverse FFT which accepts normally ordered input data, but the configuration will differ from that of Fig. 5-20.

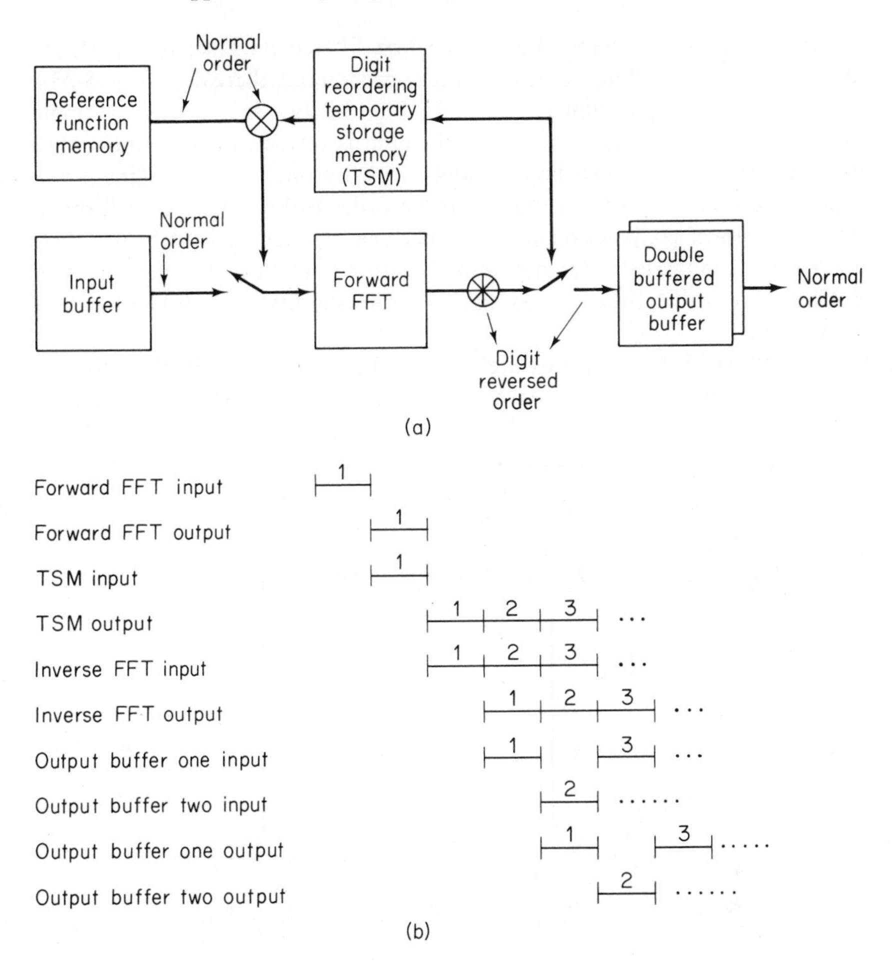

FIGURE 5-26. Single forward FFT matched filter: (a) Block diagram; (b) Timing diagram.

(implemented using a forward FFT) is performed on normally ordered data. The output is then placed in an output buffer. When this buffer is emptied, it too must be read out in digit reversed order to eliminate the digit reversing of the FFT. This output buffer is double buffered to alternately handle successive outputs.

Since the sign of the complex exponential is different for the forward and inverse transforms and since only a forward transform is available to provide both the forward and inverse transforms, it is necessary to conjugate the output of both the forward and inverse FFTs as well as modify the conjugation of the reference function.

The timing diagram for the single forward FFT system is shown in Fig.

5-26(b). Five epochs elapse before the first filtered data set is completely available, and one filtered output appears each epoch thereafter (Fig. 5-25).

The third configuration is shown in Fig. 5-27(a). In this case, a single inverse FFT is employed and the input data is read from the input buffer in digit reversed order. The data are transformed, stored in the TSM in normal order, and then read out in digit reversed order to be inverse transformed. Again, complex conjugation must be performed after each transform and a properly conjugated reference must be stored. The timing diagram, Fig. 5-27(b), shows that four epochs elapse before the first fully filtered output is available.

The above system requires digit reversed access to the IB. Since the IB in many systems can be quite large (much larger than the TSM) this address-

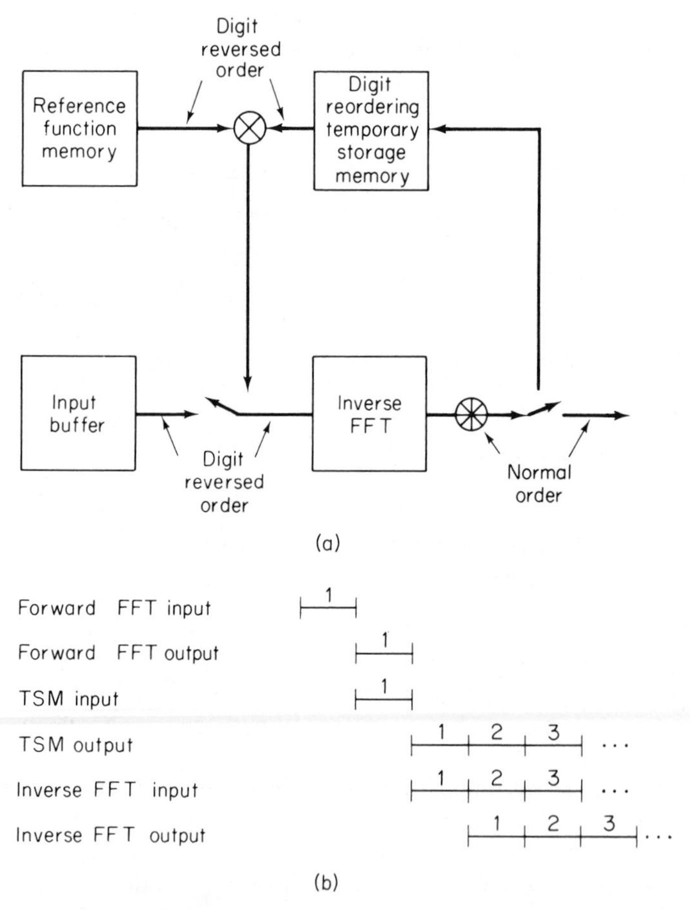

FIGURE 5-27. Single inverse FFT matched filter: (a) Block diagram, (b) Timing diagram.

ing may be difficult to implement. The fourth system we discuss avoids this problem as well as the need for a digit reordering TSM. This system uses a reconfigurable FFT as shown in Fig. 5-28(a). The FFT subsystem switches the interstage delay lines to realize both forward and inverse transforms. The switching arrangement is shown in Fig. 5-29 for a radix-4 16 K FFT. The forward transform is implemented by routing the data through the interstage delay memories (IDMs) in the order of decreasing size, while the inverse transform is implemented by sending the data through the IDMs in increasing order of size. The numbers in the delay lines memories represent the total memory for each stage. Note that it is also necessary to access the coefficient memories in a different order for the forward and inverse transforms.

The timing diagram for the reconfigurable FFT is shown in Fig. 5-28(b). As with the single inverse transform system, four initial epochs are required

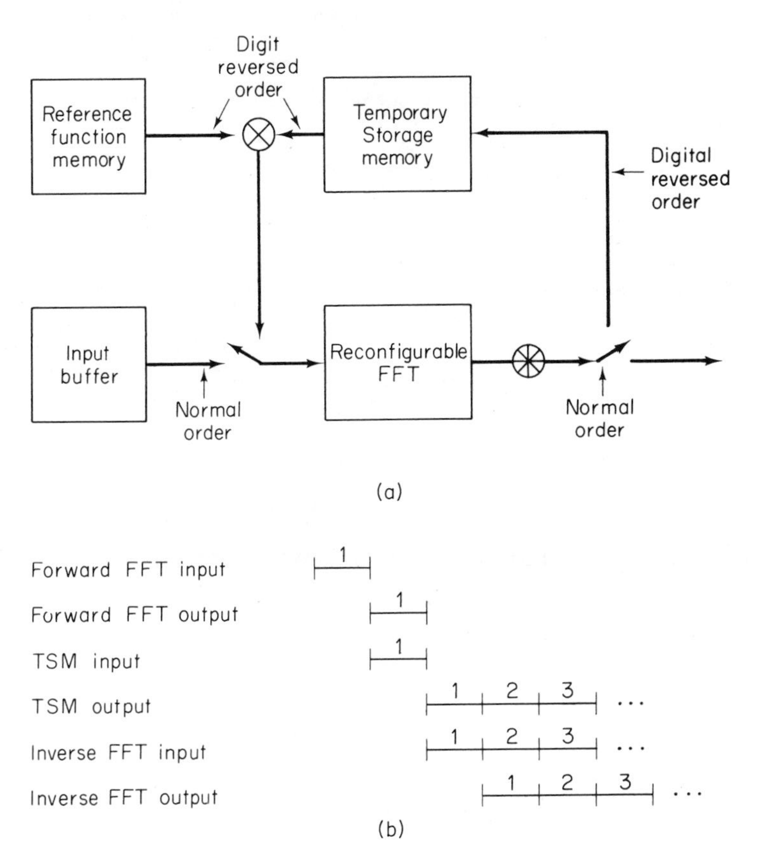

FIGURE 5-28. Reconfigurable FFT matched filter: (a) Block diagram; (b) Timing diagram.

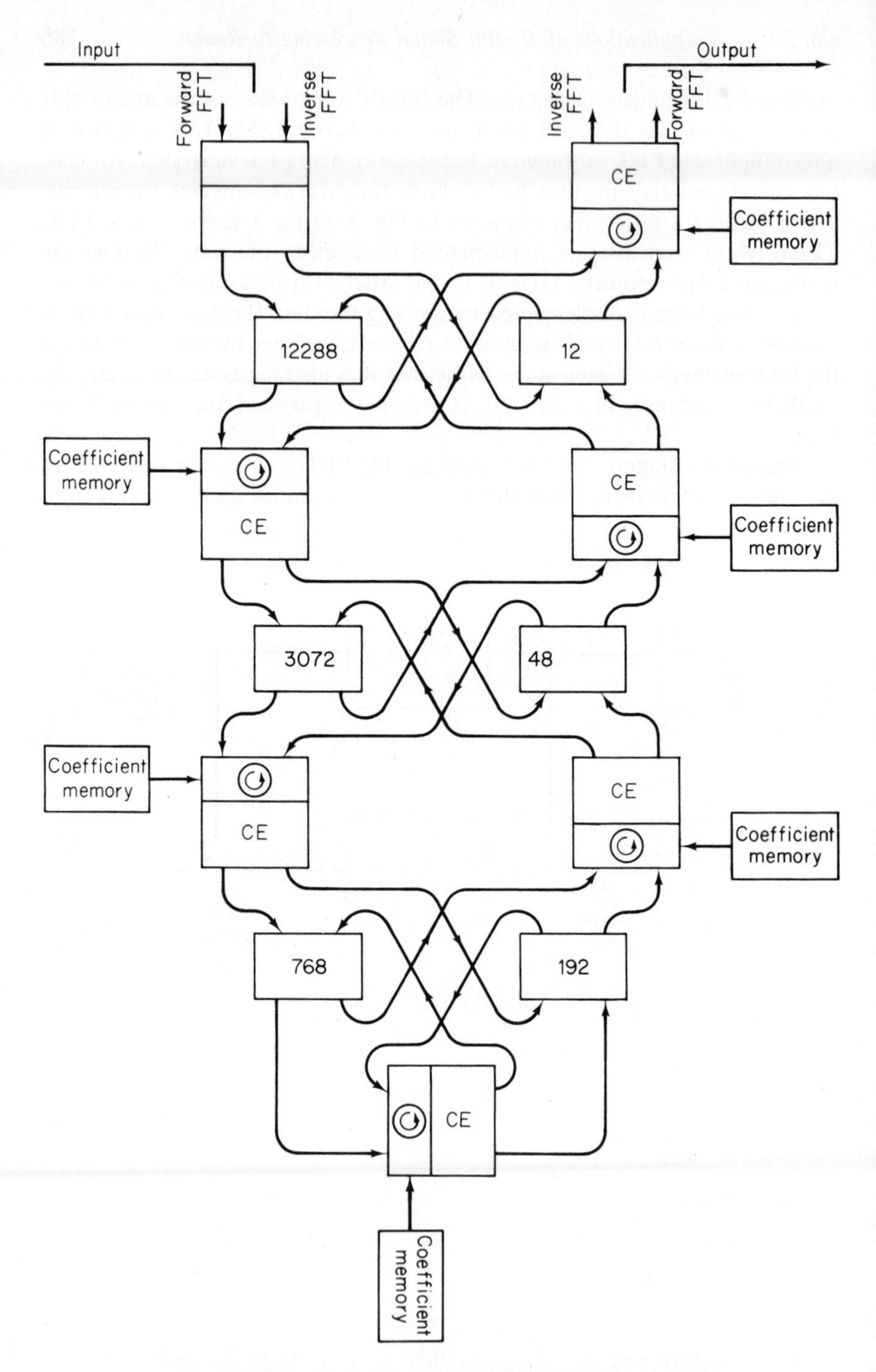

FIGURE 5-29. Radix-4, 16384-point, reconfigurable FFT.

before the first data set is completely available. It might be thought that one epoch could be saved by sending the output of the first forward FFT directly into the input of the first inverse FFT without filling the TSM. However, because the reconfiguration from forward to inverse transform requires that the short IDMs occur first in the inverse transform, the first inverse transform points would overtake the last forward transform points. This is avoided by waiting for the forward pipeline to empty completely (one epoch) before starting the inverse transform. The result is a delay of four epochs. It should be noted that the forward pipeline does not have to be completely emptied before starting the inverse and so small some savings in time could be obtained (a fraction of an epoch). However, this is not exploited in Fig. 5-28(b).

The number of epochs for each system, presented in Fig. 5-25, shows the relative performance for the four systems. The dual pipeline is the fastest but requires two complete pipeline FFTs. This can be a substantial hardware cost. The single inverse transform has better performance than the single forward transform and also does not require a double buffered output memory. The reconfigurable FFT has equivalent performance and does not require digit reversed access to the IB, nor does it require a digit reordering TSM. Moreover, for a penalty of one additional epoch of start-up delay, a complete pipelined FFT is saved over the dual pipeline system. For these reasons, the reconfigurable FFT system is an attractive choice when multiple filtering of a single input data set is required—and this is often the case in radar systems.

It is worth while to indicate the actual computation time required for the reconfigurable FFT system when the FFT is implemented with different radix FFTs and for clock rates that can be achieved with currently available digital technologies. This is done in Fig. 5-30 for a 4 K transform. The computation times are obtained from Fig. 5-25 by recalling that an epoch is N/r clock periods. Several different clock rates are shown in Fig. 5-30. It should be mentioned that clock rates of 10 MHz are reasonable. Moreover, a 16 K system is presently being built with a 30-MHz clock.

As was discussed earlier, increasing the radix decreases the computation time for a fixed clock rate. Alternatively, increasing the radix allows the same performance (computation time) to be achieved with lower clock rates. This can be an advantage since lower clock rates are easier to achieve and can allow the use of IC families with high levels of integration. This might substantially reduce hardware.

5.6.7 LFM Simulations

Any specific hardware design for a pipeline FFT requires a selection of the word lengths throughout the system. If the word lengths are too short, computation noise will degrade the resultant matched filter response. If the

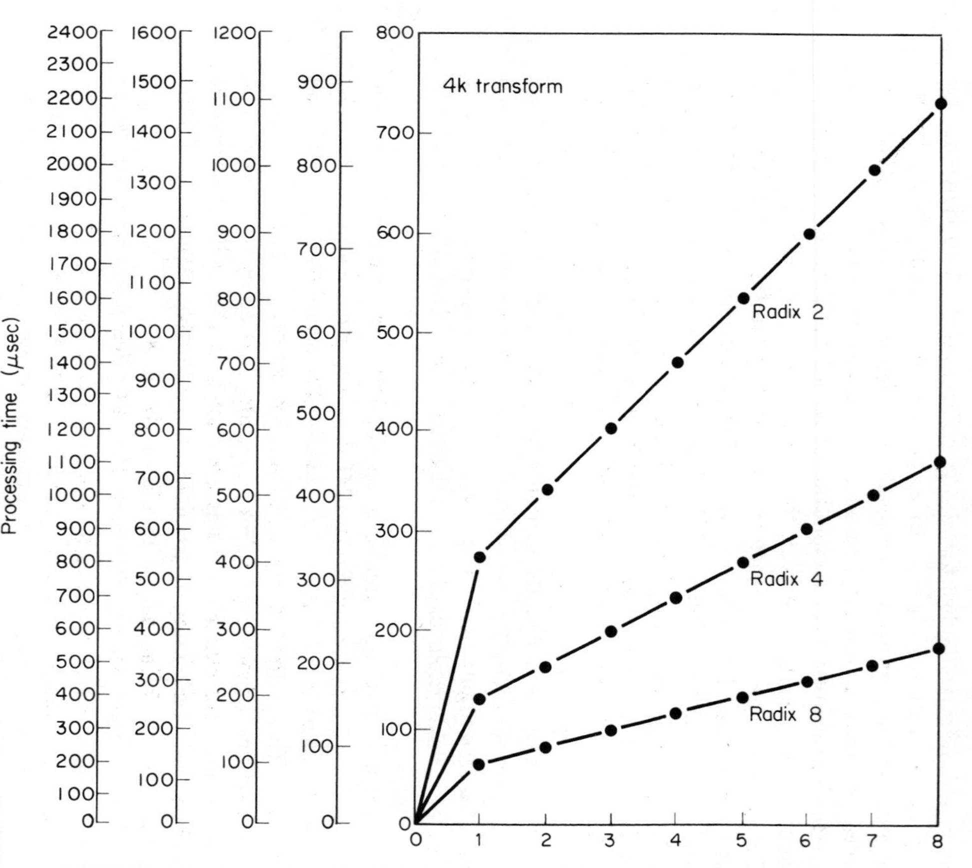

FIGURE 5-30. Processing time for the reconfigurable FFT as a function of the radix. Clock rates of 10, 15, 20, 25 and 30 MHz are shown.

word lengths are too long, the hardware required to implement the FFT may be unnecessarily large, or even prohibitive. Determining the optimal word length thus requires a compromise between hardware and performance; and determining the performance is essential. This trade off is often done through simulation, since the operation of digital hardware can be simulated exactly. This section presents some typical LFM simulations and shows the effects that will be encountered with finite precision computation.

The LFM signal is used as a baseline example because of its ease of generation and wide application. In all cases discussed here, the LFM signal has a time-bandwidth product of 2048, Nyquist rate sampling is assumed, and the processing uses a Hamming weighted matched filter. Nevertheless, the demonstrated effects on sidelobe levels and computation noise should have general applicability to all waveforms.

Figure 5-31 shows the range axis of the ideal ambiguity function. Although the output consists of 4096 points, only the middle quadrant of 1024 points

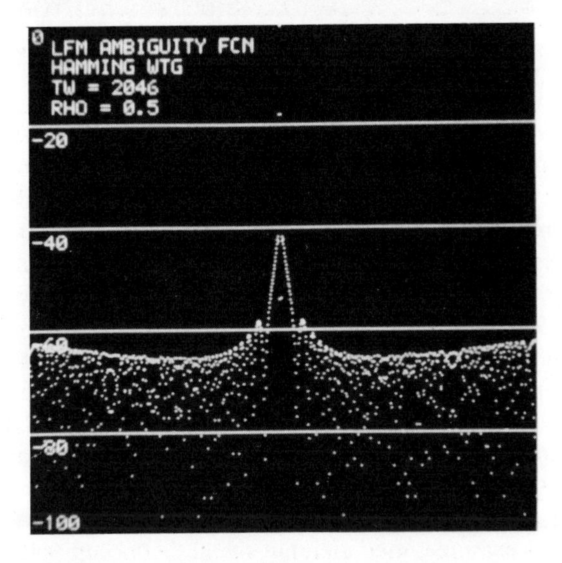

FIGURE 5-31. Middle quarter of the time axis of the ideal LFM ambiguity function. Time-bandwidth product is 2048 and the matched filter employs Hamming weighting.

centered around the main peak are shown. Also, the input to the matched filter has been shifted by one half of a range cell (the $\rho = 1/2$ case) relative to the reference signal. The result is that the peak of the matched filter output lies halfway between two discrete sample times. Thus in the output there are two samples on the main peak, one on each side of the maximum value.

Two sources of degradation are shown in Figs. 5-32 and 5-33. Figure

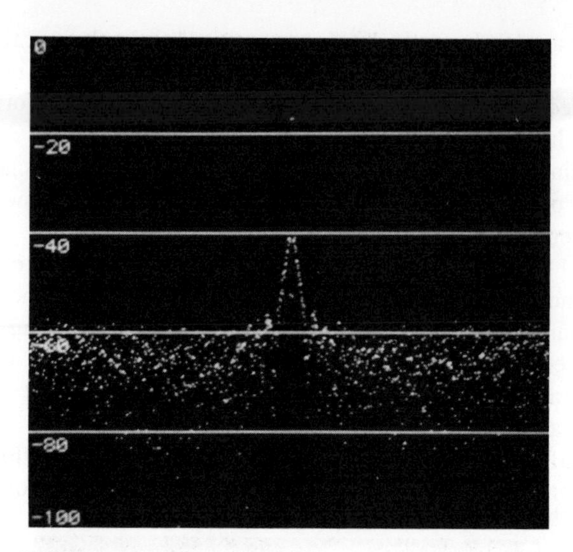

FIGURE 5-32. Ideal weighted LFM matched filter preceded by an 8-bit ADC.

5-32 shows the effect of quantizing the input signal to 8 bits. Again only the middle quadrant of the response is shown. Figure 5-32 is important for judging the further degradation introduced by the finite precision processing systems. Figures 5-33(a) and 5-33(b) show the results of using two different internal wordlengths in the FFT. The systems differ in the mantissa length carried between CEs. Figure 5-33(a) is the 11-bit mantissa system and Fig. 5-33(b) is the 9-bit system. Clearly, there are more spurious responses when using the 9-bit system. Comparison of Figs. 5-31 and 5-33(a) reveals that the 11-bit system performs nearly as well as the ideal system and most of the degradation can be attributed to the 8-bit A/D converter (ADC). However, the 9-bit system of Fig. 5-33(b) has many spurious responses above the -60 dB level. These responses are a result of further degradation when going from 11- to 9-bit internal precision.

The previous examples used an input signal that occupies the full range of the 8-bit ADC. Thus the input signal-to-noise ratio (SNR) is very high. However, a far more realistic case has an input SNR of -16.29 dB in which case the theoretical output SNR with a Hamming weighted matched filter is 15.48 dB. Figures 5-34(a) and 5-34(b) show the range axis for the 11- and 9-bit systems. The variance σ^2 of the input noise is set such that $\sigma/q = 1$ where q is the quantization step of the ADC. Section 5.8 discusses the reasons for this choice of the ratio σ/q. Notice that even though the input signal has an amplitude that is much smaller than the quantization step of the ADC, it is possible to see the main peak at 0-dB and two other points on the main lobe of the compressed pulse above the noise in Figs. 5-34(a) and 5-34(b).

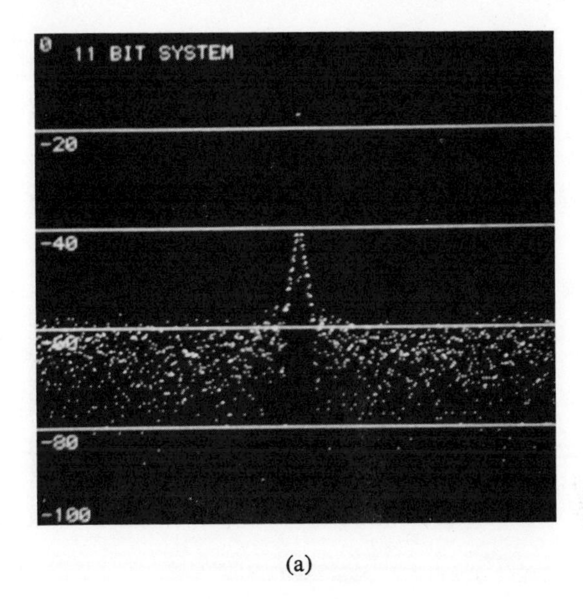

(a)

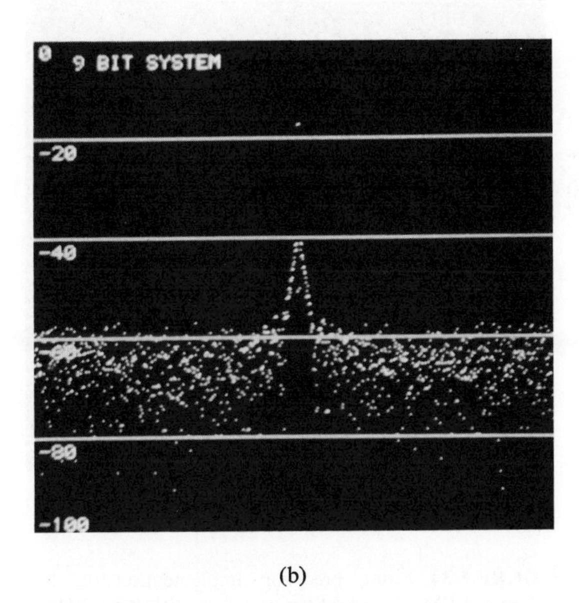

(b)

FIGURE 5-33. Finite precision implementation of the weighted LFM matched filter together with 8-bit ADC: (a) 11-bit internal precision; (b) 9-bit internal precision.

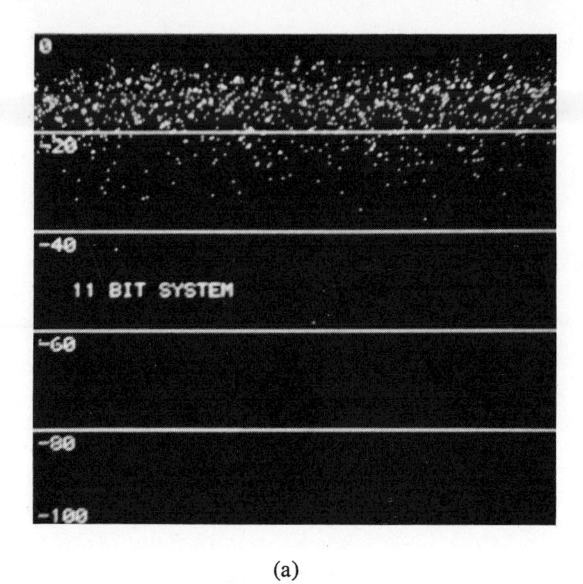

(a)

(b)

FIGURE 5-34. Finite precision implemention of the weighted LFM matched filter together with 8-bit ADC. Input SNR is −16.29 dB. (a) 11-bit internal precision; (b) 9-bit internal precision.

5.6.8 Detection Processor*

All radar systems pass the data from the matched filter to a subsystem, refered to here as the Detection Processor (DP), that accomplishes data compression by comparing the matched filtered data to a threshold. The potential targets (i.e., those that exceed the threshold) are flagged and the remaining data may be rejected. A block diagram of one possible general DP structure is shown in Fig. 5-35. At the input of the DP is a subsystem that approximates the magnitude of the I and Q samples. Next an amplitude estimator calculates an estimate of the peak amplitude of the signal based upon the amplitude of the three nearest samples. The constant false alarm rate (CFAR) subsystem provides an estimate of the ambient noise or clutter level so that the threshold can be varied dynamically to stabilize the false alarm rate. The threshold logic unit selects which of several possible thresholds is to be compared with the estimated signal amplitude. Finally, a comparator sets the ith bit of a binary channel (BC) whenever the ith amplitude estimate exceeds the selected threshold level. The DP subsystems are described further in the remainder of this section.

Theoretical calculation of the magnitude of a complex datum involves a square-root operation which is difficult to implement. An approximation to the magnitude, such as the four-region approximation[13], simplifies the hardware implementation. The four-region approximation is

$$y = \text{Max}\begin{cases} |I| \\ \frac{7}{8}|I| + \frac{1}{2}|Q| \\ \frac{1}{2}|I| + \frac{7}{8}|Q| \\ |Q| \end{cases} \tag{5.38}$$

The maximum error of the four-region approximation is 2.98% of the true amplitude.

One possible amplitude estimator uses a three-point parabolic interpolation formula to estimate the peak signal amplitude and location. This estimator is included to avoid the 1.75 dB peak loss (0.5 dB average) that can occur when sampling a Hamming weighted chirp at the Nyquist rate. The equations for this estimator are

$$\hat{\alpha} = \frac{1}{2} \cdot \frac{y_{-1} - y_1}{y_1 - 2y_0 + y_{-1}}, \qquad |\hat{\alpha}| < \frac{1}{2} \tag{5.39}$$

where y_i are magnitude samples and $\hat{\alpha}$ is the interpolated range position of the peak. The estimated amplitude is

$$A = y_0 - \tfrac{1}{4}(y_1 - y_{-1})\hat{\alpha} \tag{5.40}$$

* The material in this section was originally developed by A. E. Filip.

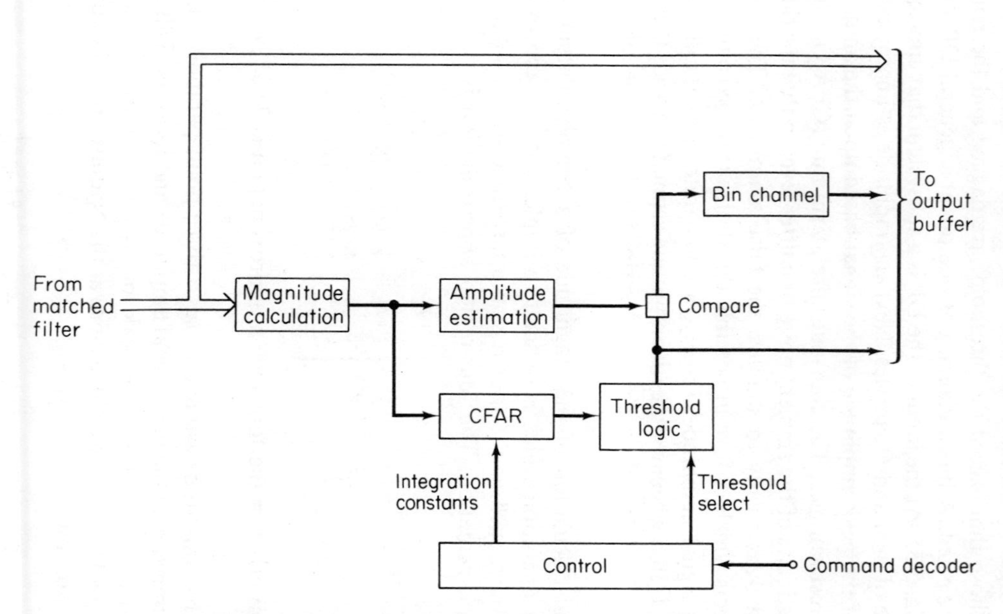

FIGURE 5-35. Block diagram of a detection processor subsystem.

Note that an alternative to the complexity of this amplitude estimator (four adds, one multiply, and one divide) is to lower the threshold by 0.5 to 1.0 dB and accept the consequences of processing eight to twelve times as many false alarms.

The CFAR integrator provides a dynamic threshold level based upon the average value of the received signal. This dynamic threshold tends to reduce false alarms caused by slowly varying changes in the background noise or clutter. The computed CFAR threshold is a function of the desired probability of false alarms (P_{FA}) and the instantaneous RMS signal level in the receive window. Under the assumptions that the received signal is corrupted by additive white Gaussian noise, the basic equation of the CFAR integrator threshold $T(n)$ is

$$T(n) = \mu_s(n)\left[2\sqrt{\frac{-\ln(P_{FA})}{\pi}}\right] \tag{5.41}$$

The bracketed factor is a constant that can be precomputed and stored. The factor μ_s is the sample mean of the received signal. It is obtained by convolving the magnitude samples with a unit-area, finite-duration window having the proper delay. The rectangular window is the obvious choice for this window but there are many other possibilities. Note that the CFAR integrator requires a finite start-up time before the computed threshold can be used. This start-up interval must be taken into consideration by the DPS in scheduling receive windows when the CFAR threshold will be used.

The binary channel (BC) is used to indicate to a postprocessor which of the samples in a receive window exceed the threshold. This data compression feature is included to simplify and speed up certain functions of the postprocessor in search and measurement modes. The control of the BC is quite simple; the ith bit is set if the ith peak amplitude estimate exceeds the ith threshold value. The BC data bits and the computed threshold values are then saved in an output buffer memory. All the complex data (I and Q channels) are usually deposited in the output buffer memory, regardless of whether of not the interpolated amplitude for each sample exceeded the threshold. Thus all data are preserved for (possible) use by a postprocessing system. This can be quite useful if complex algorithms need to be applied to improve measurements in the vicinity of detections. Extra storage is required to provide this capability but, in many systems, the additional memory requirements are acceptable.

5.6.9 Postprocessing*

The previous sections have described the basic filtering and detection operations of the signal processor and have indicated general ways in which

* The material in this section was originally reported by J. M. Frankovich.

these operations may be implemented digitally. After filtering and detection, virtually all radar systems have a subsequent stage of processing usually termed postprocessing. In general, this stage uses the matched filter output and the detections to generate target metrics and signature information (range, velocity, angle, cross section, etc.). Moreover, a special kind of postprocessor is required when the detection processor encounters a large number of targets or clutter within the volume of space being searched. In such cases, the post-processor might be required simply to define the boundaries of the target region, or it might be expected to correlate data returned from several different search pulses and to determine which, if any, of the detections correspond to a real target. This kind of computation requires that the postprocessor collect large quantities of data and process these data over an extended period of time using very complicated algorithms. In many systems these computations are left for the large, general purpose data processing system (DPS). The DPS, however, tends to be loaded to capacity by its responsibility for overall system control and management. Hence it becomes advantageous to design a high-performance postprocessor that can help map the target environment, generate the target metrics, and reduce the volume and rate of data sent to the DPS.

The postprocessor implementation may consist of hard-wired subsystems for each function, but more recently programmable units have begun to play an increasing role. This allows the user to change functions and algorithms without altering the hardware structure. Frequently, sophisticated radar systems require a variety of evolving algorithms, and so it becomes desirable to use a small, but fast, general-purpose computer as the postprocessor. In the remainder of this section, we shall concentrate on programmable postprocessors; we shall present a qualitative overview of the characteristics of both commercial and special-design programmable processors.

There are many general-purpose short-word-length (approximately 16 bits) minicomputers on the market that can execute instructions at approximately a 1-MHz rate. These have proved satisfactory for some systems, since long word lengths are usually not required in signal processing applications. Some specialized general purpose minicomputers have also been developed for signal-processing applications. Instruction rates higher than 10 MHz have been achieved using emitter-coupled logic (ECL) or Schottky transistor-transistor logic (TTL). The designs in these machines have emphasized high multiplier speeds, since multiplication is a frequent operation in many computations. However, there is also a need for a high-speed input-output system because a large amount of data often must be moved through the postprocessor. Machine designs have been proposed that use multiple processors bussed together so as to reduce both the peak-computational and input-output loads on the system(23). We next discuss the design of some of these computers more extensively.

In recent years a number of commercially available computers have become available that are intended specifically for signal-processing applications. In addition, limited-production militarized and research machines have also been developed. Although all these machines have short word lengths, the distinguishing features among these machines is the use of special architectures to achieve the fastest possible computation rate in highly coordinated memory and arithmetic units.

Commercially produced signal-processing computers began to be introduced in the late 1960s. These used early designs of medium-scale integrated (MSI) TTL memories and small-scale integrated (SSI) logic to realize cycle times as short as 300 ns. Although the available IC memory packages were limited in storage size, they were augmented by core memories to achieve larger capacity. By suitably scheduling the flow of data and programs through the fast memories, it was possible to perform signal processing far faster than with contemporary commercial general-purpose minicomputers. One important feature of these machines was the use of register file IC packages for multiple (up to 32 or more) accumulators[(17)]. This feature permitted many operands to be kept on hand rather than stored and retrieved from main memory. This is indicative of how early MSI IC developments were exploited for signal-processing uses.

In recent years the design of commercially available signal-processing computers has reflected the availability of more highly integrated circuit packages. Recent architectures have also divided the machine into several sections in order to obtain faster, parallel operations. For example, in one recent machine design[(1)] an input-output section manages the data flow and the operation of the other two sections; an arithmetic section and an index section. The arithmetic section performs the actual signal processing computations for either filtering or postprocessing. The index section manages the data manipulations within a coordinated network of separate, small IC random-access memories (RAMs) and read-only memories (ROMs) that contain the various programs and data blocks. The arithmetic section, in particular, has a parallel pipelined byte rather than word structure in order to realize the desired compromise between speed and economy. All of the logic is TTL with some Schottky elements. A complete radix-2 butterfly can be performed in 1 μs in parallel with the input-output data flow. In another commercially available machine, this butterfly can be done in 1/2 μs.

The development of MSI ECL logic and memory packages has lagged somewhat behind corresponding events in TTL technology. Early ECL machines were found mostly in the large or "super" computer area[(43)] and were characterized by very high computing speeds. Of necessity, these machines used mostly discrete or SSI IC components. More recent designs use both MSI and large-scale integrated (LSI) components. As an aside, it is interesting to note that these large machines, and the early signal-processing machines, are essentially single instruction stream, single data stream machines in which computing power is achieved primarily by the use of high-speed, highly integrated components, as well as through architectural innovations. In later signal-processing machines, the processing program functions are divided so as to realize a multiple instruction stream machine; that is, the control-type instructions in a program loop are run concurrently in a control processor, separate from the arithmetic processor.

In the Fast Digital Processor (FDP) research computer[(15)], developed at MIT Lincoln Laboratory, a different approach was taken to obtain high signal-processing speeds. Four-way parallelism in the arithmetic element and matching parallel access to the data memories yielded a single instruction stream, multiple data stream machine with exceptional signal-processing performance for a small computer. An instruction time of 150 ns was achieved using primarily SSI ECL packages for logic and MSI packages for memory. However, the available SSI technology (ECL II) and the amount of parallelism resulted in a machine of rather large physical size.

The Lincoln Digital Voice Terminal (LDVT) computer[(4)], a more recent design at Lincoln Laboratory, which uses a large percentage of MSI ECL packages, is also intended for signal processing. This computer was designed for real-time speech applications but with modifications, is directly applicable to radar processing. Separate program and data memories, realized with LSI ECL packages, are accessed in parallel to achieve a 55-ns effective instruction time. This single instruction stream, single data stream machine has a high-speed multiplier, realized with MSI ECL packages, and an instruction set designed for high-performance signal-processing applications. The parallelism in arithmetic computations used in the FDP and in other machines to achieve high performance in FFT calculations was dispensed with; the result is a dramatic reduction in the size and cost of both hardware and programs.

Many standard general-purpose mini and larger computers have been used in various parts of signal-processing systems. These efforts can be successful when the performance requirements are easily met. When they are not, one way to enhance the performance of the computer is to attach a special-purpose unit as a peripheral device. The multiplications and additions corresponding to the filter calculations can be performed in this peripheral without the compromises usually required for this hardware in a general-purpose machine. Such a unit can also make maximum use of available technology to achieve high speed and to accommodate the word lengths and data formats required for the particular application. Peripheral units of this type have been produced commercially. Early in the history of minicomputers, array processors of this type became available. A typical array processor can perform matrix as well as filter calculations rapidly and efficiently. If the host computer for such a unit does not directly enter into the primary signal data flow of the processor then we essentially have a specialized signal processor for which the host computer simply becomes a control computer.

It is often the case that these add-on processors are either not directly applicable to the postprocessing algorithms required in large radar systems, or they are not fast enough. In this case, the only recourse is a new design that is tailored to the radar environment, and which, through technology and parallelism, achieves the required performance. The design of such processors to achieve speed and programmability with simple structure is challenging. The combination is not always achievable; consequently, recent efforts in this area have been an impetus for developing advanced technologies and architectures.

5.6.10 Signal Generator

The ability to generate a wide variety of radar waveforms, including chirps and uniform or nonuniform bursts, is necessary if the flexibility of a digital signal processor is to be fully exploited. Figure 5-36 shows one structure for

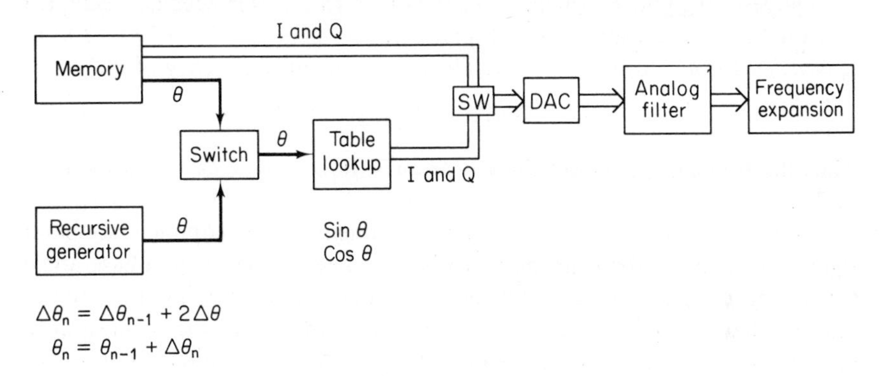

FIGURE 5-36. Block diagram of a digital signal generator.

a digital signal generator consisting of five major subsystems: (1) a memory and recursive generator, (2) a memory for table look-up, (3) a D/A converter, (4) an analog filter to smooth the output of the D/A converter, and (5) mixers to convert the signal to the proper intermediate frequency.

The function of the memory, recursive generator and table look-up is to generate digital samples of the baseband radar waveform. This can be done in one of two ways. Either the in-phase (I) and quadrature (Q) samples can be read directly out of memory and sent to the D/A converter, or if the waveform is only phase modulated, the phase can be calculated recursively or read out of memory, and the I and Q samples formed by looking up the sine and cosine of this phase angle.

Because LFM is such a widely used signal, it is usually worthwhile to provide a special module to generate the quadratic phase sequence for LFM. This can be accomplished with a second order recursion of the form:

$$\begin{aligned}
\Delta\theta_n &= \Delta\theta_{n-1} + 2\Delta\theta \\
\theta_n &= \theta_{n-1} + \Delta\theta_n \qquad n = 0, 1, \ldots, N-1 \\
\Delta\theta_0 &= -(N+1)\,\Delta\theta \\
\theta_0 &= 0 \\
\Delta\theta &= \pi TW/N^2 \\
N &= \epsilon TW
\end{aligned} \tag{5.42}$$

where

TW is the time-bandwidth product of the chirp,
ϵ the sampling frequency relative to the bandwidth W, and
$N + 1$ the number of samples generated.

The resulting phase sequence is $\theta_n = (TW/N^2)(n - N)n$ which is a sampled version of the continuous LFM phase function $\theta(t) = (W/T)(t - T)t$ at $t_n = nT/N$ for $n = 0, 1, 2, \ldots, N$. The instantaneous frequency of $\theta(t)$ is:

$$f(t) = \tfrac{1}{2}\theta'(t) = (W/T)t - W/2 \tag{5.43}$$

Thus the frequency is swept linearly from $-W/2$ to $W/2$ as t varies from 0 to T.

Since the in-phase and quadrature components of the complex baseband signal are required for transmission, the sine and cosine of the phase angle θ must be obtained, perhaps by using a table lookup. Simulation studies indicate that 10 bits of precision for both phase angle and sine-cosine amplitude would yield nearly ideal performance for a LFM waveform with a TW product of 2048.

The D/A converter (DAC) converts the sequences of digital samples to an analog waveform. Ideally, the converter is a hold circuit. However, practical DACs suffer from glitches at the transition points, and these overshoots take some finite time to die out (the settling time). A way around this problem is to sample the output of the DAC with a narrow analog pulse after the settling time has elapsed. This will create an analog pulse train in which the analog amplitude of each pulse is proportional to the amplitude of each digital sample.

Since the frequency spectrum is periodically repeated due to the sampled nature of the digital signal or the analog pulse train, it is necessary to filter out these aliased spectra with an analog filter. The signals are then mixed to an intermediate frequency (IF) required by the radar transmitter equipment.

5.7 DIGITAL TECHNOLOGY

Since any digital approach for radar signal processing would use one or more integrated circuit (IC) technologies, it is worthwhile to review some of these technologies (circa 1976). In all cases, we shall restrict ourselves to IC families that are widely available with reasonably complete product lines. Several additional technologies are emerging (e.g., integrated-injection logic (IIL) and large charge-coupled device (CCD) memories (65 K)), but are not discussed here since, at present, they either do not have a full product line or they are not available in quantity. Similarly, many high-performance specialized digital components have been developed for particular applications. These are also not discussed, again since they are not generally available.

Digital components fall in two main categories: memory and logic. The characteristics of representative devices from these two categories are shown in Table 5.2 for several IC families. The memory sizes selected are the largest presently available in each of the technologies. It should be noted that smaller sizes in each technology generally have faster speeds.

The propagation delay for the gates is indicative of the relative speeds. The device selected (either a NOR or a NAND) is comparable in complexity for all the technologies. With the exception of ECL 100 K, all families have four gates to a package (100 K has five). The power for both memory and gate circuits is given for a single complete IC package.

The most widely available and most generally used technology is transistor-transistor logic (TTL). This family, available in low-power, high-power, and Schottky versions, provides good speed with medium power dissipation. The various forms of metal-oxide-semiconductor (MOS) memories provide good levels of integration since the power dissipation levels are low. The MOS technology finds wide applicability in memory circuits and at present 4 K bits per package are available.* MOS devices can be either dynamic or static; that is, the contents of the memory may or may not have to be refreshed periodically. In general, the static devices are preferable because the control circuitry is much simpler, but the dynamic MOS devices have higher levels of integration and generally higher speeds. Charge-coupled devices (CCDs) have recently been applied to digital memory circuits and dynamic 16 K-bit ICs are presently available. This technology offers the prom-

* 16K-bit chips should be available soon.

TABLE 5-2

Memories

Technology	*Size type*	*Read or write cycle time (NS)*	*Power/package (MW)*
CCD	16K SERIAL	4 MHz*	200
MOS (static)	4K RAM	400	450
MOS (dynamic)	4K RAM	300	430
TTL	1K RAM	90	550
ECL 10K	1K RAM	35	520

Gates

Technology		*Propagation delay (NS)*	*Power/package (MW)*
CMOS	QUAD NAND	30†	6‡
TTL	QUAD NAND	10	40
TTL (LOW POWER)	QUAD NAND	33	4
TTL (SCHOTTKY)	QUAD NAND	3	76
TTL (LOW POWER SCHOTTKY)	QUAD NAND	10	8
TTL (HIGH POWER)	QUAD NAND	6	88
ECL 10K	QUAD NOR	2	100§
ECL III	QUAD NOR	1	240§
ECL 100K	QUINT NOR	0.75	250§

* Transfer rate
† With $V_{dd} = 10\text{V}$, $C_1 = 15$ pf
‡ At 2 MHz clock
§ Exclusive of Terminator

ise of being able to achieve even higher densities; 65 K CCD memories are being developed. Finally, the ECL family provides high speed but requires high power. ECL also provides complementary outputs and nearly constant power supply drain, both of which can be an advantage. The ECL 10 K line is the most complete and is being used in increasing numbers of systems. The ECL III and 100 K series are faster but the product lines are not extensive.

A traditional way of comparing technologies is by examining the speed-power product. Figure 5-37 shows the propagation delay time versus power per gate (as opposed to power per package in Table 5.2) for the IC families discussed here(37,42). Power and speed can often be traded directly; the power levels also strongly influence the amount of integration that can be achieved in a given IC. In Fig. 5-37, the values are only representative; the designer should examine individual specification sheets for precise maximum, minimum, and typical performance values.

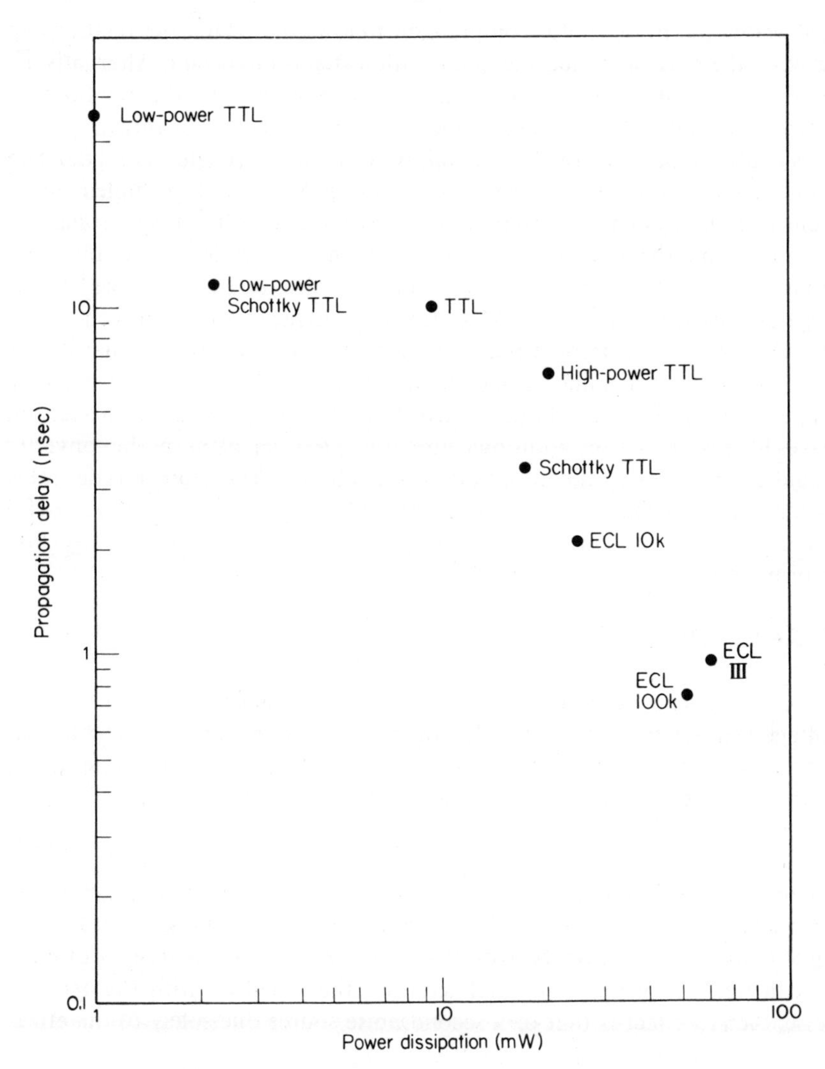

FIGURE 5-37. Delay-dissipation curve comparing several popular IC families.

5.8 ANALOG-TO-DIGITAL CONVERTERS

When the processing in a radar system is performed all digitally, some requirements develop which do not exist when the ADC occurs after the signal processing. Specifically, the received signal plus noise must be large enough to dither the lowest quantization level of the converter and must be small enough to avoid saturating the converter. Since the desired signal is often well below the noise (pulse compression follows the ADC) the relationship between the noise level and the quantization step is important. Alternatively, undesired signals (e.g., clutter) may be quite large and the relationship between the level of these signals and the total number of quantization step sizes is also important. In this section, we examine these effects and indicate some of the considerations needed in selecting ADCs and in implementing them in radar systems. Primarily, it involves a trade off between signal-to-noise ratio and the dynamic range as a function of the ratio (σ/q) of the RMS input noise level, σ, to the quantization step size, q. For example, if σ is large relative to q, then the ADC transfer characteristic approaches true linearity and errors become relatively insignificant. However, actual ADCs are comprised of a finite number of bits thus implying a limited dynamic range capacity. Setting σ large relative to q reduces quantization noise, but leaves little room for any additional amplitude. In such instances the converter would saturate or clip and information would be lost. Therefore, it is desirable to maintain as small a σ/q ratio as possible in order to maximize useful dynamic range. As will be shown, however, too low a level will also lose information.

5.8.1 ADC Level Settings*

An ADC can be modelled as a non-linear element characterized by a voltage transfer function of the general type shown in Fig. 5-38. The output consists of the true value of the input at any given sampling time and a quantization error term. The general form is given by

$$r_i = s_i + n_i + e_i \tag{5.44}$$

where s_i, n_i, and e_i are samples of the signal, receiver noise, and quantization error term, respectively. We will assume that the receiver noise is zero-mean, white Gaussian noise (WGN) with RMS value σ. If the error term is thought of as a random process, uncorrelated with the signal or with the receiver noise, then its effect is that of a second noise source due solely to quantization.

To establish a quantitative feel for the effect of quantization noise, the following experiment could be posed: drive an ADC with pure WGN of known statistics ($N(0, \sigma)$) and measure the RMS value of the output relative to σ as the ratio σ/q is varied. Deviations from σ then indicate the presence and the amount of quantization noise. The simplest analytical approach assumes that e_i of Eq. (5.44) is a random variable uncorrelated from sample to sample with itself and n_i, and uniformly distributed between plus or minus $q/2$. The variance of e_i is by inspection $q^2/12$. Given the uncorrelated assumption,

* This section was originally prepared by P. E. Blankenship.

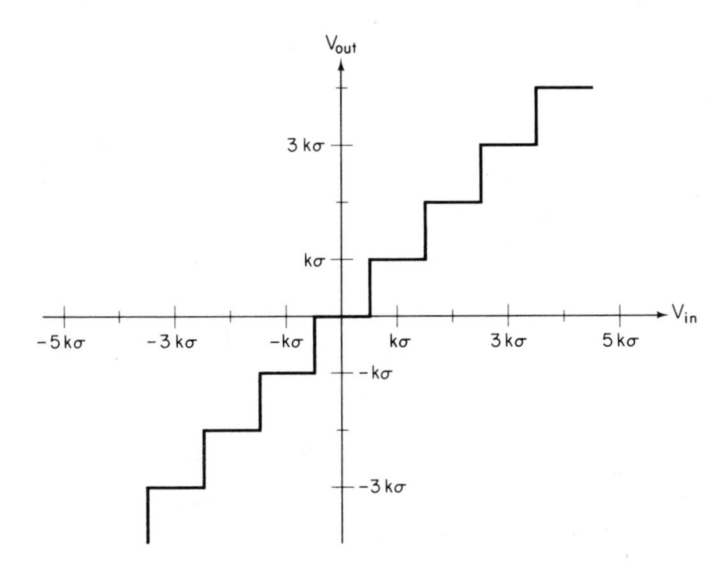

FIGURE 5-38. Input-output transfer characteristic of an ideal ADC.

the variance of the converter output σ_0^2 should be

$$\sigma_0^2 = \sigma^2 + q^2/12 \tag{5.45}$$

From Eq. (5.45) the RMS value of the output, normalized to σ, is simply

$$\sigma_0/\sigma = [1 + K^2/12]^{1/2} \tag{5.46}$$

where the parameter K is $(\sigma/q)^{-1}$. Eq. (5.46) is plotted in Fig. 5-39 and is sometimes termed the linear model since it was derived with the assumption of a purely additive, independent noise source. It predicts that quantization noise power increases without bound as σ/q decreases and the dynamic range can be exchanged for sensitivity to any desired degree. Unfortunately, the model is misleading and it is necessary to perform a more careful analysis to obtain an accurate picture of what actually occurs.

The approach taken here is to derive the exact probability density function (PDF) of the ADC output and then, using this PDF, compute the output variance as a function of σ/q. It is clear that the PDF will be discrete, and will have the general form depicted in Fig. 5-40. By direct inspection it is seen that

$$P_0 = \Pr\,[|X| \leq K/2] \tag{5.47}$$

and

$$P_{iK} = P_{-iK} = \tfrac{1}{2}\Pr\,[(i - \tfrac{1}{2})K \leq |X| < (i + \tfrac{1}{2})K] \tag{5.48}$$

where, for convenience, σ has been defined as unity and X is a sample of the WGN process. Given the Gaussian assumption Eq. (5.48) can be written as

$$P_{iK} = \operatorname{erf}\,[(i + \tfrac{1}{2})K] - \operatorname{erf}\,[(i - \tfrac{1}{2})K]$$

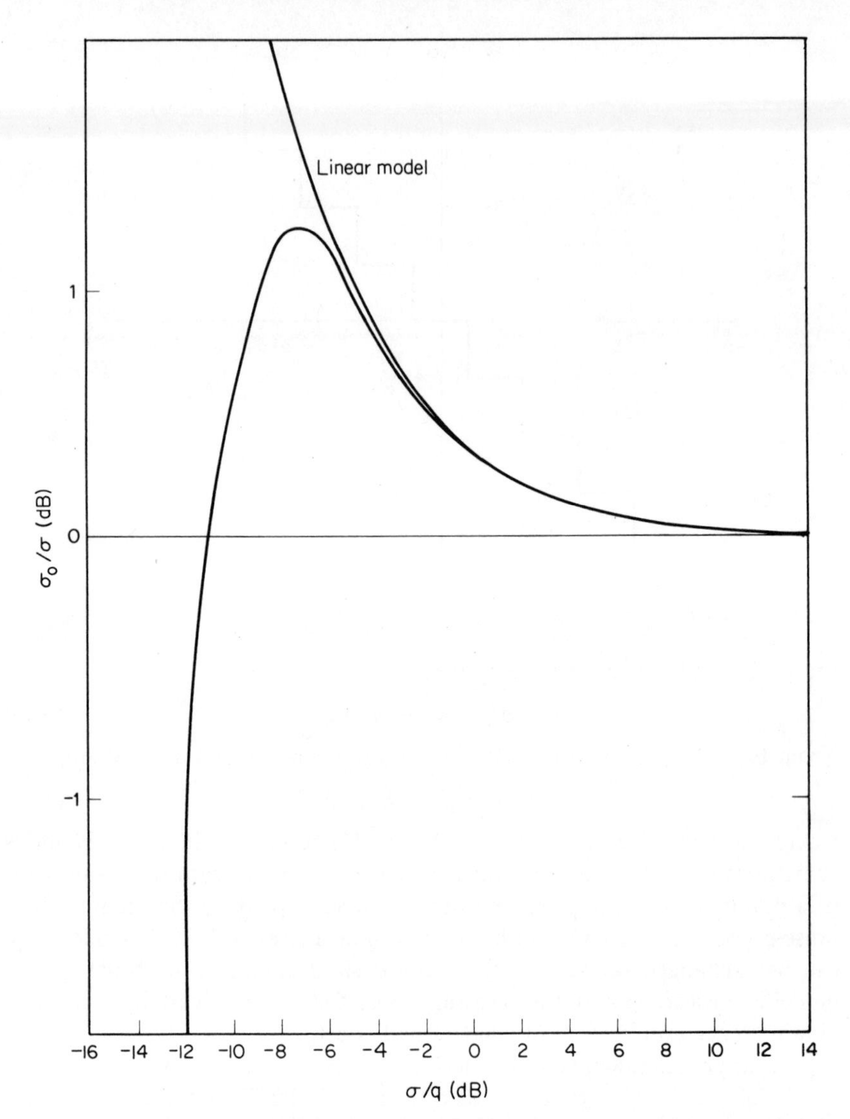

FIGURE 5-39. Quantization noise introduced by the ADC as a function of the ratio of input receiver noise standard deviation to ADC quantization step.

where

$$\operatorname{erf}(z) \triangleq \int_0^z \frac{1}{\sqrt{2\pi}} e^{-x^2/2}\, dx \tag{5.49}$$

The normalized RMS converter output is by definition

$$\sigma_0/\sigma = [2 \sum_{i=1}^{\infty} (iK)^2 P_{iK}]^{1/2} \tag{5.50}$$

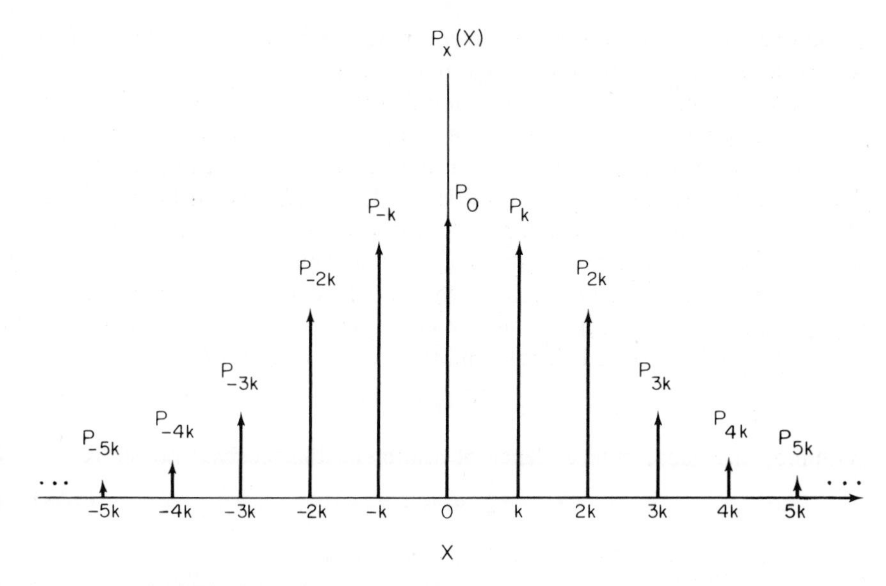

FIGURE 5-40. Discrete probability density function of the ADC output.

Equation (5.50) can be written in a more tractable form for evaluation by noting that a Gaussaian PDF is virtually zero beyond 5σ from its mean. If an integer N is chosen such that

$$(N + \tfrac{1}{2})K \geq 5 \tag{5.51}$$

then

$$\hat{\sigma} \cong K\sqrt{2}\left[N^2 \operatorname{erf}\left(N + \frac{1}{2}\right)K - \sum_{i=0}^{N-1} (2i + 1) \operatorname{erf}\left(i + \frac{1}{2}\right)K\right]^{\frac{1}{2}} \tag{5.52}$$

Equation (5.52) is also plotted in Fig. 5-39 and is seen that the linear model is a valid approximation for $\sigma/q > 1/2$ (-6 dB). Below this value, however, the RMS output falls rapidly in a manner not predicted by the linear model. The physical explanation is simple: input levels for $\sigma/q < 1/2$ are so feeble that rarely, if ever, is a quantum step boundary crossed. The ADC output is then zero for all time, in effect killing the input. The plots do indicate that σ/q can be set at a minimum of 1/2 whereupon 1.16 dB of SNR degradation is suffered. However, this is in a very nonlinear region of performance. As is also shown in Fig. 5-39, setting the ratio σ/q equal to unity represents a reasonable compromise resulting in a slight loss of dynamic range over the maximum attainable dynamic range but also operating in a linear region with a tolerable 0.34 dB degradation in sensitivity.

5.8.2 ADC Word Length

The word length of the ADCs is determined by two factors: the required quantization noise and the total received dynamic range. Simulations and the above analysis are useful for determining the quantization effects. This section addresses the dynamic range requirement.

The key to the problem lies in noting that the conversion in a fully digital processing system occurs before pulse compression. For a point target, the dynamic range is determined by the ratio of the uncompressed signal level of the largest expected target to the noise level.* However, for distributed targets, the uncompressed signals overlap and yield a potentially higher signal. This section presents a simplified analysis of this phenomenon.

The uncompressed signal level is calculated by assuming that the transmitted signal $s(t)$ is reflected from a collection of targets with amplitude a_i and delay τ_i. The received signal amplitude, as a function of time, is

$$r(t) = \sum_i a_i s(t - \tau_i) \tag{5.53}$$

Assuming the usual independence of scatterers, the received power is

$$r^2(t) = \sum_i a_i^2 |s(t - \tau_i)|^2 \tag{5.54}$$

Writing this expression in continuous form, yields

$$r^2(t) = \int C(\tau)|s(t - \tau)|^2 \, d\tau \tag{5.55}$$

where $C(\tau)$ is a cross-section density, (i.e., cross-section per unit length). If $C(\tau)$ is modeled as a constant C_0, and $s(t)$ is a pulse of duration T and amplitude A then

$$r^2(t) = A^2TC_0 \tag{5.56}$$

The noise power in the signal bandwidth W is $N_0 \cdot W$ where N_0 is the noise power per unit bandwidth (assumed constant). Then the dynamic range is

$$D = (A^2C_0)T/(N_0W) \tag{5.57}$$

This is simply the signal power per unit time duration, times the signal duration, divided by the noise power per unit bandwidth times the signal bandwidth. The required ADC word length (number of bits) is thus

$$B = (1/2)\log_2 D + 1 \tag{5.58}$$

Notice that the required number of bits increases with increasing signal energy since there is more signal overlap and/or increased cumulative signal amplitude. However, it decreases with increasing signal bandwidth since the noise level increases (σ/q is fixed at 1).

Modeling $C(\tau)$ as a constant allows Eq. (5.55) to be easily evaluated, and through Eq. (5.57) it shows the functional relationships. In general, a sophisticated model of $C(\tau)$ would have to be developed for each system and its

* This assumes, as will be assumed throughout, that the lowest level of the ADC is set at the noise level, i.e., $\sigma/q = 1$. This assures an adequate thermal noise to quantization noise level, as discussed above.

anticipated environment. The results of simulations with this model would then determine the required ADC wordlength. Some such radar simulations have been developed and indicate that word length requirements vary between 8 and 12 bits for bandwidths between 10 and 50 MHz.

Several assumptions are implicit in the above analysis and it is worthwhile examining each of them. First, the analysis assumes that the lowest quantization level equals the RMS value of the noise ($\sigma/q = 1$). One might ask whether in the presence of clutter it is necessary to preserve this ratio. However, if this ratio is lowered, thus setting the noise below the lowest quantization level, and if in the same window, but exterior to the clutter, there is a low cross-section object whose uncompressed return is below the noise, then all or a portion of the return from this target will be lost. Thus it is important to always have at least the lowest level of the ADC dithered by noise regardless of the level of the clutter. Note also that dithering more bits decreases the dynamic range. As is shown in the previous section, setting σ/q equal to one yields a small loss in SNR, but a loss that nonetheless must be taken into account in the design of a radar system.

The other parameter in σ/q is σ, the noise level. A standard technique in some radar systems is to add noise, and thus reduce the required dynamic range. This is only possible if there is adequate signal-to-noise ratio, since we are reducing the SNR to achieve reduced dynamic range requirements. However, in many situations there is adequate SNR and this practice of adding noise can be and has been employed.

5.8.3 State of the Art of High Speed ADCs

Radar systems allow high bandwidths (1 to 100 MHz) and the environment encountered generally requires large dynamic ranges (greater than 8 bits). Thus the demands on an ADC's performance can be high. It is therefore useful to summarize the characteristics of ADCs that are generally available (circa 1976) and this is done in Fig. 5-41. This plot does not include special ADCs developed for specific systems and represents only delivered devices which are generally commercially available.

5.9 MTI RADAR

The application of radar to the problem of moving target indication (MTI) in situations such as air traffic control has been fruitful. The Doppler shift phenomenon makes it possible to distinguish between moving targets and stationary objects; this fact can be exploited through signal processing to uncover targets masked by clutter (interfering signals), which may be several orders of magnitude larger than the desired signal. In an airport surveillance radar, the primary sources of clutter are ground echoes and weather returns.

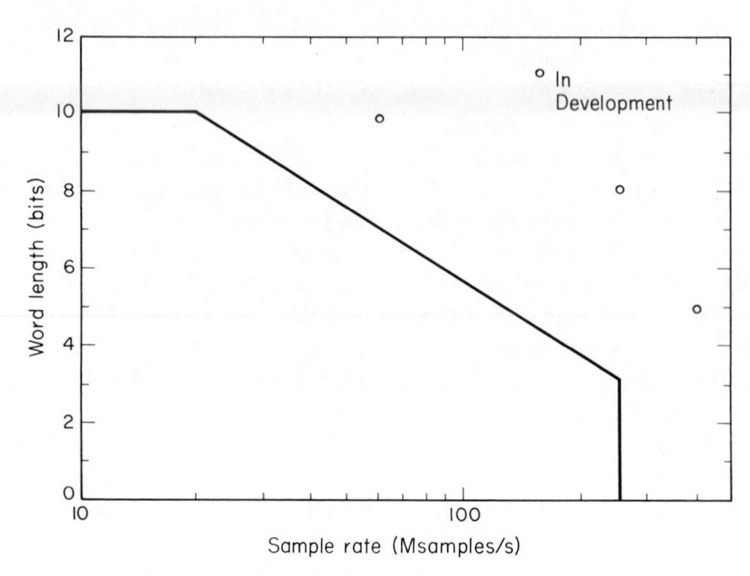

FIGURE 5-41. State-of-the-art (c. 1976) in commercially available ADCs.

Fortunately, this clutter is concentrated near zero frequency (DC) in the frequency domain, and by employing Doppler frequency sensitive filters, it is possible to enhance the signal return while attenuating the clutter component. Actually, the clutter will probably have some frequency components above DC due to antenna motion (if a rotating antenna is used), wind, or rain clouds. In practice, some assumption about the power spectrum of the clutter is made in order to design an effective MTI radar signal processor. In the following, we first consider a simple idea for clutter rejection, that is, a canceler, and then discuss "optimum" and nearly optimum processing schemes.

5.9.1 Cancelers

Since the ambiguity function of a single LFM pulse exhibits range-Doppler coupling (as we discussed in Sec. 5.4.2), it is necessary to employ a burst waveform in order to implement Doppler filtering. Of course, we must worry about the range and velocity ambiguities associated with such a waveform. As discussed above, the Doppler spectrum of a burst is periodic with a period equal to the internal PRF of the burst. The design of a clutter rejection filter amounts to the design of an FIR digital filter with stopbands to reject the clutter frequency components. If we assume that the clutter is DC only, then the clutter return in a given range bin is a constant and can be eliminated by subtracting the outputs from two successive pulses. Such a simple filter is called a two-pulse canceler (Fig. 5-42). If we represent the delay Δ as a unit delay and consider the z-transform from input to output, the two-pulse

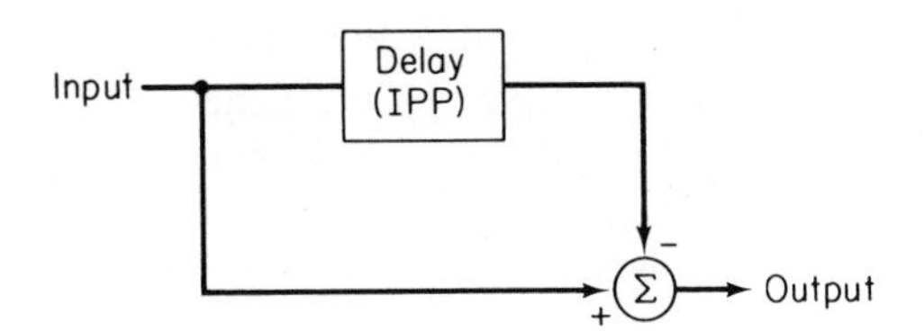

FIGURE 5-42. Two-pulse canceler.

canceler has a transfer function equal to $1 - z^{-1}$. Equivalently, it is a 2-point FIR digital filter with magnitude response $\sin(\omega/2)$ shown in Fig. 5-43. (Note that the frequency axis is normalized to the PRF.) The zero at $\omega = 0$ will completely reject the DC component of the clutter.

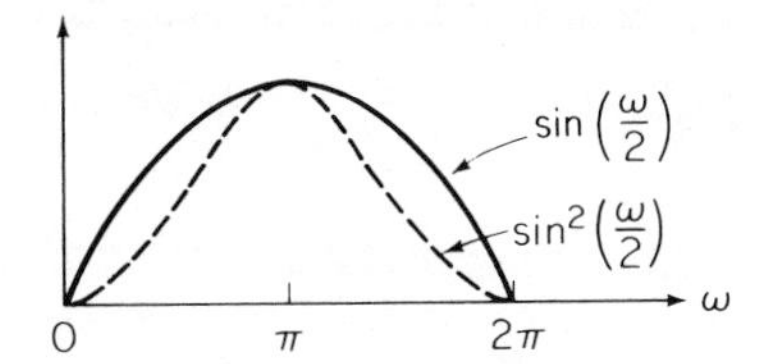

FIGURE 5-43. Doppler frequency response of the two-pulse canceler.

In practice, the clutter has a power spectrum that covers frequencies above DC. In this case, the two-pulse canceler may not adequately reject the clutter, although it will attenuate low-frequency components. The use of an additional two-pulse canceler in cascade yields a three-pulse canceler, which is equivalently an FIR filter with a transfer function $0.5 + z^{-1} + 0.5z^{-2}$. The effect of the three-pulse canceler is to attenuate further the components near DC since the magnitude of its frequency response is $\sin^2(\omega/2)$ as shown by the dashed line in Fig. 5-43. However, these ad hoc methods can be replaced by "optimum" design methods in several ways, depending on the assumptions one makes about the clutter and the desired signal. Delong and Hofstetter[10] have considered the problem of maximizing the output signal-to-interference ratio (SIR) at a given Doppler frequency under the constraint of a known clutter power spectrum. The result is a generalization of the matched filter in the sense that, if the clutter were white Gaussian noise, the optimum linear processor would be a matched filter. The optimum processor can be viewed as the cascade of a filter that prewhitens the clutter power spectrum, followed by a matched filter. Although our discussion has implicitly assumed that the clutter is uniform over all ranges of interest, the Delong-Hofstetter technique does permit a more general clutter model with range varying clutter.

Another approach to the design problem of the canceler is to formulate the problem as an equivalent FIR filter design problem. In this approach, one assumes that the Doppler spectrum of the clutter is located in frequency bands disjoint from the band occupied by the signals of interest (the moving targets). If the clutter is concentrated near DC, then the objective is to design a highpass filter whose stopband covers the range of clutter frequencies and whose passband passes the desired signals. Any of the well-known algorithms for approximating FIR digital filters can be employed for this purpose.[20] The highpass filter is then followed by a bank of narrowband bandpass filters tuned to the specific Doppler frequencies of interest.

The general clutter filter can be implemented in the following way: returns from the same range bin (i.e., the same time relative to the beginning of the transmitted pulse) over several pulses are linearly combined to form the output. Such returns are separated in time by Δ, the IPP, so the implementation requires a tapped delay line with delay Δ as in Fig. 5-44. Since there are many range bins of interest per IPP, each delay of Δ can be realized as a shift register whose length equals the number of range bins to be processed.

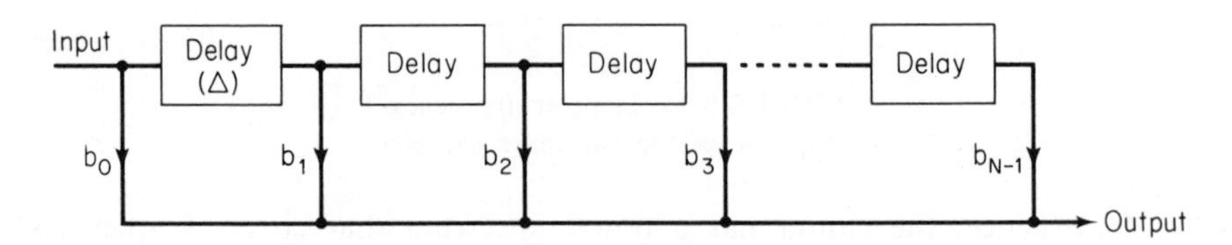

FIGURE 5-44. General digital MTI filter.

The direct implementation of the optimum linear processor with N points requires N multiplications per output point. Since a different optimum processor is designed for each Doppler channel, the filter tap weights are different for each channel. It is possible to obtain a much simpler suboptimum processor by cascading a three-pulse canceler with a bank of bandpass filters (implemented by a sliding FFT). Since the three-pulse canceler requires no multiplications (owing to its simple binary coefficients), and since the N-point FFT requires $N \log_2 N$ multiplications, only $\log_2 N$ multiplications are required per Doppler filter channel. Thus significant hardware simplifications are possible with this scheme provided its performance is adequate. The MTI system described next actually implements this suboptimum processor.

5.9.2 Specific MTI Processor*

As an example of the use of a digital MTI processor, we would like to describe a digital signal processor that has been under development at Lincoln

* The material in this section was prepared by C. E. Muehe, L. Cartledge, and R. O'Donnell and is reported in refs 25, 26 and 28.

Laboratory over the past few years. The processor, called the moving target detector (MTD), is designed to optimize the performance of the FAA's airport surveillance radars (ASRs) when used in an automated environment.

The MTD is a special-purpose, hard-wired, digital signal processor capable of processing a full 360-degree coverage in 1/16 nautical mile (nmi) steps out to a nominal range of 48 nmi. A block diagram of the MTD processor is shown in Fig. 5-45. It is preceded by a large-dynamic range receiver, which is linear over nearly the full range of the ADCs. This receiver linearity is important (and atypical), since it avoids any spectral spreading of the ground clutter returns. Such spread returns would fall into the Doppler filters and thus reduce the available MTI improvement factor. The hard limiting traditionally employed in MTI radar reduces the achievable improvement factor by about 20 dB for a three-pulse canceler.

The received signal is demodulated to baseband to obtain the complex envelope. After analog-to-digital conversion, the 11-bit words from the two quadrature channels are stored in an 8000-word core memory. Ten samples from each of 768 range gates are collected at a constant PRF. Then the PRF is changed about 20% and the data collection process repeated. During collection of 10 samples on the second PRF, data collected on the first PRF are taken out of the memory. All 10 samples from a particular range gate are read out in sequential order and fed to the rest of the processor, then 10 samples from another gate, and so on. The 10 samples from one range gate proceed into a three-pulse canceler followed by an eight-point DFT. Another parallel path contains a zero-velocity filter. This combination of filters is a good approximation to the "optimum" processor[(10)] for ground clutter. The responses of two of the near-optimum filters actually employed are shown in Figs. 5-46(a) and 5-46(b). The three-pulse canceler acts as a clutter filter and the DFT as a set of target filters. To reduce the sidelobes of the Doppler filters, the DFT filters are weighted at their output by subtracting one quarter of the output of each adjacent filter. This is simple to implement and provides a cosine on a pedestal type of weighting.

The near-optimum processor is much more easily implemented than the optimum processor. Assuming eight filters, the optimum filters would require 64 complex multiplications per range-azimuth cell. These must be performed in about 8 μs, giving a rate of 8 million complex multiplications per second or 32 million real multiplications per second. The near-optimum processor using an FFT algorithm requires four simple multiplications by $1/\sqrt{2}$. These are performed as multiplications by the constant $1/2 + 1/8 + 1/16 + 1/64 + 1/256$, thus requiring only four adders. The remainder of the MTD is also configured so that no multiplications are involved.

An empirical study of ground clutter shows that it varies appreciably from one resolution cell to the next. Many shadow areas exist within the clutter in which case the aircraft compete only with noise for detection. For thresh-

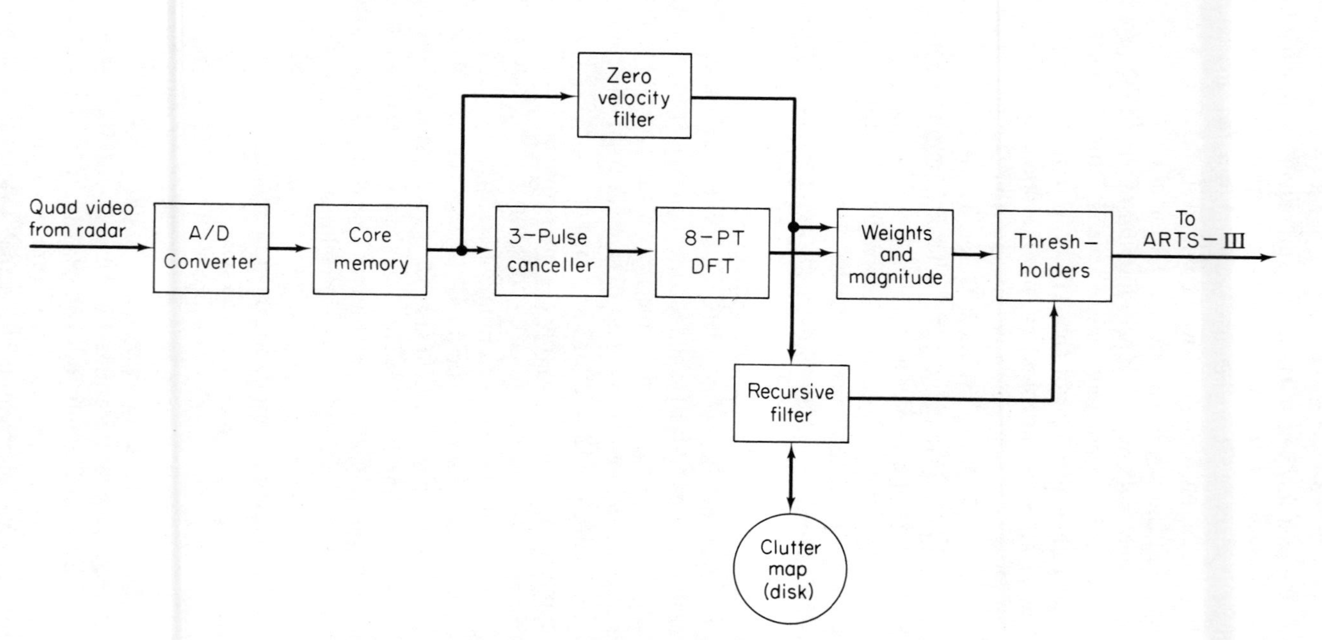

FIGURE 5-45. Block diagram of the moving target detector processor.

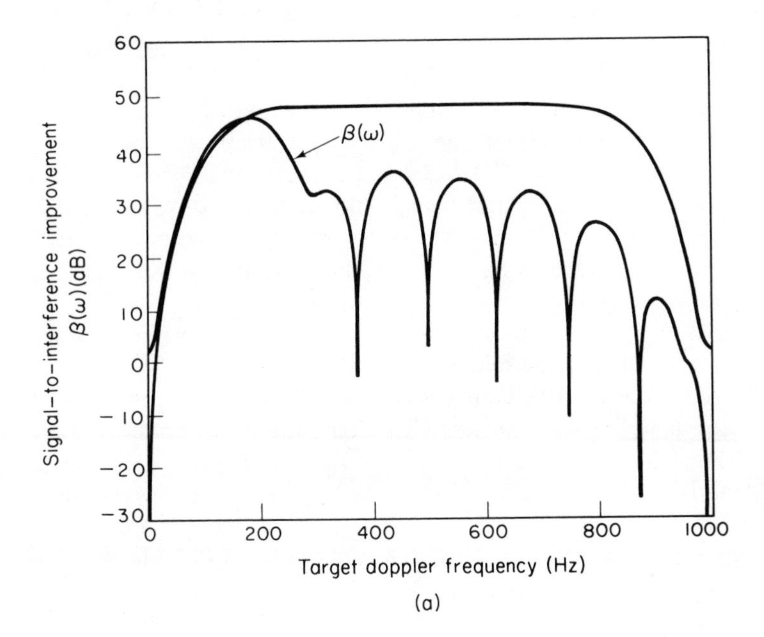

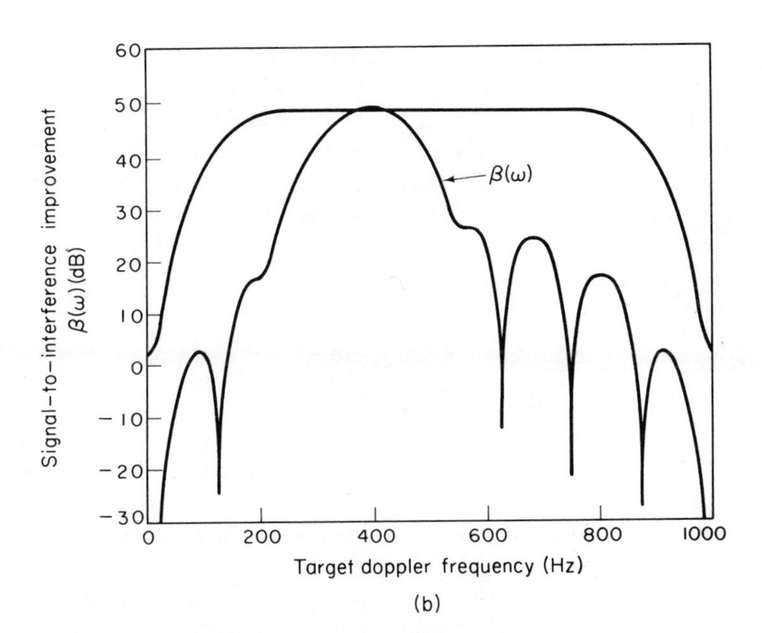

FIGURE 5-46. Signal-to-interference improvement of suboptimal M.T.I. processor using a three-point canceler and an eight-point DFT: (a) tuned to 200 Hz, (b) tuned to 400 Hz.

olding purposes, so as to get the best ground clutter estimate and the best possibility of aircraft detection, a fine-grained ground clutter map is incorporated in the MTD (see Fig. 5-45).

The azimuth coverage of the radar is broken into 480 coherent processing intervals (CPIs) each about one-half antenna beamwidth in extent. During one CPI, ten pulses are transmitted at a constant PRF. The PRF is changed for the next CPI. The clutter map is stored on a disc memory holding $480 \times 768 = 368{,}640$ clutter words, one for each range resolution cell in each CPI. The clutter words are stored in 10-bit floating-point format to preserve the large dynamic range of the clutter signals.

The ground clutter map is built up using a recursive filter, which adds, on each scan, 1/8 of the magnitude of the zero-velocity filter output to 7/8 of the value stored in the clutter map. The map requires about 10 to 20 scans to establish steady-state clutter values. To build up an accurate clutter map, careful registration of the clutter map with the true pointing direction of the antenna is essential. This is achieved by breaking up each revolution of the antenna, marked by 4096 azimuth change pulses (ACPs) into 240 units containing 17 or 18 ACPs each. Each unit contains two CPIs. The disc is accessed every two units (i.e., 34 to 36 ACPs or approximately 44 ms), during which time four CPIs worth of ground clutter data is read onto and off of the disc. The disc has a maximum access time of 18 ms. Two 3000-word MOS memories are used to buffer the data for use in the processor.

The ground clutter map values multiplied by a constant are used to establish thresholds in the zero velocity filter (number 0). A multiplier of 4 to 8 is used. In the two filters immediately adjacent (numbers 1 and 7), wherein the theory says a ground clutter residue of about -40 dB exists, another appropriate multiplier is used. These multipliers are powers of 2 or the sum of two powers of 2, so they are implemented with shifts and adds.

5.9.3 Measured Results

The MTD was connected to an ASR radar at the FAA's National Aviation Facility Experimental Center (NAFEC) at Atlantic City, New Jersey, and extensive flight testing was performed to evaluate its operational performance. This system was compared to a conventional ASR-7 equipped with a digital MTI and a modern sliding-window digitizer. The two radars were linked to a common antenna so that they were looking at the same aircraft and clutter environment at exactly the same time. The transmitter powers and receiver sensitivities were adjusted so that the round-trip sensitivities of the two systems against noise were identical to within 1 dB. What follows represents a small sampling of the rather extensive test results.

Ground clutter is not very extensive at NAFEC. The largest clutter returns are from large buildings in Atlantic City about 8 miles southeast of the radar site. The clutter level varies up to about 45 dB above noise.

The controlled aircraft was a small Piper Cherokee flying about 1000 ft above ground level. Fig. 5-47 is a 56-scan display of the tracker output. Note the meaning of the symbols employed. It was typical that the MTD produces almost 100% probability of detection over ground clutter, whereas the sliding-window detector had many radar misses in a row. If it were not for the beacon replies, the track would have been dropped several times in the sliding-window detector case.

Also in Fig. 5-47, we see the track of a non-beacon-equipped aircraft (radar only). Notice that this track is dropped in the sliding-window case

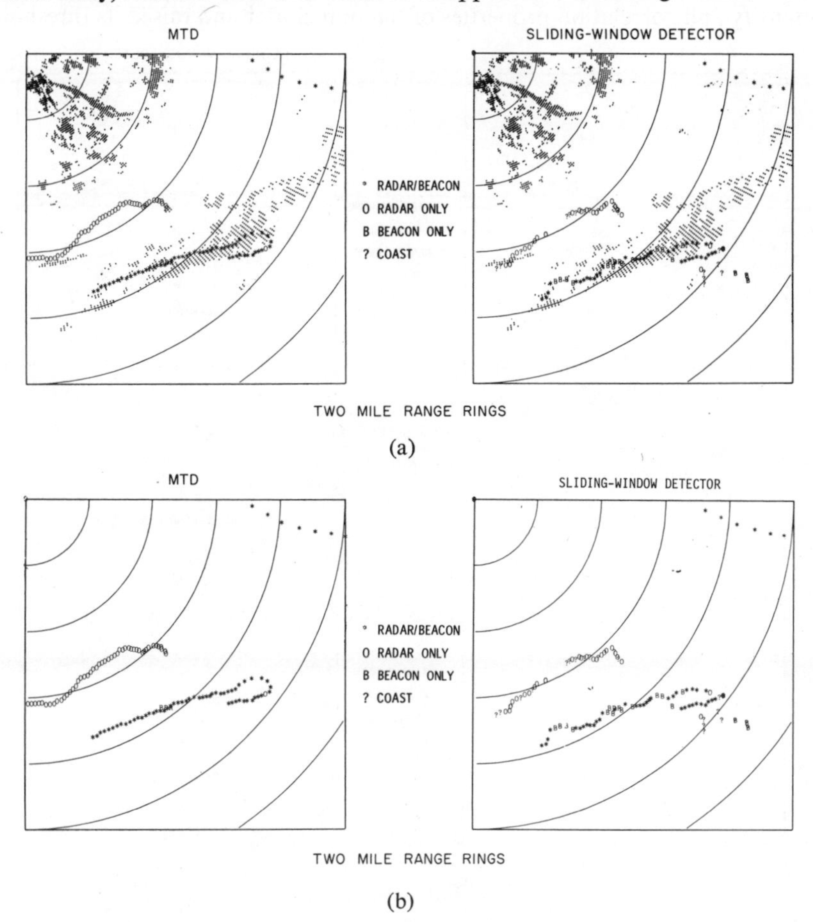

FIGURE 5-47. Tracking tests of the MTD in ground clutter: (a) With ground clutter superimposed; (b) Aircraft tracks alone.

when the aircraft flies tangential to the radar. This demonstrates the value of the ground clutter map in seeing zero-radial-velocity aircraft with the MTD because of the very large (100 to 1000 m^2) cross section presented by the side of an aircraft. Conventional MTI, however, puts an absolute null at zero velocity giving no hope for detection of tangential aircraft.

Figure 5-48 depicts what happens in a rain storm. With its subweather visibility, the MTD experiences no difficulty detecting aircraft in rain. It tracks non-beacon-equipped aircraft in rain as well as if they were in the clear. As seen in Fig. 5-48, rain causes loss of radar detection (B is for beacon detection only) for the sliding-window detector. The false alarms were controlled quite well by the sliding-window detector, because it measured the intensity and correlation properties of the rain clutter and raised its threshold

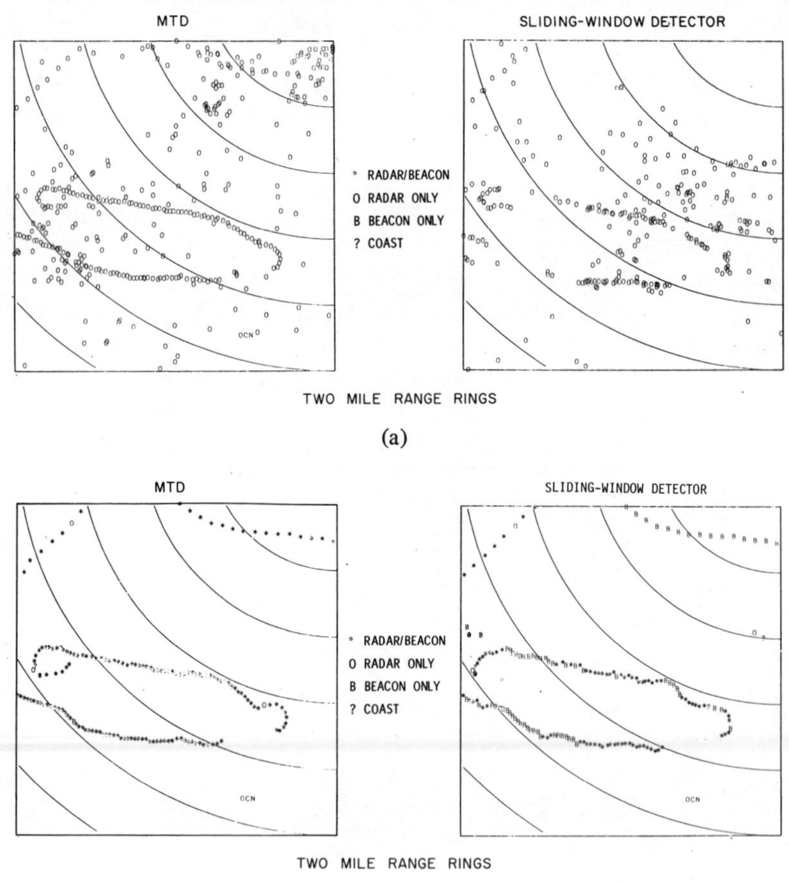

FIGURE 5-48. Detection tests of the MTD in rain: (a) 80 scans of raw data sent to tracker; (b) Tracker output.

accordingly. Unfortunately with this raised threshold, it could not detect the aircraft. Again, if the aircraft were not beacon equipped, it would not have been tracked in the region where extensive beacon only detections (Bs) occurred.

5.10 SYNTHETIC APERTURE RADAR

Synthetic aperture radar (SAR) is a signal-processing technique for improving the azimuth resolution beyond the beamwidth of the physical antenna actually used in the radar system. The azimuth resolution of a conventional antenna with physical length L is

$$\mathrm{Res}_{\mathrm{AZ}} = \frac{\lambda R}{L} \tag{5.59}$$

where R is the range from the antenna to the target and λ is the wavelength of the RF signal. This resolution is just the antenna beamwidth times the range to the target. The resolution can be improved either by increasing the antenna length or by increasing the carrier frequency of the transmitted signal. In SAR, one can view the signal processing as a means of creating an artificial antenna length much greater than L. This is accomplished by moving the physical antenna position as a function of time, and coherently processing the echoes received during the time taken to traverse the synthetic antenna length. In this way, the azimuth resolution of the SAR can exceed that of the physical antenna. Two modes of SAR will be discussed in this section: focused SAR for strip mapping and focused SAR in spotlight mode. Before considering the details of digital processing, we shall describe an idealized situation which illustrates the types of signals that arise naturally in SAR.

5.10.1 Theory

Consider the typical geometry of the SAR system for strip mapping as in Fig. 5-49. An airplane, carrying an antenna with physical length L, flies along a straight line at a constant velocity v over the ground at a height h. The antenna, pointed perpendicularly to the flight path, illuminates targets on the ground. The received signals are stored and processed to make an image of the terrain. It is also possible (and common practice) to squint the antenna, usually by looking in the forward direction. In the spotlight mode, the antenna is steered to continuously illuminate one spot on the ground for very high resolution mapping.

In order to develop the nature of the signal-processing algorithms, consider the idealized two-dimensional situation of Fig. 5-50 in which height has been suppressed. Assume for the moment that the transmitted signal is a continuous sinusoid and that there is a single point target located at coordi-

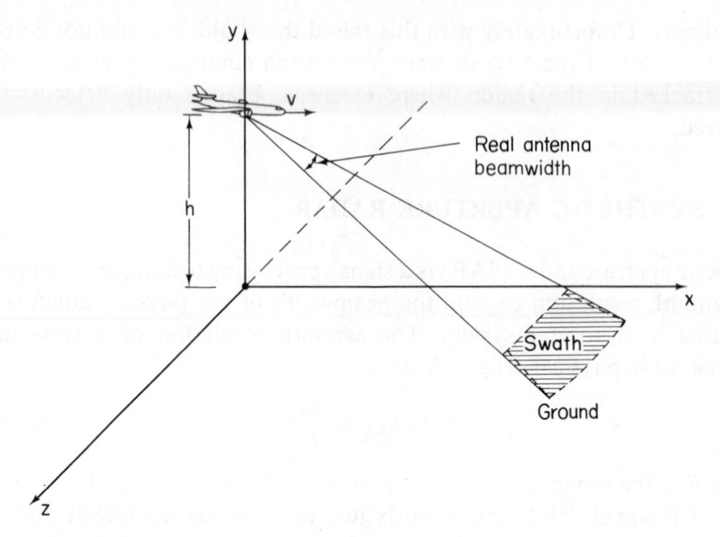

FIGURE 5-49. Typical synthetic aperture radar mapping situation.

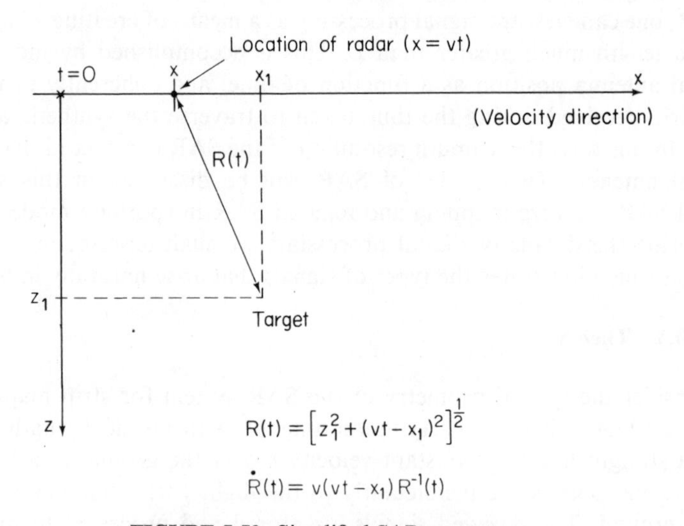

FIGURE 5-50. Simplified SAR geometry.

nates (x_1, z_1). No range resolution is possible with such a signal; we shall consider the problem of range resolution later. Since the physical antenna has a beamwidth of λ/L, the point target is illuminated when the radar antenna lies between $x_i = x_1 - \lambda z_1/2L$ and $x_f = x_1 + \lambda z_1/2L$. The relative motion between the target and the antenna introduces a Doppler shift in the received signal. It is this Doppler shift that will be exploited through coherent signal processing to obtain better azimuth information.

Since the round-trip travel time between the antenna and the target is $2R(t)/c$, the received signal is

$$r(t) = s\left(t - \frac{2R(t)}{c}\right) = e^{j\omega_0\left(t - \frac{2R(t)}{c}\right)} \tag{5.60}$$

We shall assume that the amplitude of the received signal $a(t)$ remains constant while the target is illuminated. In reality, the received amplitude will vary with position because the antenna pattern falls off to its half power points at x_i and x_f. Since the effect is known and fixed, it can be removed in the signal processing.

The instantaneous frequency of the received signal is the time derivative of the phase:

$$\begin{aligned} \omega_i(t) &= \omega_0 - \frac{2\omega_0}{c}\dot{R}(t) \\ &= \omega_0 - \frac{2\omega_0 v}{c}(vt - x_1)R^{-1}(t) \end{aligned} \tag{5.61}$$

The second term is the Doppler-shifted frequency component. By expanding $R(t)$ in a Taylor series about $t_1 = x_1/v$, we can approximate the Doppler frequency term by a linear FM term.

$$R(t) \cong z_1 + \frac{1}{2z_1}(vt - x_1)^2 \tag{5.62}$$

Thus

$$r(t) \cong e^{-j\frac{\omega_0}{cz_1}(vt - x_1)^2} e^{j\omega_0\left(t - \frac{2z_1}{c}\right)} \tag{5.63}$$

Hence the point target gives rise to a linear FM azimuth signal component at the receiver. The LFM rate is $2\omega_0 v^2/2\pi c z_1$, the time duration of the FM pulse is $\lambda z_1/vL$, the swept bandwidth is $(2v)/L$, and the time-bandwidth product is $2\lambda z_1/L^2$. This derivation suggests that the azimuth location of the point target can be determined by pulse compression matched filtering in azimuth of the received LFM signal. The azimuth resolution obtainable will be proportional to the width of the compressed pulse. Recall that the LFM compressed pulse has a compressed pulse width equal to $1/W$, where W is the swept bandwidth of the chirp. Converting from time resolution to azimuth resolution, we obtain $\text{Res}_{Az} = v(2v/L)^{-1} = L/2$. Hence we have demonstrated in this idealized case that the resolution of focused SAR is theoretically independent of both range and wavelength. It is important to note that the time-bandwidth product of the chirp depends on z_1; thus different matched filtering would have to be performed as a function of range. This results in a very complex signal processor with very severe computational requirements.

The pulse compression response can be weighted to reduce sidelobes at the cost of broadening the mainlobe just as in the case of the traditional matched filter discussed above (Sec. 5.4.2). Since, in general, the received signal will be the superposition of many echoes of varying amplitudes, the energy in the sidelobes will be an important factor in determining the dynamic range of the final map.

Even though most of the assumptions made in the preceding derivation are only approximately true in a practical situation, the essential ideas of SAR signal processing remain the same. We shall now examine the impact of some of our assumptions in more detail. First, the Taylor series approximation that leads to the LFM azimuth signal is only valid for small values of t around the time $t_1 = x_1/v$. To put it another way, the size of the synthetic antenna is limited when LFM pulse compression processing is employed. Also it is important that the target remain within the same range resolution cell while the antenna traverses its synthetic length because the azimuth pulse compression is performed for a constant range. Thus it is not possible to attain the theoretical azimuth resolution of $L/2$ in an actual system. A more serious practical problem is that the airplane does not maintain a constant-velocity linear flight as assumed and thus the received signal deviates from the ideal LFM signal. If a high-resolution map is desired it is necessary to measure the actual flight path using a high-precision inertial navigation system and then correct the phase of the received signal prior to azimuth pulse compression.

The foregoing analysis assumed the radiation of a continuous sinusoidal signal. Such a transmitted signal does not allow the determination of range, the other dimension required for developing an SAR map. Instead, a burst waveform is usually transmitted. Responses in a single range resolution cell are collected over several consecutive transmissions, and each range resolution cell is processed separately. The PRF of the radar must meet two requirements to limit ambiguities in both the range and Doppler (azimuth) dimensions. The avoidance of range ambiguities is accomplished, as before, by choosing the IPP to be less than $2R_{\text{max}}/c$. On the other hand, since pulsing the transmitter is equivalent to sampling the LFM signal returned by a point target, it is necessary to sample fast enough, in accordance with the Nyquist criterion, to avoid aliasing the azimuth frequency spectrum. Since the bandwidth of the LFM signal is $2v/L$, the PRF must satisfy

$$2v/L < \text{PRF} < c/2R_{\text{max}} \tag{5.64}$$

We have assumed that the radiation of a narrow pulse is the means for obtaining range resolution. It is more likely that pulse compression matched filtering would be used to obtain range resolution prior to performing azimuth pulse compression. Note that the pulse compression in the range dimension is independent of that in the azimuth dimension.

5.10.2 Computational Requirements

The processing technique just outlined requires matched filtering in two dimensions—range and azimuth. Traditionally, optical signal processing has been employed in SAR systems to handle the large amount of data storage and computation involved. Optical processing is usually done in nonreal time using data that have been recorded on photographic film. The processor is implemented using a highly complex lens system. Such systems have several drawbacks. They can be costly and are generally limited to making a strip map while flying in an almost straight path. Also motion compensation is difficult to implement because the optical processor is not flexible.

Although the desire for greater flexibility and real-time operation suggests digital processing, several bottlenecks are encountered: (1) the high input signal bandwidth, (2) the large storage requirement, and (3) the staggering computational load for high-resolution mapping. In all three areas, technology will have a great influence on the ultimate structure of an SAR digital processor. The storage of sufficient data for digital processing and the required processing speed are discussed next.

Data collection in a SAR involves gathering data over all ranges of interest for each transmitted pulse. If pulse compression matched filtering is used to obtain range resolution, this processing can be performed directly on the incoming data. The data from many transmissions is then stored prior to azimuth processing. If the data are arranged as in Fig. 5-51, the matrix

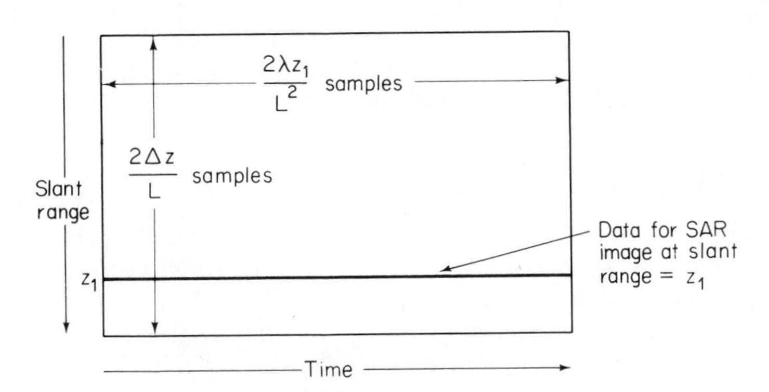

FIGURE 5-51. Two-dimensional arrangement of data storage for SAR processing.

(ordered as range versus time) is filled column by column. To perform azimuth compression, the data must be processed along one row at a time. The memory requirements are severe because a large amount of data are being collected. Furthermore, the computational requirements are also nontrivial because a different azimuth pulse compression filter is required for each range resolution cell. For example, if we assume that the range and azimuth resolutions are to be equal and that the total range interval to be mapped is Δz, then each memory column in Fig. 5-51 will contain $\Delta z/(L/2) = 2\,\Delta z/L = M$ samples, and $N(z_1) = 2\lambda z_1/L^2$ samples along each row (assuming Nyquist rate sampling) will be needed in the calculation of the final output at one point in range-azimuth space. Hence the total storage requirement is

$$\text{Memory} = \frac{2\,\Delta z}{L} \cdot \frac{2\lambda R_{\max}}{L^2} = \frac{4\,\Delta z\,\lambda R_{\max}}{L^3} \tag{5.65}$$

The total amount of calculation is that required for the operation of M azimuth pulse compression channels in parallel. Each channel requires a matched filtering operation involving $N(z_1) = 2\lambda z_1/L^2$ points, so the total computation time using direct convolution is upper bounded by

$$\frac{2M\lambda R_{\max}}{L^2} \qquad \text{multiplications per IPP} \tag{5.66}$$

Some typical numbers will show that these requirements are extremely high. Assuming that $\Delta z = 10$ km, $\lambda = 3$ cm, $R_{\max} = 10$ km, $L = 2$ m, and $v = 400$ m/s, the multiplication time must be less than 1.67 ns (6×10^8 multiplications/s) and the amount of storage must exceed 1.5×10^6 complex words.

The use of the FFT for high-speed convolution together with some parallelism might reduce the speed requirement significantly, but the rate would still be formidable. Furthermore, the storage requirements would increase because additional data buffering would be needed. In the spotlight mode for SAR the speed advantages of the FTT can be exploited further to implement a practical processor.

5.10.3 Spotlight Mode

The special form of the received azimuth signal (i.e., LFM) can be exploited to derive a processor that uses the FFT to do processing equivalent to pulse compression. Such processing is sometimes referred to as stretch* processing. The derivation rests on the fact that the azimuth resolution of the system is obtained from the different Doppler shifts of the individual targets. Consider the response from two targets, both at range z_1, one located at $x = x_1$ and the other at $x = x_2$. Using the LFM approximation derived above, the received signal can be written as the sum of two chirps.

$$r(t) = e^{-j\frac{\omega_0}{cz_1}(vt-x_1)^2} + e^{-j\frac{\omega_0}{cz_1}(vt-x_2)^2} \tag{5.67}$$

If we consider x_1 to be a reference point and multiply $r(t)$ by the conjugate of the chirp signal received from the scatterer at $x = x_1$, we obtain

$$1 + e^{-j\frac{\omega_0}{cz_1}[2(x_1-x_2)vt-x_2^2-x_1^2]} \tag{5.68}$$

* More precisely, LFM correlation and spectral analysis.

This process is sometimes called dechirping because the quadratic phase term has been removed in Eq. 5.68.

The discrimination between the two targets can now be performed with a spectrum analyzer. The reference target at $x = x_1$ will be analyzed as zero frequency while the target at $x = x_2$ will be analyzed as frequency $2\omega_0 v(x_1 - x_2)/cz_1$. This suggests a digital signal-processing structure consisting of a multiplier for dechirping followed by an FFT. The computational efficiency of the FFT is an important ingredient for obtaining a processor that can function in real time.

The FFT processing is limited to a small region around the reference point, because the LFM approximation is only valid near $x = x_1$. Thus this method of signal processing naturally lends itself to a spotlight mode for SAR. Small areas around a central reference point are imaged individually and a large map is constructed by piecing together several submaps.

Another consideration in the formation of the SAR image using the FFT is the frequency range to be analyzed after dechirping. The azimuth signal is normally sampled by the pulsed nature of the transmitter at a rate greater than the bandwidth of the chirps received from the individual point targets illuminated by the antenna beam. However, it is usually the case that the spectrum analysis to be performed by the FFT is over a much smaller frequency range (owing to the practical limitations of the LFM approximation). Therefore, it is advantageous to band limit the dechirped azimuth signal to the frequency range of interest and do a smaller FFT. In synthetic aperture radar, this operation is termed *presumming*, since one common method merely adds together adjacent points. In digital signal-processing terms, this operation can be recognized as time decimation where one is seeking to reduce the sampling rate. A digital lowpass filter is employed to avoid aliasing the frequency band of interest. The simple scheme of adding together adjacent samples is equivalent to using a lowpass filter whose impulse response is a rectangular window and whose frequency response is of the form $\sin N\omega/N \sin \omega$. This presumming operation greatly reduces the processing load for the spotlight mode of SAR. The high frequency-domain sidelobes of the rectangular window filter can be overcome by using a different window or perhaps an optimal lowpass filter. Of course, there will be additional hardware complexity, because the rectangular window requires no multiplications, only additions.

Finally, this FFT processor for spotlight SAR has another attractive feature in its implementation. Since the azimuth history is multiplied by a reference signal, it is easy to implement a motion compensation scheme by changing the reference signal to reflect the actual flight path flown. The reference azimuth signal would be LFM if the path were perfectly straight, but in other cases the phase modulation of the reference point could be derived from an inertial navigation system in order to correct for a nonlinear flight path flown at a nonuniform velocity. Such motion compensation is essential for very high-resolution mapping. The flexibility of digital processing makes it feasible to implement the motion compensation in real time as the data are being received.

5.11 SUMMARY

In this chapter we have considered the application of digital signal processing to several types of radar systems. Digital processors are attractive because of their flexibility and their ability to attain high precisions. The flexibility allows the processor to implement many different algorithms and to adapt easily to changing target environments. The precision of the processor is under the control of the system designer who can trade off the performance of the system versus the hardware cost of longer word lengths.

Our discussion of different radar systems has shown that the FFT algorithm plays an important role both in the implementation of high-speed convolution and in spectrum analysis. The presentation of a design of a pipeline FFT matched filter indicates one way in which very high speed can be obtained by using a high degree of parallelism together with very fast circuitry.

There is always a technological tradeoff to be made when designing a digital system for radar or for any other application. Many types of ICs are available, ranging from highly integrated ICs with modest speed to high-speed ICs with lower levels of integration. Given the high-speed requirements of a radar system, the recent trend has been to use high-speed technologies. However, as LSI technologies become more available, and as the levels of integration continue to increase, it is worthwhile to consider parallel implementations to achieve the same speed. The trend in IC technology to develop faster and more highly integrated ICs will encourage the design and implementation of digital systems that can process even higher bandwidth waveforms over even longer integration times than is now possible.

ACKNOWLEDGEMENT

This work was sponsored by the Department of the Air Force, the Department of the Army, and the Federal Aviation Agency. However, the views and conclusions contained in this document are those of the authors and should not be interpreted as necessarily representing the official policies, either expressed or implied, of the U.S. Government.

REFERENCES

1. J. ALLEN, "Computer Architecture for Signal Processing," *Proc. IEEE*, vol. 63, no. 4, Apr. 1975, pp. 624–633.
2. R. S. BERKOWITZ, ed., *Modern Radar: Analysis, Evaluation, and System Design*, John Wiley and Sons, Inc., New York, 1965.
3. P. E. BLANKENSHIP and E. M. HOFSTETTER, "Digital Pulse Compression via Fast Convolution," *IEEE Trans. Acoust. Speech and Signal Proc.*, vol. ASSP-23, Apr. 1975, pp. 189–201.
4. P. E. Blankenship, "LDVT: High Performance Minicomputer for Real-Time Speech Processing," *EASCON 75 Record, IEEE Pub. 75 CHO 99805 EASCON*, pp. 215A-G.
5. W. M. BROWN and L. J. PORCELLO, "An Introduction to Synthetic Aperture Theory," *IEEE Spect.*, Sept. 1969, pp. 52–62.
6. W. W. CAMP, M. AXELBANK, V. L. LYNN, and J. MARGOLIN, "ALCOR—A High Sensitivity Radar with 0.5 m Range Resolution," *IEEE 1971 Intern. Conv. Digest*, March 1971, pp. 112–113.
7. C. E. COOK and M. BERNFELD, *Radar Signals, An Introduction to Theory and Application*, Academic Press, Inc., New York, 1967.
8. L. J. CUTRONA, "Synthetic Aperture Radar," Chapter 23 in *Radar Handbook*, M. I. Skolnik, ed., McGraw-Hill, New York, 1970.
9. G. W. DELEY, "Waveform Design," Chapter 3 in *Radar Handbook*, M. I. Skolnik, ed., McGraw-Hill, New York, 1970.
10. D. F. DELONG and E. M. HOFSTETTER, "On the Design of Optimum Radar Waveforms for Clutter Rejection," *IEEE Trans. Info. Theory*, vol. IT-13, July 1967, pp. 454–463.
11. A. M. DESPAIN, "Fourier Transform Computers Using CORDIC Iterations," *IEEE Trans. Comput.*, vol. C-23, no. 10, Oct. 1974, pp. 993–1001.
12. DSP Committee IEEE ASSP Society, *Selected Papers in Digital Signal Processing, II*, IEEE Press, New York, 1976.
13. A. E. FILIP, "A Baker's Dozen Magnitude Approximation and Their Detection Statistics," *IEEE Trans. Aerospace and Electronic Systems*, vol. AES-12, no. 1, pp. 87–89, Jan. 1976.
14. B. GOLD and C. RADER, *Digital Processing of Signals*, McGraw-Hill, New York, 1969.
15. B. GOLD, I. L. LEBOW, P. G. MCHUGH, and C. M. RADER, "The FDP, A Fast Programmable Signal Processor," *IEEE Trans. Comput.*, vol. C-20, no. 1, Jan. 1971, pp. 33–38.

16. B. Gold and T. Bially, "Parallelism in Fast Fourier Transform Hardware," *IEEE Trans. Aud. and Electroacoust.*, vol. AU-21, Feb. 1973, pp. 5–16.

17. D. N. Graham, "FFT Algorithm," *Proc. of the 1971 IEEE Intern. Comput. Society Conf., IEEE Pub.*, No. 71C41-C, pp. 11–12.

18. H. L. Groginsky and G. A. Works, "A Pipeline Fast Fourier Transform," *IEEE Trans. Comput.*, vol. C-19, Nov. 1970, pp. 1015–1019.

19. R. O. Harger, *Synthetic Aperture Radar Systems Theory and Design*, Academic Press, New York, 1970.

20. R. C. Houts and D. W. Burlage, "Design Procedure for Improving the Usable Bandwidth of an MTI Radar Signal Processor," 1976 *IEEE Intern. Conf. ASSP*, Apr. 12–14, 1976, pp. 745–748.

21. J. C. Kirk, "A Discussion of Digital Processing in Synthetic Aperture Radar," *IEEE Trans. Aerospace and Electronic Systems*, vol. AES-11, no. 3, May 1975, pp. 326–337.

22. J. R. Klauder, A. C. Price, S. Darlington, and W. J. Albersheim, "The Theory and Design of Chirp Radars," *Bell System Tech. J.*, vol. 39, no. 4, July 1960, pp. 745-808.

23. Y. G. Lundh, "Multi-Processor Systems for High-Capacity Signal Processing," *Record of the International Conference on Computer Communications, Stockholm, Sweden*, Aug. 1975, pp. 325–329.

24. J. I. Marcum and P. Swerling, "Studies of Target Detection by Pulsed Radar," *IRE Trans.*, vol. IT-6, 1960, pp. 59–308.

25. C. E. Muehe, L. Cartledge, W. H. Drury, E. M. Hofstetter, M. Labitt, P. B. McCorison, and V. J. Sferrino, "New Techniques Applied to Air-Traffic Control Radars," *Proc. IEEE*, vol. 62, no. 6, June 1974, pp. 716–723.

26. C. E. Muehe, "Advances in Radar Signal Processing," Presented at ELECTRO '76, 11-14 May 1976.

27. D. O. North, "An analysis of the Factors Which Determine Signal-Noise Discrimination in Pulsed Carrier Systems," *RCA Lab. Rept. PTR-6C*, June 1943, also *Proc. IEEE*, vol. 51, 1963, pp. 1016–1027.

28. R. M. O'Donnell, C. E. Muehe, M. Labitt, W. H. Drury, and L. Cartledge, "Advanced Signal Processing for Airport Surveillance Radars," *EASCON* **74** *Record*, pp. 71A–71F.

29. G. C. O'Leary, "Nonrecursive Digital Filtering Using Cascade Fast Fourier Transforms," *IEEE Trans. Aud. and Electracoust.*, vol. AU-21, Feb. 1973, pp. 5–16.

30. A. V. Oppenheim and R. W. Schafer, *Digital Signal Processing*, Prentice-Hall, Englewood Cliffs, N.J., 1975.

31. A. V. Oppenhaim and C. J. Wienstein, "Effects of Finite Register Length in Digital Filtering and the Fast Fourier Transform," *Proc. IEEE*, vol. 60, Aug. 1972, pp. 957–975.

32. S. D. Pezaris, "A 40-ns 17-Bit by 17-Bit Array Multiplier," *IEEE Trans. Comput.*, vol. C-20, no. 4, Apr, 1971, pp. 442–447.

33. R. J. Purdy et al., "Digital Signal Processor Designs for Radar Applications," M.I.T., Lincoln Laboratory, Lexington, Mass., Tech. Note 1974–58, Vols. 1 and 2, 31 Dec. 1974.

34. L. R. Rabiner and B. Gold, *Theory and Application of Digital Signal Processing*, Prentice-Hall, Englewood Cliffs, N. J., 1975.

35. L. R. Rabiner and C. M. Rader, *Selected Papers in Digital Signal Processing*, IEEE Press, New York, 1972.

36. A. W. Rihaczek, *Principles of High-Resolution Radar*, McGraw-Hill, New York, 1969.

37. *Semiconductor Data Library, Volume 4, MECL Integrated Circuits*, Motorola Semiconductor Products Inc., 1974.

38. M. I. Skolnik, *Introduction to Radar Systems*, McGraw-Hill, New York, 1962.

39. M. I. Skolnik, ed., *Radar Handbook*, McGraw-Hill, New York, 1970.

40. "Special Issue on Modern Radar Technology and Applications," *Proc. IEEE*, vol. 62, no. 6, June 1974.

41. T. G. Stockham, "High Speed Convolution and Correlation," in *1966 Spring Joint Comput. Conf., AFIPS Conf. Proc.* vol. 28., Spartan Books, New York, 1966, pp. 229–233.

42. *The TTL Data Book for Design Engineers*, Texas Instruments, Inc., 1973.

43. J. E. Thornton, "Parallel Operation in the Control Data 6600," *AFIPS Conf. Proc.*, vol. 26, Part II, *1964 Fall Joint Comput. Conf.*, pp. 33–40.

44. J. E. Volder, "The CORDIC Trigonometric Computing Technique," *IRE Trans. Elect. Comput.*, vol. EC-8, pp. 330–334, Sept. 1959.

45. P. M. Woodward, *Probability and Information Theory, With Applications to Radar*, Pergamon Press, Elmsford, N. Y., 1953.

46. J. M. Wozencraft and I. M. Jacobs, *Principles of Communication Engineering*, John Wiley and Sons, New York, 1965.

6

SONAR SIGNAL PROCESSING

A. B. Baggeroer

Massachusetts Institute of Technology
Cambridge, Mass. 02139

6.1 INTRODUCTION

In any modern sonar system the signal processing is one of the most important components. When combined with the transducer system, it provides our senses of the ocean environment. One way by which this can be demonstrated is to examine the evolution of sonar systems. Early systems implemented crude filters and arrays. It soon became evident that a much better understanding of the ocean environment and the physics of signal propagation was required to interpret what the sonar system was indicating to an observer. When the influence of the oceanic environment was appreciated, signal-processing methods were designed that were capable of exploiting the signals at the transducers. This has continued to be a mutually stimulating process in the development of sonars: improving the signal-processing performance has required a more complete understanding of sound propagation in the ocean environment, and improving the acoustical models has led to research on more sophisticated signal processing.

As these acoustical models and signal processors evolved, the flexibility required to implement them on a sonar system has rapidly outgrown the capabilities of even the most sophisticated analog systems. These demands naturally led to employing digital systems. Initially, these were either special-purpose devices, particularly those based upon hard clipping, or general-purpose digital computers, especially those used for research on potential algorithms. At first, this research led to hard-wired digital processors; but now the trend, just as in many other fields, is toward microprocessors and minicomputers.

In this chapter our intent is to review the important concepts in sonar signal processing. Our emphasis is upon the acoustical models and the operations that the signal processing attempts to implement. We do not dwell on the details of specific implementations; we suggest that the interested reader examine Chapter 5 on radar in which the emphasis is more upon implementation issues. In many respects, particularly those concerned with matched filters, the issues and implementations are quite similar.

It is difficult to survey the entire field of sonar systems; we have, however, indicated several of the important ones in Fig. 6-1. In the past, active and passive military sonars have dominated the activity in the field. This activity was summarized in an extensive series of reports covering the intense period of research during World War II.[(1)] Presently, commerical and civilian activity in the oceans is increasing at a very rapid rate. With acoustics being the only viable means for sensing the ocean environment in most applications, nonmilitary systems are receiving much more attention.[(2)]

We have organized our material in the following manner. In Sec. 6.2, we examine some of the important characteristics of underwater acoustic signal propagation. Our intent here is not to be exhaustive, but only to be illustrative of how the oceanographic environment influences the various models of the signal propagation and subsequent signal processing. In sonar the ocean imposes some very severe limitations upon signal propagation, so it is very important to understand these influences. For the reader interested in more detail, several comprehensive treatments are available.[(3–6)] In addition, current research is published in several journals; the *Journal of the Acoustical Society of America* is the principal medium.

In Sec. 6.3, we discuss active sonar systems. In these systems the user supplies, or transmits, the acoustic energy. This energy is observed after it has propagated through the water and/or is reflected by a target. Since the user generates the energy and controls the time of transmission, he can design the signals to match the characteristics of the acoustic environment to best suit his particular needs. Examples of active sonar applications include target detection and localization, communications, navigation, and mapping and charting. The methods and implementations used in active sonar signal processing have much in common with radar and radio communications. As a result, we have employed the literature that supports both of these fields, as well as the sonar field, in our discussion. We can not emphasize, however, the mathematical details of the supporting theory in this short space; refs. 7–10 are indicative of texts where various aspects of the underlying theories can be found.

Passive sonar systems are the focus of Sec. 6.4. With these systems, one listens to signals that are radiated by the various sources of acoustic energy in the ocean. Some of these sources are natural, arising from wind, earthquakes, and marine life. The most interesting ones, however, are man-made

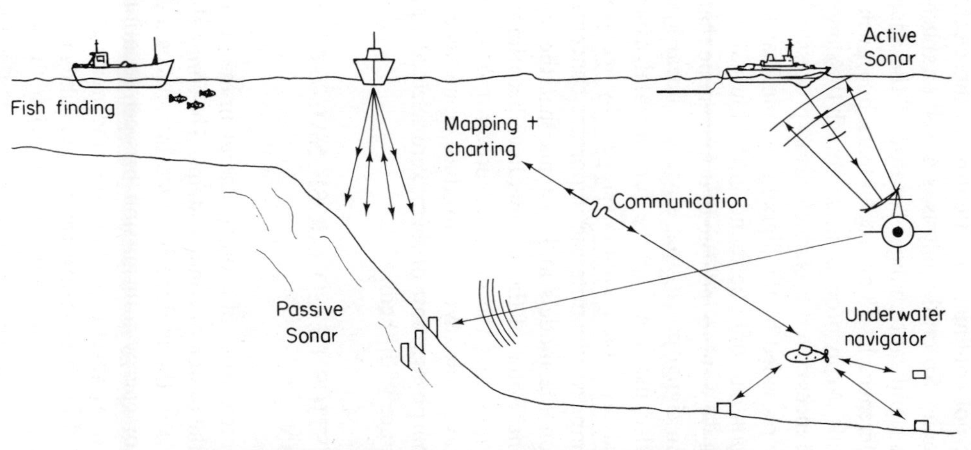

FIGURE 6-1. Several examples of sonar system applications.

by shipping and military vessels. The most important passive sonars are surveillance systems. As a result, we can discuss the principles of operation, but we cannot be too specific because of the classified aspects of the problem. The most important forms of signal processing in passive sonars are spectral analysis and array processing. We have used the literature of both these fields extensively, and have not confined our attention to those references directly applied to sonar systems. Spectral analysis is a well-established field that is currently undergoing significant changes because of the adaptability of the digital computer. References 11–13 are useful texts, but much of the material only appears in journals. Array processing is in much the same state of affairs. There are several references covering classical array-processing methods; however, the adaptability provided by the digital computer has led to many new algorithms, which again can only be found in the journals.[14–16]

Before proceeding we want to indicate that we assume that the reader has a good background in signal processing, especially linear system and transform analysis. We also use some of the fundamental aspects of random processes quite extensively. Finally, our coverage is by necessity in the nature of a survey; we attempt to illustrate the important concepts and methods. Our emphasis is upon the models and systems that the signal processing attempts to implement. Some of the systems have been implemented, while others will soon be, now that the increased power and flexibility of digital signal processing is available. We do introduce some specific applications only for illustrative purposes; most of them were selected simply because of the author's aquaintance with them.

6.2 CHARACTERISTICS OF SONAR SIGNAL PROPAGATION

The propagation characteristics of acoustic energy in the ocean have a dominant influence upon the design of sonar systems. The sonar channel is particularly difficult because of the multipath, which introduces a diverse number of modes or paths for energy transmission between locations. In addition the signals are altered by several phenomena, and noise is always present. Finally, the ocean itself imposes severe physical demands upon the platforms and devices employed in transmitting and receiving the signals. All these issues have been, and still are, the subject of an intense amount of research. Applications to military sonars have stimulated most of the research; however, other areas such as oceanography, petroleum exploration, navigation, and coastal zone investigations have contributed significantly.

Our interest here is to introduce only enough of the pertinent phenomena so that one can appreciate the motivation for the propagation models and the resulting signal-processing methods. In the development of this chapter, we examine the important propagation models, both ambient and reverberant noise, the influence of the sea-floor and sea-surface boundaries, coherence and fluctuations of the signals, and some of the constraints upon the platforms and devices. We endeavor to indicate how each of these influences the signal-processing algorithms in a sonar system. As our knowledge about these phenomena improves and the flexibility and speed of the signal-processing devices increases, more sophisticated algorithms will certainly be implemented to exploit this knowledge.

6.2.1 Propagation in the Water Column

Three important propagation factors can have a dominant influence upon the design of a sonar system: spreading, absorption, and ducting. Spreading is simply a geometrical loss; however, the structure of the ocean can lead to models different than simple *free-space* spreading. Sound absorption in the ocean can be remarkably low, especially when one compares it to acoustic transmission in the atmosphere or electromagnetic transmission in seawater, Nevertheless, some very important frequency dependences must be introduced. Finally, it is important to observe that in spite of some very impressive depths, the ocean is a comparatively thin layer on the surface of the earth. As a result, it does not require a very large horizontal distance to introduce phenomena that are commonly associated with wave guides.

Spreading is modeled either with a free-space spherical geometry leading to a $1/r^2$ dependence for the signal energy, or with a horizontally stratified geometry with a $1/r$ dependence. At short ranges the spherical model is probably more appropriate; at longer ranges, the cylindrical model is. In situations where interaction with the seabed is important, e.g., shallow water propagation, several experiments and theories suggest that a $1/r^{3/2}$ dependence is appropriate. The most important observation is that the range dependence is algebraic, so one need be concerned with it only when the range is short and absorption effects are not important.

At long ranges, losses due to absorption usually dominate system design since they have an exponential range dependence. Figure 6-2 illustrates the absorption loss versus frequency as an attenuation coefficient in decibels (dB) per kilometer (km) re 1 bar at 1 meter (m). In addition to the exponential range dependence implied, the attenuation coefficient has an approximate quadratic dependence versus frequency. (The steps in the curve of Fig. 6-2 are introduced by the onset of certain relaxation phenomena in the dissolved salts present in seawater.) This frequency dependence places some severe constraints upon several of the signal-processing parameters in a sonar system.

For active systems, high angular resolution and wide bandwidth signals suggest the use of high frequencies. The maximum usable frequency, however, is limited by absorption losses if one is to operate with transmitter sources

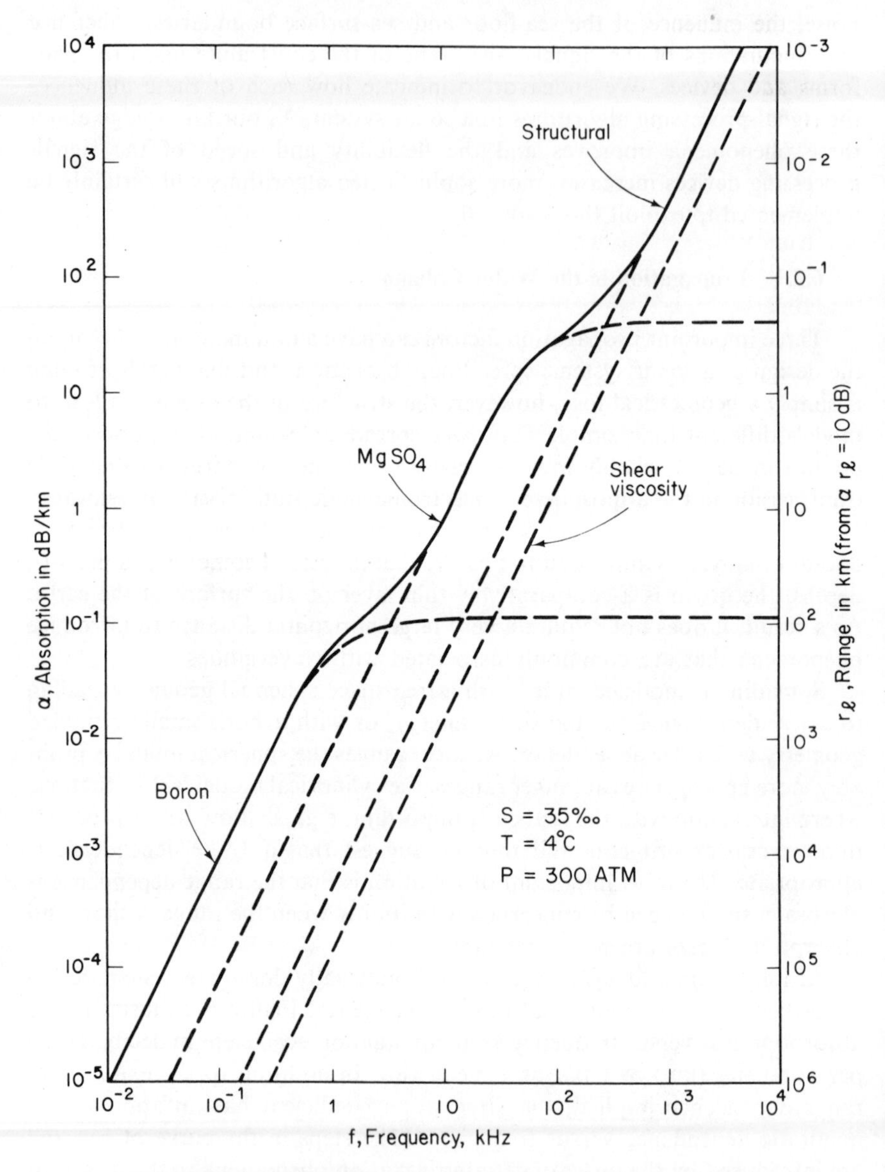

FIGURE 6-2. Volumetric absorption including all known relaxation processes (from Dyer, 122).

of reasonable power. Typically, active systems operate in the range of 2 to 40 kilohertz (kHz), with 3.5, 12, and 25 kHz being popular frequencies for which a large amount of equipment is marketed. At higher frequencies, the absorption losses are usually too high (e.g., 10 dB/km at 50 kHz). (Recreational depth sounders operate at even higher frequencies, e.g., at 200 kHz, and absorption losses make it possible for many boats to use the same frequency band without mutual interference.) At lower frequencies, the losses are lower, but the signal-processing parameters are adversely affected and propagation becomes more complex. The combination of these effects leads to maximum operating ranges of 10 to 100 km for most active systems.

Passive sonar systems operate within the low-frequency portion of the spectrum simply because absorption losses attenuate the higher components. At very low frequencies, signals can propagate over very long distances. Many experiments have demonstrated that propagation over global distances is possible.(17) Two phenomena make this possible, the low absorption losses and the stratification of the ocean, which guides the acoustic energy within various paths and ducts. Thus it is the low absorption losses in the ocean that both makes passive sonar possible and leads to the interest in the low-frequency portion of the spectrum.

The guiding of acoustic energy is the result of the comparative thinness of the ocean depth and its internal stratification. Figure 6-3 illustrates a typical bathymetric profile of the ocean where there is no vertical exaggeration. In addition, we have superimposed the wavelengths across the sonar band for geometrical comparison. The depths commonly extend from relatively shallow ones of 100 m or less on the continental shelves to 4 to 5 km typical of the ocean basins. In shallow water, it is easy to observe that the water depth is of the same order as the low-frequency wavelengths. In deep water, the water depth is usually quite large compared to the wavelengths; however, the stratification of the ocean leads to a sound velocity profile as a function of depth, which acts as a waveguide by introducing refraction. In studying ocean propagation, the wavelength of the acoustic energy along a particular propagation direction sets an important scale. In its simplest form, when the oceanic waveguide is comparable to this wavelength, phenomena unique to trapped energy, such as modes and dispersion, must be introduced in the signal-processing models.

We can illustrate some of the phenomena associated with shallow- and deep-water propagation with examples. In the first, the *Pekeris* wave guide is used to model shallow-water propagation where there is interaction at the ocean–bottom interface. The model is illustrated in Fig. 6-4. (Shallow water is often delimited by considering the water depth to be less than ten times the acoustic wavelength, but the division is not that sharp.) When one solves the Helmholtz wave equation by separation of variables, and imposes the appropriate boundary conditions at the sea surface and sea bottom, one finds that there are several modes in which energy propagates. These are summarized in Fig. 6-5, where we indicate temporal frequency–horizontal wave number dependence of the modes and the group velocity dispersion curves of the modes. This leads to the following type of propagation when a broad-

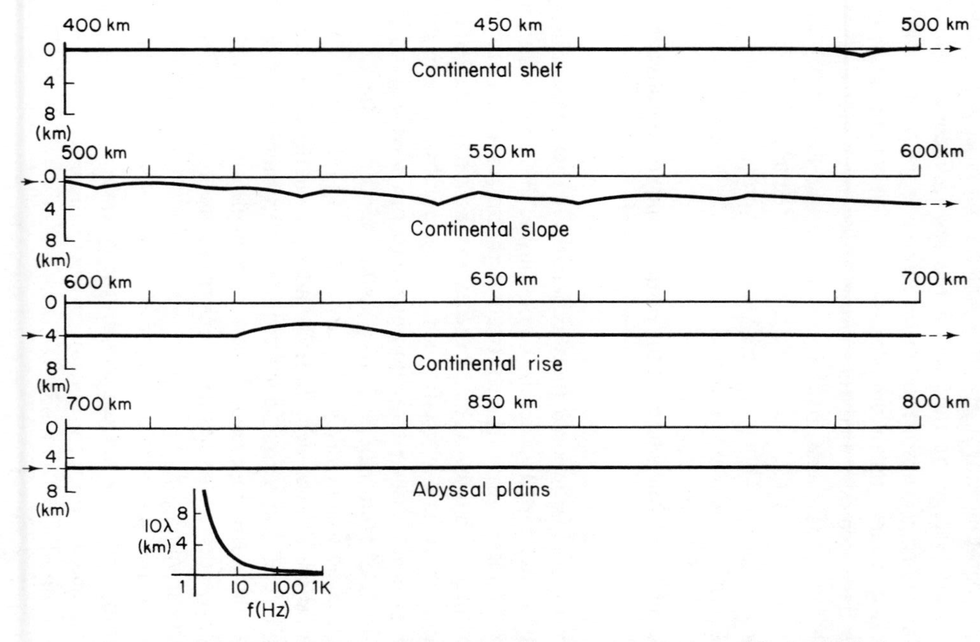

FIGURE 6-3. Bathymetric section across Georges Bank drawn with no vertical exaggeration (from Emery and Uchupi, 18).

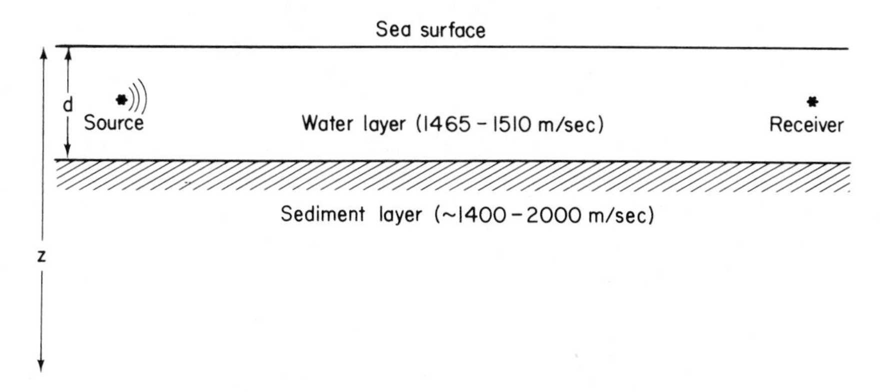

FIGURE 6-4. Pekeris waveguide model for sound propagation in shallow water overlying a sediment layer.

band impulsive signal such as a charge, or explosive, is transmitted. The transmitted energy is selectively filtered into modes. For each mode the first arrival is a low-frequency component that travels primarily as a refracted path through the higher-velocity seabed; next a waterborne, direct-path, high-frequency component arrives; finally, these two paths merge at an intermediate-frequency, low-velocity component corresponding to the minimum in the group velocity curves. In any experiment, one needs to convolve the excitation waveform with the impulse response of the channel to account for how each mode is excited. In addition, the absorption of the higher frequencies must be introduced, and this leads to stronger responses at low frequencies, especially for the refracted paths. As far as the signal processing is concerned, one should note from this example that shallow-water propagation is quite complex, involving much frequency-dependent multipath. Even though we understand the essentials of the propagation, especially by transforming into the frequency domain and using spectrograph analysis, any successful sonar system must explicitly account for this propagation phenomena by appropriate signal design at the transmitter and environmental signal processing at the receiver.

Deep-water acoustic propagation is strongly influenced by the speed of sound as a function of depth in the ocean. Some typical profiles and paths are illustrated in Fig. 6-6 for various locations and seasons of the year. Extensive tabulations of these sound-speed profiles have been compiled in atlases.[19] The relative influence of temperature and pressure upon the sound speed leads to a profile that usually contains two relative minima. Although there are some very sophisticated methods for calculating the acoustic propagation paths, a simple Snell's law analysis indicates that the sound is refracted toward these minima so that they act as wave guides, or ducts, of acoustic energy. The two minima of concern are the SOFAR

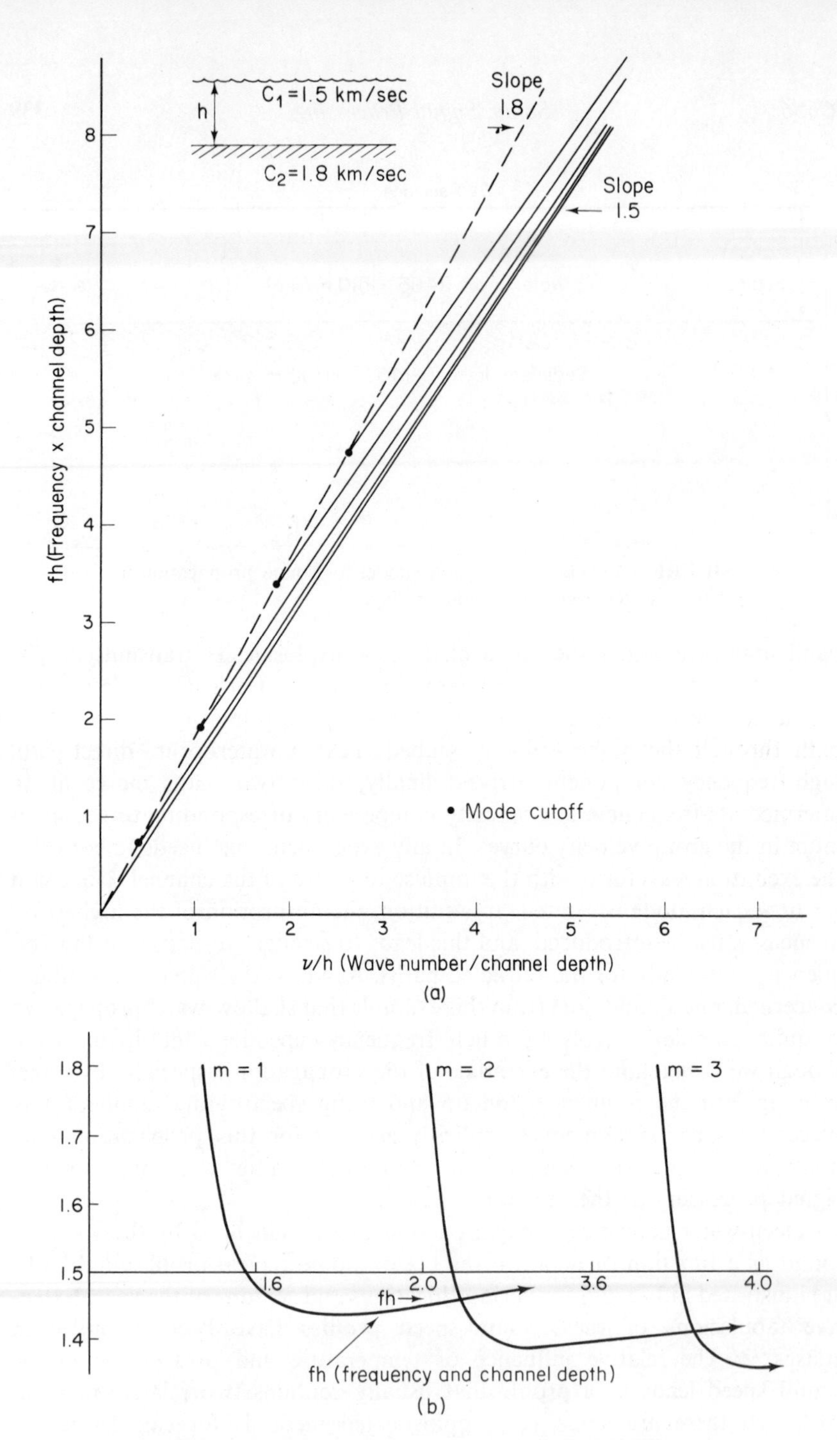

FIGURE 6-5. (a) Wave characteristics for Pekeris guide—normalized frequency vs normalized wave number for the first four modes; (b) Group velocity vs frequency for Pekeris guide for the first three modes.

channel and surface duct. The SOFAR (sound fixing and ranging) channel appears persistently in the world's oceans, usually at a depth of about 1 km, although it may appear at the surface in high latitudes. Energy, when trapped within this channel, can propagate over very long ranges, particularly at low frequencies where the absorption losses are low. This has many implications for passive sonar systems; thus numerous investigations and many signal-processing systems are closely coupled to this type of propagation.[17,20–22]

We can observe the gross effects of refraction by the SOFAR channel

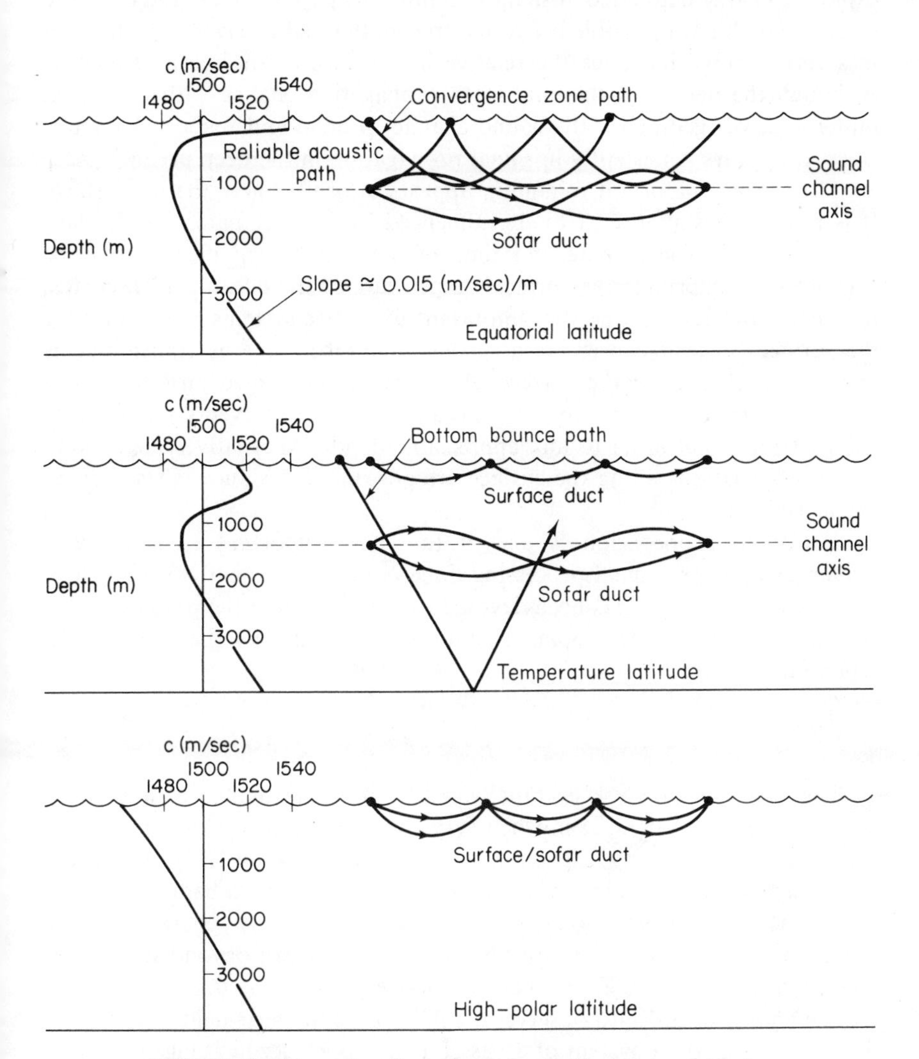

FIGURE 6-6. Examples of sound speed vs depth at various latitudes with several possible transmission paths.

(Fig. 6-7) by examining some ray paths for typical midlatitudes for several depths of the transmitter. First, the curvature of the rays toward the sound channel minimum, or axis, is evident, so that the energy is continually guided back toward this axis. Second, when the transmitter is off axis, the rays focus, or collect in a concentrated bundle, at intervals of approximately 60 to 70 km. These are known as *convergence zones.* Third, several more paths may be introduced when there are reflections off the sea floor (*bottom-bounce paths*) or refractions within it. Figure 6-8 illustrates these phenomena for an explosive charge detonated at shallow depth. The path structure indicates the extensive multipath possible between a transmitter and receiver, and the path loss versus range indicates the relative importance of the received energy. Although the details of this multipath propagation change with the transmitter–receiver geometry, the sound-channel profile, and seabed characteristics, it appears persistently in signal propagation in the deep ocean.

The second important minimum appears at or very near the sea surface. It is much less stable, is strongly influenced by local climate, and displays diurnal effects. The climate and time of day have a significant influence because they determine the surface temperature of the ocean, which creates the duct. This duct can be very important for active systems operating near the surface, since it can create a *shadow area* where energy transmitted is refracted back toward the surface. As a result, submerged targets at even modest ranges cannot be observed because this duct prevents the acoustic energy from penetrating to any appreciable depth. Alternatively, very long range propagations along the surface are possible if the duct exists over an extensive area.

Multipath propagation is one of the most important environmental influences upon the signal processing done in a sonar system. As a result, a great deal of attention has been devoted to understanding the physics of this propagation and to developing signal-processing models that adequately represent its effects.

6.2.2 Noise

The presence of noise eventually limits the performance of any sonar signal-processing system, even when very good environmental information is available for predicting the characteristics of the acoustic propagation. Generally, noise is classified as either ambient, "self," or reverberant. Ambient noise is the background noise, generally additive, that is generated by the numerous other sources of acoustic energy in the ocean beyond the ones of interest to an operator. It is observed passively, since one does not need to radiate energy in order to observe it. Self-noise is the near-field noise introduced into the sonar system of a vessel as it propels itself. It may consist of machinery, cavitation, or flow noise generated by the propulsion plant and

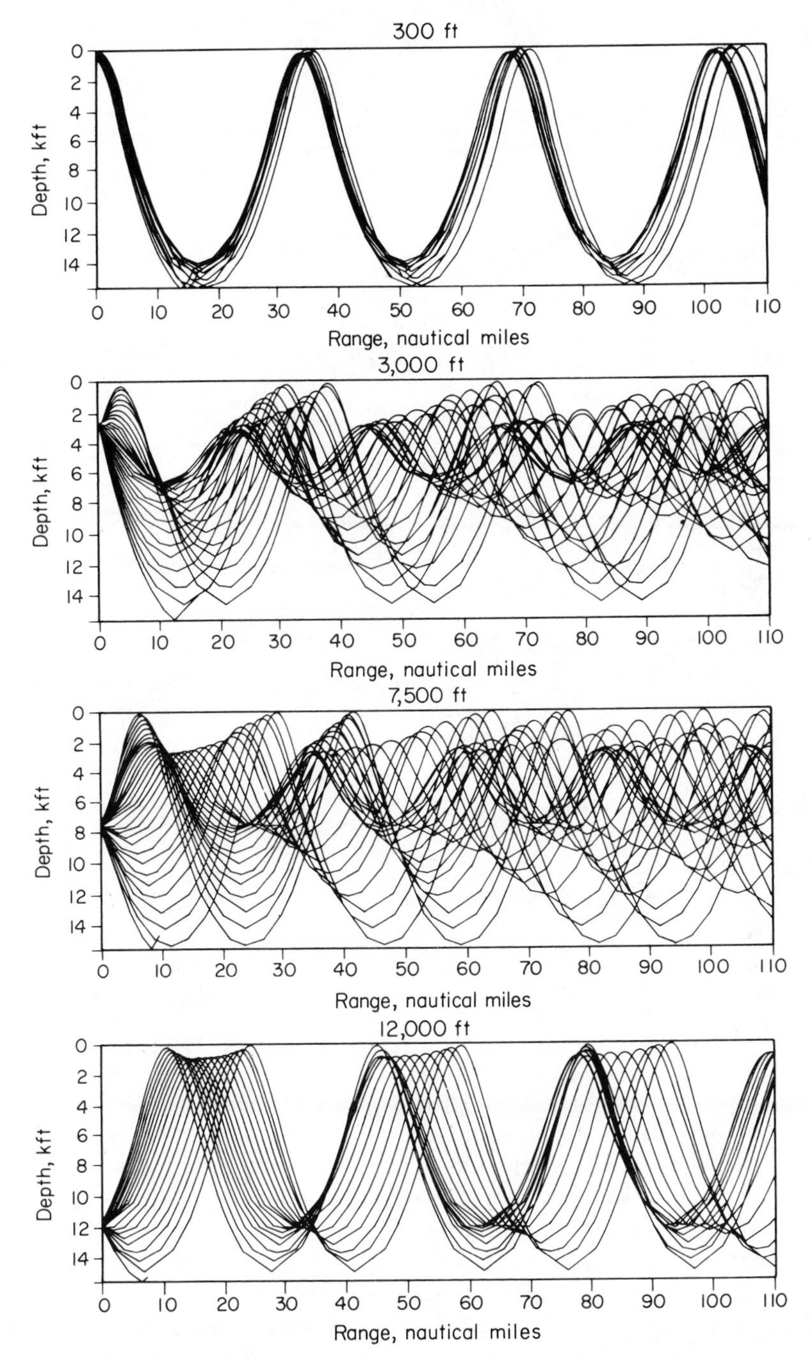

FIGURE 6-7. Ray diagrams for a source at four different transmitter depths in the deep sea (from *Principles of Underwater Sound for Engineers* by Urick; McGraw-Hill, 1967. Used with permission of McGraw-Hill Book Company).

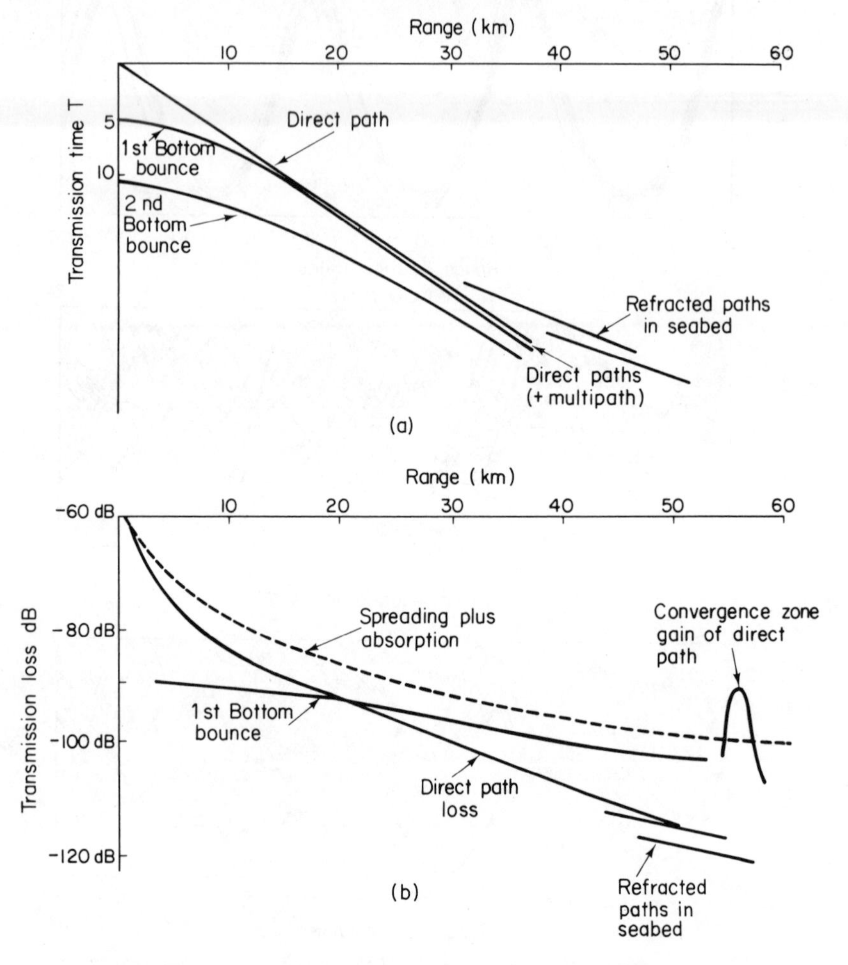

FIGURE 6-8. Long range transmission time and path loss at low frequency in the deep sea.

hydrodynamic forces. Reverberant noise is generated by spurious reflectors, or targets, in an active system. One must radiate energy in order to observe it, so its characteristics are very dependent upon the transmitted waveforms.

There are two major considerations in characterizing ambient noise, the dependence in space and in time. Although there are appreciable variations in measurements taken at different seasons and at various locations, the overall dependene upon the various sources is well understood. Temporal dependence is often summarized by Wenz's curve, which is illustrated in Fig. 6-9.[23] In the 1- to 10-hertz (Hz) range, oceanic turbulence and seismic activity at the sea-floor are often suggested as its source. In the 10- to 300-Hz range, shipping is the dominant source; and, in fact, the average noise level has been

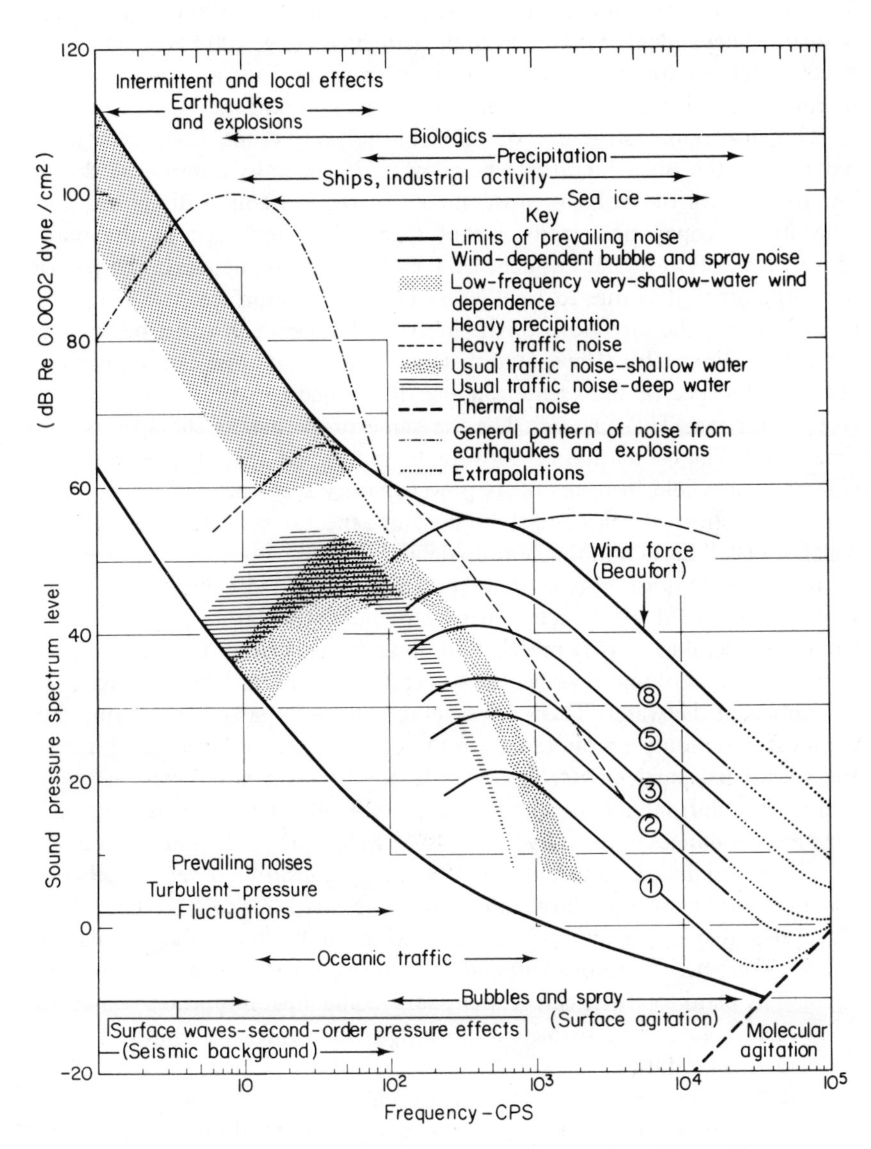

FIGURE 6-9. Ambient noise spectra in the ocean (from Wenz, 23).

increasing in recent years due to activity on the sealanes. The spectra in this range are very dependent upon one's proximity to the sealanes; they also can be very tonal since a major portion of the excitation is the rotation machinery of the propulsion systems. As an example, there are typically on the order of two to three thousand ships in the North Atlantic at any one time. Above

300 Hz, wind forces and sea-surface activity become important, so this range is weather dependent. Finally, at high frequencies (e.g. 100 kHz), which are beyond the spectral region of interest for most but not all sonar systems, thermal excitation of water molecules becomes important.

The directional structure of ambient noise has not been extensively reported in the open literature; however, it is generally considered that at low frequencies the noise is most intense in the horizontal direction, since long-range propagation from distant shipping is regarded as the major source, whereas at higher frequencies, the noise is more intense in the vertical direction, since it is due to weather activity on the surface.[24-26] If one is near shipping, the directions of the individual vessels can often be resolved as well as their spectra. The spatial structure of ambient noise can be described in terms of a spectral representation, usually termed a wave-number of wave-vector function. This function has the same properties as the spectra of a time series. The estimation of both the temporal and spatial structure of an ambient noise field in terms of its power-density spectrum and/or its wave-number function is a particularly important aspect of passive sonar systems, which we shall discuss in much more detail later.

Reverberant noise is generated when the transmitted energy of an active system is reflected by objects other than the particular target of interest. Often it is useful to classify reverberant noise as being either from a boundary or volumetric. Volume reverberation is caused by the scattering of particles or bubbles in the water. There are several sources of particle scattering, but the most commonly encountered are the deep scattering layer, gas bubbles, and suspended particulants. The deep scattering layer is biologic in origin. It migrates diurnally between 200 and 1000 km, and displays frequency selective resonances in the range of 1.5 to 25 kHz. Gas bubbles are introduced by shipping and surface turbulence. While they constitute a very small component in the water, their large difference in density leads to large reflections. They also display strong resonances owing to bubble pulse oscillation. Finally, there are numerous suspended particles, objects as large as fish and as small as plankton, that can scatter energy. The most important consideration is that volume reverberation is always present; and it can impose serious performance limitations.

Boundary reverberation results from reflections off the sea surface and seabed. At long horizontal ranges, boundary reverberation is a dominant concern, since the acoustic energy undergoes several reflections off these boundaries. Since the characteristics of these boundaries are such an important aspect in the propagation of sonar signals, we consider them separately.

Although the actual reverberation mechanisms are important in understanding its nature, the most important considerations as far as signal processing is concerned are how reverberation scatters the energy as a function of range, Doppler, and space. This scattering is most conveniently modeled as a time varying, possible spatially dependent, random linear filter. The statistical properties of these filter models are often described in terms of a scattering-function representation. These functions, or their close counterparts, attempt to quantify the range and Doppler scattering of the medium statistically. They play an important part in effective signal design for an active sonar system, and we examine them in the section on active systems.

6.2.3 Interaction with the Boundaries

The simplest description of the seabed and sea surface is that they confine acoustic energy within the water column. For example, the sea surface is often modeled as a pressure release boundary with a reflection coefficient of -1, and the sea floor as a partially rigid boundary with a reflection coefficient typically in the range of 0.01 to 0.3 (i.e. -40 to -10 dB). These descriptions may be adequate for many purposes; however, the boundary interactions are much more complicated than this, so it is important to comment on their characteristics in more detail.

The seabed can be described by how strongly it reflects acoustic energy, grossly termed its reflectivity, and how it disperses, or spreads, this energy in time. Both the reflectivity and time spreading are frequency dependent. The reflectivity has been measured in a variety of experiments, which suggest the following types of dependences.[27-30] Rocky and highly compacted, fine sand bottoms are most reflective, whereas mud and silted ones are least. The topographic relief is also important, for if one has relief larger than one eighth of a wavelength, constructive and destructive interference of the reflected wavefronts can occur. This can be of particular importance at high frequencies, or short wavelengths, where it does not require very much relief for this to occur. Grazing angle is also an important parameter; it is often approximated by a Lambert's law relationship implying a $\sin^2(\theta)$ dependence, where θ is the grazing angle. At low grazing angles, refraction phenomena can occur that mitigate this relationship.

The time dispersion, or spreading, is strongly dependent upon the frequency of the signal. At high frequency (e.g., at the 12 kHz of a fathometer), the spreading introduced is typically less than 10 milliseconds (ms) of two-way travel time, roughly 8 m. At 3.5 kHz, probably the most commonly used sonar frequency, the penetration is typically in the range of 50 to 150 ms, or 100 m. One can often infer the near surficial characteristics of the seabed from the structure of a 3.5-kHz echo sounder. As one approaches the seismic frequency range of 100 Hz and below, very deep penetration of acoustic energy into the seabed is possible. Reflections from strata 10 to 20 km and refractions from 30 to 50 km are routinely detectable with existing seismic systems. (For further details, see Chapter 7). The most important consideration about sonar systems operating with energy propagated through the seabed, is that the earth strongly attenuates high-frequency energy. Often a dependence of αf^1 is advocated with a spread of 0.5 to 10 dB/km-Hz for α.[30a] As a result, although very deep penetrations into the seabed are possible, the shallower ones account for the dominant portion of the energy used by the sonar system. As a means of comparison, Fig. 6-10 illustrates reflections from the seabed at the same location at several frequencies.

The sea surface also introduces several important phenomena when energy is reflected off it. The water–air boundary leads to a high reflectivity simply because of the large difference in the acoustic impedances. Roughness can, however, introduce time dispersion, and wave motion leads to some Doppler spreading. Refraction phenomena can also be present at low grazing angles. The three most important parameters influencing sea-surface scatter-

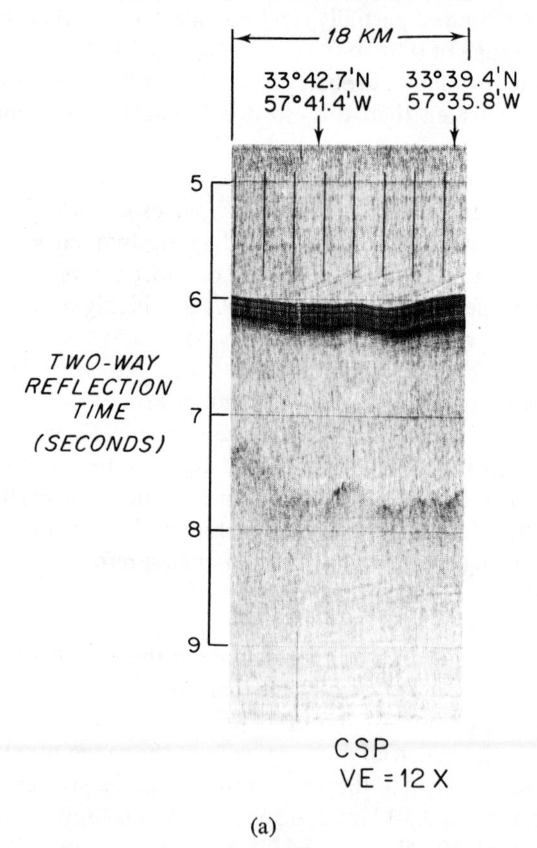

(a)

FIGURE 6-10. Examples of reflections from the seabed at the same location at three frequencies: (a) Seismic 0–100 Hz; (b) Echosounder at 3.5 kHz; (c) Fathometer at 12 kHz (from KNORR cruise #31, leg #3 of the Woods Hole Oceanographic Institute, photographs courtesy of Mr. Jamie Austin).

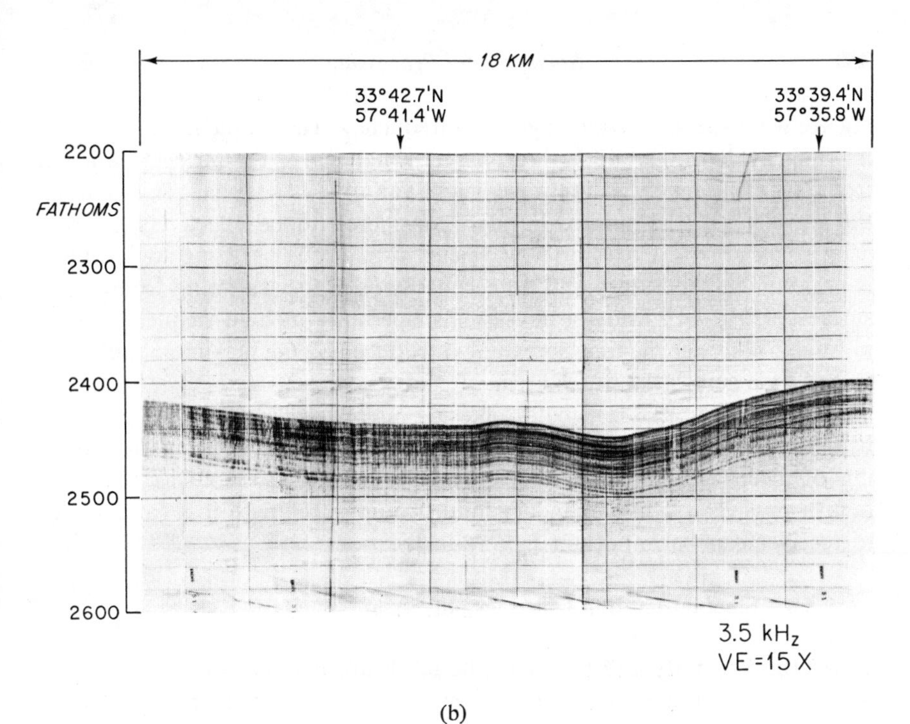

(b)

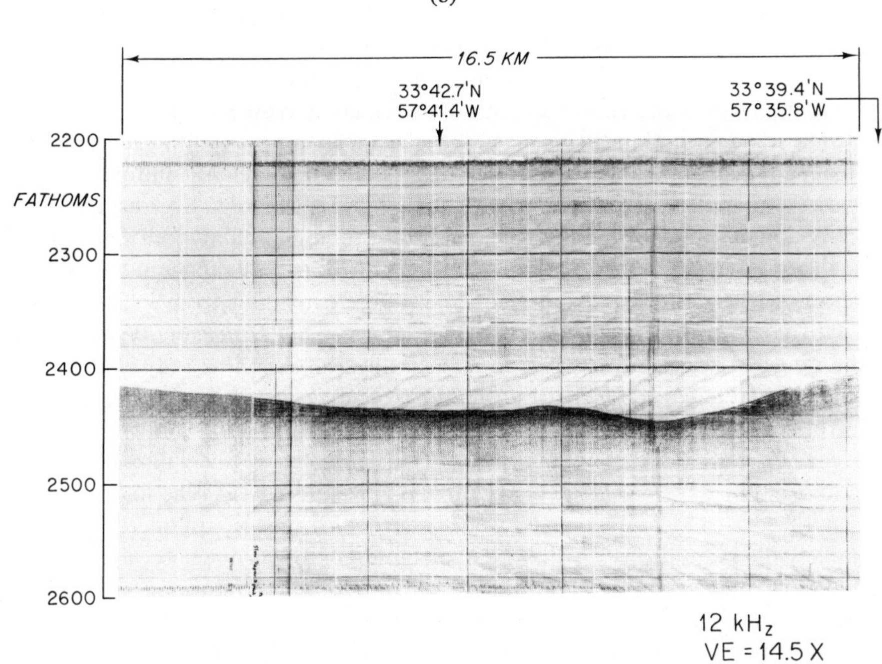

(c)

FIGURE 6-10. (*Continued*)

ing are wind speed, grazing angle, and frequency. The frequency dependence is primarily dependent upon the apparent roughness of the surface compared to the incident energy. If the wave-height variation is less than $\sin(\theta)/8$ times the wavelength, the surface may be considered to be effectively smooth, acting as a mirror. For shorter wavelengths and lower grazing angles, there is an abundant literature and numerous theories for scattering from rough surfaces.[31-33] For wide-beam systems operating at high frequencies, one must also consider the time spreading introduced by the maximum multipath transit time across the rough area.

The implications for signal processing introduced by the boundaries can be quite significant. Often they introduce the dominant spreading, or scattering, of energy in an active system. Although they are probably not as important in passive systems, boundary influence upon the path loss from source to receiver can be important in certain locations.

6.2.4 Fluctuations and Coherence of Sonar Signals

As the flexibility and power of the hardware and software available for processing sonar signals have shown dramatic growth in the past few years, it has become particularly important to determine just what limits are imposed by the ocean itself. For example, one cannot design arbitrarily narrow filters or integrate over extremely long durations. Nor can one build very large arrays and expect to achieve extremely narrow resolution. There is always some motion in the propagation paths that imposes a finite amount of frequency spreading, and there is always some incremental multipath that causes a signal wavefront to decorrelate across the face of a large array.

A major source of temporal fluctuations (i.e., finite bandwidth signals) is conjectured to be the result of internal waves introducing changes in the sound velocity. These waves have a frequency range extending from inertial and tidal frequencies of one cycle per day to the *Väisälä frequency*, typically in the range of several cycles per hour. The physical oceanography that leads to these waves can be found in a number of texts, but their net effect is to introduce fluctuations into tonal signals propagated over long ranges in the ocean.[34-36] Since these signals are of particular concern to passive sonar systems, they have received considerable emphasis recently. The fluctuations that they introduce are typically on the order of tens of millihertz, which implies filter durations on the order of hundreds of seconds. At higher frequencies, one encounters turbulence effects and the frequency spreading introduced by the motion of the sea surface.

The spatial coherence of sonar signals is still a subject of active investigation. Typical measures of coherence distances (i.e., for the wavefronts of a plane wave) are 10 to 100 wavelengths for horizontal separations and less than 10 wavelengths for vertical separations.[37] Given the stratification of the ocean, this anistrophy should not be surprising. In assessing the performance of a sonar array-processing system, the degree of spatial coherence is quite important. If a plane wave decorrelates across the face of an array, there is a limit upon the attainable resolution regardless of the size of the array; in addition, the decorrelated field implies that one should use a form of signal processing that is different from the usual forms of direct beam forming.

6.2.5 Platform and Medium Considerations

The ocean is a very demanding environment in which to work, whether upon a vessel or within it. As a result, relatively simple transducer and sensing systems require considerable skill and planning in their implementations. Numerous experiments involving very sophisticated, state of the art signal-processing concepts have failed because of inattention to the mechanical integrity required of systems when operating in the ocean. We do not intend to stress this topic as there are several texts on it; however, we want to note some of the more significant considerations that affect signal processing.[38]

The wave forces that can be generated at or near the ocean surface will eventually test the mechanical integrity of any transducer system. On any moving platform flow noise past the hydrophones may be very important. They also introduce accuracy limitations into data observations through noise, vibration, and positional uncertainty. One of the most difficult environmental aspects of working in the ocean is the extreme pressures encountered. Ambient pressure increases at 1 atmosphere (atm)/30 ft, so operating in the deep ocean may require equipment to withstand pressures in excess of 2 tons/in.^{2}. These pressures also affect instrument sensitivities. Finally, seawater is a very corrosive medium, which requires very special packaging considerations.

The overall effect of all these factors is that, more so than in most environments, one cannot always design ideal transducer systems because of the trade offs that must be made to accommodate the environment. In addition, extreme pressures make the deep ocean inaccessible for all but very specialized submersibles, so one must rely upon remote-sensing systems for which implacement costs are very high. Signal processing must incorporate all these limitations if it is to operate successfully. One significant benefit of the great flexibility of the digital systems now evolving is that they can adapt to and compensate for the environmental effects to a degree that was not possible with analog systems.

6.3 ACTIVE SONAR SYSTEMS AND DIGITAL SIGNAL PROCESSING

In exploiting the ocean for commercial, scientific, or military purposes, the principal means of acquiring information is through systems that actively transmit acoustic energy into the water and subsequently interpret the received echoes. In this context, we can identify two related aspects in active sonar signal processing. In the first, one simply wants to process the received signals in the most effective manner in order to extract the desired information. In the second, one wants to design the transmitted signals such that the receiver can optimize its desired performance.

In this section we examine the signal-processing aspects of active sonar systems. We have indicated our organization of the section in Fig. 6-11, where we show a horizontal and a vertical organization of our material. Horizontally, we show the way in which information usually flows in a sonar system. First is the front end, where the actual operations upon the received waveforms are done. This typically involves some type of matched filtering operation or a combination of them. In the next section we have environmental processing. This section takes data samples from the front end and incorporates the effects of the environment, such as the multipath and Doppler spreading. The final section is a target analysis section. Here information gained from a sequences of pulses is analyzed in order to determine such quantities as target motion and structure. Vertically, we show three levels related to the implementation of various systems. At the top is the underlying theory and models that guide the structure of the implementation. Next are some implementations of the respective theories, which inevitably involve some compromise with the theory, since the "optimum" processor can seldom be practically realized. The theory, however, serves as a guide to structure and sets performance benchmarks to which the loss due to the suboptimum realization can be compared. Finally, we have various applications. These must synthesize the theory and implementations at each stage in the processing in order to create the overall system.

In our discussion we successively examine each row in Fig. 6-11. We start

	Front end processing	Environmental processing	Pulse sequence processing
Theory and models	Correlators, Matched filters, Ambiguity functions	Point reflectors, Range and doppler Spread environments, Scattering and two Frequency correlation Functions	Swerling models, Tracking algorithms and Kalman filters
Examples of implementations	DELTIC systems, Digital FIR's, FFT based algorithms	Tapped delay lines, Frequency domain Proc.,	Pulse integrators, Minicomputers and Microprocessor appl.
Examples of applications	Target detection and range/ doppler estimation Tracking and navigation Communications Mapping and charting		

FIGURE 6-11. Section organization of digital signal processing for active sonars.

with a brief survey of the theories and models that have motivated the design of sonar signal processors. Next we examine some of the more important and extensively used implementations. Finally, we discuss representative applications in detection systems, range and Doppler estimation, and tracking and navigation.

6.3.1 Active Sonar Signals, Correlators and Matched Filters, and Ambiguity Functions

The various forms of active sonar signals have been strongly influenced by the technology available to generate them at the transmitter and to process them at the receiver. In spite of continual evolution with the technology, the basic structure of the system has remained virtually the same as a correlation-based operation. The effort has been placed on effective signal design consistent with the technology available to take advantage of the correlation properties of the signals.

The simplest sonar system simply consists of a transducer excited near its resonant frequency (the transmitter) and a narrowband energy detector (the receiver). The more sophisticated systems now employ programmable, adaptive modulators with multielement, steerable arrays for transmitters and digital processors often in the form of microprocessors and minicomputers for receivers.

The transmitter–receiver system and the associated waveforms are usually designed to accomplish two tasks: (1) detection of targets, and (2) estimation of their range, Doppler, and bearing. These are closely coupled both in terms of effective signal design and receiver processing. Figure 6-12 illustrates the system and the important components of the front end of a receiver. They essentially consist of an appropriately designed transmitted signal, the sonar channel, a quadrature demodulated correlator (often termed a matched filter for reasons that we subsequently explain), a square-law device, and possibly a threshold device for the detector and a range, Doppler, bearing scanners for the estimator. In this section we examine some of the popularly used signals, the properties of the correlation operation, and the performance aspects of the system as encompassed in the structure of the ambiguity function for the waveforms employed.

The waveforms, or signals, used still consist of a few elementary forms in spite of the application of some very elaborate theories. They also share much in common in their development with radar, and the interested reader is urged to consult Chapter 5.[39-41] The simplest waveform is simply a gated continuous wave (*cw*) tone. Its representation is given by

$$s_t(t) = \begin{cases} \sqrt{\dfrac{2E_t}{T}} \cos(2\pi f_c t) & 0 \leq t \leq T \\ 0 & \text{elsewhere} \end{cases} \tag{6.1}$$

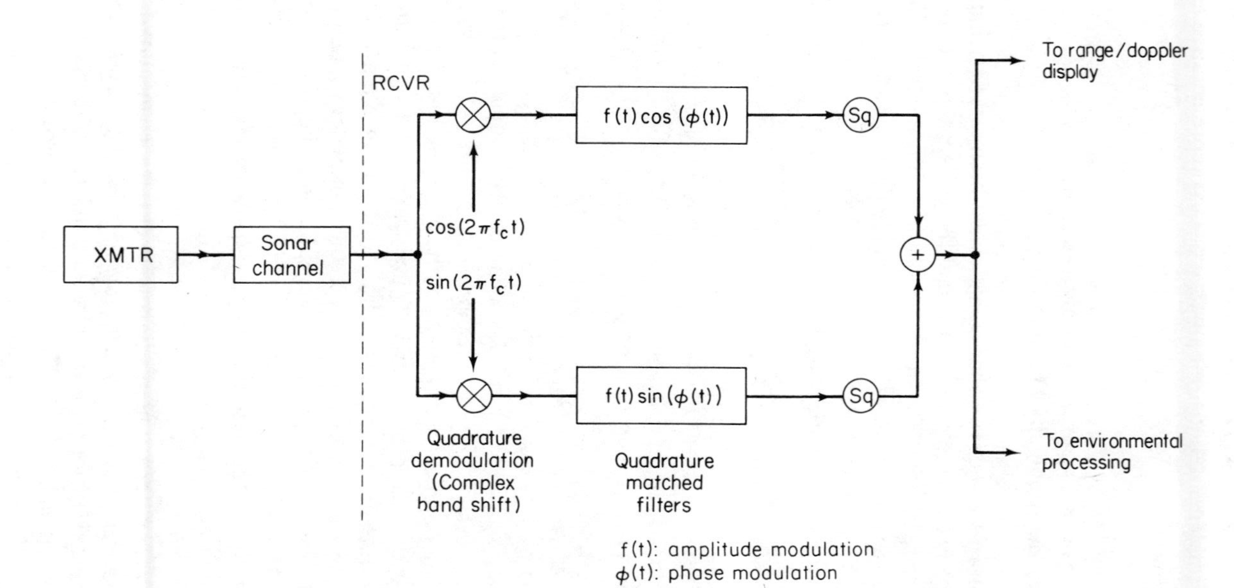

FIGURE 6-12. Front end processing structure for a quadrature matched filtering of a sonar signal.

where

$$E_t = \text{transmitted energy}$$
$$T = \text{pulse duration}$$
$$f_c = \text{center frequency}$$

Durations can range from a fraction of a millisecond to 1 to 2s, and center frequencies can range from a few hundred hertz to tens of kilohertz, depending upon the system and its desired operating range. This is illustrated in Fig. 6-13(a). The pulse can be generated by simply gating or shocking a transducer; however, one usually obtains a slightly different signal in the water because of the finite bandwidth of a transducer. A more realistic model of the transmitted signal is given by

$$s_t(t) = \begin{cases} \sqrt{\frac{2E_t}{T}}\, a(t) \cos(2\pi f_c t) & t > 0 \\ 0 & t < 0 \end{cases} \tag{6.2}$$

where $a(t)$ is a modulation which depends upon the bandwidth of the transmitter. As we shall observe shortly, there are a number of limitations intrinsic with the simple gated tone; however, gated tones are quite easy to implement.

The next class of signals consists of a sequence of gated tones whose polarity, or more generally phase, has been modulated according to some signal-design algorithm. These have been popularly termed pseudorandom (PR) or pseudorandom noise (PRN) sequences. The rationale for their use will become more apparant when we discuss ambiguity functions. An example of a PR sequence is illustrated in Fig. 6-13(b).

A signal set that has some very good properties for range resolution and that can be implemented comparatively easily are chirped or frequency-modulated (FM) signals. The general form of this class of signals is given by

$$s_t(t) = \sqrt{\frac{2E_t}{T}}\, a(t) \cos[2\pi f_c t + \varphi(t)] \tag{6.3}$$

where

$$\varphi(t) = 2\pi\mu\frac{t^2}{2}$$

and μ is termed the chirp rate and $\varphi(t)$ is the amplitude modulation which is used. If the instantaneous frequency is defined as the derivative of the phase modulation, one can see that with this type of modulation the instantaneous frequency is swept linearly during the transmission interval. Other types of phase modulation, such as hyperbolic cosine as well as frequency stepping using a programmable frequency synthesizer, can also be employed; however, the most important parameters are the bandwidth and the chirp rate of the signals.

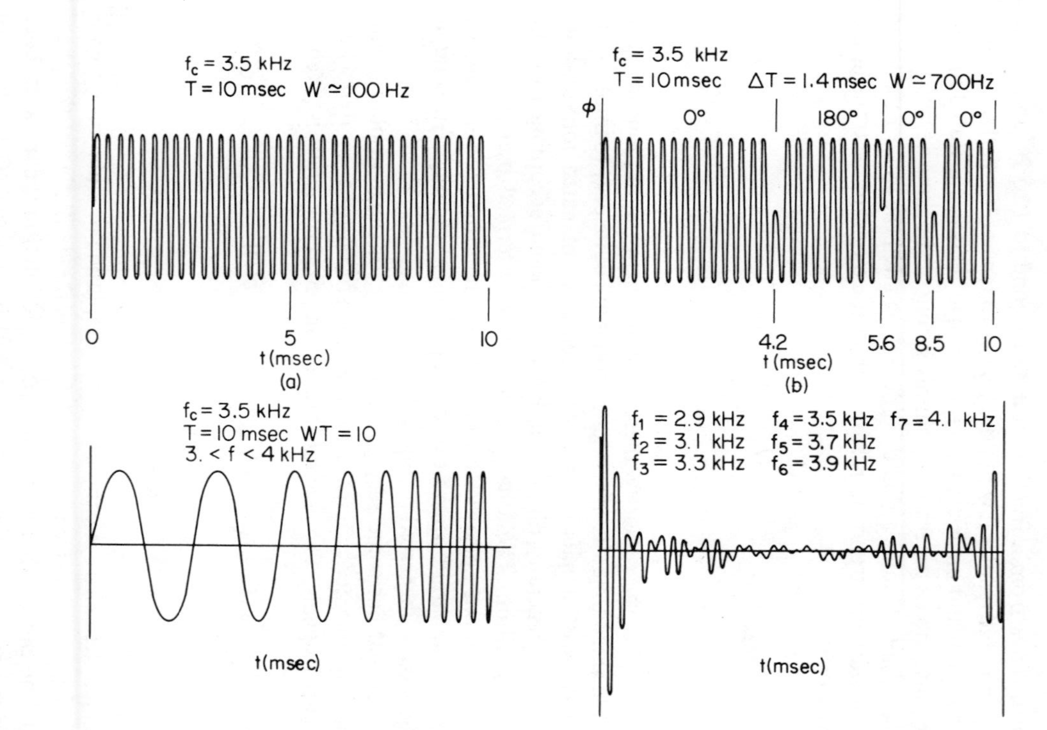

FIGURE 6-13. Examples of commonly used sonar signals at the center frequency: (a) Rectangular CW pulse; (b) Coded pulse; (c) Frequency modulated chirp; (d) Frequency shifted coded pulses.

In sonar communications some of the more advanced systems may employ frequency diversity coded signals. These are a frequency-domain dual of PR sequences. In this signal set, a message and its associated redundancy coding, if any, are encoded as a set of tones at different frequencies and transmitted simultaneously. The time-domain structure of the waveform may appear to be quite complex, but the frequency-domain structure consists of a set of weighted, finite-duration tones at different frequency cells. The signals may be represented as

$$s_t(t) = a(t) \sum_{n=-N}^{N} a_n \cos [2\pi(f_c + n\,\Delta f)t] \tag{6.4}$$

and an example is illustrated in Fig. 6-13(d).

In describing active sonar systems, it is useful to make a distinction between broadband and narrowband systems. In a broadband system a direct, or baseband, representation of the waveforms must be used; in a narrowband system a complex, or quadrature, representation of the waveforms is more convenient. The distinction between the two classes is somewhat arbitrary, but a useful criterion is that the bandwidth of a narrowband system is less than 10% of the carrier, or center, frequency. The complex, or quadrature, representation for narrowband signals is very convenient for describing the Doppler and phase shifts introduced by the propagation and reflection processes. Since most active systems are narrowband, we employ the complex representation in our discussion.

The complex representation essentially defines the quadrature components of a signal with respect to a center frequency. The quadrature components can be determined by demodulating the narrowband signal with respect to cosine and sine references of the center frequency, as illustrated in Fig. 6-14. It is easily verified that the original narrowband signal can reconstructed according to

$$f(t) = f_c(t) \cos (2\pi f_c t) + f_s(t) \sin (2\pi f_c t) \tag{6.5}$$

as illustrated in the figure. Both the quadrature components and the narrowband signal itself may be conveniently represented in terms of a quantity called the complex envelope, or

$$f(t) = \text{Re}[f(t)e^{j2\pi f_c t}] \tag{6.6a}$$

$$\tilde{f}(t) = f_c(t) - jf_s(t) \tag{6.6b}$$

It is easy to observe that the magnitude of the complex envelope is the real envelope, and that its phase is the shift of the narrowband signal relative to the cosine reference of the center frequency. The complex envelope for a signal can often be identified directly from an analytic description of it by using these observations. For example, the chirp signal of Eq. (6.3) has the representation

$$s_t(t) = \sqrt{2E}\, \text{Re}[\tilde{s}_t(t)e^{j2\pi f_c t}] \tag{6.7}$$

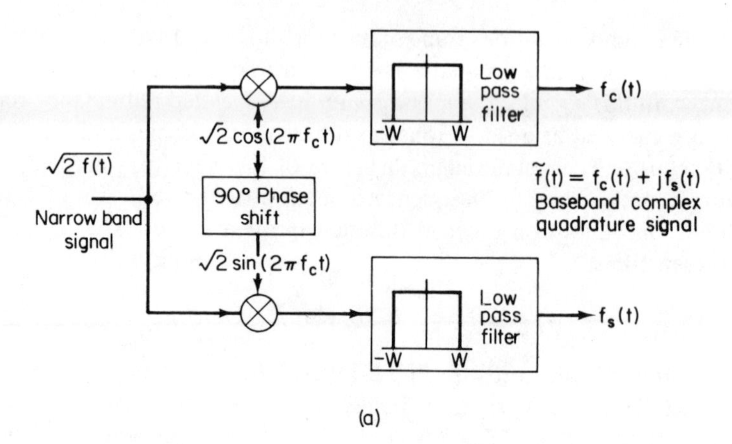

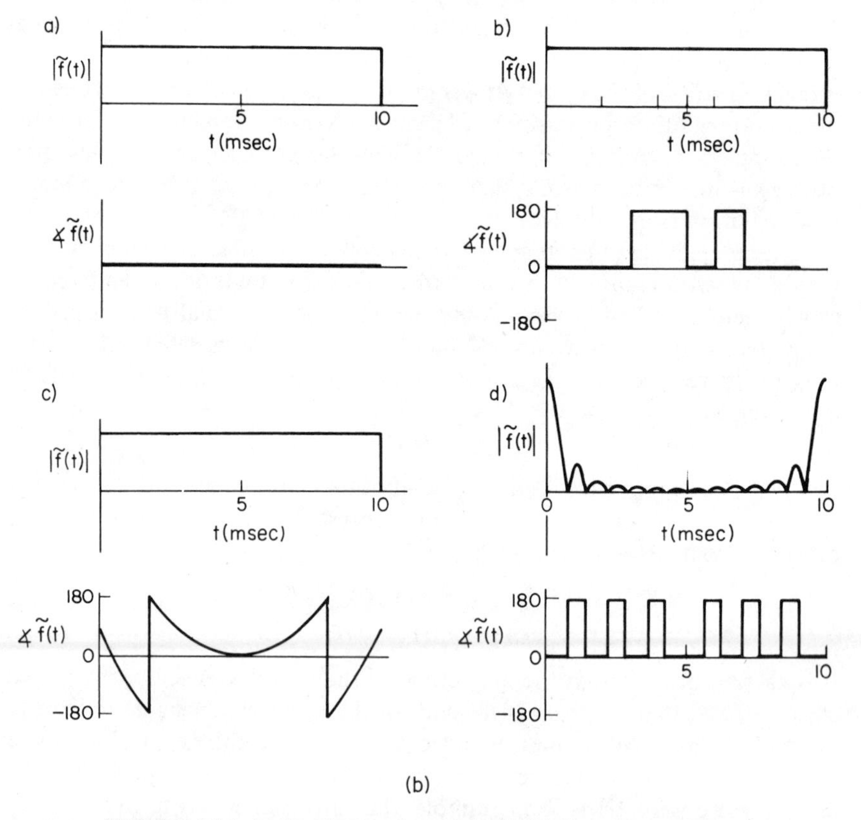

FIGURE 6-14. (a) Quadrature demodulator, (b) Complex envelope (magnitude and phase representation) of the sonar signals illustrated in Figure 6-13.

where

$$\tilde{s}_t = \frac{1}{\sqrt{T}} a(t) e^{j2\pi\mu\frac{t^2}{2}}$$

The complex notation simplifies many of the equations needed to describe the signal processing done in the front end of an active sonar system. Probably the most important of these operations is the correlation between two narrowband signals. It is easy to demonstrate that under the narrowband assumption one has

$$\int x(t)y(t)\,dt = \tfrac{1}{2}\,\mathrm{Re}\left[\int \tilde{x}^*(t)\tilde{y}(t)\,dt\right] \tag{6.8}$$

One should also note the implementation aspects suggested by the complex representation. The complex envelope is a lowpass representation, so the sampling rates required for a digital implementation are significantly less than those which would be needed if a direct sampling of the observed narrowband signal were done. As a result, some active sonars essentially implement the quadrature demodulation suggested by Fig. 6-14, and process the signals according to the formulas indicated by the complex representation.

Once the signal is transmitted, several phenomena in the transmission path and the reflection process can have an influence upon what finally is observed at the receiver. The effects of these phenomena are introduced by channel models of varying degrees of complexity; nevertheless, one of the simplest models, the slowly fluctuating, point target model, leads to the most important receiver structure, the correlation receiver. In this model the receiver observes a delayed and Doppler-shifted version of the transmitted signal in the presence of white, or broadband, noise. A random attenuation and phase shift is also often introduced, where the attenuation has a Rayleigh distribution and the phase a uniform one. The received signal may be represented as

$$r(t) = \sqrt{2E_t}\,\mathrm{Re}[\alpha e^{j\theta}\tilde{s}_t(t - T)e^{j2\pi f_d t}e^{j2\pi f_c t}] + w(t) \qquad (6.9)^*$$

Physically, the delay and Doppler shift are introduced by the range and range rate of the target. The attenuation includes several effects, such as the transmission loss and the target cross section; the random phase arises because of the range uncertainty compared to a wavelength. This is probably the simplest of useful models, and we discuss more complex ones after we introduce the concepts of scattering functions and random channels and targets.

The spatial aspects and array processing for an active system have a lot in common with that of passive systems. At this point we simply note that the spatial processing consists of some type of beam forming and defer more detailed discussions on array signal processing until Sec. 6.4.

* The range delay may be computed using a sound speed of 1490 to 1500 m/s, and the Doppler shift for two-way transmission is given by 0.68 to 0.69 Hz/knot-kHz (range rate in knots, carrier in kilohertz).

The correlation receiver performs a very intuitive operation, which can be substantiated theoretically in numerous ways. It is important, however, to distinguish two separate criterion: detection of a target and estimation of its range and Doppler. In the detection problem the correlation receiver is used to generate a sufficient statistic, which is the basis for a threshold comparison in making a decision if a target is present within a certain range and Doppler region. The performance of the detector is customarily presented in a receiver operating characteristic (ROC) that plots the detection probability as a function of the false alarm probability. In the estimation problem the correlation receiver is used to generate a statistic that is a function of the possible range and Doppler values of the target. This statistic is then maximized versus these values in making the estimate. The performance of the estimator is typically represented as the variance in the estimated parameters. This performance analysis generally has two facets to it, the variance under high signal-to-noise ratios and the probability of an ambiguous, or globally anomalous, estimate. The first is specified by a linearized analysis (Cramer–Rao bound) involving the behavior of the ambiguity function of the signals at the origin; the second is indicated by the sidelobe structure of the ambiguity function. We discuss these aspects in more detail when we examine the ambiguity function of the signals previously introduced; several extensive texts are available on detection and estimation theory.[7,10]

The basic correlation operation indicated in most signal-processing texts has the forms

$$\ell = \int_{-\infty}^{\infty} r(t)s(t)\,dt \tag{6.10a}$$

$$= \int_{-\infty}^{\infty} R^*(f)S(f)\,df \tag{6.10b}$$

where we have used Parseval's theorem in relating the time- and frequency-domain realizations. Because of the random phase present in most received sonar signals, one must account for the correlation that is present in both quadratures of the signals. It is straightforward to demonstrate that this random phase leads to a complex correlation operation involving the complex envelope representation of the quadrature components. Several derivations based upon a number of criteria and methods all indicate that the receiver front end should implement the following operation:

$$\ell = \left| \int \tilde{r}^*(t)\tilde{s}(t:\underline{a})\,dt \right|^2 \tag{6.11a}$$

$$= \left| \int \tilde{R}^*(f)\tilde{S}(f:a)\,df \right|^2 \tag{6.11b}$$

where

$\tilde{r}(t)$ = complex representation of the received signal

$\tilde{s}(t: \underset{\sim}{a})$ = replica of the transmitted signal where one includes via the parameter $\underset{\sim}{a}$ any modifications in the waveform due to the propagation–reflection process, *eg* range delay and Doppler rate

In a detection system, the parameter $\underset{\sim}{a}$ sets the range and Doppler value in the particular resolution cell of concern. If the range and Doppler are unknown, or more generally if one wishes to estimate these parameters, then the correlation operation in Eq. (6.11) is done as a function of the a, the unknown range and Doppler parameters. (More generally, parameters in addition to range and Doppler can be included in the parameter $\underset{\sim}{a}$.)

One can observe how the operation expressed by Eq. (6.11) is realized in the receiver front end of Fig. 6-13. The first part is essentially a quadrature demodulator for extracting these components of the signal. (It also rejects out-of-band noise.) The next operation is simply the complex correlation for each quadrature. Finally, there is the magnitude square operation.

There are several issues that become relevant when we discuss implementations, which we should note here. First, an alternative realization at the carrier frequency, or any other one obtained by heterodyning the signals, is possible. These realizations have the form of a bandpass correlator followed by a square-law envelope detector. This is probably the most convenient analog implementation, but the high sampling rates required do not make it attractive for digital implementations.

The concept of matched filters is closely coupled to correlator receivers. The matched filter is simply a convenient realization of the operation in terms of a linear time-invariant system. The impulse response of this system is specified as the reflected image of the image of the transmitted signal with which one is correlating the observed signal. There is one particular advantage of the matched-filter realization when one is concerned with range estimation. The receiver operation of Eq. (6.11) indicates that one should perform the correlation as a function of the range delays of interest. It is easy to observe that this is simply the output of the matched filter during the delay epoch of concern; that is,

$$\ell(\hat{T}) = \int \tilde{r}^*(t)\tilde{s}_t(t - \hat{T})\, dt = \int \tilde{r}^*(t)\tilde{h}(\hat{T} - t)\, dt \qquad (6.12)$$

with

$$\tilde{h}(t) = \tilde{s}_t(-t)$$

If one uses a frequency-domain realization, the dual statement can be made regarding the scanning across a Doppler interval by using a matched filter in the frequency domain.

Several important points should be made regarding the characteristics of matched filters. First, when operating against a background of white noise and detecting a point target in a given range–Doppler resolution cell, the detection signal-to-noise ratio depends only upon the average received energy-to-noise ratio. It does not depend upon the shape of the signal. The waveform becomes a factor when there is a reverberant environment, as we shall subsequently discuss, and when one is concerned with estimating target range and Doppler. In both of these problems the influence of the wave shape is determined by the ambiguity function of the transmitted signal, and this function is strongly influenced by the available time-bandwidth of the signal.

The ambiguity function is proportional to the noise-free output of the receiver in Fig. 6-13 when the correlation replica is scanned as a function of range and Doppler. We have

$$E[\ell(\hat{T},\hat{f}_d)]\Big|^2_{\substack{\text{signal}\\\text{only}}} = 2E_t\sigma_t^2\theta(\hat{T}, T_t, \hat{f}_d, f_{d_t}) \tag{6.13a}$$

where

$$T_t, f_{d_t} = \text{true target range and Doppler}$$

$$T, f_d = \text{scanning estimate of range and Doppler}$$

and

$$\theta(\hat{T}, T_t, \hat{f}_d, f_{d_t}) \propto \left| \int \tilde{s}_t(t - T_t)\tilde{s}_t(t - \hat{T})e^{j2\pi(f_{d_t} - \hat{f}_d)t}\, dt \right|^2 \tag{6.13b}$$

is the ambiguity function. By convention, the ambiguity function is normalized to have a value of unity when the true target and scanning parameters are equal. This is most conveniently done by scaling the energy out as a factor in describing the transmitted signal.

Sketches of the ambiguity functions for some of the signals previously introduced are indicated in Fig. 6-15. They are plotted as a function of the scanning parameters $\hat{T}$, $\hat{f}_d$, which is how the output would appear in estimating the range and Doppler of a target. We can observe several important properties of these signals. First, the simple gated pulse has an extent along the range axis that is determined by its length, or duration, T, while along the Doppler axis its extent is determined by the reciprocal of the duration. Obviously, with this one degree of freedom, one cannot obtain arbitrarily good range and Doppler resolution. The PR sequences can obtain good range and Doppler resolution by suitable coding of the sequences. The objective is to make the sequence appear as broadband noise for a long duration. If this is done, the wide bandwidth produces a narrow extent along the range axis, and the long duration and constant amplitude produces a narrow extent along the Doppler axis. There are a number of signal design, or coding, algorithms for doing this.[42,43]

The important issue is to control the appearance of spurious sidelobes in the ambiguity function. Figure 6-15(b) illustrates the general shape of an ambiguity function for a PR sequence. One should note that the range resolution is determined by the reciprocal of the bandwidth and the Doppler resolution by the reciprocal of the duration, while the approximate level of the sidelobes with good signal design is determined by the reciprocal of the

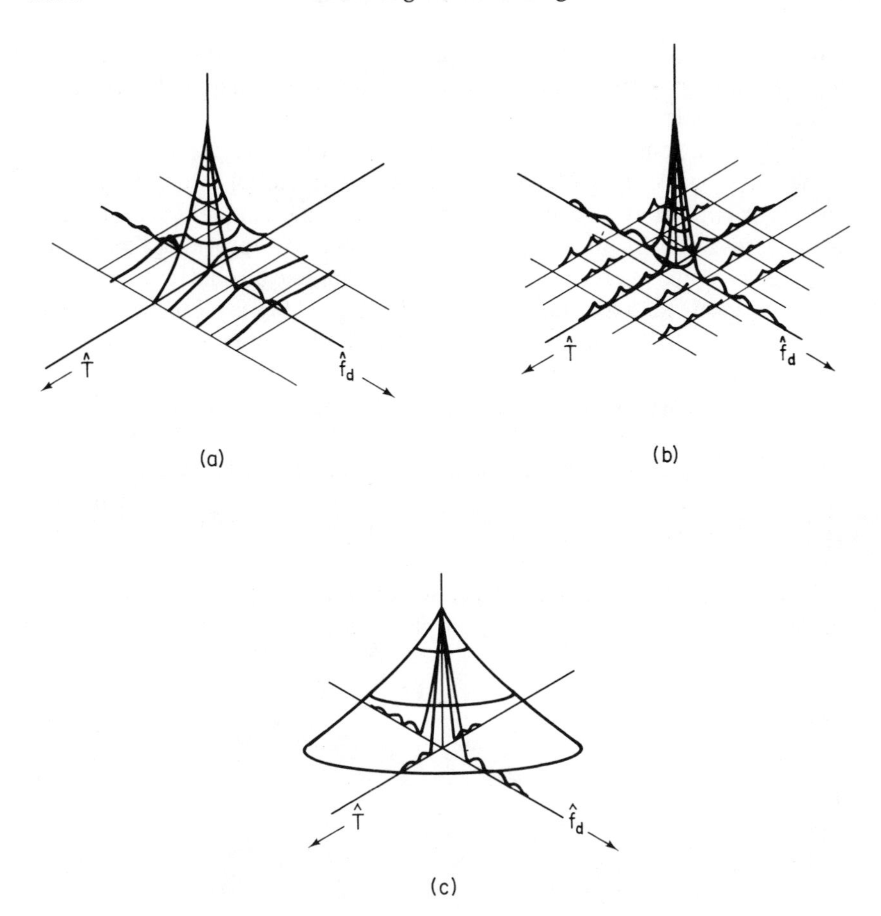

FIGURE 6-15. Ambiguity functions for several sonar signals: (a) Rectangular pulse; (b) coded pulse; (c) chirped FM pulse.

time–bandwidth product. The chirp, or FM, signal has an ambiguity function that is a "sheared" version of the ambiguity function of the unchirped signal if one uses a linear FM sweep. Figure 6-15(c) illustrates the ambiguity function for a chirped rectangular pulse. The important observation is that there is an ambiguous strip in the range–Doppler plane whose slope is determined by the chirping rate. The advantage of the chirped signals, and the PR sequences as well, is that they are basically phase modulations, so one can increase the energy in a transmitted signal without changing the peak power. This is a particularly important consideration in many sonar systems, since the transducers often have peak displacements of the hydrophone faces. The chirping of the signal removes some of the ambiguity function, which is distributed along the range axis and displaces it to the strip indicated. This improves one's range resolution capabilities only if one has a priori knowledge of the Doppler. Fortunately, this is often the case. We note that the extent along the range axis for a chirped signal is again determined by the reciprocal of the bandwidth.

These are the ambiguity functions of some of the more popular signal sets for active sonar systems. There has been a substantial amount of effort devoted to optimizing the signals; however, the fundamental issues are that the range resolution is primarily dependent upon the reciprocal of bandwidth, the Doppler upon the reciprocal of the duration, and the sidelobes must be carefully examined since the volume under the ambiguity function is conserved.[39] We shall now comment upon how the ambiguity function can specify these performances analytically.

In analyzing the range and Doppler resolution performance, it is important to distinguish the high and low signal-to-noise ratio (SNR) situtations. In the high SNR case, the performance is specified by the structure of the ambiguity function at the origin (i.e., where the true target and estimated parameters are equal). This structure can be related to the mean square duration and bandwidth of the transmitted signal. These results are derived using the Cramer–Rao bound of estimation theory; using these results one can demonstrate that the variances on the range and Doppler estimates are given by[41]

$$\mathrm{Var}[\hat{T} - T_t] \simeq \frac{2\sigma_b^2 E_t}{N_0}[J^{-1}]_{11} \tag{6.14a}$$

$$\mathrm{Var}[\hat{f}_d - f_{d_t}] \simeq \frac{2\sigma_b^2 E_t}{N_0}[J^{-1}]_{22} \tag{6.14b}$$

where

$$[J] = \left[\begin{array}{c|c} \dfrac{\partial^2\theta}{\partial\hat{T}^2} & \dfrac{\partial^2\theta}{\partial\hat{T}\,\partial\hat{f}_d} \\ \hline \dfrac{\partial^2\theta}{\partial\hat{f}_d\,\partial\hat{T}} & \dfrac{\partial^2\theta}{\partial\hat{f}_d^2} \end{array}\right]_0 = \left[\begin{array}{c|c} \int f^2\,|\tilde{S}_t(f)|^2\,df & \int t\tilde{s}_t(t)\,\dfrac{\partial s_t^*(t)\,dt}{\partial t} \\ \hline \int t\tilde{s}_t^*(t)\,\dfrac{\partial\tilde{s}_t(t)}{\partial t}\,dt & \int t^2\,|s_t(t)|^2\,dt \end{array}\right]$$

and the $|_0$ specifies that the ambiguity function is evaluated at the true target parameters T_t, $\hat{f}d_t$. When the off-diagonal terms in this matrix are zero, we find the dependences indicated above where the sharpness, or convexity, of the ambiguity function along the range axis is determined by the reciprocal of the signal bandwidth and correspondingly for the Doppler axis. When the off-diagonal terms are nonzero as is the case with a chirped signal, then the estimates are coupled.

In the low SNR situation, the sidelobes of the ambiguity function become important. When the noise is significant, spurious peaks may be generated as the range–Doppler plane is scanned with the correlator receiver. If there is a large sidelobe, this noise may cause the value at the sidelobe to exceed that at the origin. Consequently, the presence of large sidelobes in the ambiguity function can lead to globally erroneous estimates of the range and Doppler in the low SNR situation. The role of a priori information about the possible values of the target's range and Doppler is particularly important in this situation, since one can automatically reject spurious sidelobes if they are at unreasonable values of range and Doppler.

6.3.2 Environmental Processing, Spread Channels and Targets, and Scattering Functions

The ocean is a very reverberant environment with extensive multipath and some Doppler spreading in which to propagate acoustic energy. We described some of the physical phenomena that lead to these characteristics. For signal processing, however, it is important to have a means of modeling the effects of the environment upon signal propagation. In this section we shall introduce such models and indicate some of their implications upon signal processing.

There are two types of phenomena that we are interested in modeling at this point. (A third type is related to array processing, and we defer introducing it until the Sec. 6.4.) The first is associated with the multipath and spurious targets distributed in range; it is termed *range spreading*. If an impulse is transmitted, one receives a signal that has its energy spread out over a finite extent. This may be the result of energy propagating over different paths from the target or from reflectors in the vicinity of the target. The second phenomenon is the dual of range spreading and is associated with the energy being reflected by a moving target or propagating through a moving medium. It is called *Doppler spreading*, since if one transmits a tonal signal, the received energy is spread across frequency when this phenomenon is present.

The principal means of characterizing spread sonar channels is to model them as a random, time-varying filter. The randomness is introduced because one seldom has enough detailed knowledge of the acoustic environment and the relative geometries of the transmitter, channel, target, and receiver to use a deterministic description. As a result, one accepts the randomness in the channel or targets, and attempts to describe the average properties of the spreading. This is usually done by employing a *scattering-function* description, or one of its related descriptions such as the *two-frequency correlation function*.[41,44-47] The scattering function indicates how a transmitted pulse statistically redistributes its energy in range and Doppler; the two-frequency correlation function indicates the correlation that is introduced among adjacent frequencies when a tone is transmitted.

The model for the spread sonar channel or target is given by

$$\tilde{s}_r(t) = \sqrt{2E_t} \int \tilde{b}(t, \lambda)\tilde{s}_t(t - \lambda)\, d\lambda \tag{6.15}$$

where

$\tilde{s}_t(t)$ = transmitted signal
E_t = transmitted signal energy (normalization of the signal energy is still convenient to do)
$\tilde{s}_r(t)$ = received signal
$\tilde{b}(t, \lambda)$ = random, time-varying impulse response of the channel

(Note that both T and λ correspond to range delay variables.) The linearity assumption is usually well satisfied, since the signal pressure levels are quite small compared to the static conditions. The duration of the impulse response is determined by the multipath structure of the ocean and how the sonar system is using it; the time variation is determined by any Doppler shifts encountered in the propagation and reflection process. The uncertainties in prediction make specification of the impulse response, $\tilde{b}(t, \lambda)$, very difficult; so it is treated as a random process and characterized in terms of its first and second moments. (It is also modeled as a Gaussian process, which makes the moment description complete.) Generally, the first moment, or mean, is zero because of the phase uncertainties. The second moment, or covariance function, incorporates the important information regarding the statistical nature of the channel; this is usually described in terms of a scattering function.

The scattering-function method of characterizing random channels and propagation media has been used in several fields and has had numerous inventors.(44–50,41,51–53) Nevertheless, the assumptions made have much in common in spite of descriptions in various domains (see ref. 47 for a unifying treatment). The most important are the concepts of stationarity and uncorrelated scattering. The stationarity assumption asserts that the scattering, or reflection, at a range interval is a stationary random process (it may be a constant if the channel is not Doppler spread). The uncorrelated assumption asserts that the reflection, or scattering, is uncorrelated as a function of range. Under these two assumptions we can model the covariance of the reverberation signal, $\tilde{b}(t, \lambda)$, in the following form:

$$\tilde{K}_{\tilde{s}_r}(t_1, t_2, \lambda) = E_t \int \tilde{K}_b(t_1 - t_2, \lambda)\tilde{s}_t(t_1 - \lambda)\tilde{s}_t^*(t_2 - \lambda)\, d\lambda \qquad (6.16)$$

where

$\tilde{K}_{\tilde{s}_r}$ = covariance of the scattering as a function of the range delay, λ
$\tilde{K}_b$ = covariance of the observed reverberation, or scattering

Generally, it is more convenient to use the power density spectrum of the reflection process as a function of range, and this is the scattering function

$$S_{\tilde{b}}(f, \lambda) = \int \tilde{K}_{\tilde{b}}(\tau : \lambda)e^{-2\pi f \tau}\, d\tau \qquad (6.17)$$

It physically represents the statistical redistribution of energy as a function of range and Doppler. If one also Fourier transforms with respect to the range variable, we obtain the two-frequency correlation function

$$R_b(f, v) = \int S_b(f, \lambda)e^{j2\pi\lambda v}\, d\lambda \tag{6.18}$$

This function describes the correlation as a function of frequency in the scattered process.

The sonar channel is principally spread in range, although Doppler spreading from the sea surface and moving targets can be important as well. There is a very strong dependence upon the manner in which the channel is used and, in particular, upon what path structures are excited. Recently, there has been extensive experimental work directed toward measurement of the sonar scattering function; some illustrative results which have been published for a relatively short one way path are indicated in Fig. 6-16.

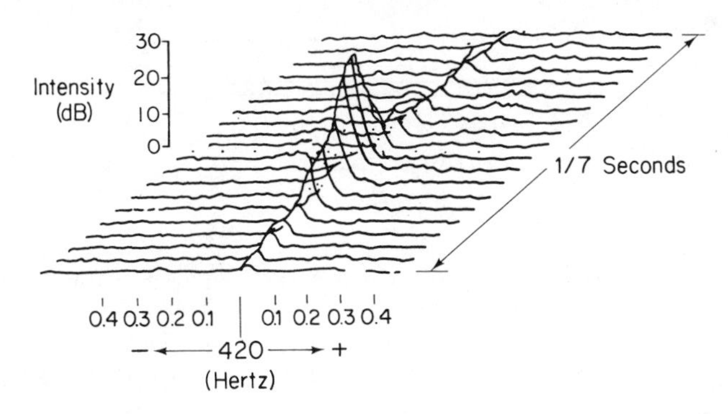

FIGURE 6-16. Scattering function—typical summer conditions seas 2 to 4 ft (from DeFerrari, 53).

In using a scattering-function model to design the signal processing for an active sonar system, several important performance measures are useful. One needs to be careful to specify models for both the scattering of the unwanted reverberation noise and the target reflection process. In addition, one wants to distinguish among the problems of target detection, range–Doppler estimation, and communications. The signal processing related to the environmental modeling for each of these problems is illustrated in Fig. 6-17.

In detecting a slow point target in a reverberant environment using a correlator or a matched filter, which is generally a suboptimal processor, the SNR is given by

$$\text{SNR} = \frac{\dfrac{2E_t\sigma_t^2}{N_0}}{1 + \dfrac{E_t}{N_0}\displaystyle\iint S_t(f, \lambda)\theta(\lambda, \lambda_t, f, f_{d_t})\, df\, d\lambda} \tag{6.19}$$

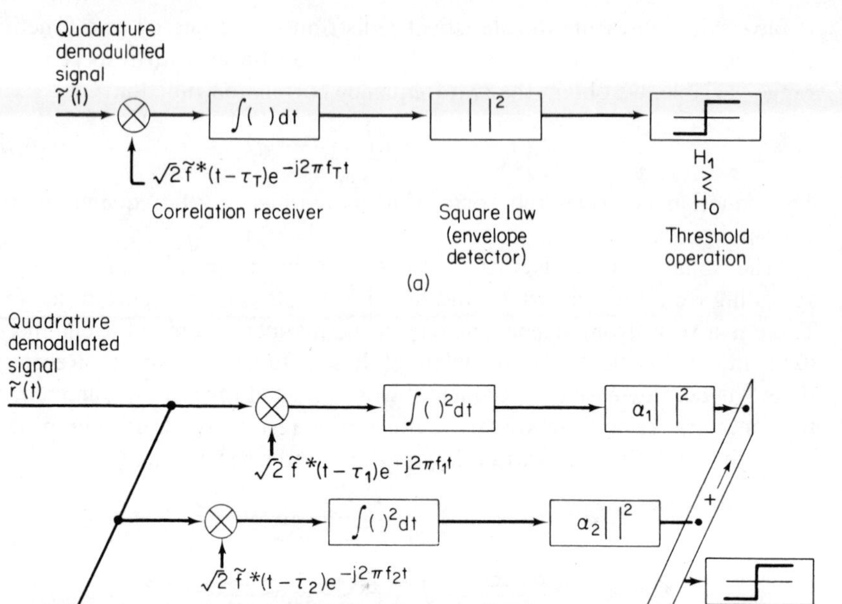

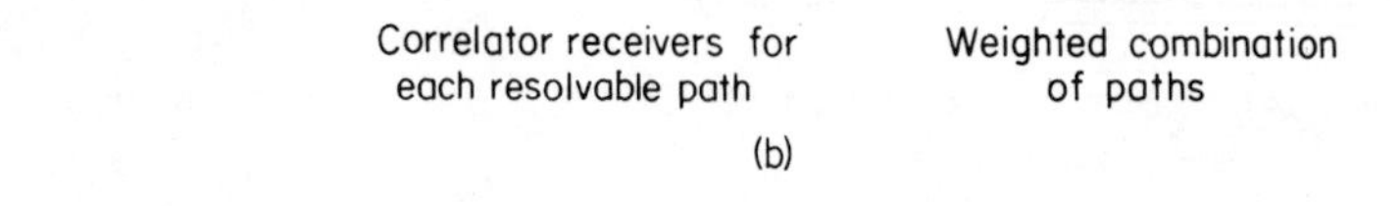

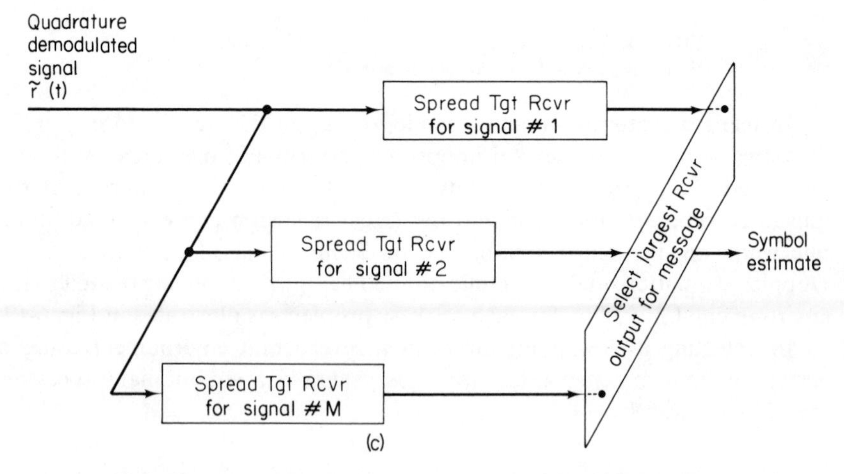

FIGURE 6-17. Structure of sonar receivers for several applications: (a) Detection of point targets; (b) Detection of spread targets in reverberation; (c) Communication receiver.

where

$S_b(f, \lambda)$ = scattering function of the reverberation

N_0 = background white noise level

E_t = transmitted energy

σ_t^2 = target strength

θ = ambiguity function of the transmitted waveform

There are several issues that relate to the performance and effective signal design for sonar systems operating in a reverberant environment encompassed in this expression. First, there are two cases of concern. If the integral in the denominator is much less than unity, then the situation is said to be background, or white, noise limited. Physically, the reverberation is not significant in the detection, and received energy is the primary concern. Conversely, if the integral is much greater than unity, the system is reverberation limited. Here an increase in the transmitted energy simply increases the reverberation noise as fast as it increases the reflected signal from the target, so no net gain is achieved. In this situation the performance of the system can only be improved by effective signal design. This is the second important issue. The performance is specified by the overlap of the ambiguity function of the target displaced to its range and Doppler and the scattering function. Effective signal design concentrates upon minimizing this overlap. This is usually done by making the ambiguity function very narrow either along the range axis, which is termed *range gating*, or along the Doppler axis, which is termed *Doppler gating* on Fig. 6-18. One can see the advantages of a chirped waveform when operating in a reverberant environment. Figure 6-18(a) illustrates a comparison of the overlap for a pulsed continuous-wave (CW) signal and a chirped FM signal of the same duration. There is an approximate reduction of the overlap by a factor of μT^2, the time bandwidth product of the chirped signal. The performance loss due to reverberation as a function of chirp bandwidth to reverberation bandwidth is indicated in Fig. 6-18b. High chirp can minimize the loss in this situation. Chirped signals are not always an advantage. If there is a Doppler displacement between the target and the reverberation, then we want to maintain the narrow Doppler axis width of the unchirped signal. Reference 41 describes the parameter optimization for this problem in detail.

In detecting a spread target or in an underwater communication system, the processing should be somewhat more complicated than the simple correlator receiver used for point targets. In these situations an approximately optimal receiver combining the outputs consists of a correlator receiver at each resolvable range–Doppler cell. The resolution and number of these cells can be determined using the scattering function and the two-frequency correlation function of the transmitted waveform. The correlation between two

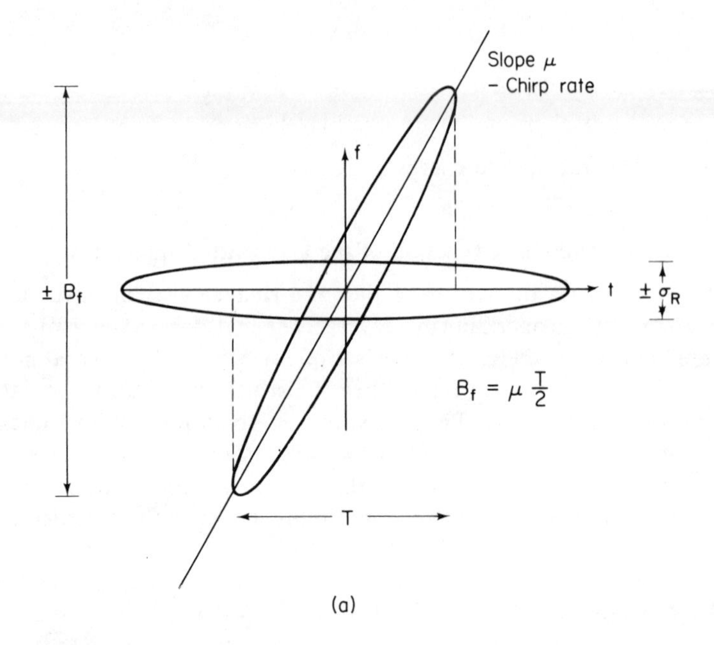

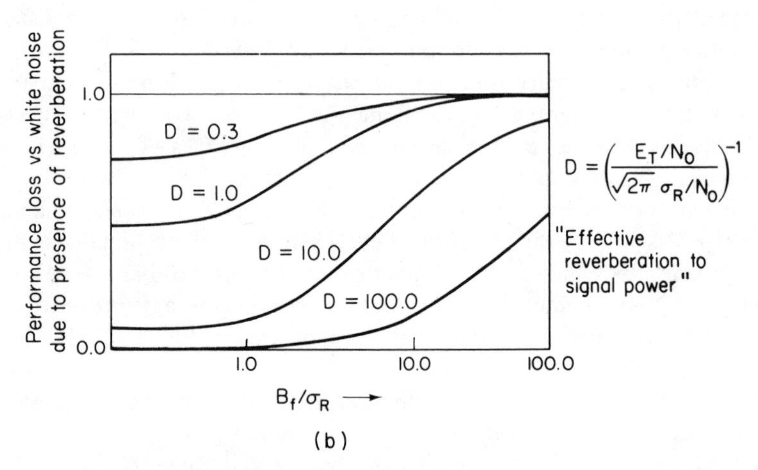

FIGURE 6-18. (a) Overlap of ambiguity function and scattering function for range spread reverberation and an FM chirp; (b) Performance loss for Figure 6-18(a) as a function of signal bandwidth to reverberation bandwidth (from VanTrees, 42).

matched filters, or correlators, centered at (λ_1, f_{d_1}) and (λ_2, f_{d_2}) in the range–Doppler plane is given by

$$\gamma(\lambda_1, \lambda_2, f_d, f_{d_2}) = E_t \int S_{\tilde{b}}(f, \lambda)\tilde{\phi}(\lambda_1, \lambda, f_{d_1}, f)\tilde{\phi}^*(\lambda_2, \lambda, f_{d_2}, f)df\,d\lambda \quad (6.20)$$

The environmental processing can be seen in Fig. 6-17b in conjunction with this formula. It simply consists of an incoherent combination of each resolvable range–Doppler cell weighted by the strength of the scattering at the cell. This is then thresholded to make the actual decision for detection.

Both the detection of spread targets and underwater acoustic communications involve combining the resolvable paths to account for the environmental effects of the target and propagation medium. The major differences between these systems appear in the manner in which the channel is used and in how the paths are identified. In the detection problem the same signal is transmitted, and one wants to identify just those cells that are associated with the reflected energy of the target. In the communication problem, different signals, usually some form of shifted frequency, are transmitted depending upon the message, and one wants to identify all the paths that were excited by the signal.

It is easy to observe that good signal design can improve the performance of these systems significantly. Generally, one wants to use signals that resolve as much detail of the target as possible in a detection system. In communications the issue is somewhat more subtle. The key to good signal design is to excite the *diversity* of the channel in an optimal manner. This implies that there is an optimum number of resolvable paths that should be generated for a given amount of transmitted energy. (See refs. 41 and 48 for a complete discussion on optimal diversity.) A quantitative analysis of the performance and the calculation of an SNR or error probability can be a difficult task. Many models lead to a chisquared type of analysis that emphasizes degrees of freedom; however, more complex methods have been extensively developed.[41,48]

Before proceeding to Sec. 6.3.3, which covers the processing of sequences of pulses, we shall comment on the environmental processing that is associated with range–Doppler estimation. The estimation problem is closely coupled to the detection one. An operator is generally scanning the range–Doppler plane when he is attempting to locate targets. The issue is to identify a target reflection using whatever features that one can associate with such reflections. Once located, the issue is to separate the target signal from the reverberation so that its range–Doppler and strength may be determined. One wants to map the target features, so signals with good resolution properties should be used. Unfortunately, there is not a simple analysis available for this with spread targets, such as there was with the Cramer–Rao bounds and the signal ambiguity function for slow point targets.

In summary, the acoustic environment sets important limitations. In most systems it is quite complex and changes appreciably, so it requires a great deal of flexibility in the processing done to account for it. The capabilities of digital signal processors are and will be employed extensively for obtaining this flexibility.

6.3.3 Processing of Pulse Sequences

Up to now we have concentrated upon the processing that is done upon a single pulse transmission. One seldom, if ever, uses a single observation in an active sonar system. There is a continual probing of the environment and monitoring of the observed signals. In all cases a sequence of observations is used, and the capabilities of digital processors are employed. In discussing this area, however, it is important to note that there are many more methods of processing the data sequences in this section than in the previous ones, and almost all involve some type of computer program. We shall concentrate upon some of the methods that are closely connected with signal processing, but we emphasize that there are many more methods.

In target detection the processing of a sequence of pulses usually consists of summing the correlator output or its square root from each pulse. There are two important issues that lead to four different possibilities in this summation process. These are generally referred to as *Swerling target models.*[54] In this classification of targets one considers two different probability density models for the target strength and the fluctuation among the reflected pulses. In all cases the phase of each pulse is modeled as a uniformly distributed random variable, which is statistically independent among the pulses. The Swerling models for the correlator output statistic are summarized in Fig. 6-19. The first density model is appropriate for a reflection produced by a large number of small scattering centers; the second is an approximation to a single specular component with a random phase plus a large number of smaller scattering centers. The fluctuation models correspond to the magni-

Probability density of receiver output statistic for a single pulse

Fluctuation rate	Exponential	Rayleigh
– slow	Case #1 – constant cross section of a target with a small extent compared to signal wavelength	Case #3 – constant cross section of a target with a large extent compared to signal wavelength
fast	Case #2 – fluctuating cross section of a target with a small extent compared to signal wavelength	Case #4 – fluctuating cross section of a target with a large extent compared to signal wavelength

FIGURE 6-19. Swerling target model classifications for combining pulses.

tude of the correlator output remaining constant, but random, in the slowly fluctuating case, and the output fluctuating statistically independently from pulse to pulse in the fast case. The optimum method of combining the correlator outputs can be derived using transcendental functions; however, in most cases the outputs are simply added.[41] The performances of the various cases have been extensively tabulated, and they are used as a basis of performance prediction in many situations.

The details of the analysis are not particularly important for active systems since other effects tend to obscure them. The important issue is that they incorporate two important physical considerations in combining the pulses: (1) the model for the output of the correlator receiver, and (2) the amount of fluctuation from pulse to pulse. As far as the active sonar detection problem is concerned, the signal processing for combining pulse sequences usually consists of some form of averaging the correlator output in an attempt to obtain the required SNR by the averaging process. If the environment and reflector were completely stationary, and if there were sufficient processing capability to scan all range–Doppler cells over long sequences of pulses, one could in principle achieve any desired receiver operating characteristic. This is seldom the case, so one must be content with combining a limited number of pulses, typically less than 10–100. In addition, the combination can be done far more efficiently with considerably less memory requirements by using a fading memory averaging instead of the exact averaging. The performance loss is negligible.

In processing a sequence of active pulses to estimate the range and Doppler of a target, the capabilities of both modern estimation theory and digital hardware technology are used extensively. Essentially, one wants to smooth a sequence of raw range–Doppler estimates from the correlator output to produce a track of the target. This smoothing should incorporate the dynamics of the target motion, such as the coupling of the Doppler to the sequence of range estimates, as well as any constraints on the target, such as maximum accelerations or speeds. The most commonly used procedures employ a form of least-squares analysis. For example, curve fitting with polynomials with some form of outlier rejection is one of the simplest methods. At the other end in complexity, the target motion is modeled in terms of a dynamical system with unknown excitation forces and noise corrupting the range–Doppler observations. This is most conveniently done in terms of a state equation model of the form

$$\mathbf{x}(n+1) = \mathbf{f}(\mathbf{x}(n), n) + \mathbf{g}(\mathbf{x}(n), n)\mathbf{u}(n) \qquad \text{state equation} \tag{6.21}$$

$$\mathbf{y}(n) = \mathbf{h}(\mathbf{x}(n), n) + \mathbf{w}(n) \qquad \text{observation equation}$$

where the state matrices $(\mathbf{f}, \mathbf{g}, \mathbf{h})$ incorporate the track dynamics and observations method, and the random processes $(\mathbf{u}, \mathbf{w})$ model the uncertainties in the track and observations of the target. For example, in a very simple model one may have

$$
\begin{aligned}
x_1 &= \text{north/south position} \\
x_2 &= \text{north/south velocity} \\
x_3 &= \text{east/west position} \\
x_4 &= \text{east/west velocity} \\
y_1 &= \text{range observation} \\
y_2 &= \text{range rate observation} \\
y_3 &= \text{bearing observation}
\end{aligned}
$$

$$
\mathbf{f}(\mathbf{x}) = \begin{bmatrix} 1 & \Delta T & 0 & 0 \\ 0 & 0 & 0 & 0 \\ 0 & 0 & 1 & \Delta T \\ 0 & 0 & 0 & 0 \end{bmatrix} \qquad \mathbf{g} = \begin{bmatrix} 0 \\ 1 \\ 0 \\ 1 \end{bmatrix}
$$

$$
\mathbf{h}(\mathbf{x}) = \begin{bmatrix} \sqrt{x_1^2 + x_3^2} \\ \dfrac{x_1 x_2 + x_3 x_y}{\sqrt{x_1^2 + x_3^2}} \\ \tan^{-1}(x_1, x_3) \end{bmatrix}
$$

The advantages of this type of modeling are that the constraints of the dynamics and the uncertainties in the observations can be introduced directly into the estimation algorithms; and the resulting estimation algorithms are recursive in nature, which leads to very convenient implementation on a digital processor. The difficulty, however, is that the estimation equations are nonlinear, so analytical solutions cannot be obtained. In addition, all the details of the estimation algorithms are not currently understood, particularly the global aspects and the performance at low SNRs. Nevertheless, several investigations involving extensive simulations have been made.[55-57]

Most of these investigations have involved some type of nonlinear estimation theory oriented around Kalman filtering. The nonlinear aspects are incorporated by linearizing about an estimated track, which is generally referred to as extended Kalman filtering (EKF). When the SNR is sufficient and certain constraints (observability) in the dynamical modeling are satisfied, some of these algorithms have worked well using range and Doppler observables. If the performance of the estimator is too low, the linearization becomes too noisy and the algorithms degrade rapidly.

These discrete time filters (pulse-by-pulse processing) are usually not classified as digital filters; however, when one examines their recursive nature, they are really a generalization of infinite impulse response filters. The major difference is that the feedback matrices are not constant, but functions of time and possibly the observed and estimated data. This is an area where implementation issues, which have been the focus of the digital signal-processing field, and processing algorithms, which have been the focus of the estimation theory literature, are starting to interact to solve a common problem of designing efficient and accurate tracking algorithms and systems.

We have elaborated briefly upon just two aspects of processing sequences of active pulse reflections. Several of the systems that we mention in the sequel also process entire sequences in their operation. To discuss the details of this processing requires too much space, so we shall simply comment on some of their more important aspects.

Communication systems often encode their messages into a sequence of transmissions. There are several reasons for doing this. First, the encoding can provide more efficient communication in terms of energy per bit, as well as provide increased reliability by introducing redundancy in the code sequence. Second, the encoding can provide a degree of security by using coding methods that introduce some form of cryptographic concealment. In employing encoding procedures, both the transmitter and receiver must have a modest amount of signal-processing capability. The transmitter must implement the coding algorithm, and the receiver must decode the observed data sequence to make a minimum-error probability estimate of the message. As one might suspect, the more advanced procedures can become quite involved with the details of coding theory and information theory.

Maps and charts of the seabed are generated by plotting estimates of depth obtained by a sequence of reflections from the seabed. A single sounding does not convey an appreciable amount of information; one needs a grid of soundings to determine the topographic relief that is usually required. This necessitates a variety of processing methods; some simply compensate for the finite beamwidth of the sounding system; others compile the grid of points, interpolate the data, and contour the relief. Again, all these tasks require a considerable amount of signal-processing capability.

We have now completed our discussion, brief as it may have been, on some of the theory and models that underlie active sonar signal processing. We shall next turn our attention to some implementations that have been used to realize the operations suggested by the theory and models.

6.3.4 Implementation of Correlation Receivers and Matched Filters

The use of correlation receivers and matched filters has been an important issue ever since the role of signal processing was recognized in sonar. It is important to realize, however, that the matched filter is an optimum receiver only under some very special situations; in particular, it assumes that the interfering noise is effectively white across the band of interest. Nevertheless, when one combines it with the appropriate environmental processing, as indicated in Sec. 6.3.2, it forms an important component for many sonar receivers that are essentially optimum.

In its simplest form a correlation receiver for a pulsed CW signal may be realized by a bandpass filter whose frequency and damping are matched to the signal and its duration. As soon as the advantages of large time–bandwidth signals such as PR sequences and chirped FM became evident, considerable effort was directed toward implementing receivers capable of exploiting these advantages. Initially, almost all these receivers were directed toward time-domain realizations, since large time–bandwidth Fourier transformation was not practical in real time.

One of the first systems, and certainly the most extensively used, was the DELTIC (delay line time compression).[58-59] It is essentially a digital technique for realizing a tapped delay line implementation of the convolution operation at the signal bandwidths commonly used in active sonar systems. It uses a recirculating delay line to compress the long-duration, small-bandwidth sonar signal into a short-duration, wide-bandwidth signal in the delay line. The basis system is illustrated in Fig. 6-20(a). The delay line has a length of $(N - 1) \cdot \Delta T$ with $(N - 1)$ taps spaced every ΔT seconds. The data are sampled every $N \cdot \Delta T$ seconds. In its operation a data sample is taken, progresses through the entire length of the delay line, and recirculates to the second tap before the next sample is taken. The process repeats, and eventually the delay-line taps contain the last N samples of the data. The data are shifted out of the lines by overwriting the contents of the first tap with a new sample.

The DELTIC system can be used to implement a correlation operation by combining two of them and a set of multipliers and a summing bus, as indicated in Fig. 6-20(b). Both the top and bottom DELTICs are loaded with the signals to be correlated, and the multiplier–adder network carries out the correlation operation. If one of the signals is a stored reference, as it usually is in an active sonar system, one of the lines is modified such that, once a signal is loaded, it recycles continuously. The multiplier–adder network is then a sampled version of the matched filter.

In principle, this system could operate in a completely analog manner. It was quickly recognized, however, that there were a number of advantages in hard clipping the signals, so that one was actually performing a polarity coincident correlation. The actual performance loss incurred by this digital version of a linear correlator varys according to the particular environmental conditions, but in many cases the loss is quite small compared to the advantages offered. The digital realization did not require shaping circuits to maintain the fidelity of the recirculated signal, the hardware savings in multiplier–adder network was considerable, and this was a very important factor during the era of the development of the systems. An operational advantage is that the hard clipping eliminated the sensitivity to the dynamic range of the signals,

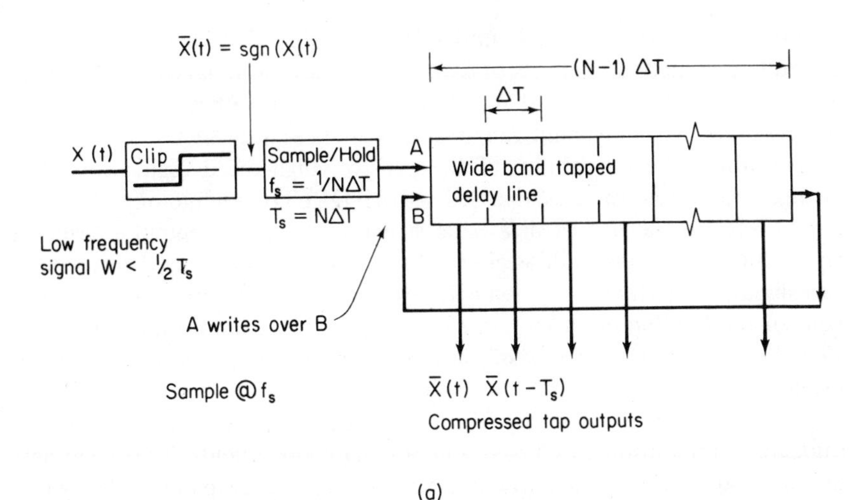

(a)

(b)

FIGURE 6-20. (a) DELTIC system structure; (b) Cross correlation with DELTICs.

which as we comment on later has been an important practical aspect of sonar signal processors.

Much research has been done upon determining the loss in performance due to the hard clipping and upon evaluating just how many quantization levels are really needed.[60-62] The most frequently stated result is that when the system is operating against a white noise background and when the SNR is above threshold, there is a 2-dB performance loss compared to the ideal linear implementation. Several other situations have also been investigated; for example, when operating against a narrowband jammer, there is a loss of $\frac{1}{2}$ dB for each decibel of interference-to-signal ratio (ISR). Additional results are available, but it is important to realize that they must be very carefully qualified. The effects of quantization beyond binary have also been analyzed, and they suggest that a 4-bit quantization suffices for most situations, providing the dynamic range is adequately scaled.

Naturally, the availability of inexpensive, high-speed digital logic has changed the circumstances that led to the DELTIC technology, We now would refer to the tapped delay line structure as a finite impulse response (FIR) filter. We have multilevel shift registers that can be used to quantize the signals to virtually any realistic level of accuracy, and adaptive gain control and floating-point formats ease many of the dynamic range problems. Figure 6-21 illustrates some of the details of a correlator receiver using recent technology.

The advent of fast Fourier Transform (FFT) technology is having the same impact upon sonar as it has had upon many other fields. The advantages in active sonar are particularly evident with large time–bandwidth signals, for it is here that the computational advantages offered by the FFT excel. The crossover point in terms of efficiency in the number of operations for a frequency-domain realization is commonly quoted at 32 length points, which is below the time–bandwidth products now being considered in proposed systems. FFT implementation is discussed in detail in Chapter 5, to which we refer the reader for the hardware aspects of the FFT.

The front-end processing required for estimating the range and Doppler of a target benefits significantly from the current technology available for high-speed correlation operations. In scaning the range–Doppler plane, the receiver must implement a correlation operation at every resolution cell for all the bearings of interest. This can imply performing a very large number of correlations. Figure 6-22 illustrates two alternatives for scanning the range–Doppler plane. The first realization consists of a bank of Doppler frequency shifters (or a time compressor for a broadband system), followed by a network of correlators, or matched filters. The correlators themselves may have an FFT realization, but they must return to the time domain to produce the range estimate. In some cases the Doppler bank can be eliminated by exploiting some of the properties of the transmitted waveform; for example, when an FM chirp is employed, an uncompensated Doppler shift leads to a range offset under the narrowband signal conditions. The second realization is a dual to the first, as we have used a Parseval realization to transform the two-frequency correlation operation of the front end into the frequency domain. This system consists of an FFT operation, a bank of linear phase shifters to incorporate the time shift for the range delays, and a correlation operation, which itself may be implemented by an FFT (i.e., a pass through the time domain). The output emerges as a function of the Doppler shift for each of the range delays used. The relative efficiencies of these two realizations of the same estimator, or range–Doppler scanner, depends upon the number

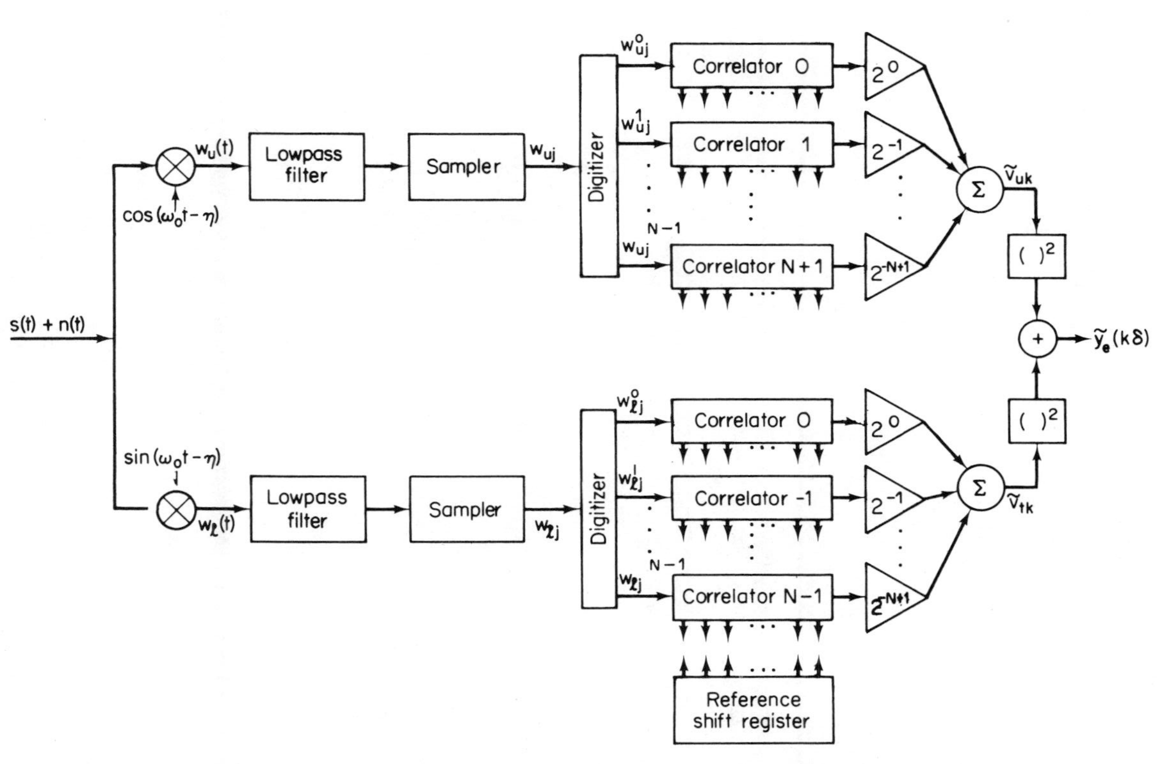

FIGURE 6-21. Example of matched filtering with current digital technology (from Turin, 62).

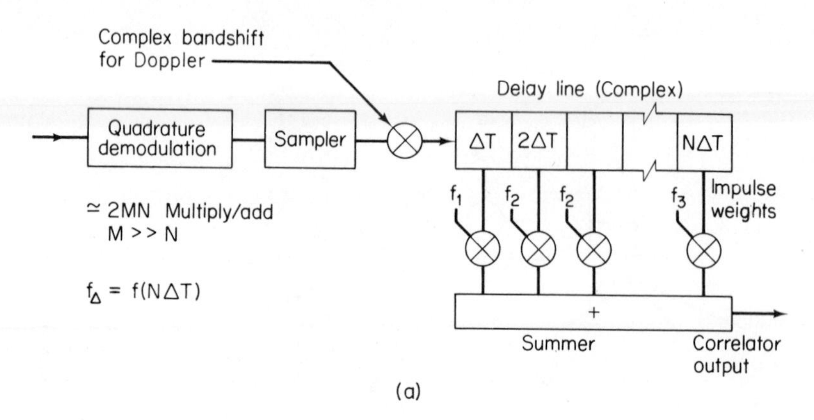

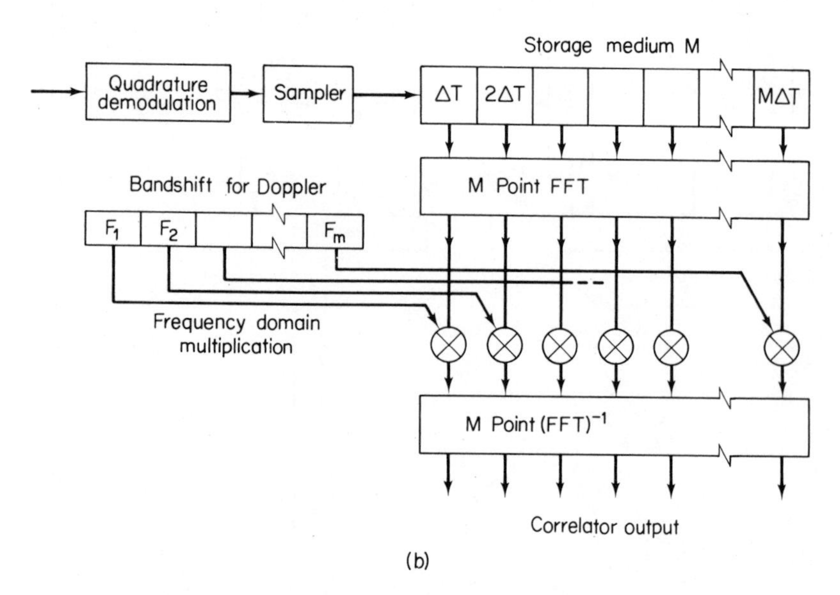

FIGURE 6-22. (a) Time domain realization of correlation operation at the front end; (b) Frequency domain realization of correlation operation at the front end.

of resolution cells required along both axes; this is determined by the resolution of the signals compared to the region of uncertainty that must be scanned.

In most range–Doppler estimation procedures the scanning operation must be followed by an operation that selects the location of the maximum correlation. This operation must incorporate all a priori information about the target location, as well as the relevant environmental factors.

One other important consideration in the structure of the front-end processing is whether the operations are implemented at the center frequency of the sonar system or demodulated to quadrature components at baseband. In a digital realization there are a number of advantages to the demodulated signals because of the lower sampling rates required. To implement the operations at the center frequency, a sampling rate of four times the center frequency is required. This can lead to some very long data arrays. In comparison, the demodulation to baseband requires two complex samplings, both quadratures, at the signal bandwidth, in effect a savings of f_c/W in the dimensions of the array storage. Accurate beamforming requires sampling well in excess of the Nyquist rate, typically 4-5 times, even with baseband, quadrature demodulation.

There are several tasks that the front end of an active sonar must implement. Two of the more important are dynamic range compression and compensation for the Doppler motion of one's own ship. The dynamic range compression involves some form of automatic gain control (AGC), time-varying gain (TVG), or logarithmic amplification. These compensations are particularly important for analog or fixed-point systems, both of which have a limited dynamic range compared to that which can appear in a reflected sonar signal. As digital processing using floating-point hardware becomes more prevalant, these factors will probably be less important. They will not disappear, since array elements will still need to be equalized, and output data will need to be compressed for presentation on most displays. Doppler compensation simply consists of frequency shift operation. It is usually done at the operating, or center, frequency, but there is no difficulty in doing it digitally at the sampled baseband signals.

6.3.5 Implementations of Environmental Processing

The sonar channel and reflection processes are generally so complex that they are modeled in terms of random filters and scattering functions, which were described earlier. Effective processing using these environmental models depends upon how well the scattering function, or its equivalent, can be determined and the flexibility of the signal processing in adapting to the inevitable changes in the environment. The estimation of the scattering function can often be done by a combination of determining the oceanographic variables, such as the sound-velocity profile, from either tabulated data or in situ measurements, and performing some form of adaptive processing. Once the scattering process has been characterized, this information must be cast in a form such that the signal processing can implement the equations specifying the receiver. We cannot comment upon the estimation problem, since this becomes coupled to specific situations, and the flexibility now available in the processing is still stimulating work on this topic; however, we can briefly indicate some of the implementations that can be used in processing reverberant signals.

In discussing these implementations it is important to qualify the remarks and indicate some perspective. First, there are a number of constraints and interrelationships among the duration of the signal, T, the duration of the reverberation, L, the bandwidth of the signal, W, and the bandwidth of the Doppler spread, B. In addition, the sonar channel often tends to have concentrated scattering due to the dominance of some of the paths. Second, spread channels have a long and extensive literature not only in sonar, but also in radar, radar astronomy, and ionospheric and tropospheric communications. In fact, the applications in these fields are probably more advanced than in sonar.

Many implementations for environmental processing revolve around a tapped-delay-line model for Eq. (6.15), the fundamental random filter model. If the signal is band limited to W and the reverberation is duration limited to L, the structure indicated in Fig. 6-23(a) can be used to model the reverberant signal.[63-65] In this tapped-delay-line structure, the signal is modeled in the form

$$\tilde{s}_r(t) = \sum_{i=1}^{2WL+1} \tilde{b}_i(t)\tilde{s}_t\left(t - \frac{i}{2W}\right)$$

where the $\tilde{b}_i(t)$ are a set of random, time-varying tap gains whose spacing is determined by the reciprocal of the bandwidth of the transmitted signal. Under some mild assumptions, it can be demonstrated that the tap gains are uncorrelated stationary random processes whose power spectral densities are proportional to a slice of the scattering function at range

$$\lambda = i/2W, \qquad \text{i.e., } S_b(f, i/2W)$$

as indicated in Fig. 6-23(b). One should further note that, if one Fourier transformed the tap gains over the duration of the data, the coefficients in the frequency domain are approximately independent. The important implementation issue with this structure, however, is that it is a very convenient representation of the reverberant signal, especially from a digital-hardware viewpoint. As a result, one can focus his attention on a number of algorithms, which suggest that one should be estimating the tap gains in a receiver implementation.

Before discussing the use of the tapped-delay-line model in some receivers, it is useful to make a few observations. If the channel or reflector being modeled has no or negligible Doppler spread, which is often the case in sonar, the tap gains are constant and are effectively the sampled impulse response of the channel. Several efforts have been made to measure this impulse response.[66-68] There is also an obvious dual to this representation, which involves Fourier transforming the signals. This leads to a transfer function description.

Many receiver structures have been derived using the tapped-delay-line representation of a reverberant signal. In Fig. 6-24 we illustrate a receiver

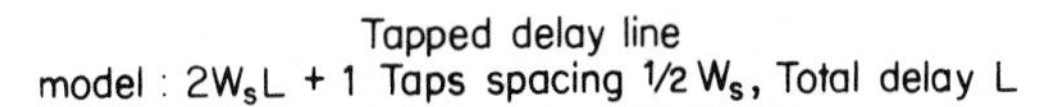

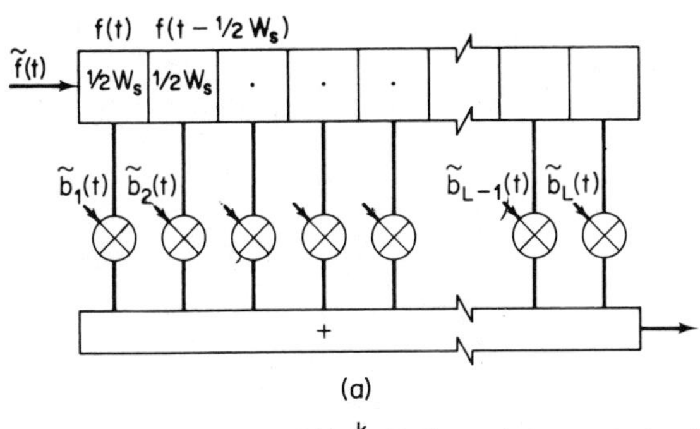

(a)

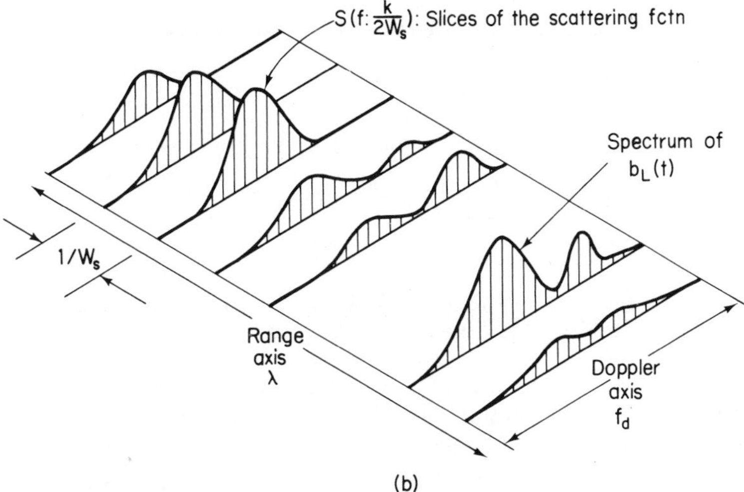

(b)

FIGURE 6-23. (a) Tapped delay line model for spread channels; (b) Spectra of the tap weights in the delay line model as slices of the scattering function.

implementation for detecting a spread target and estimating its range and Doppler. It involves some modeling assumptions, so it is not universally applicable for all reverberant signals. There are two inputs to the environmental processing section of the receiver. In one input, the output of the correlator–matched filter at the various Doppler offsets is input to a set of tapped delay lines. The second input consists of estimates of the scattering functions of both the target and the reverberation noise. These must be obtained from either knowledge of the environmental conditions or from some form of adaptive estimation. The tapped-delay-line outputs are magnitude squared and weighted according to a ratio determined by the intensity of the scattering functions of both the target and the noise, as well as the level

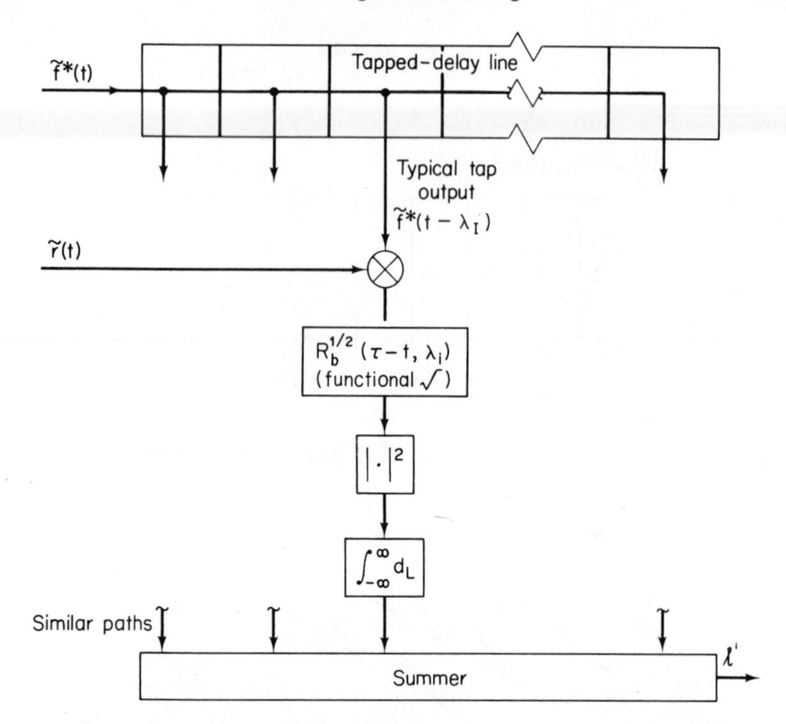

FIGURE 6-24. Approximation of the sonar receiver in the situation of low signal to noise in a reverberation spread environment.

of the ambient background noise. These weighted outputs are finally added to produce the detector output, which may drive a display or serve as an input to a pulse sequence processor. The same system can be used as an estimator by scanning the outputs as a function of the range–Doppler values of the target (i.e., by changing the weights of the tapped-delay-line outputs as a function of the target parameters).

The most important aspect of the digital filtering role in these systems is that to operate successfully the environmental information must be combined with the correlator receiver outputs. This requires a great deal of flexibility, especially because of the changing nature of the channels and targets. The digital processors appear to be the only ones with this flexibility.

There are several other representations available for modeling the environmental processing (e.g., those coupled to two-frequency correlation functions). The literature on spread channels is extensive, and we have suggested only one model that is well suited for digital processing. Almost all the models, however, are similar in one respect. They attempt to characterize the random nature of the channel and targets in terms of a set of variables or processes that can be resolved by using the correlation properties of the transmitted waveforms. As a result, the ability to perform rapid correlations at the front end of the receiver is particularly important.

6.3.6 Processing of Pulse Sequences

The impact of the minicomputer and microprocessors have been particularly important in many active sonar systems, yet nothing has been especially unique in these implementations to the sonar environment. Most applications involve either machine-coded or Fortran or Algol programs on machines with a 16- to 24-bit word length and 32 K of memory. These are often supplemented by disc memories. The minicomputers are also very useful and adaptable in driving operator-interactive displays. In this section we shall comment upon some of the situations where the use of a minicomputer for processing sequences of pulses is especially advantageous.

In target detection and range–Doppler estimation, a sequence of pulses must often be combined simply to build up the signal-to-noise ratio. This combination can be done in a number of ways, such as a direct sum of the magnitude or the squared magnitude, or lowpass filtering a sequence from a range–Doppler resolution cell. In doing this combination, it is important to accumulate running statistics such as the mean, variance, and possibly a histogram of the detector output. These are useful in adaptively setting thresholds and rejecting spurious outliers, or anomalous outputs, in accumulating the sequence of pulse outputs for both detection and range–Doppler localization.

Tracking and navigation systems typically involve the solution of relatively complicated sets of equations. A number of target motion algorithms are used, and nonlinear estimation methods such as the extended Kalman filters are becoming more popular. Essentially, one needs a modest capability in general-purpose numerical analysis for these algorithms, and it was not until the advent of the minicomputer that this could be done without the use of a large, general-purpose digital computer.

Communication systems employ coding for both improved reliability and cryptographic security. The decoding algorithms are often complicated and operate on a long sequence of pulses, and again a modest capability in numerical processing is required. A number of coding, and decoding algorithms have been developed, such as the Viterbi algorithm, that can yet make an impact upon efficient underwater communication systems.

Finally, mapping and charting systems accumulate an enormous amount of data, which undergo a significant amount of processing before the final map or chart is produced. The pulse sequences must be edited to remove spurious returns, corrected for source–receiver geometry, collated on a navigational coordinate system, and contoured. Without the minicomputer it would be virtually impossible to handle the enormous sequences of pulses that are generated with today's mapping systems.[123]

6.3.7 Examples of Sonar Signal Processors

In previous sections we discussed some of the aspects of active sonar signal processing from both a theoretical and modeling viewpoint, as well as implementation. In this section we shall briefly introduce examples of several active sonar that are illustrative of recent systems. Classification issues do, however, prevent us from discussing the most sophisticated systems. We introduce three examples: (1) one of the recent military active sonar for detection and tracking of targets from a hull-mounted array, (2) a navigation and positioning system for tracking a submerged research vessel, and (3) a configuration for a communication system. By using these examples one can observe the flow of signal processing and the interconnection of the various operations.

In an active detection and tracking system one is principally interested in locating submarines and determining their course. A variety of propagation paths are used in this—direct, bottom bounce, convergence zone, or surface duct—all depending upon the environmental conditions. Figure 6-25(a) is the block diagram for a relatively low cost, simple active sonar, the DE-1160-AN/SQS-56 System developed by the Raytheon Company.[68] Some specifications for the system are also shown.

One can observe the sequence of the signal processing as follows. First, a controller, or operator, selects an appropriate waveform for transmission; this waveform is determined by the environmental features of the target and reverberation, the propagation paths to them, and the desired operation of the system. The waveform is then delayed, weighted, and amplified before being applied to the individual transducer elements of a hull-mounted array to form a beam. If the system is operating in a search mode, the beam direction may be stepped, or swept, on successive transmissions in order to scan a desired area. The transmitted energy enters the sonar channel and is reflected back to the array, as well as any reverberant energy that may be present. At the array the transducer elements are switched to the receive mode. Their outputs are preamplified, bandpass filtered, and gain compensated to condition the signals. These signals are delayed, weighted, and summed to generate a received beam. This signal processing is contained in the block "search/track active processing."

An expanded version of this section is indicated in Fig. 6-25(b). The first operation is a matched filtering, or correlation receiver, operation. This is done at baseband by quadrature demodulation followed by a sampling for A/D conversion. The correlation operation is generated as a function of range after Doppler compensation for one's own ship motion; in addition, Doppler shifts for the target can be introduced. The correlator outputs are then magnitude squared and averaged as a function of beam direction over a sequence of pulses if the system is operating in a search mode. The number of pulses in the averaging operation is a function of stationarity of the environment. If the system is operating in a tracking mode, the correlator output is maximized as a function of range and Doppler, and is then used as an input for a target motion analysis algorithm. The details of the processing are indicated in the Fig. 6-25(b).

Some very interesting applications of signal processing for tracking and navigation concern the problem of locating an underwater platform, or submersible. There are several oceanographic systems that employ this technology; examples include the positioning, or control, of the drilling head in the Deep Sea Drilling Program and tracking the research submersible Alvin. This is quite different than tracking a hostile submersible by observing reflections, for in these applications the target operates cooperatively within an array of transponders. The basic operating principle of these systems is the use of an array of three or more transponders to generate lines of position with respect to the tracked vehicle or platform. Figure 6-26(a) indicates the geometry used for tracking a vessel, whether submerged or on the surface, using such a transponder array.(69)

In this system there are two modes of operation, pulsed and Doppler. In the pulsed mode the platform emits an omnidirectional waveform. This waveform is received by the transponder array; each of these elements heterodynes the signal to a different frequency and rebroadcasts it back to the platform. Each frequency-shifted channel is then processed using a correlation receiver to produce an estimate of the round-trip time to each element in the transponder array. The round-trip times are then used to generate lines of position, really surfaces of position since the system operates in three dimensions, from which the platform location relative to the tranponder array can be determined. As one might suspect, the accuracy of this system is very dependent upon the estimates of the round-trip times, so accurate range–Doppler estimation is imperative. This is basically implemented with a correlation receiver, as suggested in Fig. 6-26(b). Accuracies on the order of of 2 to 3 m have been reported using this type of system in conjunction with a precision survey of the transponder array locations.

In the Doppler tracking mode, a CW signal is transmitted from each element in the transponder array to the platform. The signal processing at the platform demodulates each signal and applies a digital phase-locked loop to integrate the observed Doppler shift for an estimate of the phase difference between the platform and the transponder. These phase differences are then used to generate surfaces of position from which the location of the platform is determined. This system needs to have the phase differences initialized, so it is used in conjunction with the pulsed system. In addition, it must operate

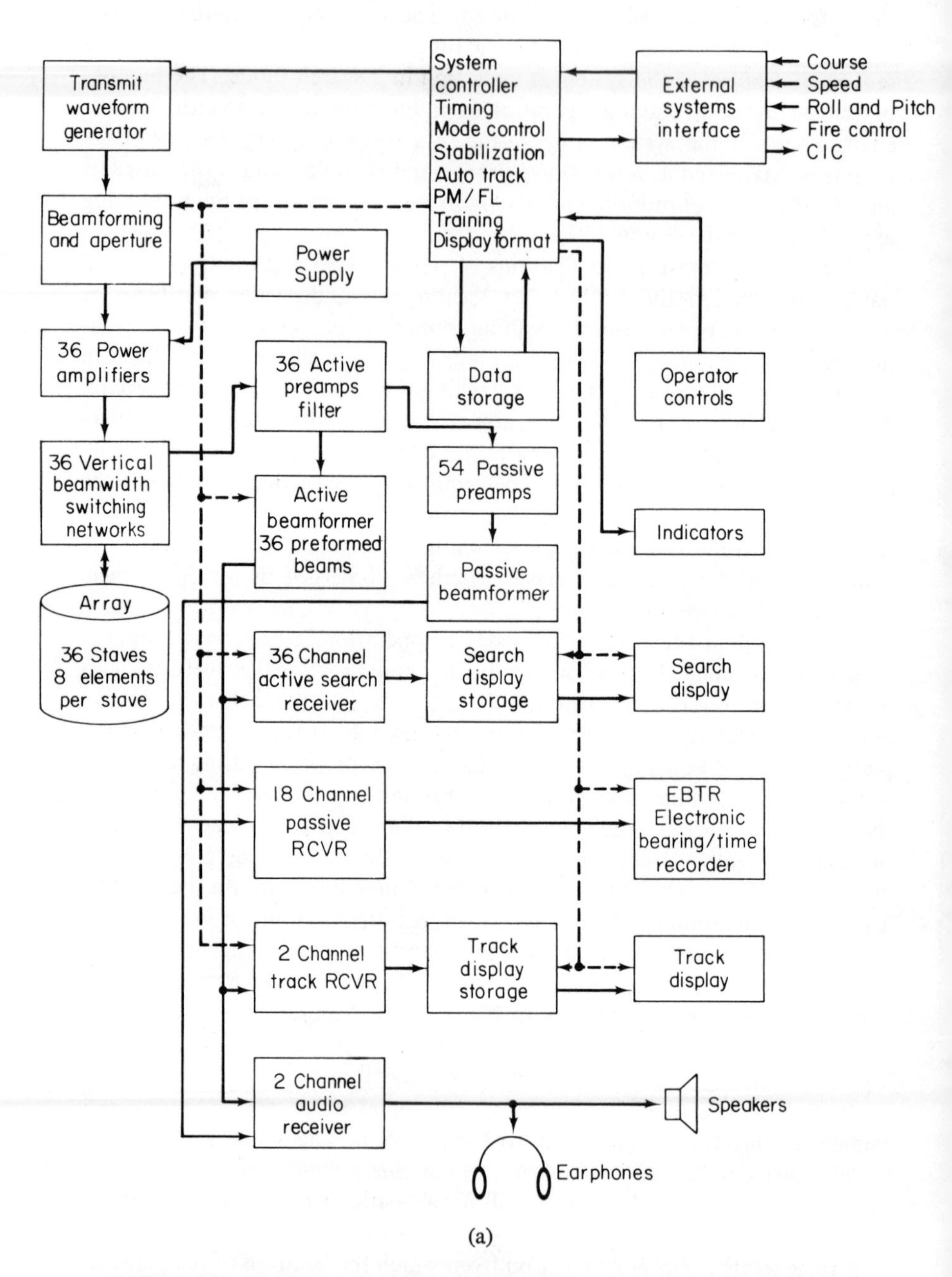

(a)

FIGURE 6-25. (a) Signal flow diagram of the Raytheon DE-1160-AN/SQS-56 sonar system; (b) Structure of the signal processing of important components of the Raytheon active section DE-1160-AN/SQS-56 sonar system (from Skitzki, 68).

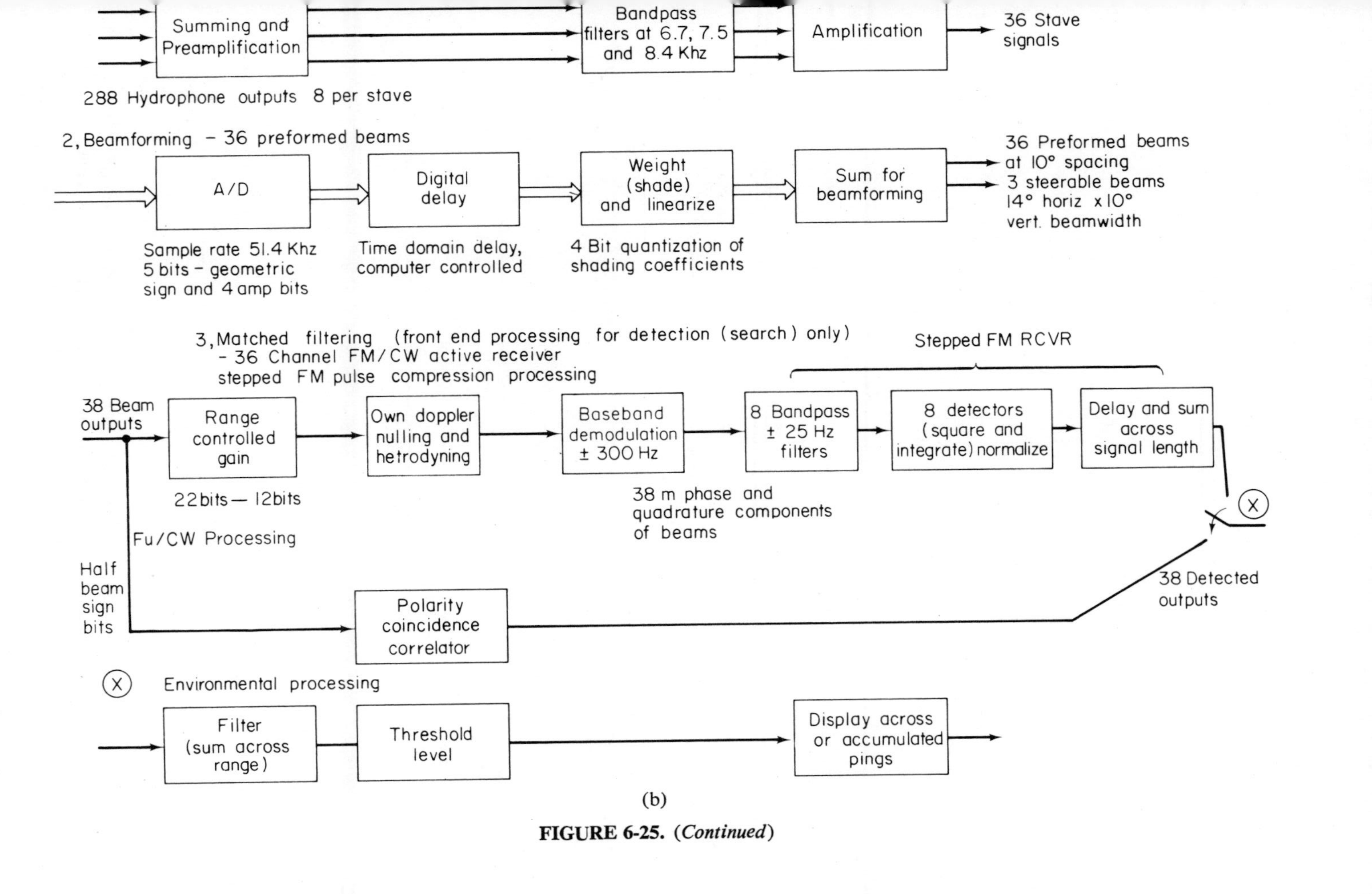

(b)

FIGURE 6-25. *(Continued)*

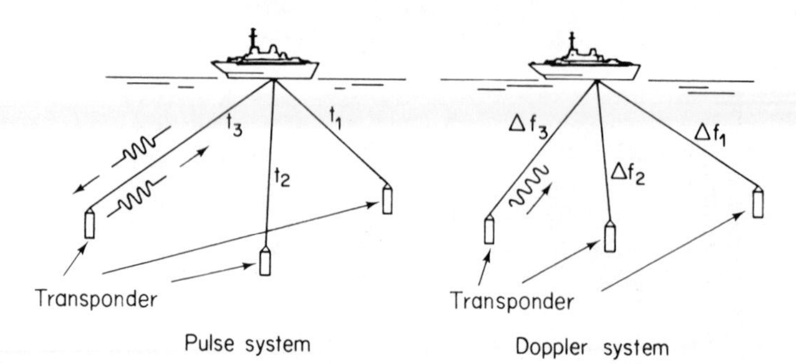

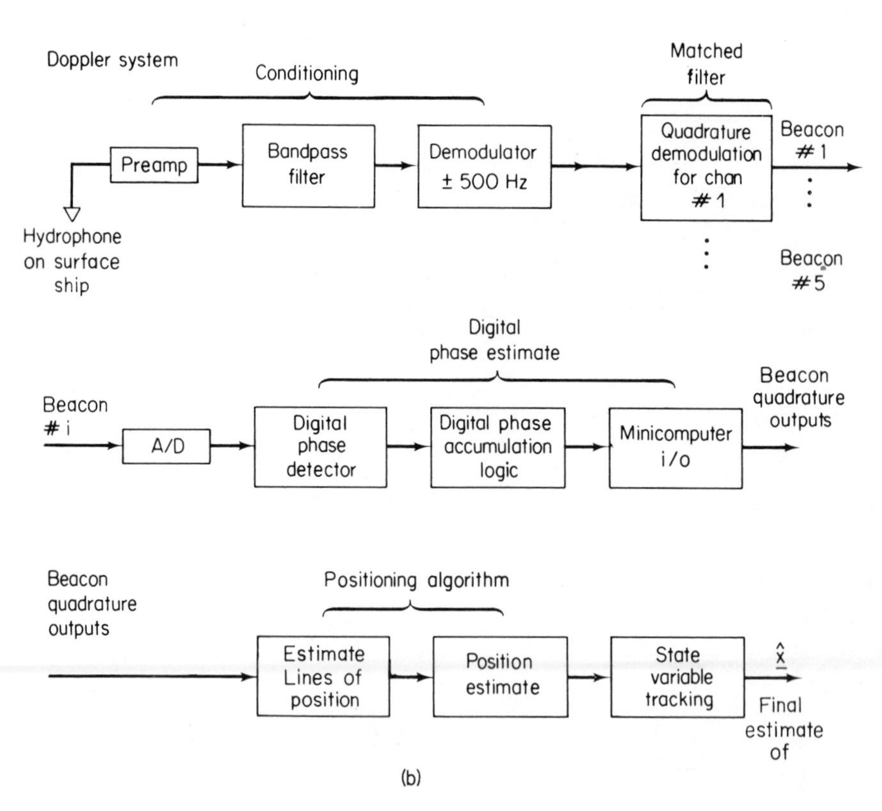

FIGURE 6-26. (a) Woods Hole Oceanographic Institute acoustic navigation system; (b) Signal processing in the Woods Hole Oceanographic Institute acoustic navigation system (from Spindel, 69).

at a sufficiently high SNR such that cycle skipping in the phase-locked loops is not present; otherwise, one is going to have lane offsets in the lines of position. Centimeter accuracies have been reported.[69]

This system is very strongly oriented toward digital processing. In working with the research, submersible navigation algorithms are implemented in real time upon a small minicomputer contained within the submersible itself. The pulsed-operating mode uses digital logic to produce its estimates of range; the Doppler-operating mode essentially employs a digital phase-locked loop to generate the estimated phase differences due to the length of the various transmission paths. One needs to be somewhat careful regarding the environmental situation in using this system, since surface-reflected, bottom-reflected, and direct paths can be present. As a result, these considerations must enter the range and Doppler estimation algorithms.

Underwater communication systems can be either very simple or exceedingly complex depending upon the environment and the needs of the user. At short ranges and vertical propagation paths, underwater communication is readily adaptable to voice transmission and low data rates; for example, 2.5 kbits/s can be sustained with a few watts of power. All one really needs is some type of amplifier–transducer that can convert speech energy to acoustic energy in the water. Modulation to a carrier frequency may not even be required, since sound propagates very effectively in the spectral band occupied by the human voice; however, one may want to use some type of modulation to form directive signals that can be transmitted more efficiently by small transducers.

The major difficulties in underwater communications arise when one wants to communicate at high data rates over long distances. The reverberant multipath structure of the ocean and its strong attenuation at high frequencies leads to a range spread, bandwidth limited channel, so that high data rates for digital communication encoding either voice or data information are severely limited at long ranges, owing to this complex propagation medium.

The most effective digital communication systems are usually based upon spread channel models and explicitly incorporate the range spread reverberation in both the signal design and the receiver structure. Doppler spreading can be a factor, but the channel is usually much more spread in range than in Doppler when one considers reasonable pulse durations in the signaling. Figure 6-27 illustrates some of the important aspects of an underwater communication system that is based upon a frequency-shift keying form of signaling. (Differential phase shift keying, DPSK, is also an effective signaling method in many environmental situations where the channel has short-term phase stability.)

Several important signal-processing considerations in the design of an underwater communication system are illustrated in Fig. 6-27. At the transmitter it is often useful to employ some form of digital encoding from a

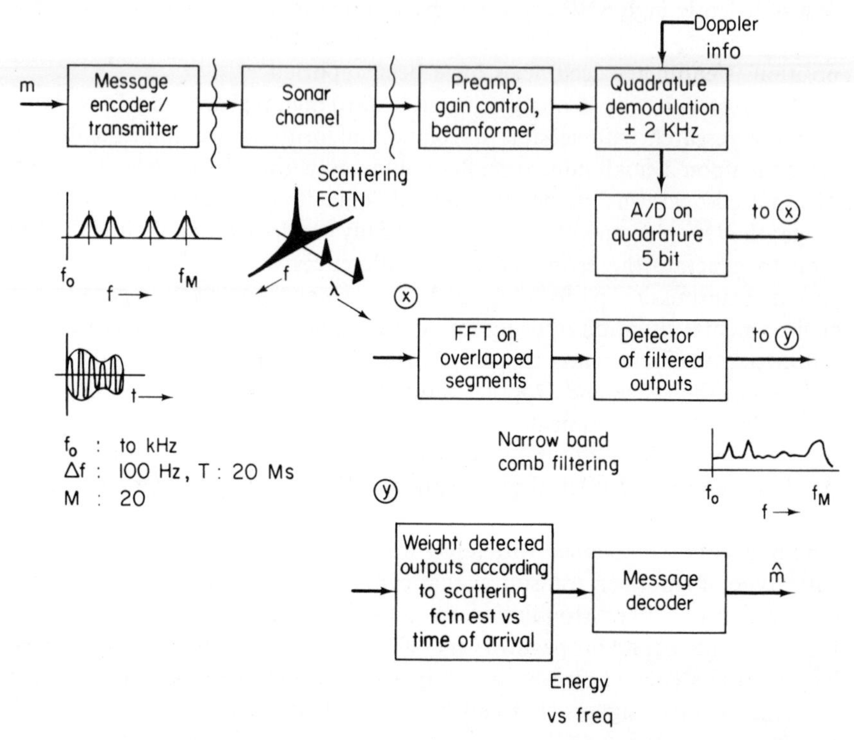

FIGURE 6-27. FSK sonar communication system structure.

binary message sequence into a transmitted signal set. This is done to introduce some redundancy in the signals, to optimize the diversity in the transmitted signals, and to provide some form of cryptographic security if desired. In exploiting the underwater channel, most scattering functions have narrow Doppler spreads and possibly extensive range spreading, so this suggests a signaling set based upon narrowband elements. In addition, the extensive range spreading requires some form of frequency hopping to clear the channel of the reverberant energy. The combination of the scattering function and the signaling set defines an implicit diversity of the channel, and this should be optimized for maximum performance. The optimal diversity is a function of the transmission rate compared to the channel capacity. As a result, some form of adaptive estimation of the scattering function of the channel is often useful. At the receiver, the front end consists of a bank of correlation receivers, or matched filters, for producing estimates of received energy at each resolvable multipath in range and Doppler. These resolved outputs are the inputs of a set of tapped delay lines. The outputs of these delay lines are weighted according to the scattering function resolved range–Doppler cells; thus again some form of adaptive measurement of this function is useful so that the receiver can make the optimum combination of the observed signal energy. If one has not identified the multipath structure fairly accurately, the incorrect range–Doppler cells are used, which naturally degrade the system performance. Finally, the resolved paths are weighted and magnitude squared to form a statistic that is used to decide which message was transmitted; that is, the same form of output from all possible transmitted messages is compared in making an optimum (minimum error probability) decision.

The digital signal processing in all the components of such a system is extensive, and probably not possible to implement without it. From the matched-filter implementations at the front end to the extensive logic required in the encoding and decoding procedures, the flexibility of the digital hardware pervades the system implementations.

This completes our discussion of examples of active sonar systems. We have emphasized the structure of some recently developed systems. We have not focused upon the details of the individual implementations, but rather upon what they were attempting to do. Just as in many fields, the full impact of the capabilities of digital signal processing has yet to absorbed in sonar. Nevertheless, the flexibility demanded by the ocean acoustical environment combined with the advantages of speed, weight, power, and cost ensures that the bulk of systems developed in the future will be centered about digital technology.

6.4 PASSIVE SONAR SYSTEMS AND DIGITAL SIGNAL PROCESSING

In passive sonar one is principally concerned with estimating both the temporal and spatial structure of an observed signal field; by in large, this involves some form of spectral and/or wave-vector analysis. The most important application of passive sonar is for military surveillance systems that detect and track submarines, so it is difficult to become too specific about these systems because of security classification considerations. There are, however, applications in seismology in studying submarine earthquakes and in some types of navigation systems.

A variety of sensor systems are used in the acquisition of passive sonar data; their configurations determine the capabilities and processing of the data. Figure 6-28 illustrates some of the systems that have been used for passive sonar. In general, there are two categories of sensing systems, point sensors and arrays. Examples of point sensors include omnidirectional or directional hydrophones mounted on a ship and sonobuoys, which are deployed from the air or a ship. Array systems include ship-mounted transducer arrays, moored arrays, and towed arrays. Point sensors are capable of producing a temporal analysis of the ambient wave field, so they employ only spectral analysis methods. Arrays are capable of both temporal and spatial

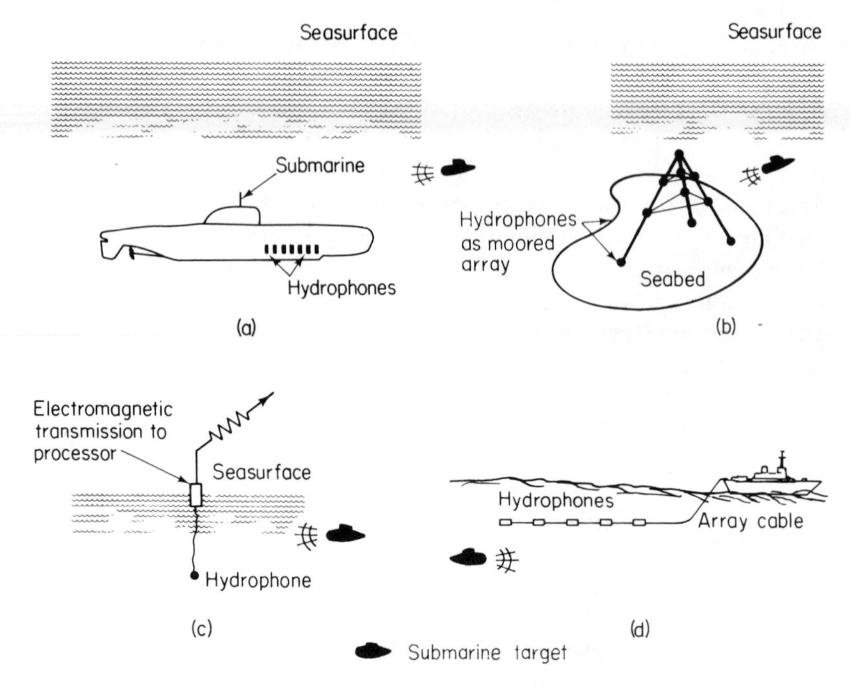

FIGURE 6-28. Examples of several passive sonar systems; (a) Ship mounted hydrophone arrays; (b) Moored surveillance arrays; (c) Sonobuoy systems; (d) Towed array systems sections.

analysis, so they employ both spectral and wave-vector analysis. Often, however, these two problems are factored by performing some form of array processing, or wave-vector analysis, upon narrowband components of the signals.

In this context we have organized the discussion in this section as illustrated in Fig. 6-29. Our discussion sequence proceeds horizontally. First, we shall discuss the underlying theory and models used to represent and processing the signals. We shall review some of the aspects of spectral analysis and its application to passive sonar. This primarily concerns the concepts of a spectral representation and the properties of typical sonar power spectra. Next we introduce wave-vector representations and array processing. Wave-vector representations generalize the concepts of spectra in time series, and array processing concerns the aspects of spatial filtering and beam patterns. Then we shall discuss some processing methods and their implementations for performing spectral and wave-vector estimation. For time series this consists of a brief review of some of the more established methods of spectral estimation and a survey of some recently developed adaptive methods. For spatial processing we examine the conventional methods of estimating, or detecting, power as a function of bearing, or wave vector, and then discuss

	Temporal analysis	Spatial analysis
Theory and models	Spectral representations, properties of passive sonar spectra	Wavector representations, array processing, spatial filtering and beam patterns
Examples of processing methods and implementations	Spectral analysis methods — indirect (Blackman–Turkey), direct (FFT) and adaptive	Beamforming – fixed, preformed, adaptive and multiplicative; Wave vector estimation – conventional and adaptive
Examples of Applications	Passive sonar microearthquake analysis	Passive sonar

FIGURE 6-29. Organization of digital signal processing for passive sonars.

some adaptive methods that have recently appeared. We have categorized these as either closed or open loop because of their implementation structure. Finally, we shall examine to a very limited extent some applications.

6.4.1 Power Spectra Representations for Passive Sonar Signals

In passive sonar one is usually attempting to detect and clarify the various sources of ambient noise in the ocean. In some spectral bands a large component of this noise is generated by ships, and some of this power is concentrated in regions where long-range propagation is possible. The spectra generally consist of two major components.(3) The first is a broadband component in the 100- to 1000-Hz band, which is caused by the cavitation of the propellers and the hydrodynamic forces acting upon a vessel as it moves through the water. The second consists of the narrowband tonal lines generated by the propulsion and the rotating machinery on a vessel. These narrowband components are often synchronized since they are on the same power train, but this need not be a harmonic relationship. As a result they shift as a function of the rotation speeds when radiated, and relative motion between the source and receiver may introduce additional Doppler shifts. The tonal components are distributed across a broadband region, so the Doppler shifts must be treated proportionally. There is often a template for this tonal structure that is unique to each class of vessel because of its power plant construction. In addition, the tonal lines are often very stable with very small bandwidths, because their sources are physically large with considerable amounts of inertia.

The overall impact of this structure of the passive sonar spectrum has been an extensive amount of effort devoted to high-resolution spectral analysis and tonal line tracking. This has led to the use of adaptive spectral estimation algorithms and tracking loops, both of which can only be conveniently implemented using digital technology.

Since passive sonar primarily involves spectral analysis, it is useful to review some of the more essential aspects of this method of representing signals. The power density spectra of a wide sense stationary random process is defined as the Fourier transform of the autocorrelation function of the process, or

$$S_x(f) = \int_{-\infty}^{\infty} R_x(\tau)e^{-j2\pi f\tau}\, d\tau \tag{6.23}$$

where

$$R_x(\tau) = \text{autocorrelation function of the process.}$$

The autocorrelation function and the power density spectra are ensemble properties of a hypothesized model for the sonar signals. These quantities must be estimated from the observed data, and this usually involves some kind of time averaging, either directly in the time domain or indirectly in the frequency domain. The issue then becomes the determination of when these time averages converge in some probabilistic, or statistical, sense to the hypothesized ensemble averages. This is addressed by a combination of realistic modeling and invoking an *ergodic* hypothesis. Usually these issues are not raised directly, but obviously the signals can only be considered to be stationary over limited durations, and the ergodic theory is automatically introduced with the substantive mathematical questions it implies deferred.

The most important aspect of the spectral theory in its use in passive sonar is that it is an uncorrelated representation. This implies that disjoint spectral bands of a stationary random process are statistically uncorrelated, so that by transforming to the frequency domain, for example, by an FFT, the resulting data are statistically uncorrelated. The advantages of this are very significant, for it implies that one can treat each frequency region separately, for example, with parallel processing, and that one does not need to consider the effects of correlation among disjoint spectral bands.

A few additional properties of power spectra are often employed in passive sonar signal processing. They are related to the input–output properties of these signals when applied to linear, time-invariant systems. The input–output relationships for the autocorrelation and the power density spectra are

$$R_y(\tau) = \int \left(\int h(t_1)h(t_1 - t)\, dt_1 \right) R_x(\tau - t)\, d\tau \tag{6.24a}$$

$$S_y(f) = |H(f)|^2 S_x(f) \tag{6.24b}$$

where $h(t)$ and $H(f)$ are the impulse response and transfer function of the linear system.

The spectral relationship is particularly important, for it asserts that one can "shape" the spectra of a time series by passing broadband, white noise, that is, constant spectral level through an appropriate filter. Alternatively, one can represent the process in terms of a shaping filter.

6.4.2 Wave-Vector Representations for Passive Sonar Signals and Array Processing

In spatial signal processing it is useful to have a representation that is analogous to the frequency-domain representation of temporal signal. This requires the generalization of the power density concept for stationary processes to the wave-vector function for homogeneous spatial processes. We define a space–time correlation function of a spatial process as

$$E[X(t_1, \mathbf{z}_1)X(t_2, \mathbf{z}_2)] = R_x(t_1, t_2 : \mathbf{z}_1, \mathbf{z}_2) \tag{6.25a}$$

If the space–time correlation function is a function of only the time differences and spatial separations of its arguments, then it is respectively stationary and/or homogeneous. Many passive sonar fields are well modeled by the homogeneous assumption over a limited areal extent. The basic concept of the wave-vector representation is that a stationary, homogeneous process can be represented as a sum of uncorrelated plane waves whose power distribution is specified by the wave-vector function.[70–72] We have

$$X(t, \mathbf{z}) = \sum_{f, \mathbf{v}} X(f, \mathbf{v}) e^{j2\pi(ft - \mathbf{v}\cdot\mathbf{z})} \tag{6.25b}$$

where

$$E[|X(f, \mathbf{v})|^2] \propto P_x(f, \mathbf{v})$$

The wave-vector function $P_x(f, \mathbf{v})$ is determined by a sequence of Fourier transforms of the space–time correlation function.

If the space–time correlation function is first transformed with respect to the temporal variable, one obtains a cross-spectral covariance function. If this function is evaluated at the locations of an array, it specifies the cross-spectral covariance matrix of the array outputs. This matrix is a fundamental quantity in many adaptive array-processing algorithms, for many of them use estimates of it. If we next transform with respect to the spatial separation, we obtain the wave-vector function, or

$$S_\lambda(f : \mathbf{z}_1 - \mathbf{z}_2) = \int R_x(\tau : \mathbf{z}_1 - \mathbf{z}_2) e^{-j2\pi f\tau}\, d\tau \qquad \text{spectral covariance function} \tag{6.26a}$$

$$P_x(f, \mathbf{v}) = \int S_x(f : \mathbf{z}) e^{+j2\pi\mathbf{v}\cdot\mathbf{z}}\, d\mathbf{z} \qquad \text{wave-vector function} \tag{6.26b}$$

This function has all the attributes of a power density spectrum, and plays a similar role in spatial filtering. The most important aspect, however, of this representation is that it asserts that disjoint wave-vector regions are uncorrelated, so that each region can be processed separately. This is usually consistent with the physical processes generating the signals. Disjoint wave-vector regions correspond to different source directions, and the sources are seldom coupled in their operation. The only notable exception is when there is correlated multipath present.[73]

Since this function is quite important in the subsequent discussion of signal processing, it is useful to comment upon some of its properties. There is usually a dispersion constraint imposed by the wave equation of the propagation medium between the temporal frequency, f, and the magnitude of the wave vector $|\mathbf{v}|$.[4,5] This constraint can often be exploited to reduce the signal processing required, since some regions of the frequency–wave vector space do not correspond to physically attainable propagating signals.

The signals of most interest are point sources in direction. Ideally, these signals have an impulsive wave-vector function and a spectral covariance function whose magnitude is constant. In reality, there is always some spatial spreading of the impulse owing to the microstructure and incremental multipath. Usually, the spatial spreading of the wave vector is larger in the vertical direction than in the horizontal because of the stratification of the ocean.[37]

For a given temporal frequency, all propagating signals in the ocean have a spatially band-limited wave-vector function. This leads to spectral covariance functions, which are often oscillatory in their form.[74-75] Many measurements of the covariance of ambient noise have been reported, and it is now routinely measured by adaptive systems. One of the more commonly used covariance is for isotropic noise, which has a $\sin(x)/x$ where x is $2\pi|\mathbf{v}|/\lambda$ form for its spectral covariances.

There are several situations where the homogeneity of the signal field is not satisfied; in these cases, one has the same set of difficulties that are usually associated with nonstationary processes. Probably the most common examples of nonhomogeneity of a signal field occur when there is wavefront curvature across an array and when mode effects are present. Mode effects are important in shallow water and low frequency deep water signal propagation.[124] Such arrays are usually distributed over extensive areas so as to introduce as much curvature as possible in relation to a spherically radiating source. This curvature can be exploited to produce a focus and passive ranging algorithms, but the signal processing must incorporate the nonhomogeneity explicitly.

Having introduced a method of representing the spatial signal, we now turn our attention to describing some of the fundamental aspects of the signal processing that is performed in a passive sonar array.

In many situations the acoustic environment can be very directional, especially the targets or sources of interest. Consequently, significant gains can be obtained by spatial filtering and array processing. In its simplest form, spatial filtering consists of inserting delays at the transducer outputs to account for the relative propagation delays of a signal from a particular location, and then summing the results. The signals arriving from the direction of interest are then added coherently, or in phase; those from other directions do not add coherently, so they cancel. The processing, however, is usually more sophisticated, with different weights and phase shifts being applied to each array transducer output. Often these are adaptively set in sonar signal processing as a function of the observed data. The two basic methods of beamforming are illustrated in Fig. 6-30.

The fundamental concept for array processing that is based upon a plane wave representation for the signals is the Fourier transform relationship between the transducer weightings and the beam pattern of an array. This is given by

$$\mathcal{G}(f, \mathbf{v}) = \sum_{i=1}^{N} G_i(f) e^{j2\pi \mathbf{v} \cdot \mathbf{z}_i} \tag{6.27}$$

where

$G_i(f)$ = weightings applied to the array transducers located at z_i as a function of frequency

$\mathcal{G}(f: \mathbf{v})$ = response of the array to a plane wave propagating across it with wave vector $\mathbf{v}$ and frequency f.

We can observe that Eq. (6.27) is fundamentally a discrete Fourier transform operation. The beam pattern at a particular frequency can be found by fixing the magnitude of the wave vector to the value determined by the dispersion relationship of the propagation medium, for example, in free-field acoustic propagation in the ocean, and then evaluating the wave-vector response of Eq. (6.27) as a function of the bearing angle of the plane wave.

The Fourier relationship of Eq. (6.27) suggests an uncertainty type relationship between the physical extent of the array and the beamwidth, so that for narrow beamwidths one wants as large an array as possible. Certainly, this is the case, but practical considerations usually limit the overall dimensions of an array. In addition, the coherence of the propagation medium sets a point of diminishing return on extremely large array geometries. This inverse relationship suggests that, since the wave-vector magnitude decreases with increasing frequency, one should operate a sonar system at as high a frequency as possible if small beamwidths are desired. Unfortunately, propagation medium effects, particularly absorption, limit one's ability to do this.

The Fourier relationship of Eq. (6.27) also indicates that by adjusting the weighting pattern one can tailor the beam pattern according to various criteria. These procedures share much in common with filter design for finite impulse response filters, especially when the array elements are equally spaced within a linear array. A variety of criteria can be adopted for the beam pattern design, including minimum beamwidth, controlled sidelobe levels, and null placement. Minimum-beamwidth arrays were introduced early in the sonar signal processing literature by Pritchard with a quadratic optimization procedure.[76] The concept of controlling sidelobes is virtually identical with Chebychev filter design, and Dolph–Chebychev design has been extensively

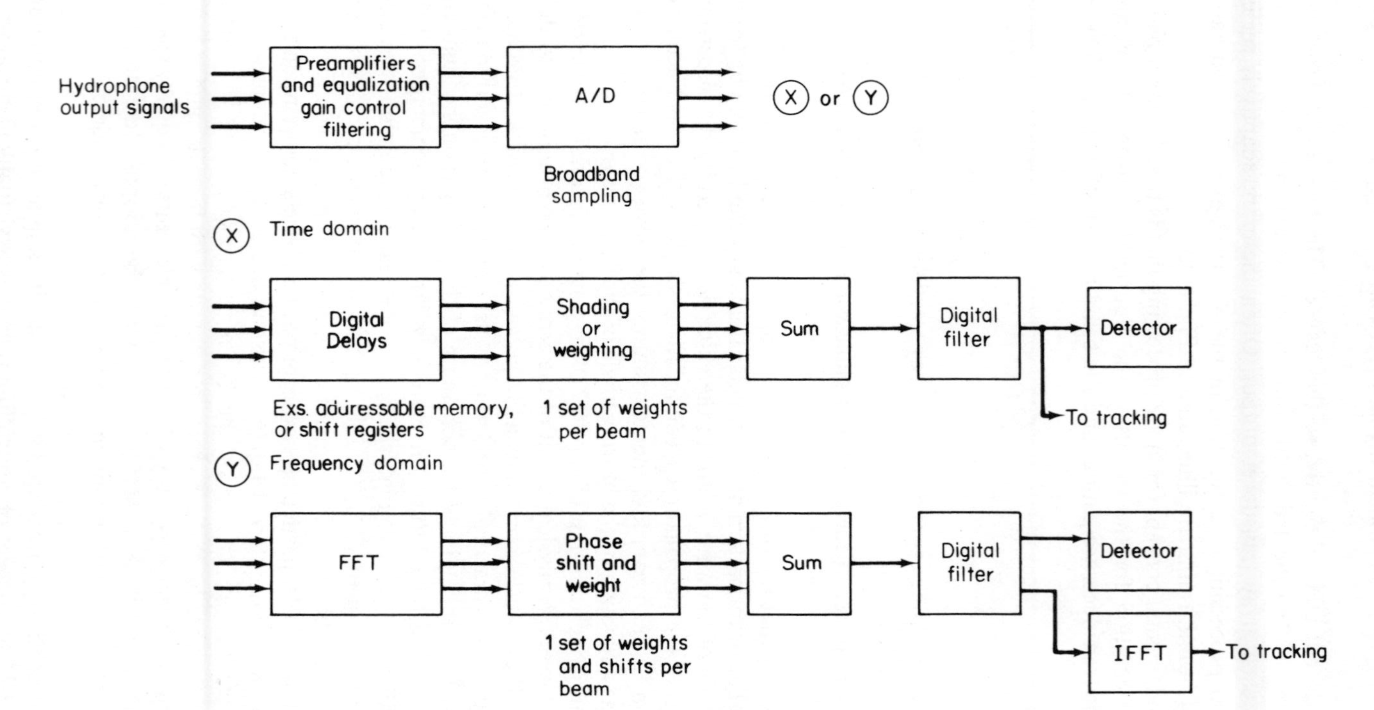

FIGURE 6-30. Basic methods of time domain and frequency domain beamforming for passive (and active) systems.

discussed.[77,3] Null-placement procedures have been particularly important, since strongly interfering directional noise fields often occur in the sonar environment.[78] We subsequently discuss null-placement arrays in the section on implementations. Figure 6-31 illustrates several array designs based upon a variety of array signal-processing algorithms.

In many situations, complete coverage across the full extent of an array geometry is not possible. For example, the maximum array separation suggested by the spatial version of sampling theorem is one half of a wavelength, so for large array geometries this can lead to a tremendous number of transducer elements, which simply overwhelms the processing capabilities of an array processor. Under these circumstances, *thinned arrays* are often used. These may take the form of local clusters of array elements or simply isolated transducer elements with large interelement spacing. Thinned arrays generally have good resolution compared to a completely filled array of the same extent; they do, however, have some serious limitations because they generate large sidelobe levels. If the interelement spacing is at regular intervals, then periodic grating lobes are introduced into the beam pattern. If the spacing is irregular, the average sidelobe level increases, and large lobes may appear in directions that are difficult to predict analytically. Array design for these situations is in general a difficult task, since there is not much structure to exploit in the transform relationships. There has been, however, some very interesting work based upon randomization arguments.[14] Figure 6-32 illustrates some important sonar array situations where the irregular and large interelement spacing appear.

The array processing encompassed by linear weighting and phase shifting of the transducer outputs leads to a theory strongly coupled to Fourier transform analysis. This is a natural result, given the plane wave representation of the signal field. However, a number of important applications employ correlation operations among the array outputs. The basic concept is well illustrated by the split-beam correlator, which is illustrated in Fig. 6-33. In this array processing, the transducer elements are partitioned into two sections, each of which is beamed as a linear array. The outputs of these two sections are then cross correlated. The concept of operation is that, if there is a signal source in the direction of the beam, it will be focused by each partition and produce identical, or similar, outputs that will display a high degree of correlation. Conversely, if there is no energy in the signal field in the direction of the beams, the array outputs will be dissimilar and will not be strongly correlated. Although the split-beam correlator is used extensively, product or multiplicative arrays have not seen a significant amount of use beyond this.[79-80] One of their major limitations is that the product operation introduces a definite threshold in the performance of the system if the SNR at the beam outputs is not sufficient. They can also introduce some spurious sidelobes, which are essentially the spatial equivalent of modulation products.

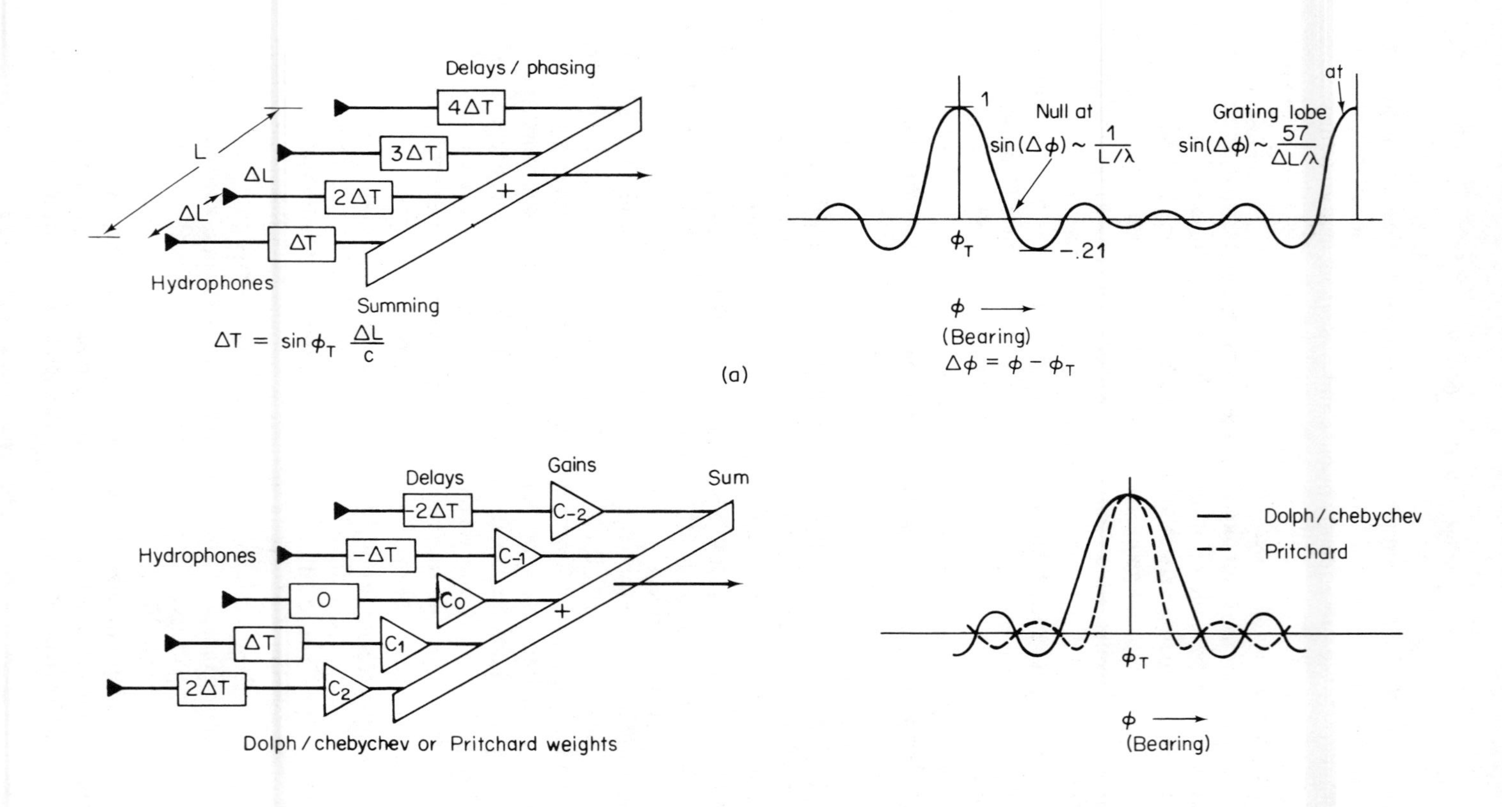

Delays / phasing
4ΔT
3ΔT
2ΔT
ΔT
L
ΔL
ΔL
+
Hydrophones
Summing
$\Delta T = \sin \phi_T \frac{\Delta L}{c}$
(a)
1
Null at
$\sin(\Delta\phi) \sim \frac{1}{L/\lambda}$
Grating lobe at
$\sin(\Delta\phi) \sim \frac{57}{\Delta L/\lambda}$
ϕ_T
−.21
ϕ ⟶
(Bearing)
$\Delta\phi = \phi - \phi_T$
Delays
Gains
Sum
−2ΔT
−ΔT
0
ΔT
2ΔT
C_{-2}
C_{-1}
C_0
C_1
C_2
+
Hydrophones
Dolph / chebychev or Pritchard weights
(b)
Dolph / chebychev
Pritchard
ϕ_T
ϕ ⟶
(Bearing)

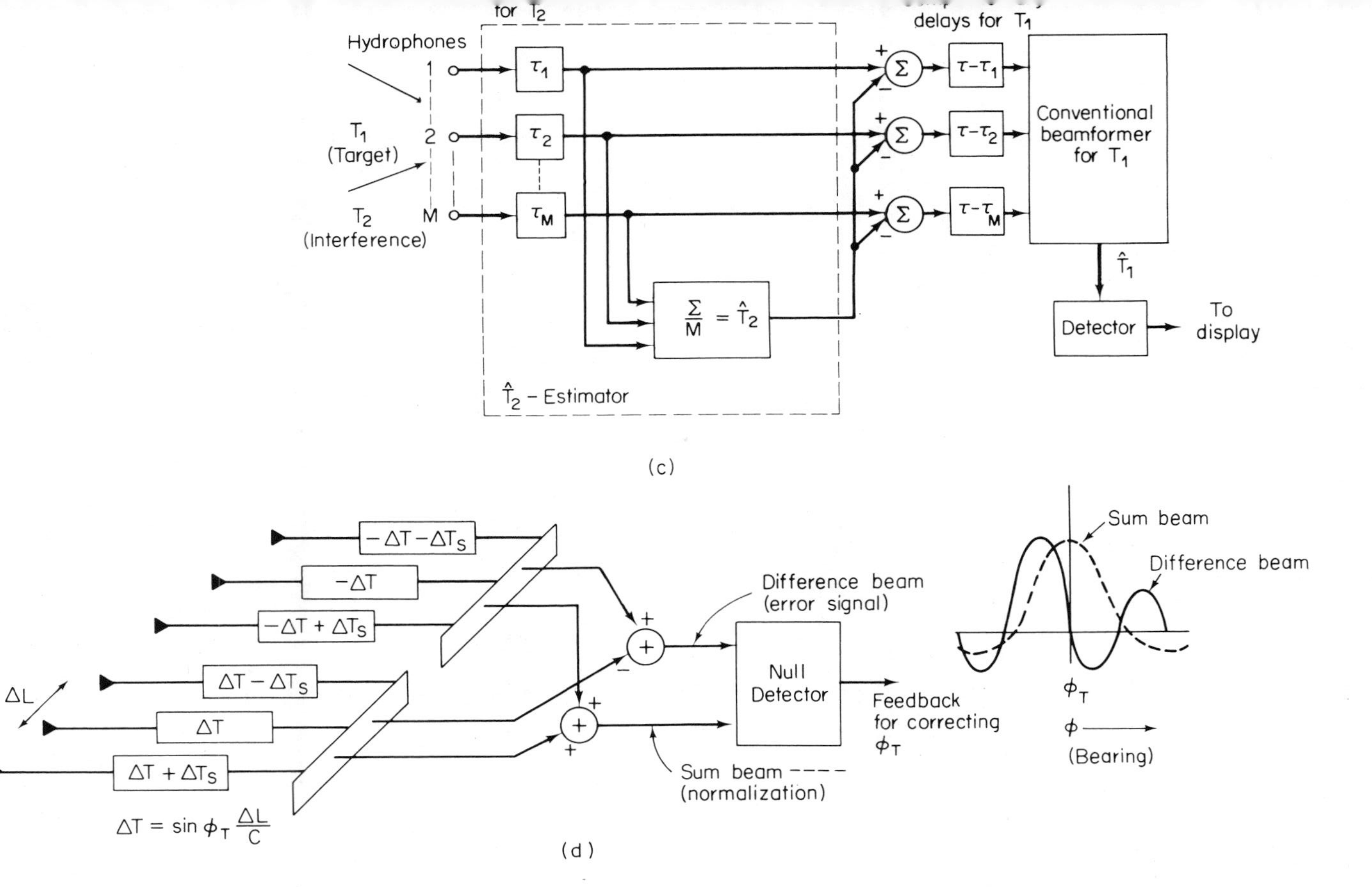

FIGURE 6-31. Examples of beamforming in several sonar systems; (a) Uniformly weighted and phased (steered)array; (b) Shading (or weighting) of array; (c) Null steering processor (from Anderson, 105); (d) Split beam null placement for tracking.

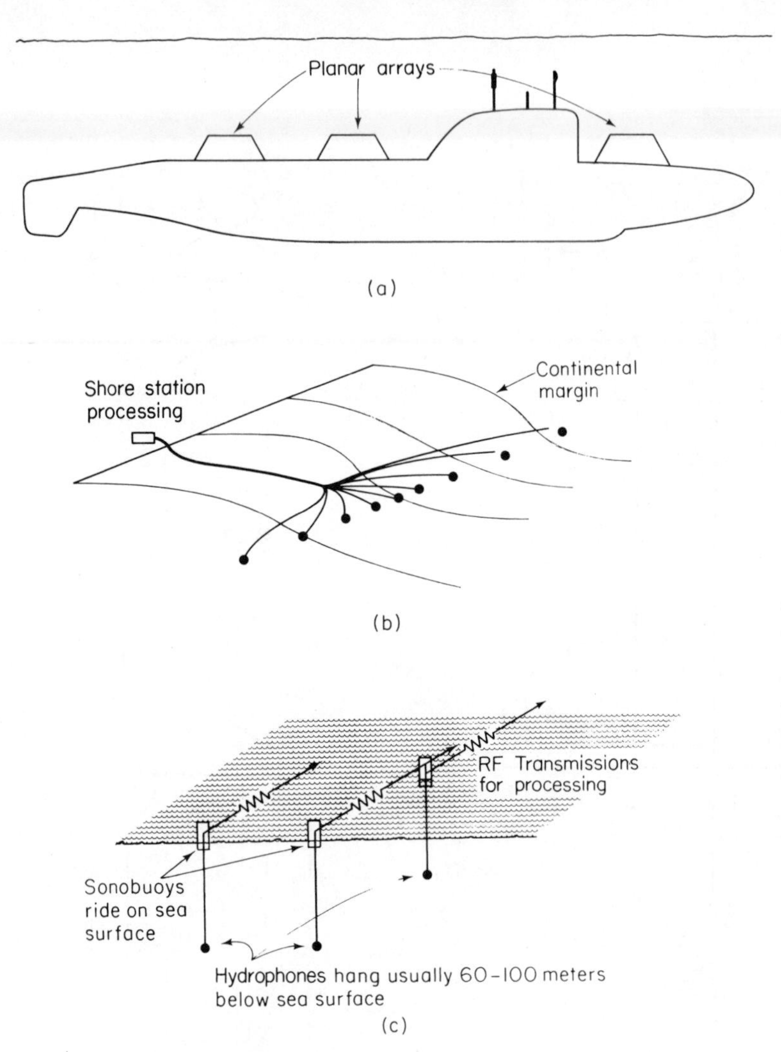

FIGURE 6-32. Examples of distributed, or widely separated, passive sonar systems: (a) Three planar arrays widely separated along a submarine (used to measure wavefornt curvature for passive ranging); (b) Nonuniformly spaced bottom moored surveillance arrays; (c) Distributed sonobouy arrays.

Sonar array processing encompasses a large number of algorithms and systems that are quite diverse in concept; however, a few basic concepts underlie much of the processing. This is very analogous to time-series analysis; there are many signal-processing methods, but spectral representations and linear system theory form the basis for most of them. At this point we want

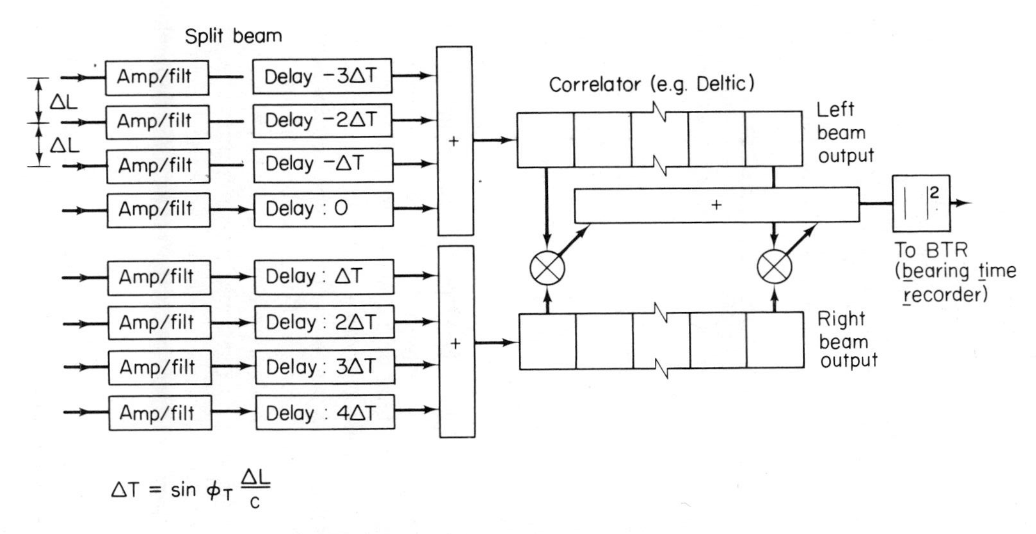

FIGURE 6-33. Split beam correlator array processor.

to discuss an extension of this to array processing, the coupling of wave-vector spectral representations and array beam patterns.

The response of a sonar array to a single plane wave signal is indicated by the Fourier relationship of Eq. (6.27), It is easy to derive that the response of the array to an entire signal field can be found in terms of the spectral covariance of the field and the array weightings, or

$$\sigma_G^2(f) = \sum_{i=1}^{N} \sum_{j=1}^{N} G_i^*(f) S_x(f : \mathbf{z}_i, \mathbf{z}_i) G_j(f) \tag{6.28}$$

If the signal field is also homogeneous, then the wave-vector interpretation is possible; this is given by

$$\sigma_G^2(f) = \int |\mathcal{G}(f, \mathbf{v})|^2 P_x(f, \mathbf{v})\, d\mathbf{v} \tag{6.29}$$

where

P_x = wave-vector function describing the distribution of power of the signal field

$\mathcal{G}$ = wave-vector response of the array system

This result is completely analogous to the input–output relationship of the power density spectra with linear systems.

The most obvious and important aspect of this equation is that the intensity of signal at the output of an array-processing system is determined by the overlap of the wave-vector function of the signal field and the wave-vector response of the array. A very large part of the theory of array processing centers upon this equation. For example, in many situations, one component of the ambient acoustic signal field consists of interfering noise and another is a signal from a source in a particular direction. An efficient array-processing algorithm minimizes the response due to the noise component by minimizing the overlap of its beam pattern with the intense portions of the noise field, while simultaneously maintaining a beam in the target, or source, direction. The algorithms suggested by these minimization procedures require a great deal of flexibility since the properties of the acoustical environment must be estimated in real time. It is very doubtful if they could be implemented without today's digital technology.

6.4.3 Processing Methods for the Temporal Analysis of Passive Sonar Signals

The processing methods for analyzing the temporal structure of passive sonar signals by in large consist of the signal processing of spectral analysis. At present, three general categories of spectral estimation algorithms are used or are being investigated in passive sonar systems: (1) the "classical" indirect procedures of Blackman and Tukey, (2) frequency-domain, or FFT, direct procedures, and (3) data-dependent adaptive methods. All are becoming strongly dependent upon the capabilities of digital implementations, either as hard-wired devices, microprocessors, or as software on a minicomputer. Many of the recently developed algorithms, especially those with high-resolution capabilities in the third category, could not be realized without digital technology.

In spectral analysis there is a recurring trade off that must be made. Because of the finite length of the data, which may be the result of recording limitations or the short-term stationarity of the signals, we cannot achieve high resolution, or low bias, and high stability, or low variance, spectral estimates. The principal design issue in selecting an algorithm and adjusting its parameters is to establish the best trade off within the limitations of the data.

We shall now briefly discuss each of these categories. We shall not emphasize the first two categories since they are fairly well established in the signal-processing literature. It does appear though that the sonar signal processing community is providing much of the research on adaptive, high resolution spectral analysis algorithms and their implementations.

The Blackman–Tukey, or indirect procedure, follows the definition of a power spectra.[81,82] The signal processing associated with it is illustrated in Fig. 6-34(a). One first forms an estimate of the autocorrelation function of the observed signal. The estimated correlation function is next multiplied by a window function, which is introduced to provide some statistical stability in the spectral estimate. Finally, the windowed correlation function is transformed to produce the spectral estimate. It is also often useful to prewhiten the data so as to minimize the leakage created by the intense regions of the spectra. (This is often an iterative procedure in which one uses an estimated spectra to design a whitening filter, passes the observed data through it a second time, and then repeats the estimation algorithm.)

The two performance parameters of interest are the bias and variance of the estimate. Under some assumptions, these are given by

$$\Delta\hat{S}_x(f) = \int S_x(f - v)W(v)\,dx - S_x(f) \qquad \text{bias} \tag{6.30a}$$

$$\sigma^2_{\hat{S}}(f) = \frac{1}{T}\left[\int |W(x)|^2\,dx\right]S^2_x(f) \qquad \text{variance} \tag{6.30b}$$

The well-established trade off between these two performance parameters can easily be seen with these two formulas. Low bias demands that the transform of the window function approximate an impulse such that the spectrum is not smeared by the convolution operation. This requires a long-duration window function. Low variance, however, demands that the window duration be as small as possible.

There is an extensive literature on the theoretical aspects of this algorithm (e.g., the choice of window function). Its implementation requires a significant

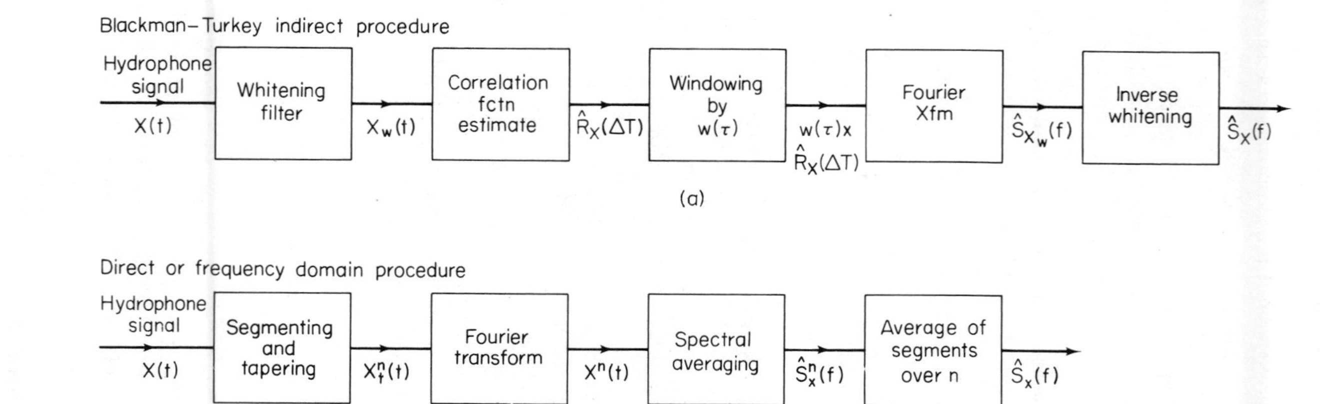

FIGURE 6-34. (a) Blackman–Tukey indirect method of spectral analysis; (b) Direct, or frequency domain, method of spectral analysis.

amount of signal processing. The whitening filter must be iteratively tuned in those applications where there is a large range in the spectral levels. This may necessitate the use of some form of filter design. The correlation operation dominates the computations required by the algorithm; this suggests that a fast multiply–add capability is needed for rapid implementations. When the hardware for this is not available, the process is often hard clipped and a polarity coincidence correlation of the lagged product is implemented. Under a Gaussian process assumption, it can be demonstrated that the correlation of the clipped process is given by $\sin^{-1}[R(\tau)/R(0)]$, which usually represents an acceptable distortion of the true correlation function. The windowing is a comparatively simple operation, and the final transformation to the frequency domain is easily accomplished with the FFT.

The FFT, or direct, method of spectral estimation is illustrated in Fig. 6-34(b).[83,84] It is termed direct because one transforms the data directly to the frequency domain without first forming an estimate of the autocorrelation function. The transformation to the frequency domain yields a representation of the signal in which the Fourier components are uncorrelated, so each frequency band can be analyzed separately. The advent of the FFT algorithm has been very significant in passive sonar; many signal-processing systems are now evolving based upon it, particularly those requiring a real-time implementation.

There are two categories of FFT spectral estimation, the unsegmented and segmented versions. Most passive sonar systems employ a modified segmented version. In the unsegmented version the data are simply transformed to form a periodogram, and the magnitude squared Fourier coefficients are simply averaged by convolving them with a window in the frequency domain. The averaging introduces statistical stability, or low variance; it does, however, degrade the resolution. In the segmented version, the data are partitioned into segments, or blocks, which are often overlapped. These segments are tapered to reduce sidelobes and the correlation introduced by the overlap. These are then transformed with a shorter-length FFT, and the magnitude squared Fourier coefficients are averaged across the segments for each frequency. This can be averaged further across frequency if more stability is required. In a system that operates in real time, the data are being observed continuously. As a result, most systems simply partition the data into blocks as they are observed and then average the most recently acquired blocks. The extent of the averaging is an important parameter in that short-term are noisy and have high resolution averages, while long-term averages are stable but can encounter a nonstationary environment.

The performance of the bias and variance of the FFT methods encounters the same trade off as the indirect methods. The flexibility in the algorithm in terms of possible tapers and averaging make the format of general results quite complicated. We indicate these results for a relatively simple version:

$$\Delta \hat{S}_x(f) = \sum_n W_n S_x\left(f + \frac{n}{T}\right) - S_x(f) \qquad \text{bias} \tag{6.31a}$$

$$\sigma_S^2(f) = \frac{1}{T} S_x^2(f) \sum_n |W_n|^2 \qquad \text{variance} \tag{6.31b}$$

where W_n are the set of frequency-domain averaging weights.

The spectral analysis system employed in most passive systems before the introduction of the FFT was basically a narrowband spectrograph. A narrowband spectrum analyzer would simply scan the frequency range of interest and indicate the spectral intensity by a stylus marking. This would be done on a repetitive basis, and eventually a spectral history of the acoustical environment would be generated. Such systems are called low-frequency analysis record (LOFAR) "grams." The detection and classification of ships on these systems primarily consisted of looking for line structures generated by the tonal components. An example of a "gram" based on a minicomputer system is indicated in Fig. 6-35.[125] (The tonal components in the spectral estimate are due to the machinery of a ship.) A variation on this type of processing can be used to improve the resolution of frequency bands of particular interest. In this method of processing the band of interest is demodulated to baseband, and the resolution is increased by a slower sweep and a narrower filter in the spectral analyzer.

The FFT, or direct, method of spectral analysis is changing LOFAR analysis significantly. A typical system is indicated in Fig. 6-36. It consists of a signal conditioner for gain control, an A/D converter, an FFT system, and an averaging of the squared magnitude of the Fourier coefficients. In addition a *zoom* option can be employed in which a frequency band is demodulated and then transformed over a longer duration to increase the resolution. The duration of the averaging is a compromise between the stability needed for detecting the line and the stationarity of the environment.

In passive sonar there is a very high premium upon the resolution capabilities of any spectral analysis method. As a result, a considerable amount of attention has been devoted to some of the adaptive, high-resolution algorithms that have been developed recently in sonar and several other fields.[85–87] Several of these adaptive algorithms have appeared; we shall comment, however, only upon the three that have attracted the most attention.

Both the indirect and direct methods suffer from several shortcomings. Although they are now very easy to implement with several special-purpose systems commercially available, they often have high sidelobe leakage and insufficient resolution. Two methods, the maximum entropy method (MEM) or autoregressive (AR) procedure and the maximum likelihood method have been extensively investigated; a third, the method of Pisarenko, is starting to attract some interest.

The *maximum entropy method* is probably the most popular terminology for a spectral estimation procedure that also is known as autoregressive spectral analysis, linear prediction, and whitening and innovations processing.

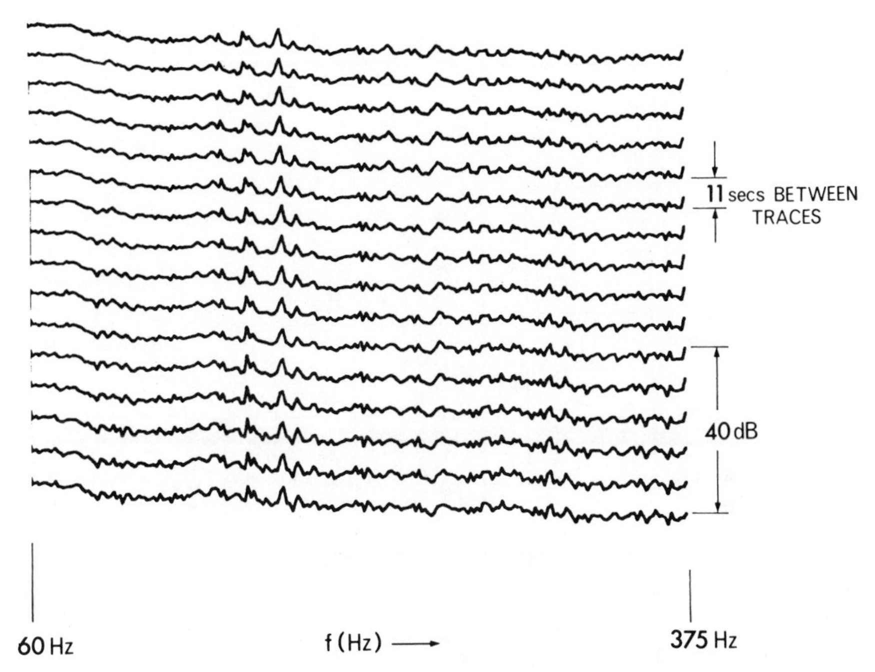

FIGURE 6-35. Spectral analysis of shipping noise at successive time segments (note the presence of prominent stable peaks corresponding to machinery noise), (Courtesy, R. Seynave, SACLANT Center), (16 averaged FFTs, 1.5 Hz resolution).

It is popularly attributed to Burg, who introduced the "entropy" description, but it was certainly suggested and employed by several others before this introduction.[88–90] There are several implementations; one uses the following analytic definition for the spectral estimate:

$$\hat{S}_{\text{MEM}}(f) = \frac{\hat{P}_E(N)}{|\hat{H}(f)|^2} \qquad \text{MEM eq.} \tag{6.32}$$

where

$$P_E(N) = R(0) - \underline{r}^T \underline{h} \qquad \text{prediction error}$$

$$H(f) = \begin{pmatrix} 1 \\ -\underline{h} \end{pmatrix}^T \underline{E}(f) \qquad \text{transfer function}$$

$$E^T(f) = (1, e^{j2\pi f \Delta T}, \cdots e^{j2\pi f(N-1)\Delta T})$$

$$\begin{bmatrix} \hat{R}(0) & & & \hat{R}(N-1)\Delta T \\ & \hat{R}(1)\cdot & & \\ & & \cdot & \\ & & & \cdot \\ \hat{R}(N-1)\Delta T & R(0) & & \cdot R(0) \end{bmatrix} \underline{h} = \begin{bmatrix} R(\Delta T) \\ \cdot \\ \cdot \\ \cdot \\ R(N\Delta T) \end{bmatrix} = \underline{r} \qquad \text{"normal" equation}$$

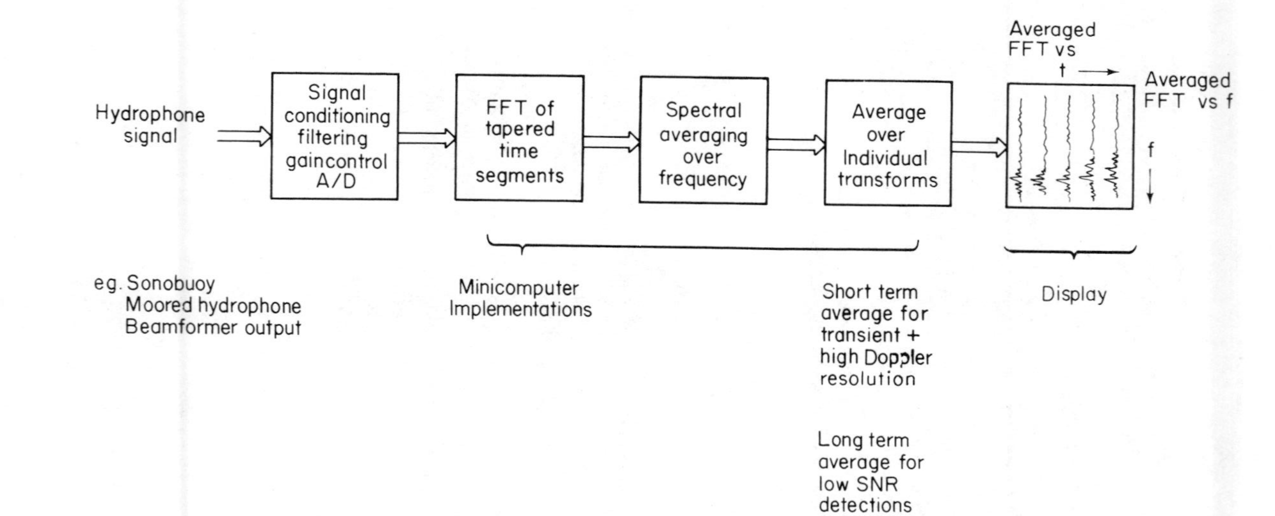

FIGURE 6-36. Sonobouy system based upon FFT processing.

The interpretation of this method of spectral analysis is that the filter is an all-pole network that attempts to whiten the observed data. The spectral estimate is then proportional to the inverse of the magnitude squared of the transfer function in accordance with the input–output properties of power spectra, as indicated by Eq. (6.24a). There are also several other useful interpretations of the algorithm. The whitening filter has the form of a prediction-error operation in which one uses a linear combination of the previous N samples to predict the next sample, and then subtracts this estimate from the sample itself to generate an error process. One can demonstrate that for an Nth-order prediction filter, the error signal is uncorrelated with itself for separations up to length N. The prediction-error process has a long and rich history in time-series analysis, ranging from some of the original work of Kolmogorov, through such diverse fields as seismic processing and information theory, to its extensive use in many fields at the current time. It was the capabilities of digital signal processing, however, that permitted this powerful theoretical tool to fulfill its potential.

The actual implementation of the MEM spectral estimation algorithm can be done in several ways. The most strightforward procedure is to estimate the autocorrelation function just as in the indirect method, set up the "normal" equation expressed by Eq. (6.32), solve the "normal" equation for the impulse response of the prediction error filter, and determine the estimate from these coefficients according to the MEM formula. In this formulation, the Toeplitz form of the "normal" equation leads to an efficient method of computing the filter coefficients known as the Wiener–Levinson algorithm. Several variations of this algorithm have also appeared; all are motivated by the need for efficient and numerically stable solutions to the normal equations. An alternative procedure for calculating the coefficients is to calculate recursively the prediction error filter's impulse response directly from the data.(91) (This is known as the Burg formulation.) Finally, the original formulation for the entropy terminology suggests a method based upon the concept of positive definite extensions of covariance functions.(88,91,92)

The *maximum likelihood method* was originally developed in the context of seismic and sonar array processing.(93,94) It was subsequently extended to time-series analysis, and it is this application that we discuss here.(87) The MLM spectral estimate is defined by

$$S_{\mathrm{MLM}}(f) = \left(\underline{E}^{T}(f) \begin{bmatrix} R(0) & R(\Delta T) \cdot & & R(N-1)\Delta T \\ & & \cdot & \\ & & \cdot & \\ & & \cdot & \\ & & & \cdot R(0) \end{bmatrix}^{-1} \underline{E}(f) \right)^{-1} \tag{6.33}$$

The interpretation of this method of spectral analysis is that the estimate is the noise power output of a finite-length filter, which is constrained to have a unity, or distortionless, response for the frequency of interest and a minimum response from all the remainder of the spectral band. The spectra is simply the noise power estimate as a function of the frequency of interest. The actual implementation of the method consists of estimating the correlation function of the data, forming the Toeplitz correlation matrix, inverting it, and computing the quadratic form in Eq. (6.33) as a function of frequency. The correlation function can be computed using all the previously mentioned techniques, such as the lagged product, clipped lagged product, or FFT methods; in addition, the Toeplitz form of the correlation function can be exploited in inverting it. Most of the computation, however, is involved in calculating the quadratic form as a function of frequency.

In comparing the performance of the MEM and MLM algorithms, several issues are important.[95,96] Figure 6-37 illustrates a comparison using ensemble (i.e., perfectly estimated) correlation functions, which is typical of the results one obtains.[86] The MEM probably has the better resolution of the two methods; however, it is prone to spurious spectral peaks due to high sidelobes. It is also a much noiser estimate, although it has been reported that the local area in a tonal peak is quite stable.[86] The estimate is consistent in that one can demonstrate asymptotic unbias for increasing order and length of data. This property, however, is strongly dependent upon uniform sampling, which limits its extension to array processing. The MLM is a much more stable estimate with far fewer spurious sidelobes. It does, however, have less resolution than the MEM; and it is a biased estimate, although this can be compensated for in terms of a factor dependent upon the filter length.

The Pisarenko method has not attracted as much attention as the MEM and MLM.[85,97,98] It is in some respects a combination of MEM and eigenvalue analysis, which is a generalized spectral analysis. Essentially, one attempts to identify the additive white, or uncorrelated, noise contribution to the autocorrelation of a process and then subtract it from it. This in itself can increase the resolution of the MEM; the important consideration is not to remove too much since it could destroy the positive definiteness of the correlation function. Next, one attempts to identify the spectral content of the modified correlation function as a sum of tonal signals. It should be noted, however, that much more experience with this algorithm is needed before it can be used with confidence, even by the research community.

This completes our discussion of the temporal analysis methods that are used with passive sonar systems. As we mentioned earlier, it is largely a story in spectral analysis. In this context it benefits extensively from several fields, and research on passive sonar systems is contributing significantly to the spectral analysis literature. The capabilities and flexibility provided by digital signal processing motivate much of this effort, for they can implement the the extensive amount of processing required. We now turn our attention to

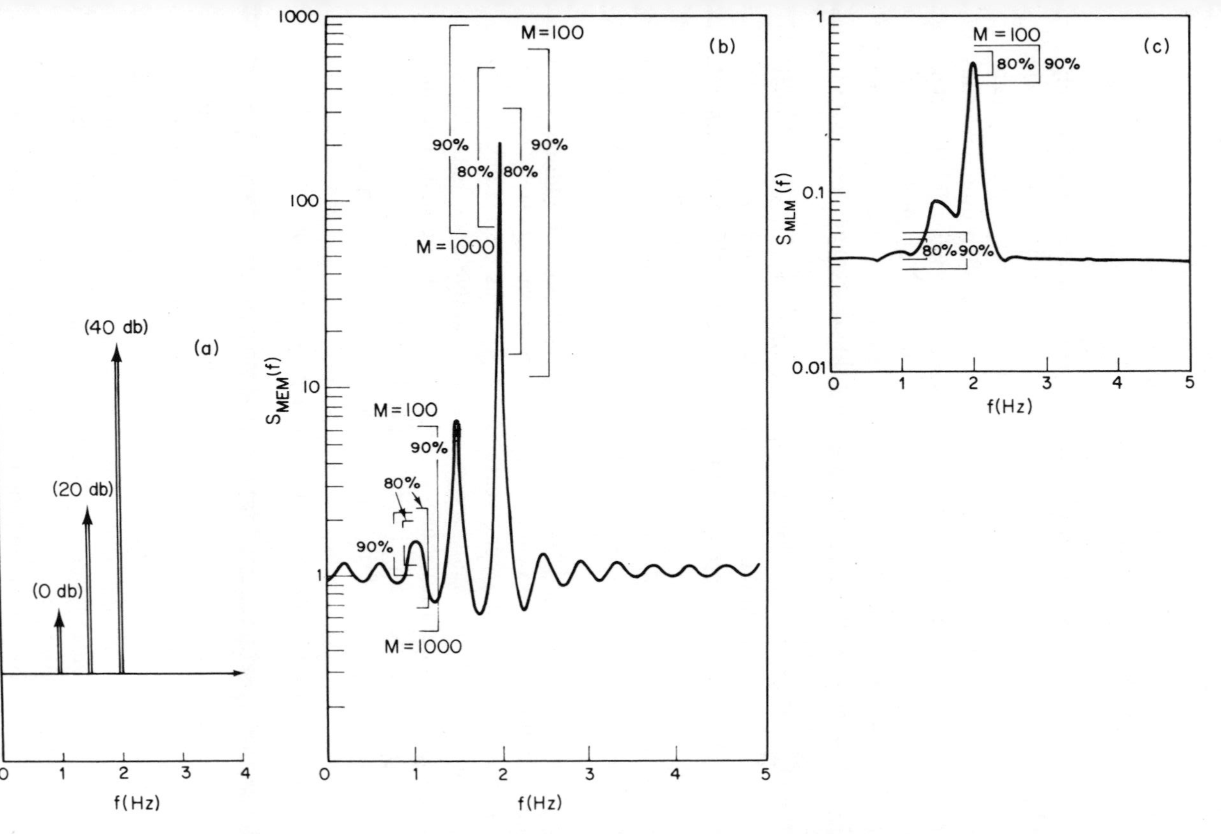

FIGURE 6-37. Comparison of maximum entropy method and maximum likelihood methods of spectral analysis: (a) True spectra consisting of three tones in white noise; (b) Maximum entropy estimate based upon a 2.4 sec correlation lag; (c) Maximum likelihood estimate based upon a 2.4 sec correlation lag (from Baggeroer, 96).

the area where passive sonar has made the most gain in recent years, the topic of array processing.

6.4.4 Array Processing Methods for the Spatial Analysis of Passive Sonar Signals

Passive sonar systems use arrays and spatial signal processing extensively, and efficient, high-speed processing has always been particularly important because of the requirements for high-resolution, large array element systems operating in real time. In passive sonar the SNRs are often quite low, thus requiring high resolution in both the time and spatial domains for the detection of the narrowband, directional signals of interest. This high resolution implies that a large number of frequency bands and beams must be scanned, and it is the capability of the signal processing that determines how well this can be done in real time.

A number of array processing systems have been used in passive sonar. In Sec. 6.4.2 we introduced the concept of beam forming and element weighting to control the beam-pattern direction and shape. This suggests a straightforward implementation consisting of a delay and weighted summation of the outputs of the transducers of an array. When one wants to form several beams, this becomes quite difficult, because the number of delay elements required can be very large. This has led to the use of shift registers and hard-clipped beam formers, popularly known as DIMUS (digital multiple beam steering) systems. Adaptive beam forming has been an important research topic in passive sonar in recent years. These systems have ranged from relatively simple null-placement devices to fully adaptive ones that estimate the spatial correlation of the acoustic environment and then use this information to implement some form of optimal processing.

Before discussing some of these array processors, it is useful to distinguish two types of processing. In the first, one simply wants to steer a beam in a particular direction to filter spatially the signals, or plane wave, coming from that direction. This is termed *beamforming*, and it may be adaptive. In the second, one is concerned with estimating the intensity of the acoustic environment as a function of the frequency and wave vector, or bearing. This takes the form of a parameter estimation problem; it is very similar in many respects to spectral estimation.

Direct beamforming consists of implementing the operation indicated in Eq. (6.27), and several systems for doing this are illustrated in Figure 6-38. Each element is delayed, or phase shifted, weighted, and summed. Generally, the weights are not frequency dependent so these systems often have a broadband capability. The weightings are designed to realize a desired beam pattern, such as those indicated in Fig. 6-38(b), where one incorporates the trade offs among beamwidth, sidelobe levels, and null locations. The delays can be introduced by mechanically steering the array by pointing it in the particular direction of interest, or by electronic steering by introducing delays or phase shifts. The electronic delays can be implemented in several ways. In the first, one uses a delay-line network. The delay lengths are difficult to change rapidly as one scans in bearing, so a number of ingenious switching networks have been designed for connecting the array elements and the weighted summing junction. The connection, or switching, can be realized either mechanically or electrically. Alternatively, a preformed beam system can be used; one forms several beams simultaneously using separate summing junctions for each beam direction. By overlapping the beams and interpolating between them, one can cover, or scan, an entire bearing sector with a sufficient number of beams. A frequency-domain alternative consists of successively transforming a segment of each array element into the frequency domain, possibly with a multidimensional FFT, and then combining each frequency with the proper phase and amplitude weighting. Inverse transformation can be then performed, although it is often useful to remain in the frequency domain for further processing. Several array transform processing systems that are able to implement this type of beam forming have been constructed recently.[99,100] Figure 6-38(c) illustrates several of these beamforming systems.

The DIMUS class of systems is by far the most popular type of electron-

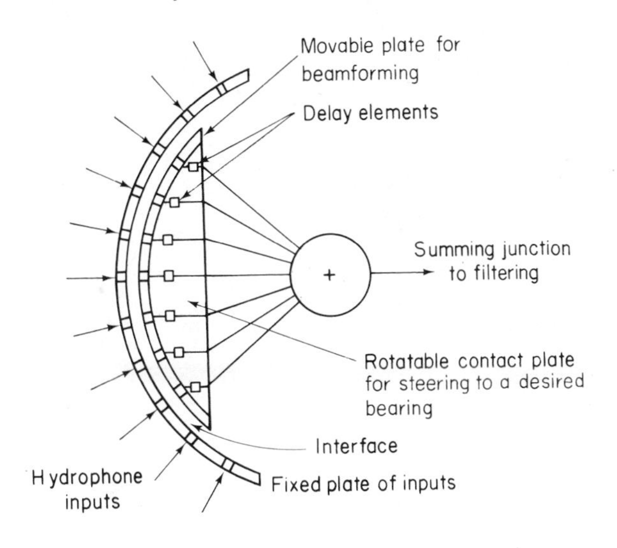

FIGURE 6-38. (a) Mechanical steering of a cylindrical (or spherical) array using a delay compensation plate; (b) Preformed beam steering in time domain; (c) Array transform, or frequency domain, beamforming network.

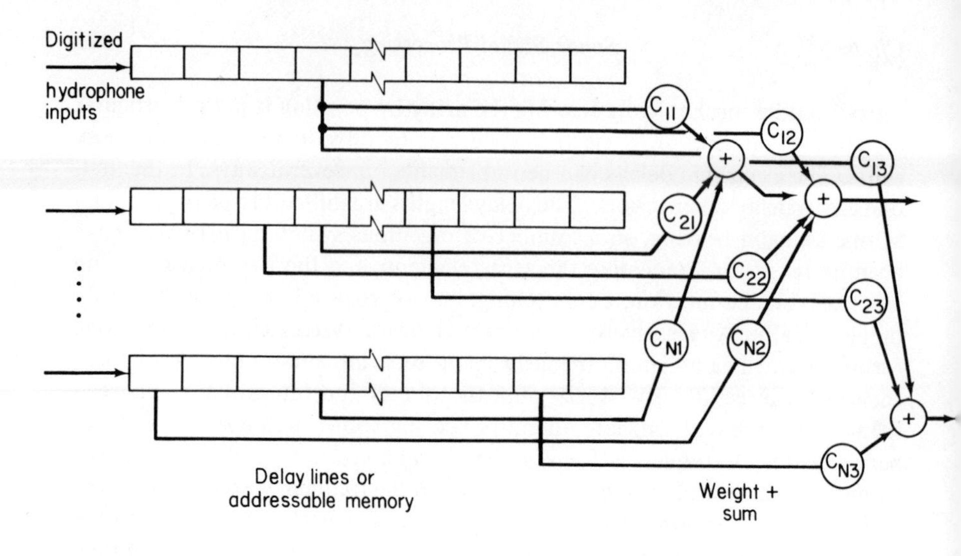

(b)

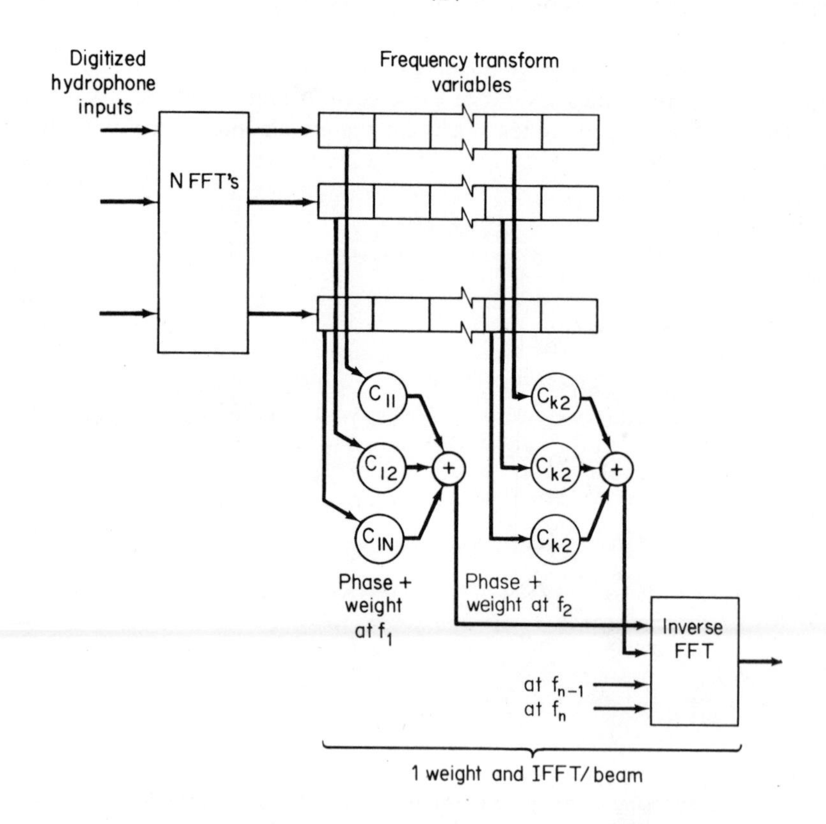

(c)

FIGURE 6-38. (*Continued*)

ically steered beamforming.[101–103] It essentially consists of connecting a tapped delay line to each array element, so that the desired delay for any direction can be achieved by selecting the proper tap. The steering operation consists of specifying the appropriate tap connection. Since each tap can be selected more than once, one can design as many beams as desired and achieve multiple steering. In addition, several taps can be made on the same delay line, or array output. This leads to a frequency selective response; however, it is not commonly implemented in direct beamforming. The implementation issues then focus upon two aspects, the construction of the delay line and the connections for the taps. The delay lines have often been implemented with DELTIC systems, which were described in Sec. 6.3.4. This implied a hard-clipped system in which just the polarity of the array output signals is used in the beamforming operations. The performance loss in comparison to linear beam forming depends upon the acoustical environment. Against a spatially white background, it is a relatively minor 2 dB, which is a tolerable loss considering the beamforming capability provided; however, strong directional interference introduces a much higher loss in performance, as well as the possibility of "capturing" the beam. With the introduction of inexpensive shift registers, clocked parallel shift registers are now being used to implement the delay operation. These systems have also been hard clipped, although multilevel systems are now becoming more feasible. The hard clipping eliminated the problems of dynamic range adjustment and equalization of the response of the array transducers. The latter is a particularly troublesome problem on moving platforms, such as a submarine. The tap connections for a multiple-beam system can be very messy and complex, so it is useful to attempt to organize the memory, or shift registers, efficiently. Figure 6-39 illustrates a DIMUS system for a simple array that forms multiple beams. The number of connections required can be substantial; for example, if it used 12 input phones and formed 10 beams it would require at least 120 separate connections, and current technology has implemented a 256-element, 80-beam systems.[104] Quite often one staves the cylindrical and spherical arrays so as to use only those elements and their associated delay lines that are directly facing the beam direction. The class of DIMUS-type systems is used extensively; it is easy to observe that their general nature leads to digital signal-processing implementations.

Adaptive systems have the capability of adjusting their weights to minimize the effects of interfering noise. Equation (6.29) summarizes the essence of much of what adaptive array processing is about. One attempts to minimize the overlap between the beam pattern, or wave-vector response, of an array and the wave-vector function of ambient acoustical noise environment. This minimization must be done within the constraints of the array geometry and the direction of the beam. In some environments, such as those with directional interfering noise sources, adaptive array processing can offer distinct advantages and significant performance gains.

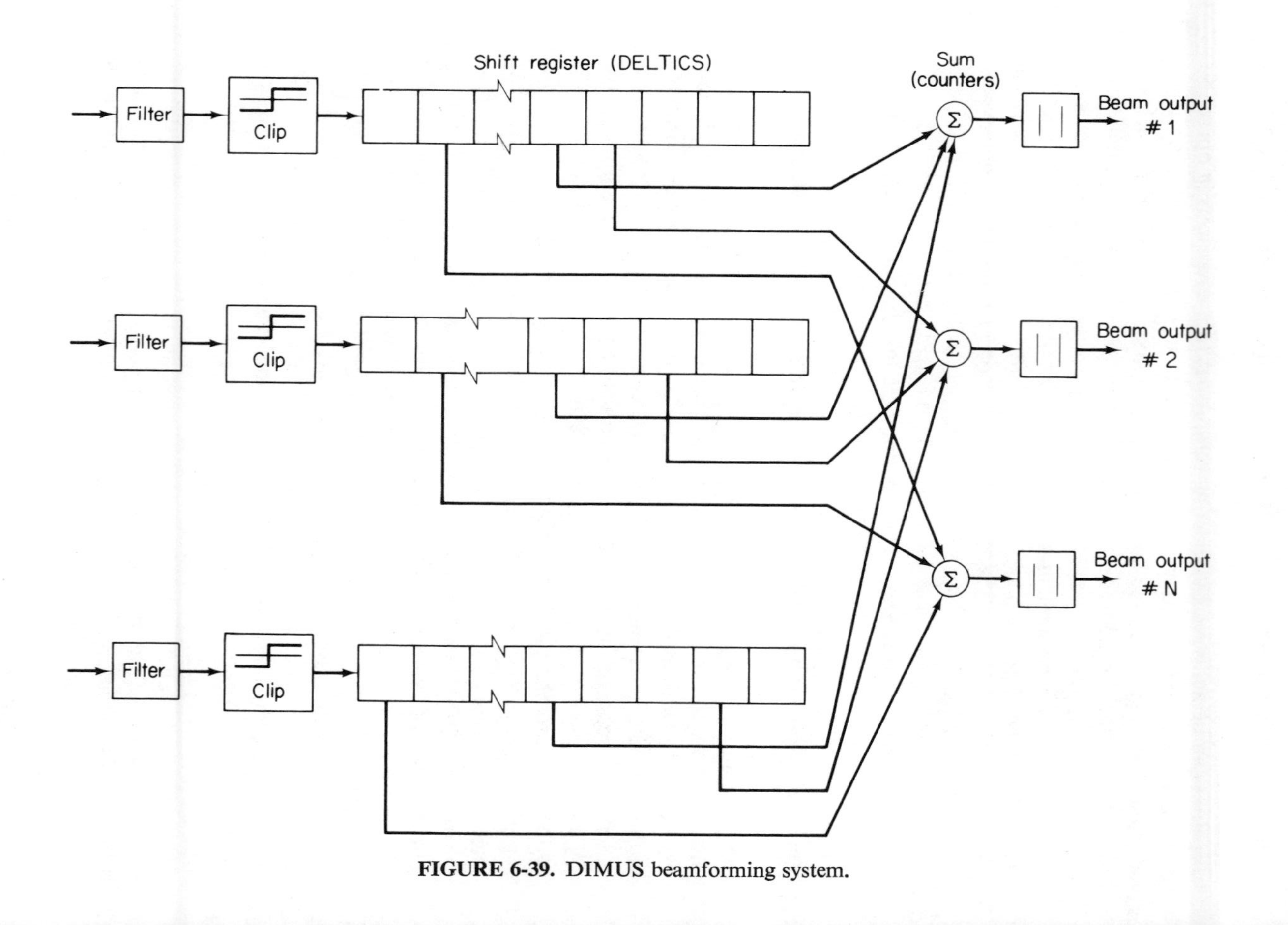

FIGURE 6-39. DIMUS beamforming system.

The structures of adaptive beamforming systems are quite varied; however, we have attempted to organize them into three categories. The first is operator adaptive; the operator has the capability of adjusting the weights dynamically in response to his perception of the environmental noise field. Adjusting the weights directly is not very efficient, so this class of systems often takes the form of a parameter adjustment, such as the placement of a collection of nulls. The second class consists of a system for estimating the environmental noise structure, for example by measuring the spectral covariance $S_x(f: \mathbf{z}_i, \mathbf{z}_j)$ in Eq. (6.28) of the signals at the array element locations, and a subsequent system for implementing an optimal beamformer algorithm. We call these open-loop systems since the performance of the beamformer is not fed back to the beamformer design system. Finally, we have closed-loop adaptive arrays. In these systems one is implementing some form of stochastic gradient algorithm in which a performance measure for the beamformer, such as minimum mean square error or minimum noise output power, is iteratively optimized within the constraints of the array geometry and target direction. The iterative optimization uses the performance of the beamformer for adapting the array weights according to the stochastic gradient algorithm employed.

In describing optimal beamformers and adaptive arrays, it is useful to distinguish two operations that are often confused. These are beamforming and wave-vector-function, or bearing intensity, estimation. In the beamforming operation, one wants to form a beam in a particular direction that minimizes the effects of interfering noise sources. The output of a beamformer is a waveform, which is often subsequently subjected to some form of spectral analysis. It is usually important to classify systems as being either broad- or narrowband in their operation. In the wave-vector-function estimation operation, one wants to measure the intensity of the acoustic environment as a function of frequency and/or wave vector, or direction. The output of the estimator is a function of these parameters, and not a waveform.

The operator-adaptive systems attempt to adjust a set of parameters to improve, or optimize, the performance of a beamformer. The parameters are usually the location and intensity of interfering directional noise sources. Figure 6-40 illustrates the DICANNE (digital interference canceling adaptive null network equipment) system as an example of this class of adaptive arrays.[105,106] In this system an operator steers a beam in the direction of an interfering noise source. This beam output is then used to subtract an estimate of the interference from each array output. This is followed by a conventional beamforming operation in the direction of the target. It is straightforward to extend this version of the system to operate against several directional interferences. This system is an optimal structure for array processing against directional interference.[78] The major difficulty is estimating the location and intensity of the interferences, and implementing the extensive

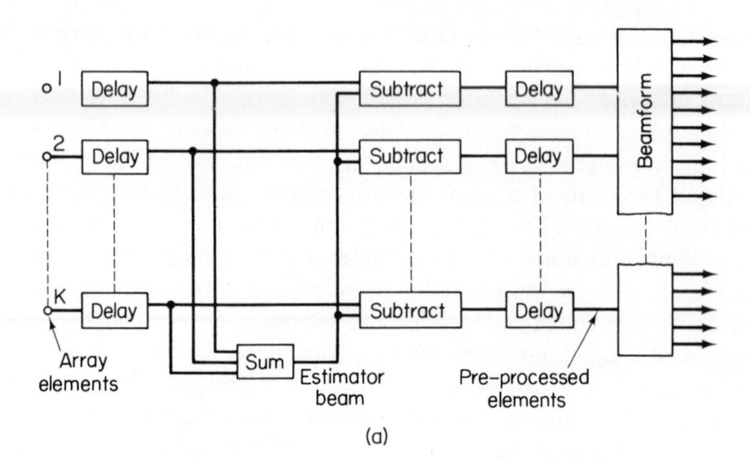

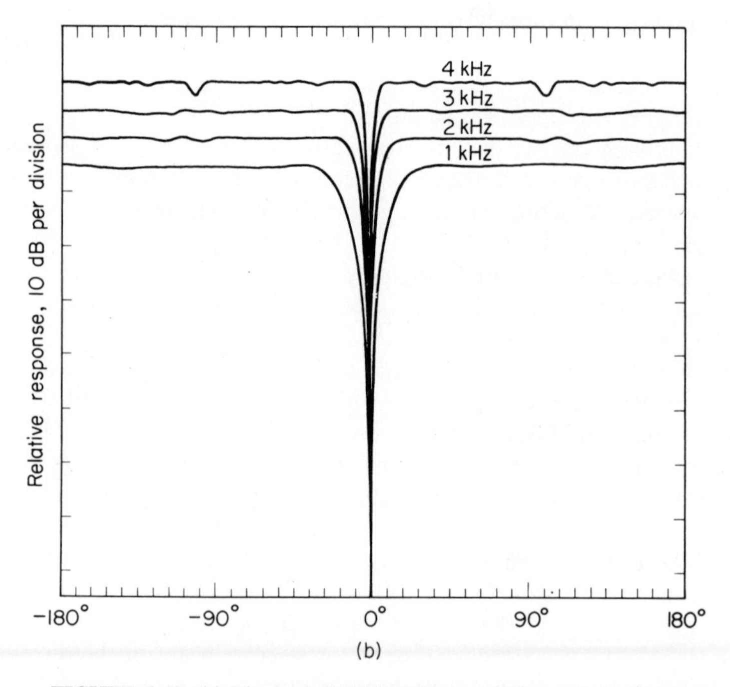

FIGURE 6-40. (a) The basic DICANNE process for null steering; (b) Null placement pattern response for a noise source at 0 degrees at four frequencies (from Anderson, 105).

number of delays required. An operator can continually track the location parameters, and a number of feedback algorithms for automatic estimation and tracking have been suggested. The delays have led to hard-clipped implementations using DELTICs or shift registers. In most applications, these beamformers are broadband and are often followed by some form of spectral analysis algorithm.

The open-loop adaptive array systems can be used for both optimal beam forming and wave-vector-function estimation. They operate by first measuring the second-moment statistics of the acoustic environment, usually the spectral covariance matrix across the array elements, and then using this estimated information to design both beamformers and estimators. They usually operate in the frequency domain by Fourier transforming segments of the input data, usually by parallel FFTs; thus one may consider them to be a collection of parallel narrowband processors.

The most commonly employed open-loop adaptive array processor is a minimum variance, unbiased or distortionless, beamformer and the spatial version of the maximum likelihood method for wave-vector-function estimation.[107–111] It is based upon the following optimization problem: find the optimal set of weights that minimizes the variance of the output a linear beam former that has a unity, or distortionless, response in the direction of a target; that is,

$$\operatorname*{Min}_{[G_i]}\left[\sum_{i,j} G_i^*(f)S_x(f:\mathbf{z}_i,\mathbf{z}_j)G_j(f)\right] \tag{6.35}$$

where

$$\mathcal{G}(f,\mathbf{v}_T)=\sum_i G_i(f)e^{-j2\pi\mathbf{v}_T\cdot\mathbf{z}_i}=1$$

The solution to this problem is given by

$$\mathbf{G}(f)=\mathbf{S}_x^{-1}(f)[E(f,\mathbf{v}_T)]\sigma_G^2(f,\mathbf{v}_T) \qquad \text{beamformer} \tag{6.36a}$$

$$\sigma_G^2(f,\mathbf{v}_T)=[E^t(f,\mathbf{v}_T)\underset{\sim}{S}_x^{-1}(f)E(f,\mathbf{v}_T)]^{-1} \qquad \text{output variance} \tag{6.36b}$$

We can observe that the optimal beamformer uses only the spectral covariance matrix and a steering vector in its design, and that the output variance is the expression for the maximum likelihood method of wave-vector estimation. This suggests the implementation indicated in Fig. 6-41. It consists of three components: (1) an accumulation of the spectral covariance matrix, (2) a beamformer based upon this estimated matrix, and (3) a MLM algorithm for the estimate of the frequency wave-vector function of the acoustic environment.

Several other algorithms have been proposed for estimating the wave-vector function. If the array elements have a uniform spacing, which is often the case in practice, one can employ the spatial version of the maximum entropy method. The extensions to nonuniformly spaced linear arrays and arrays distributed in two or three spatial coordinates is still a topic of active research. One can also perform an eigenvalue–eigenvector decomposition of the spectral covariance matrix.[112] This is equivalent to factor, or principal component, analysis in several other fields.

One needs to be aware of the limitations of these algorithms, for they can lead to a sensitive situation if pressed too much. To achieve the potentially

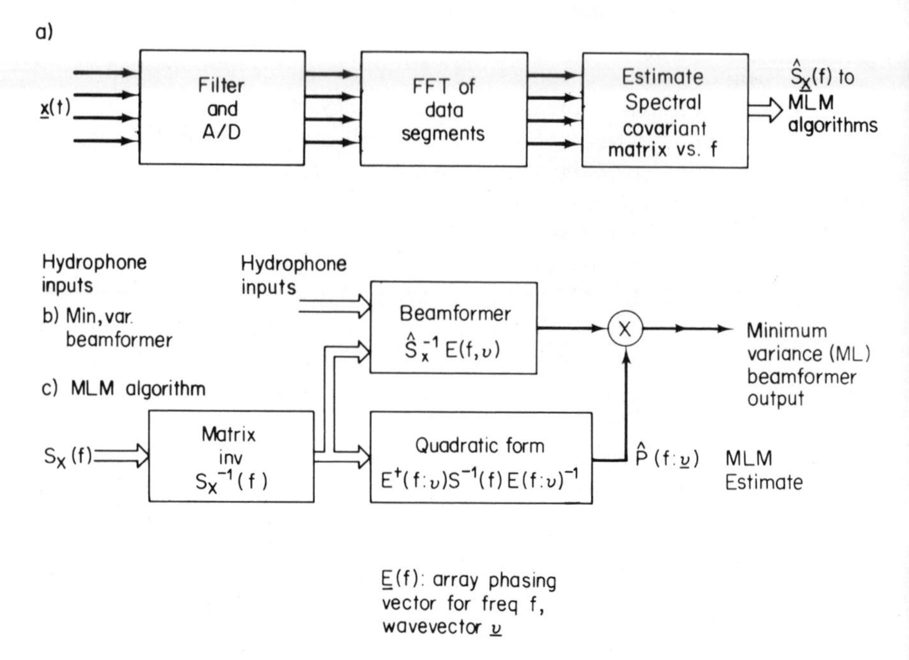

FIGURE 6-41. Adaptive array processing methods for frequency wavevector estimation: (a) Maximum likelihood method; (b) Maximum entropy method.

high array gains indicated by this algorithm in some acoustical environments, one needs to have very accurate estimates of statistics, precise knowledge of array locations, and high-resolution processing. Several sensitivity analyses have been discussed in the literature.[113]

Adaptive array processors based upon performance feedback involve stochastic approximation methods for solving the Wiener–Hopf equation, or some equivalent equation, for the optimal beamformer weightings. One can motivate their operations as follows: a linear beamformer is specified in terms of a solution to a Wiener–Hopf, or normal, equation, and a gradient algorithm for solving this equation is specified. The ensemble quantities in these equations are replaced by estimates derived from the data, and the gradient algorithm is iterated using these estimated quantities. In interpreting these algorithms it is useful to cast the structure of the optimal beamformer in terms of the collection of tapped delay lines, as indicated in Fig. 6-42. In this operation the array outputs are passed through a tapped-delay-line combiner to produce an estimate of the desired signal, usually a beam in a particular direction. This estimate is compared with a model of the desired signal, and the error in this estimate is fed back to adjust the weights.

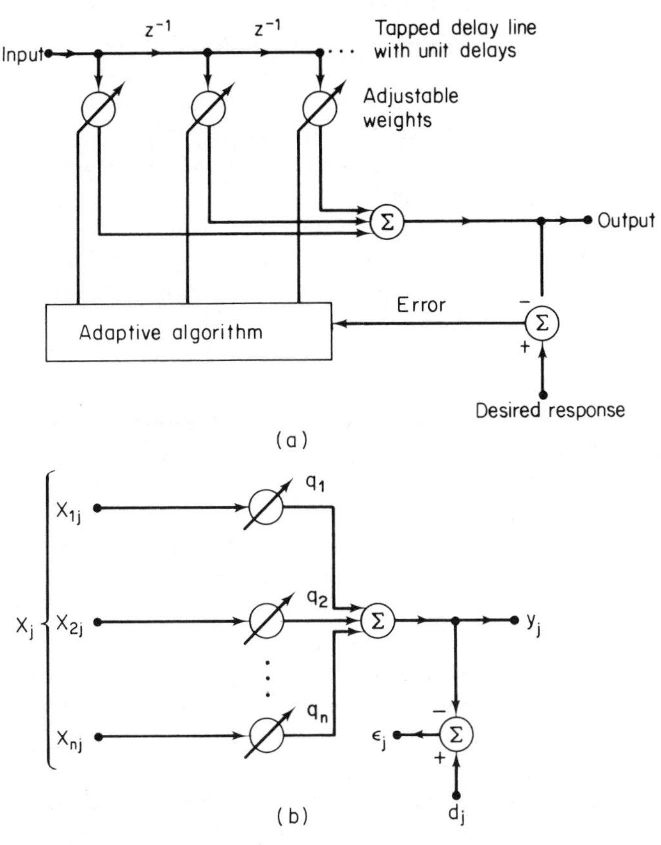

FIGURE 6-42. Adaptive least mean square (LMS) array processor system structure. (from Widrow, 114)

Over the last decade several stochastic approximation algorithms have been proposed for adjusting the weights.(114–119) The least mean square (LMS) algorithm was the first of these; it has the form

$$\mathbf{G}(N+1) = \mathbf{G}(N) - \mu\{[d(N) - \mathbf{G}^T(N)\mathbf{R}(N)]\mathbf{R}^T(N)\}^T \qquad (6.37)$$

where

$\mathbf{G}(N)$ = weights at the Nth iteration,

$\mathbf{R}(N)$ = array outputs at the Nth iteration

$d(N)$ = model of the desired signal

The major differences in the earlier algorithms involved the model for the desired signal. The first version injected a pilot tone to simulate the arrival of a signal from a particular direction; the second used a steering vector model for the desired signal.(114,115) Later versions of the algorithms introduced the concept of constraining the response of the beamformer in certain directions, in effect producing a stochastic approximation version of the maximum likelihood method.[116–119] An example of the adaption of the beam pattern for such a constrained response is indicated in Fig. 6-43. In this sequence one can observe that the algorithm adapts to the noise field by placing nulls in the direction of strong interference while maintaining a unity response in the target direction.

Several interesting aspects associated with this class of adaptive arrays are worth commenting on. The adaption rate is strongly influenced by the parameter in the stochastic approximation algorithm. It essentially determines the amount of feedback for the error signal. There are bounds on it for simple stability of the feedback, and within the stable region large amounts of feedback lead to rapid adaption by noisy performance, whereas small feedback leads to slow adaption and smoother performance. A trade off must be effected in which one considers the performance of the system versus the stability of the acoustic environment.

Most of the applications published have used only a single tap at each array output, although the algorithm can work with more than one tap and introduce some frequency selectivity. One also needs to specify either a narrowband or baseband implementation. In the narrowband operation, the array outputs are block transformed, and the input to the tapped delay lines is the Fourier coefficient in the frequency band of interest. In the baseband operation, the array outputs are demodulated, and the input to the delay lines is the quadrature components of the demodulated signal.

Almost all the implementations of this class of adaptive array processing have been done on general-purpose minicomputers. The algorithms demand a great deal of flexible processing capability, and they could not be implemented without a significant amount of digital signal processing. These algorithms have, however, become quite popular, and several applications to adaptive radar arrays have recently appeared.[120]

6.4.5 Examples of Applications of Passive Sonar Signal Processing

As mentioned earlier, examples of passive sonar systems, particularly those involving the latest research, usually become entangled in security classifications. This makes it quite difficult to give examples that contain field data and actual systems. In this section we shall briefly indicate two examples of passive sonar signal processing; we emphasize, however, that there is a very extensive classified literature on this topic.

One of the most extensively used operational systems is the split-beam correlator discussed in Sec. 6.4.2 and illustrated in Fig. 6-43. As this system successively scans in bearing, it generates a picture of the acoustic environ-

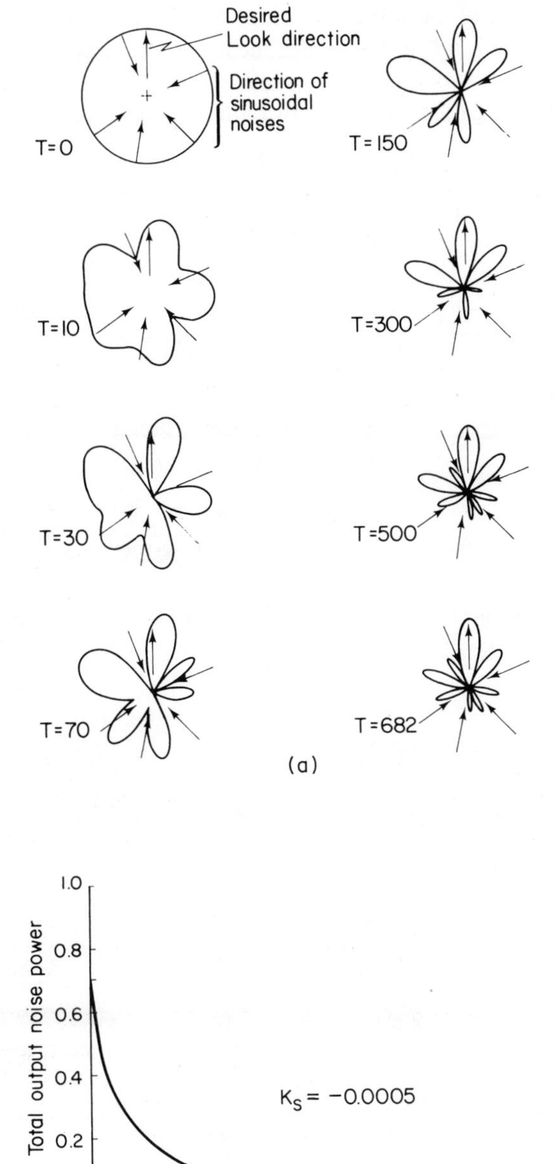

FIGURE 6-43. Beam pattern response of LMS array processor during the adaption process. (from Widrow, 114)

ment. An example of an implementation of the split-beam correlator and its output is illustrated in Fig. 6-44.[121] One can observe how the history of the target environment evolves on the output by the track positions. When coupled with the navigation of one's own ship, target motion analysis can be

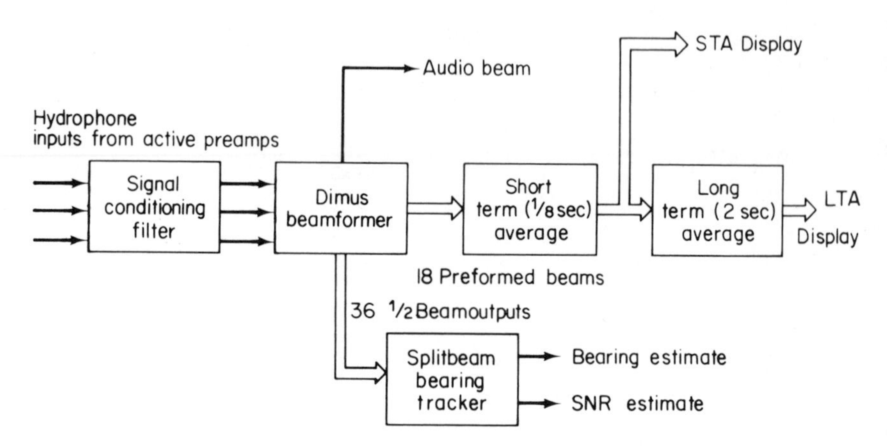

FIGURE 6-44. Bearing vs time recording of a clipper correlator output (from Winder, 121).

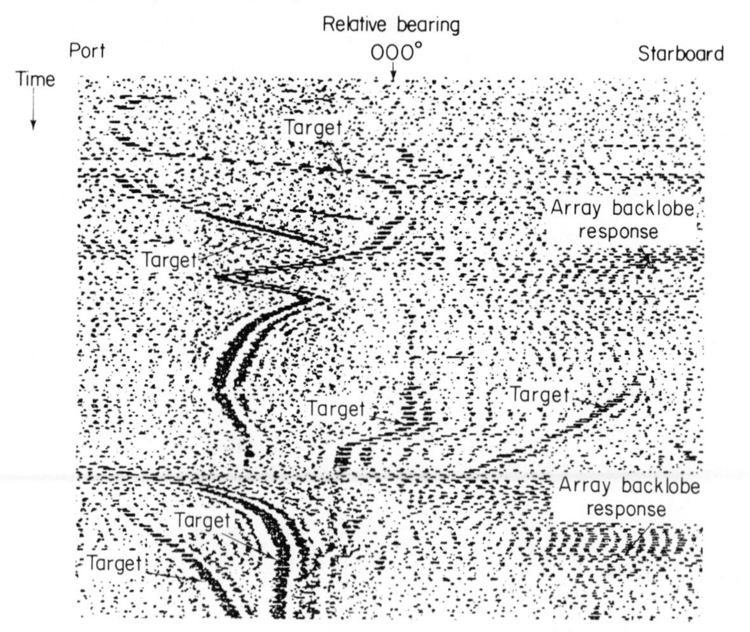

FIGURE 6-45. Structure of the signal processing components of the passive section of the Raytheon DE-1160 sonar system (from Skitzki, 68).

done to generate estimates of the range to the various targets as well as their bearing. One can also observe the appearance and disappearance of several of the targets, as well as the possibility for false alarms.

Most passive sonar systems use a number of sensing devices and processing options. Figure 6-45 illustrates the organization of an entire system.(68) One can observe that there are a number of options for both temporal and spatial analysis of the signals, depending upon the capabilities of the sensing system and the information desired by the operators. We have discussed most of the individual components in this section. Additional processing may be done within the general-purpose minicomputers that control the signal flow within the system. Future implementations will undoubtedly use digital signal processing more extensively as some of the algorithms discussed in this section find their way into operational systems.

6.5 CONCLUDING REMARKS

In this chapter we have attempted to survey some of the important aspects of sonar signal processing. It is important to remember that the signal processor is only one of the components of a sonar system. While we have discussed the ocean environment because of its fundamental influence upon the signal processing, we have omitted comments upon the transducers, displays, and operators—all of which interact significantly with the signal processing.

The material chosen emphasized signal-processing concepts, not whether the implementations were analog or digital. Digital signal processing, however, has been used extensively in sonar systems in the past, and the true impact of the modern technology of digital hardware has yet to be felt in its entirety. This is the rationale for our emphasis upon the underlying concepts and models, for it is only now that the hardware capability exists to implement the processing suggested by them. We may soon arrive at the point where the capabilities of the signal processing are adequate for most sonar applications. At this point we couple it to our rapidly increasing knowledge of the acoustics of the ocean so that we operate more effectively in the medium which occupies over seventy percent of the earth's surface.

REFERENCES

1. *Principles and Applications of Underwater Sound*, Summary Tech. Rept., Div. 5, National Defense Research Council, vol. 7; reprinted by the Department of the Navy, HQ Naval Material Command, Washington, D.C.

2. *Present and Future Civil Uses of Underwater Sound*, Committee on Underwater Telecommunication, National Research Council, National Academy of Sciences, Standard Book No. 309–01771–8, Washington, D.C.

3. R. J. Urick, *Principles of Underwater Sound for Engineers*, McGraw-Hill, New York, 1967.

4. C. B. Officer, *Introduction to the Theory of Sound Transmission*, McGraw-Hill, New York, 1958.

5. I. Tolstoy and C. S. Clay, *Ocean Acoustics: Theory and Experiment in Underwater Sound*, McGraw-Hill, New York, 1966.

6. V. O. Albers, *Underwater Acoustics Handbook*, Pennsylvania State University Press, University Park, Pa., 1960.

7. H. L. VanTrees, *Detection, Estimation, and Modulation Theory*, Part I (1969), Part II (1970), Part III (1971), Wiley, New York.

8. J. Wozencraft and I. Jacobs, *Principles of Communication Engineering*, Wiley, New York, 1965.

9. M. Skolnik, *Introduction to Radar Systems*, McGraw-Hill, New York, 1962.

10. C. W. Helstrom, *Statistical Theory of Signal Detection*, Pergamon Press, Elmsford, N.Y. 1960.

11. R. B. Blackman and J. W. Tukey, *The Measurement of Power Spectra from the Point of View of Communications Engineering*, Dover, New York, 1958.

12. G. M. Jenkins and D. G. Watts, *Spectral Analysis and Its Applications*, Holden-Day, San Francisco, 1968.

13. D. R. Brillinger, *Time Series Data Analysis and Theory*, Holt, Rinehart and Winston, New York, 1975.

14. B. D. Steinberg, *Principles of Aperture and Array Systems Design*, Wiley-Interscience, New York, 1976.

15. H. L. VanTrees and A. B. Baggeroer, *Array Processing* (in press), 1977.

16. C. W. Horton, Sr., *Signal Processing of Underwater Acoustic Waves*, U.S. Government Printing Office, Washington, D.C., 1969.

17. R. A. Frosch, "Underwater Sound: Deep-Ocean Propagation," *Science*, Vol. 146, 13 Nov. 1964, pp. 889–893.

18. K. O. Emery and E. Uchupi, *Western North Atlantic Ocean: Topography, Rocks, Structure, Water, Life and Sediments*, American Association of Petroleum Geologists Memoir 17, 1972.

19. M. Ewing and J. L. Worzel, "Long-Range Sound Transmission," in *Propagation of Sound in the Ocean*, Geological Society of America Memoir 27, 1948.

20. R. J. Urick, "Long-Range Deep Sea Attenuation Measurements," *J. Acoust. Soc. Amer.*, vol. 39, 1966, pp. 904–906.

21. J. C. Steinberg, J. G. Clark, H. A. DeFerrari, M. Kronengold, and K.

Yacoub, "Fixed System Studies of Underwater Acoustic Propagation," *J. Acoust. Soc. Amer.* 1972, pp. 1521–1536.

22. L. L. Booda, "*Top Level Defence Guidance in ASW Grows*," Sea Technol., vol. 16, no. 11, 1975, pp. 16–18.

23. G. M. Wenz, "Acoustical Ambient Noise in the Ocean: Spectra and Sources," *J. Acoust. Soc. Amer.*, vol. 34, 1962, pp. 1936–1956.

24. V. O. Knudsen, R. S. Alford, and J. W. Emling, "Underwater Ambient Noise," *J. Marine Res.*, vol. 17, 1948, pp. 416–420.

25. G. R. Fox, "Ambient Noise Directivity Measurements," *J. Acoust. Soc. Amer.*, vol. 36, 1964, pp. 1537–1540.

26. E. H. Axelrod, B. A. Schoomer, and W. A. VonWinkle, "Vertical Directionality of Ambient Noise in the Deep Ocean at a Site Near Bermuda," *J. Acoust. Soc. Amer.*, vol. 40, 1966, pp. 205–211.

27. R. J. Urick and D. S. Saling, "Backscattering of Sound from the Deep-sea Bed," *J. Acoust. Soc. Amer.*, vol. 34, 1961, pp. 158–160.

28. O. F. Hastrup, "Digital Analysis of Acoustic Reflectivity in the Tyrrhenian Abyssal Plain," *J. Acoust. Soc. Amer.*, vol. 47, 1970, pp. 181–190.

29. H. P. Bucher et al., "Reflection of Low Frequency Sonar Signals from a Smooth Ocean Bottom," *J. Acoust. Soc. Amer.*, vol. 37, 1965, pp. 1037–1051.

30a. E. L. Hamilton, "Geoacoustical Models of the Sea Floor," in *Physics of Sound in Marine Sediments*, L. Hampton, ed., Plenum, New York, 1974, pp. 181–223.

30b. H. P. Bucher, "Sound Propagation Calculation Using Bottom Reflection Functions," ibid., pp. 223–240.

30c. P. F. Lonsdale, R. C. Tyce and F. N. Speiss, "Near Bottom Acoustic Observations of Abyssal Topography and Reflectivity," ibid., pp. 293–318.

30d. D. L. Bell and W. J. Porter, "Remote Sediment Classification of Reflected Acoustic Signals," ibid., pp. 319–337.

30e. C. S. Clay and W. K. Leong, "Acoustic Estimates of the Topography and Roughness Spectrum of the Seafloor Southwest of the Iberian Peninsula," ibid., pp. 373–446.

30f. T. Akal, "Acoustical Characteristics of the Sea Floor: Experimental Techniques and Some Examples from the Mediterranean Sea," ibid., pp. 447–480.

30g. R. R. Goodman and A. Z. Robinson, "Measurements of Reflectivity by Explosive Signals," ibid., pp. 537–564.

31. C. Eckart, "Scattering of Sound from the Sea Surface," *J. Acoust. Soc. Amer.*, vol. 25, 1953, pp. 566–570.

32. H. W. Marsh, M. Schulkin, and S. G. Kneale, "Scattering of Underwater Sound by the Sea Surface," *J. Acoust. Soc. Amer.*, vol. 33, 1961, pp. 330–334.

33. C. S. CLAY and H. MEDWIN, "Dependence of Spatial and Temporal Correlation of Forward Scattered Underwater Sound on the Surface Statistics, Pt. I, Theory, Pt. II, Experiment," *J. Acoust. Soc. Amer.*, vol. 47, 1970, pp. 1412–1429.

34. I. DYER, "Statistics of Sound Propagation in the Ocean," *J. Acoust. Soc. Amer.*, vol. 48, 1970, pp. 337–345.

35. F. DYSON, W. MUNK and B. ZETLER, "Interpretation of Multipath Scintillation Eleuthera to Bermuda in Terms of Internal Waves and Tides," *J. Acoust. Soc. Amer.*, vol. 59, 1976, pp. 1121–1133.

36. R. E. WILLIAMS and C. H. WEI, "Spatial and Temporal Fluctuation of Acoustic Signals Propagated over Long Ocean Paths," *J. Acoust. Soc. Amer.*, vol. 56, 1976, pp. 1299–1309.

37. R. E. WILLIAMS and C. H. WEI, "The Correlation of Acoustic Wavefront and Signal Time-Base Instabilities in the Ocean," *J. Acoust. Soc. Amer.*, vol. 59, 1976, pp. 1310–1316.

38. J. F. BRAHTZ, ed., *Ocean Engineering*, Wiley, New York, 1968.

39. P. M. WOODWARD, *Probability and Information Theory*, McGraw-Hill, New York, 1953.

40. W. M. SIEBERT, "A Radar Detection Philosphy," *IRE Trans. Inform. Theory*, vol. IT-2, 1956, pp. 204–221.

41. H. L. VANTREES, *Detection Estimation and Modulation Theory, Pt. III*, Wiley, New York, 1971.

42. H. L. VANTREES, "Optimum Signal Design and Processing for Reverberation Limited Environments," *IEEE Trans. Military Electron*,, vol. MIL-9, 1965, pp. 212–229.

43. R. S. KENNEDY and I. L. LEBOW, "Signal Design for Dispersive Channels," *IEEE Spectrum*, Mar. 1964, pp. 231–237.

44a. P. E. GREEN, Jr., "Radar Measurements of Target Scattering Properties," in *Radar Astronomy*, J. V. Evans and T. Hagfors, eds., McGraw-Hill, New York, 1968.

44b. R. PRICE, "Detectors for Radar Astronomy," ibid., pp. 547–614.

45. E. C. WESTERFIELD, R. H. PRAGER and J. L. STEWART, "Processing Gains Against Reverberation (Clutter) Using Matched Filters," *IRE Trans. Inform. Theory*, vol. IT-6, 1960, pp. 342–348.

46. P. BELLO, "Characterization of Randomly Time-Variant Linear Channels," *IEEE Trans. Communi. Systems*, vol. CS-11, 1963, pp. 360–393.

47. D. MIDDLETON, "A Statistical Theory of Reverberation and Similar First-Order Scattered Fields, Parts I and II," *IEEE Trans. Inform. Theory*, vol. IT-13, 1967, pp. 372–414.

48. R. S. KENNEDY, *Fading Dispersive Communication Channels*, Wiley, New York, 1969.

49. T. KAILATH, "Measurements in Time-Variant Communication Channels," *IRE Trans. Inform. Theory*, vol. IT-8, 1962, pp. 229–236.

50a. R. LAVAL, "Sound Propagation Effects on Signal Processing," in *Signal Processing*, J. W. R. Griffiths, P. Stocklin and C. Van Schoonveld, eds., Academic Press, New York, 1973, pp. 223–242.

50b. R. S. THOMAS, J. C. MOLDON, and J. M. ROSS, "Shallow Water Acoustics Related to Signal Processing," ibid., pp. 281–297.

50c. P. H. MOOSE, "Signal Processing in Reverberant Environments," ibid., pp. 413–428.

51a. A. WASILJEFF, "The Influence of Time and Space Variant Random Filters on Signal Processing," *Proceedings of the NATO Advanced Study Institute on Signal Processing with Emphasis on Underwater Acoustics*, paper no. 2, Reidel Publishing Co., Dordrecht, The Netherlands, 1976.

51b. R. LAVAL, "Some Remarks on Sound Propagation in a Variable Ocean," ibid., paper no. 4.

52. R. C. SPINDEL and P. M. SCHULTEISS, "Acoustic Surface-Reflection Characterization Through Impulse Response Measurements," *J. Acoust. Soc. Amer.*, vol. 51, 1972, pp. 1812–1818.

53. H. A. DEFERRARI and L. NGHIEM-PHU, "Scattering Function Measurements for a 7-NM Propagation Range in the Florida Straights," *J. Acoust. Soc. Amer.*, vol. 56, 1974, pp. 45–52.

54. P. SWERLING, "Probability of Detection for Fluctuating Targets," *IRE Trans. Inform. Theory*, vol. IT-6, 1960, pp. 269–308.

55. J. M. F. MOURA, H. L. VANTREES and A. B. BAGGEROER, "Space–Time Tracking by a Passive Observer," *Proceedings of the 4th Symposium on Nonlinear Estimation Theory and Its Application*, San Diego, Calif. 1973.

56. M. A. GALLOP and L. W. NOLTE, "Bayesian Detection of Targets of Unknown Location," *IEEE Trans. Aerospace Electron. Systems*, vol. AES-10, 1974, pp. 429–435.

57. C. N. PRYOR, "An Adaptive Kalman Filter for Multimode Range-Doppler Sonar," *Proceedings of the NATO Advanced Study Institute on Signal Processing with Emphasis on Underwater Acoustics*, paper no. 36, Reidel Publishing Co., Dordrecht, The Netherlands, 1976.

58. V. C. ANDERSON, "Deltic Correlator," *Harvard Acoust. Lab. Tech. Memo. No. 37*, Jan. 5, 1956.

59. W. B. ALLEN and E. C. WESTERFIELD, "Digital Compressed-Time Correlators and Matched Filters for Active Sonar," *J. Acoust. Soc. Amer.*, vol. 36, 1964, pp. 121–139.

60. J. I. MARCUM, "A Statistical Theory of Target Detection by Pulsed Radar," *IRE Trans. Inform. Theory*, vol. IT-6, 1960, pp. 145–267.

61. P. BELLO and W. HIGGINS, "The Effect of Hard Limiting on the Propabilities of Incorrect Dismissal and False Alarm at the Output of an Envelope Detector," *IRE Trans. Inform. Theory*, vol. IT-7, 1961, pp. 60–66.

62. G. Turin, "An Introduction to Digital Matched Filters," *Proc. IEEE*, vol. 64, 1976, pp. 1092–1112.

63. T. Kailath, "Sampling Models for Linear Time-Variant Filters," Technical Report 352, Research Laboratory of Electronics, MIT, Cambridge, Mass., May 1959.

64. R. Price and P. E. Green, Jr., "A Communication Technique for Multipath Channels," *Proc. IEEE*, vol. 46, 1958, pp. 555–570.

65. R. M. Lerner, "A Matched Filter Detection System for Complicated Doppler Shifted Signals," *IRE Trans. Inform. Theory*, vol. IT-6, 1960, pp. 380–385.

66. R. W. Lucky, "Techniques for Adaptive Equalization of Digital Communication Systems," *Bell System Tech. J.*, Feb. 1966, pp. 255–286.

67. H. L. Groginsky, L. R. Wilson and D. Middleton, "Adaptive Detection of Statistical Signals in Noise," *IEEE Trans. Inform. Theory*, vol. IT-12, 1966, pp. 337–348.

68. P. Skitzki, "Modern Sonar Systems," Raytheon (description of the Raytheon DE 1160 Destroyer Sonar system, personal communication from R. T. Karon, Raytheon Co.).

69. R. C. Spindel, R. P. Porter, W. M. Marquet and J. Durham, "A High Resolution Pulse-Doppler Underwater Acoustic Navigation System," *IEEE J. Oceanic Eng.*, vol. OE-1, Aug. 1976, pp. 6–14.

70. A. M. Yaglom, "Second Order Homogeneous Random Fields," *Proceedings of the 4th Berkeley Symposium on Mathematical Statistics and Probability*, vol. 2, 1961, pp. 593–620.

71. E. Wong, *Stochastic Processes in Information and Dynamical Systems*, Chap. 7, McGraw-Hill, New York, 1971.

72. A. B. Baggeroer, "Space–Time Processes and Optimal Array Processing," Technical Publication 506, Navy Undersea Center, San Diego, Calif. Dec. 1976.

73. N. L. Owsley, "Adaptive Space–Time Processor Performance Considerations," *Proceedings of the NATO Advanced Study Institute on Signal Processing with Emphasis on Underwater Acoustics*, paper no. 42, Reidel Publishing Co., Dordrecht, The Netherlands, 1976.

74. B. F. Cron and C. H. Sherman, "Spatial Correlation Functions for Various Noise Models," *J. Acoust. Soc. Amer.*, vol. 34, 1962, pp. 1732–1742.

75. H. Cox, "Spatial Correlation in Arbitrary Noise Fields with Application to Ambient Sea Noise," *J. Acoust. Soc. Amer.*, vol. 54, 1973, pp. 1289–1301.

76. R. L. Pritchard, "Optimum Directivity Patterns for Linear Point Arrays," *J. Acoust. Soc. Amer.*, vol. 25, 1953, pp. 879–890.

77. C. L. Dolph, "A Current Distribution of Broadside Arrays Which Optimizes the Relationship Between Beamwidth and Side-lobe Level," *Proc. Inst. Radio Engr.*, vol. 34, 1946, pp. 335–356.

78. V. C. Anderson, "DICANNE, A Realizable Adaptive Process," *J. Acoust. Soc. Amer.*, vol. 45, 1969, pp. 398–405.

79. A. Berman and C. S. Clay, "Theory of Time-Average Product Arrays," *J. Acoust. Soc. Amer.*, vol. 29, 1957, pp. 805–820.

80. D. C. Fakley, "Time Averaged Product Array," *J. Acoust. Soc. Amer.*, vol. 31, 1959, pp. 1307–1310.

81. E. J. Hannan, *Time Series Analysis*, Methuen, London, 1961.

82. J. S. Bendat and A. Piersol, *Measurement and Analysis of Random Data*, Wiley, New York, 1966 (first ed.); 1973 (second ed.).

83. P. D. Welch, "The Use of the Fast Fourier Transform for Estimation of Spectra: A Method Based Upon Averaging over Short, Modified Periodograms," *IEEE Trans. Electroacoust.*, vol. AU-15, 1967, pp. 70–76.

84. J. W. Cooley, P. A. W. Lewis, and P. D. Welch, "Historical Notes on the Fast Fourier Transform," *IEEE Trans. Electroacoust.*, vol. AU-15, 1967, pp. 76–79.

85. O. L. Frost, III, "Power Spectrum Estimation," *Proceedings of the NATO Advanced Study Institute on Signal Processing with Emphasis on Underwater Acoustics*, paper no. 7, Reidel Publishing Co., Dordrecht, The Netherlands, 1976.

86. R. T. Lacoss, "Data Adaptive Spectral Analysis Methods," *Geophysics*, vol. 36, 1971, pp. 661–675.

87. T. J. Ulrych and T. N. Bishop, "Maximum Entropy Spectral Analysis and Autoregressive Decomposition," *Reviews of Geophysics and Space Physics*, vol. 13, 1975, pp. 1–19.

88. J. P. Burg, "Maximum Entropy Spectral Analysis," presented at the 37th Meeting of the Society of Exploration Geophysicists, Oklahoma City, Okla., 1967.

89. E. Parzen, "Multiple Time Series Modelling," in *Multivariate Analysis*, P. Krishnaiah, ed., Academic Press, New York, 1970, pp. 389–407.

90. H. Akaike, "Power Spectrum Estimation Through Autoregressive Model Fitting," *Ann. Inst. Statist. Math.* Tokyo, 1969, pp. 407–419.

91. J. Claerbout, *Fundamentals of Geophysical Data Processing*, McGraw-Hill, New York, 1976.

92. L. C. Pusey, "The Role of the Stationarity Equation in Spectral Analysis and Wave Propagation," Ph. D. Thesis, Department of Electrical Engineering, Massachusetts Institute of Technology, Cambridge, Mass., 1976.

93. J. Capon, "High Resolution Frequency Wavenumber Spectral Analysis," *Proc. IEEE*, vol. 57, 1969, pp. 1408–1418.

94. D. J. EDELBUTE, J. M. FISK and G. L. KINNISON, "Criteria for Optimum Signal Detection Theory for Arrays," *J. Acoust. Soc. Amer.*, vol. 41, 1967, pp. 199–205.

95. J. CAPON and N. R. GOODMAN, "Probability Distributions for Estimators of Frequency Wave Number Spectra," *Proc. IEEE*, vol. 58, 1970, pp. 1785–1786.

96. A. B. BAGGEROER, "Confidence Intervals for Regression (MEM) Spectral Estimators," *IEEE Trans. Inform. Theory*, vol. IT-22, 1976, pp. 534–545.

97. D. R. BRILLINGER, "Fourier Analysis of Stationary Processes, *Proc. IEEE*, vol. 62, 1974, pp. 1628–1642.

98. V. E. PISARENKO, "On the Estimation of Spectra by Means of Nonlinear Functions on the Covariance Matrix," *Geophys. J. Roy. Astron. Soc.*, vol. 28, 1972, pp. 511–531.

99. Texas Instruments, "A Description of the Array Transform Processor," Austin, Tex.

100. SPS-42, *Users Manual*, Signal Processing Systems, Waltham, Mass., 1974.

101. V. C. ANDERSON, "Digital Array Phasing," *J. Acoust. Soc. Amer.*, vol. 32, 1950, pp. 867–870.

102. V. C. ANDERSON and P. RUDNICK, "Small Signal Detection with the DIMUS ARRAY," *J. Acoust. Soc. Amer.*, vol. 32, 1960, pp. 871–877.

103. W. R. REMLEY, "Some Effects of Clipping in Array Processing," *J. Acoust. Soc. Amer.*, vol. 39, 1966, pp. 702–707.

104. V. C. ANDERSON, "Current Major Projects of the Marine Physical Laboratory," by MPL Staff, Scripps Institute of Oceanography, San Diego, Calif., 1975.

105. V. C. ANDERSON, "DICANNE, A Realizable Adaptive Process," *J. Acoust. Soc. Amer.*, vol. 45, 1969, pp. 398–405.

106. V. C. ANDERSON and P. RUDNICK, "Rejection of a Coherent Arrival at an Array," *J. Acoust. Soc. Amer.*," vol. 45, 1969, pp. 406–410.

107. F. BRYN, "Optimum Signal Processing of Three Dimensional Arrays Operating on Gaussian Signals and Noise," *J. Acoust. Soc. Amer.*, vol. 34, 1962, pp. 289–297.

108. H. MERMOZ, "Matched Filters and Optimum Use of an Array," *Proceedings of the NATO Advanced Study Institute on Signal Processing with Emphasis on Underwater Acoustics*, Grenoble, France, Sept. 1964.

109. D. MIDDLETON and H. GROGINSKY, "Detection of Acoustic Signals by Receivers with Distributed Elements," *J. Acoust. Soc. Amer.*, vol. 38, 1965, pp. 727–737.

110. J. B. LEWIS and P. SCHULTEISS, "Optimum and Conventional Detection Using a Linear Array," *J. Acoust. Soc. Amer.*, vol. 49, 1971, pp. 1083–1091.

111. H. L. VanTrees, "A Unified Approach to Optimum Array Processing," *Proceedings of the 1966 Hawaii Symposium on Signal Processing.*

112. N. L. Owsley, "A Recent Trend in Adaptive Spatial Processing for Sensor Arrays: Constrained Adaption," in *Signal Processing*, Academic Press, New York, 1973.

113. H. Cox, "Resolving Power and Sensitivity to Mismatch of Optimum Array Processors," *J. Acoust. Soc. Amer.*, vol. 54, 1973, pp. 771–780.

114. B. Widrow, P. Mantey, L. Griffiths, and B. Goode, "Adaptive Antenna Systems," *Proc. IEEE*, vol. 55, 1967, pp. 2143–2159.

115. L. J. Griffiths, "A Simple Algorithm for Real-Time Processing in Antenna Arrays," *Proc. IEEE*, vol. 57, 1969, pp. 1696–1704.

116. O. L. Frost, III, "An Algorithm for Linearly Constrained Adaptive Array Processing," *Proc. IEEE*, vol. 60, 1972, pp. 926–935.

117. J. H. Chang and F. B. Tuteur, "A New Class of Adaptiver Array Processors," *J. Acoust. Soc. Amer.*, vol. 49, pp. 639–649.

118. R. T. Lacoss, "Adaptive Combining of Wideband Array Data for Optimal Reception," *IEEE Trans. Geoscience Electron.*, vol. GE-6, 1968, pp. 78–86.

119. S. P. Applebaum, "Adaptive Arrays," Syracuse University Research Corp., Rept. SPL TR 66–1, 1966.

120. W. F. Gabriel, "Adaptive Arrays, An Introduction," *Proc. IEEE*, vol. 64, 1976, pp. 239–271.

121. A. A. Winder, "Sonar System Technology," *IEEE Trans. Sonics Ultrasonics*, vol. SU-22, 1975, pp. 291–329.

122. I. Dyer, Class notes for subject 13.85, Fundamentals of Underwater Acoustics, Mass. Inst. of Tech.

123. H. K. Farr, "BO'SUN, A High Resolution Automatic Charting System for the Continental Shelf," *Proc. of the IEEE OCEAN '74 Conference*, vol 1, pp. 10–14 (1974).

124. C. S. Clay, "Waveguides, Arrays, and Filters, *Geophysics*, vol. 31, pp. 501–505, June 1966.

125. R. Seynave, SACLANT ASW Center, La Spezia, Italy, personal communication.

7

DIGITAL SIGNAL PROCESSING IN GEOPHYSICS

E. A. Robinson

100 Autumn Lane
Lincoln, Mass. 01773

and

S. Treitel

Amoco Production Company
Tulsa, Okla. 74102

7.1 INTRODUCTION

Exploration geophysics has a dual scientific role, the *search* for new petroleum and mineral deposits and the *research* required to develop new methods and to improve scientific understanding. Progress in either role is dependent upon the other. The exchange of ideas and the communication needed to attain these goals depend upon the use and development of general scientific models that are broad enough to cover and bridge the gap between these two roles. Perhaps the most important thing that must be recognized is the integrity of the geophysicist working on the problems, whether he be in the field, office, or research laboratory. In this respect, it is necessary that the working geophysicist have a broad scientific objective under which to operate, and at the same time that he be given full freedom and opportunity to develop his own creativity and imagination.

Let us now look at some of the fundamental notions of model building. First, a model is simply a representation of something else. Typically, it is a representation in which details that appear unessential for the intended use of the model are omitted. A geophysical model is supposed to represent the real earth in certain significant respects. However, the advantage of a model over the real earth is the relative ease with which the model may be manipulated and experimented with, as well as the conceptual understanding of the earth made possible by the model. With these purposes in view, it is clear that the objective in model building cannot be to create an exact duplicate of the real earth. In summary, models are, or should be, useful substitutes for what is modeled.

If, in fact, there exists a useful connection between the behavior of the earth and the corresponding behavior of the model, the model can be useful in analyzing data from the earth and making geophysical and geological decisions.

Models may be described verbally, graphically, physically, or by mathematical functions and equations. Models may also take the form of computer programs. Some modes of describing models might be more convenient than others. However, most models can be expressed in several different modes. A summary of models of interest in exploration geophysics has recently been given by Wood and Treitel.[1]

In geophysics, it is essential to have several expressions of the same model, but at different levels of sophistication. The approach of model building is broad and flexible, and a host of model types can be constructed. At the highest level, any given model should be connected with other models, both for geophysical and geologic purposes. At the intermediate level, the model should take into account all the important geophysical parameters, so that the geophysicist may have the flexibility and scope to experiment with and to adjust the model to the field situation. At the deepest level, the model should be built on a firm mathematical and physical foundation in order to support the ultimate consequences resulting from it.

In essence, model building is a systematic coordination of theoretical and empirical elements of knowledge into a joint construct. It is important to discern between, on the one hand, the hypothetical assumptions that constitute the theoretical part of the model and, on the other hand, the empirical observations that the model serves to interpret. Empirical observations enter in different ways in the model construction: at an early stage when observations and experience are accumulated, whether a tentative model has or has not been formed; at more advanced stages, in assessing the parameters of the model on the basis of observations and in testing the theoretical model against the empirical evidence.

The search aspect of geophysics represents empirical research. More specifically, a seismic crew exploring for oil may be regarded as a scientific organization carrying out empirical research about the structure of the sedimentary layers. On the other hand, the research aspect of geophysics represents research of a more theoretical nature. Empirical and theoretical research (or search and research) are to some extent independent vehicles for geophysical progress. In some respects, theoretical research is ahead of what has been verified empirically; in other respects, the converse is true. To put it otherwise, it is not true that the theoretical approach has general priority over the empirical approach. However, matters differ when it comes to the reporting of fresh research results or to expository treatments in articles and textbooks, for there it is often appropriate to present the theoretical results first and only then follow with the empirical aspects. This sequence does not necessarily represent the order of discovery, especially in geophysics, for which the empirical development of the exploration seismic method in the 1920s preceded much of the corresponding theoretical work.

7.2 PREDICTIVE DECONVOLUTION

The method of predictive deconvolution[2,3] constitutes an early success in the application of communication theory to geophysics. It is based on a statistical, minimum-delay model of the observed seismogram; the approach has, moreover, found widespread use in other disciplines, such as speech processing. For these reasons we shall begin this chapter with a detailed description of the appropriate theoretical framework.

There are two basic approaches to seismic data processing: the deterministic approach and the statistical approach. The deterministic approach is concerned with the building of mathematical and physical models of the layered earth to better understand seismic wave propagation. These models involve no random elements; they are completely deterministic. The statistical approach is concerned with the building of seismic models involving random components. For example, in the statistical model that we shall discuss, the depths and reflectivities of the deep reflecting horizons are considered to have a random distribution. A major justification for using the statistical approach in seismology is due to the fact that large amounts of data must be processed; any data in large enough quantities take on a statistical character, even if each individual piece of data is of a deterministic nature.

The model required for the application of predictive deconvolution is a statistical one. It depends upon two basic hypotheses: (1) the statistical hypothesis that the strengths and arrival times of the information-bearing events on the seismic trace can be represented as a random spike series, and (2) the deterministic hypothesis that the basic waveform associated with each of these events is minimum delay. There are various ways of checking a model to see if it conforms with the physical situation.

The method of predictive deconvolution has been successful in many instances in deconvolving reverberations from field records, both recorded on land and offshore. The general success of the method suggests that the random-event hypothesis and the minimum-delay hypothesis are valid under a wide range of field situations.

Let us designate the received seismic trace by x_t (where it is assumed that the filtering effect of the recording instruments has already been removed from x_t by some restoring filter operation). The received seismic trace x_t is the resultant of many deep reflections, each of which contributes a pulse-train waveform of shape b_t. However, because all the pulse-train waveforms are overlapping with each other to various degrees, it is not possible to obtain a direct measurement of the individual waveform shape b_t. As we shall see, the fact that the waveform b_t is minimum delay allows us to obtain an indirect measurement of its shape.

The predictive deconvolution method is based on the following statistical model. The received seismic trace x_t (included within an appropriately chosen time gate) is considered to be the result of convolving the waveform b_t with a random spike series ϵ_t; that is, $x_t = b_t * \epsilon_t$. The spike series ϵ_t represents the reflections from the deep reflecting horizons in the sense that the timing of a spike represents the direct arrival time of a reflection, and the amplitude of the spike represents the strength of the reflection. For data-processing purposes, this spike series ϵ_t is considered as a random uncorrelated (i.e., white noise) series. Hence the autocorrelation of the received seismic trace x_t is the same as the autocorrelation of the individual waveform b_t, except for a constant scale factor. This scale factor will not affect the final results, so it may be neglected. We can, therefore, compute the autocorrelation of the waveform b_t from the received seismic trace x_t.

The autocorrelation coefficients ϕ_j are computed from the sampled trace by means of the formula

$$\phi_j = \sum x_{i+j}x_i$$

where the summation runs over all the time indexes i within the time gate. Because $\phi_{-j} = \phi_j$, the autocorrelation coefficients need only be computed for nonnegative values of the time-shift index j.

Often it is useful to weight the autocorrelation by some set of tapered weighting factors w_j in order to obtain the weighted autocorrelation

$$r_j = w_j\phi_j$$

A typical set of weighting factors would be the triangular weighting coefficient set given by

$$w_j = \begin{cases} 1 - \dfrac{|j|}{N} & \text{for } |j| = 0, 1, 2, 3, \ldots, N \\ 0 & \text{for } |j| = N+1, N+2, \ldots \end{cases}$$

Here N represents the time index at which the autocorrelation is truncated; the value of the parameter N must be specified.

From the autocorrelation, we can compute the *prediction operator* for the waveform b_t. Because the waveform b_t is minimum delay, a prediction operator for prediction distance α will predict the waveform α time units ahead; that is, the operator predicts the tail of the waveform from time α onward. (This would not be true if the waveform were not minimum delay. For an illustration of this principle, see ref. 4, pp. 90–91.) A delay of α time units will line up this predicted tail with the tail portion of the waveform b_t; by subtracting this delayed predicted tail from the waveform b_t, we obtain the prediction error. Because the delayed predicted tail cancels the tail part of the waveform b_t, it follows that the prediction error represents the initial part of the waveform b_t. Hence the prediction-error operator eliminates the tail part of the waveform b_t. Let us designate the coefficients of the prediction operator by

$$\begin{array}{rl} & (k_0, k_1, k_2, \ldots, k_m) \\ \text{time:} & \;\;0,\;\; 1,\;\; 2, \ldots,\;\; m \end{array}$$

There are various methods of computing this operator; for numerical work, the Gauss method of least squares has certain advantages. According to the method of least squares, the prediction operator is determined by minimizing the mean-square prediction error. The minimization leads to the set of simultaneous linear equations called the normal equations, which involve the autocorrelation coefficients that we have just computed as the knowns and the prediction operator coefficients as the unknowns. The normal equations are

$$\begin{aligned} k_0 r_0 + k_1 r_1 + k_2 r_2 + \cdots + k_m r_m &= r_\alpha \\ k_0 r_1 + k_1 r_0 + k_2 r_1 + \cdots + k_m r_{m-1} &= r_{\alpha+1} \\ &\cdots \\ k_0 r_m + k_1 r_{m-1} + k_2 r_{m-2} + \cdots + k_m r_0 &= r_{\alpha+m} \end{aligned}$$

In this set of equations, the positive integer α denotes the prediction distance; we must specify some value for α.

These equations may be solved by an efficient recursive procedure due to Levinson.[5] Using this procedure, the machine time required to solve the normal equations for a digital filter with m coefficients is proportional to m^2, as compared to m^3 for the conventional methods of solving simultaneous equations. Another advantage of using this recursive method is that it requires computer storage space proportional to m, rather than m^2 as in the case of the conventional methods. A simplified treatment of the recursive method suitable for writing a computer subroutine may be found in ref. 6, and is also discussed later in this chapter.

Once we have computed the prediction operator coefficients $(k_0, k_1, k_2, \ldots, k_m)$, we then know in effect the coefficients of the prediction-error operator, for the prediction-error operator coefficients for prediction span α are given by

$$\begin{array}{lcccc} & (1, 0, 0, \ldots, 0, & -k_0, & -k_1, & -k_2, \ \ldots, \ -k_m) \\ \text{time:} & 0, 1, 2, \ldots, \alpha - 1, & \alpha, & \alpha + 1, & \alpha + 2, \ldots, \alpha + m \end{array}$$

It is the prediction-error operator that is the required inverse operator for deconvolving the seismic trace x_i. Since the prediction-error operator is linear, we can apply it to the received seismic trace $x_t = b_t * \epsilon_t$ (which represents many overlapping waveforms b_t with arrival times and strengths given according to the spike series ϵ_t). In so doing, we eliminate the tails from each of the waveforms b_t but leave intact their initial portions, thereby increasing seismic resolution. If more resolution is desired, the prediction distance can be lessened, which will have the effect of further compressing the energy in the waveform.

Accordingly, we convolve* the prediction-error operator with the seismic trace x_i; this computation is carried out according to the discrete convolution formula:

$$y_i = x_i - k_0 x_{i-\alpha} - k_1 x_{i-\alpha-1} - k_2 x_{i-\alpha-2} - \cdots - k_m x_{i-\alpha-m}$$

The result y_i is the prediction-error series (for prediction span α), which represents the deconvolved seismic trace.

In the special case when the prediction distance α is chosen to be unity, the prediction-error operator is the least-squares inverse of the (minimum-delay) waveform b_t, and the prediction-error series represents the random spike series ϵ_t (i.e., the series designating the strengths and arrival times of the deep reflections).[3]

All the above holds within the limitations of statistical errors imposed by noise, computational approximation, and the finiteness of the data, and within the limitations of specification errors imposed by the model. The success of the method of predictive deconvolution, as we have discussed, depends largely upon the validity of the basic hypotheses as to the minimum-delay nature of the waveform b_t and the random nature of the spike series ϵ_t. The beauty of the predictive method is that the only data required in order to perform the deconvolution are the received seismic trace.

To smooth the output prediction error series, we can apply a postfiltering operation to it. This postfilter can be some type of digital bandpass filter or a digital waveform shaping filter (see Sec. 7.5). Instead of applying the postfilter to the prediction error series, we can instead first cascade the prediction-error filter with the postfilter, and apply the cascaded filter to the received seismic trace. The final output will be the same in either case: the smoothed seismic trace without reverberations.

* The *deconvolution* of the trace is accomplished by *convolving* the trace with the *inverse* operator, i.e., with the prediction-error operator.

7.3 DYNAMIC DECONVOLUTION

The method of predictive deconvolution makes only minimal assumptions about the detailed structure of the layered earth. In particular, it becomes desirable to establish more quantitative relationships between the reflection and transmission coefficients that describe a layered medium and the characteristics of the ideal seismogram that this stratification produces. An approach to the deconvolution problem based on such a more explicit treatment of the layered structure is called *dynamic deconvolution.*

Dynamic deconvolution makes use of an entire seismic trace, including all primary and multiple reflections, to yield an approximation to the subsurface structure. We consider plane-wave motion at normal incidence in a horizontally layered system sandwiched between the air and the basement rock. Energy degradation effects are neglected so that the layered system represents a lossless system in which energy is lost only by net transmission downward into the basement or net reflection upward into the air; there is no internal loss of energy by absorption within the layers. The layered system is frequency selective in that the energy from a surface input is divided between that energy which is accepted over time by net transmission downward into the basement, and the remaining energy that is rejected over time by net reflection upward into the air. Thus the energy from a downgoing unit spike at the surface as input is divided between the wave transmitted by the layered system into the basement and the wave reflected by the layered system into the air. This reflected wave is the observed seismic trace resulting from the unit spike input. From surface measurements we can compute both the input energy spectrum, which by assumption is unity, and the reflection energy spectrum, which is the energy spectrum of the trace. But, by the conservation of energy, the input energy spectrum is equal to the sum of the reflection energy spectrum and the transmission energy spectrum. Thus we can compute the transmission energy spectrum as the difference of the input energy spectrum and the reflection energy spectrum.

Furthermore, we know that the layered system acts as a pure feedback system in producing the transmitted wave, from which it follows that the transmitted wave is minimum delay. Hence, from the computed energy spectrum of the transmitted wave, we can compute the prediction-error operator that contracts the transmitted wave to a spike. We also know that the layered system acts as a system with both a feedback component and a feed-forward component in producing the reflected wave, that is, the observed seismic trace. Moreover, this feedback component is identical to the pure feedback system that produces the transmitted wave. Thus we can deconvolve the observed seismic trace by the prediction-error operator computed above; the result of the deconvolution is the waveform due to the feed-forward component alone. The feed-forward component represents the wanted dynamic structure of the layered system, whereas the feedback component represents the unwanted reverberatory effects of the layered system. Because this deconvolution process yields the wanted dynamic structure and destroys the unwanted reverberatory effects, we call the process dynamic deconvolution. The resulting feed-forward waveform in itself represents an approximation to the subsurface structure; a further decomposition yields the reflection coefficients of the interfaces separating the layers.

Let us now discuss in some detail a typical and well-known model used in geophysics, the familiar model of a flat-layered earth. A horizontal line represents the surface, and below the surface there are media whose interfaces are parallel to the surface. The media have various thicknesses and compressional velocities, as depicted in Fig. 7-1. Although most workers number the layers from the top downward, we shall find it more useful to do the opposite, number them from the bottom upward. This is done for notational simplicity in the derivations that follow.

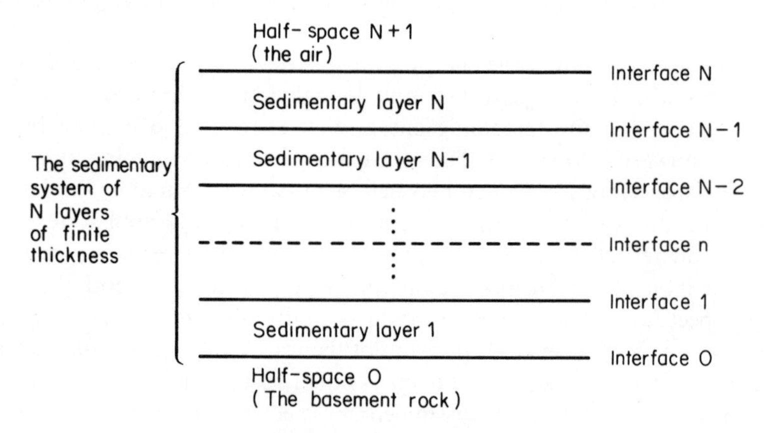

FIGURE 7-1. The layered system.

The lowermost layer or basement is a half-space denoted by index 0. On top of the basement lie N proper layers of finite thickness, which represent the sedimentary column, and which we denote by indexes running from 1 at the bottom to N at the top. In other words, layer 1 is the first sedimentary layer in geologic time and is the deepest layer, whereas layer N is the last layer in geologic time and represents the surface layer. Of course, the surface layer would be water in the case of marine exploration. The topmost layer (or air) is a half-space and is denoted by the index $N + 1$. Thus the stratified system consists of N sedimentary layers of finite thickness sandwiched in between the basement rock and the air. The terms *layered system* or *sedimentary system* refer to the N layers of finite thickness and include neither the air nor the basement rock.

There are $N + 1$ horizontal interfaces. The lowest interface is denoted by index 0 and represents the top of layer 0, that is, the top of the basement. The highest interface is denoted by index N and represents the top of layer N, that is, the surface of the ground or the water, as the case may be. Generally, we may say that interface n is the top of layer n, where the integer n runs from 0 through N.

We restrict ourselves to plane compressional wave motion at normal incidence to the horizontal interfaces. To satisfy appropriate boundary conditions, two plane compressional waves can exist within each layer, one a compressional wave traveling vertically upward and the other a compressional wave traveling vertically downward. For definiteness, let us measure the wave motion itself in terms of particle velocity.

If a downgoing unit spike is incident on the top of interface n, then the reflection coefficient r_n is equal to the resulting upgoing spike reflected from the top of interface n, and the transmission coefficient t_n is equal to the resulting downgoing spike transmitted through interface n. If an upgoing unit spike is incident on the bottom of interface n, then the reflection coefficient r'_n is equal to the resulting downgoing spike reflected from the bottom of interface n, and the transmission coefficient t'_n is equal to the resulting upgoing spike transmitted through interface n (see Fig. 7-2).

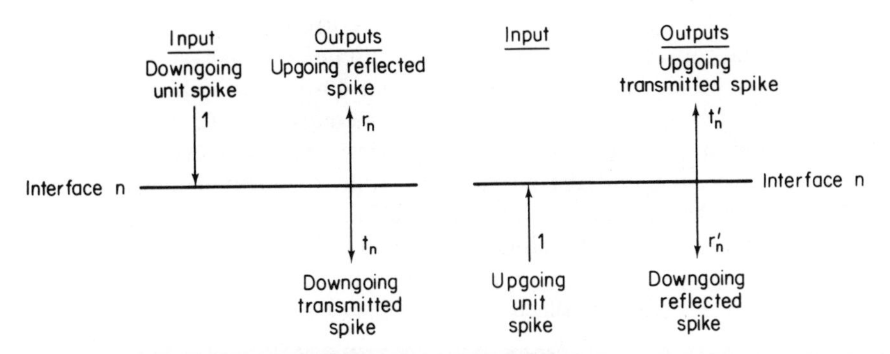

FIGURE 7-2. Schematic diagram illustrating the reflection and transmission coefficients for an interface.

The last three coefficients can be expressed in terms of the first as follows:

$$t_n = 1 + r_n$$

$$r'_n = -r_n$$

$$t'_n = 1 - r_n$$

These coefficients are real numbers. The reflection coefficients are always in the range $(-1, 1)$, and the transmission coefficients are always in the range $(0, 2)$. The two-way transmission factor of interface n is equal to $t_n t'_n = 1 - r_n^2$.

We assume that the layered system is lossless; that is, there are no energy degradation effects such as absorption within the layers. Thus the only way that the sedimentary system can lose energy is by having a downgoing wave travel into the basement never to return, and/or by having an upgoing wave travel into the air never to return.

For mathematical simplicity, it is convenient to add hypothetical interfaces where necessary, so as to make the two-way travel time in each layer equal to the same quantity, which we shall define as one time unit. Of course, the reflection coefficients are zero and the transmission coefficients are unity for any such added hypothetical interfaces.

We let z represent the unit delay operator. Any wave train of spikes $a_0, a_1, a_2, \ldots$, where a_s denotes the amplitude of the spike at discrete integer time s, can be represented by its generating function,

$$A(z) = a_0 + a_1 z + a_2 z^2 + \cdots$$

Often we shall refer to such a wave train simply as the wave A, where the capital letter A actually denotes the generating function $A(z)$. The quantity $\bar{A}(z)$ is often denoted simply by $\bar{A}$, and represents the z transform of the wave train $a_0, a_1, a_2, \ldots$,

$$\bar{A}(z) = a_0 + a_1 z^{-1} + a_2 z^{-2} + \cdots$$

In other words, a generating function and its corresponding z transform can be obtained from each other by the substitution of z by z^{-1}.

Because we always let the input be a unit downgoing spike at time zero incident at the topmost interface of a sedimentary system, we can simply refer to the output wave train reflected up into the upper half-space as the *reflection response*, and the output wave train transmitted down into the basement as the *transmission response* (see Fig. 7-3).

Let us now consider a sedimentary system of $n - 1$ layers with reflection coefficients $r_0, r_1, \ldots, r_{n-1}$. Let us also consider another sedimentary system of n layers with the same reflection coefficients $r_0, r_1, \ldots, r_{n-1}$, plus the additional reflection coefficient r_n. For these reflection coefficients to be the same, layer n of the second system must be of the same material as the half-space n of the first system, and all the layers below must have the same impedances in the two systems (see Fig. 7-4).

We now wish to combine the reflection response R_{n-1} of the $n - 1$ layer system with the reflection coefficient r_n in such a way as to give the reflection response R_n of the n-layer system. If we refer to Fig. 7-5, we see that the reflection response R_n is made up of an infinite series of components, as follows:

1. The spike r_n resulting from the reflection upward of the source spike from the nth interface.
2. The spike train $t_n R_{n-1} t'_n$ resulting from the transmission downward of

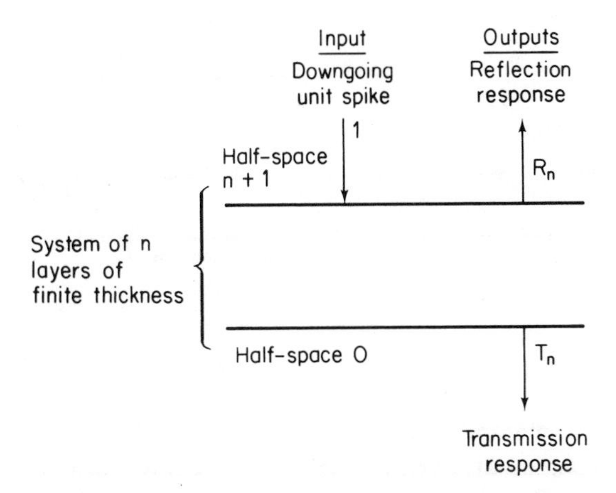

FIGURE 7-3. Schematic diagram illustrating the reflection and transmission responses for a system of n layers of finite thickness.

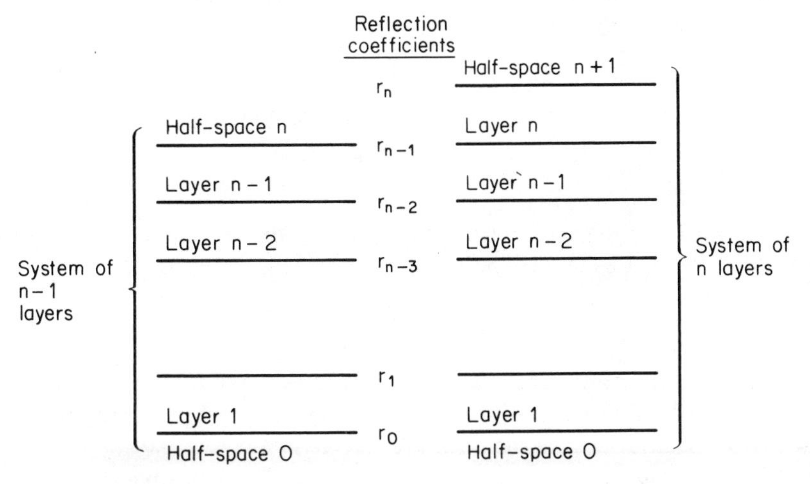

FIGURE 7-4. Two layered systems with the same reflection coefficients $r_0, r_1, \ldots r_{n-1}$. The system of n layers has an additional reflection coefficient r_n.

the source spike through the nth interface, reflection upward from the $n - 1$ layer system, and transmission upward through the nth interface.

3. The spike train $t_n R_{n-1} r'_n R_{n-1} t'_n$ resulting from the transmission downward of the source spike through the nth interface, reflection upward from the $n - 1$ layer system, reflection downward from the nth interface, reflection upward from the $n - 1$ layer system, and transmission upward through the nth interface, and so on.

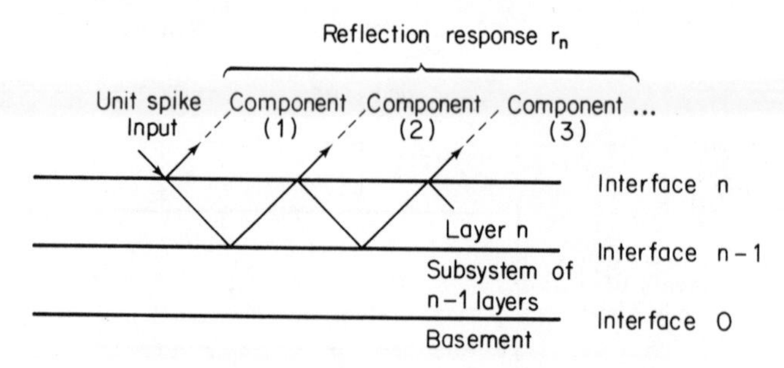

FIGURE 7-5. Schematic make-up of the reflection response R_n.

Spike 1 occurs at the time of the source spike, spike train 2 occurs at a delay of one time unit (i.e., at a delay of the two-way travel time through the nth layer), spike train 3 occurs at a delay of two time units, and so on. Summing all these contributions, we have

$$R_n = r_n + t_n R_{n-1} t'_n z + t_n R_{n-1} r'_n R_{n-1} t'_n z^2 + \cdots$$

This expression may be factored as

$$R_n = r_n + t_n R_{n-1} t'_n z[1 + r'_n R_{n-1} z + (r'_n R_{n-1} z)^2 + \cdots]$$

which, if we sum the geometric series in brackets, becomes

$$R_n = r_n + \frac{t_n R_{n-1} t'_n z}{1 - r'_n R_{n-1} z}$$

Using the relationships given earlier between the reflection and transmission coefficients, this expression becomes

$$R_n = \frac{r_n + R_{n-1} z}{1 + r_n R_{n-1} z}$$

This equation for combining r_n and R_{n-1} to form R_n is of the same mathematical form as the Einstein addition formula in the theory of relativity to combine two velocities to give the resulting velocity.[7]

In a similar manner, the transmission response T_n can be obtained in terms of the reflection coefficient r_n and the transmission coefficient t_n of the nth interface, and the reflection response R_{n-1} and the transmission response T_{n-1} of the $n - 1$ layer system. If we refer to Fig. 7-6, we see that the transmission response T_n of the n-layer system is made up of an infinite series of components, as follows:

1. The spike train $t_n T_{n-1}$.
2. The spike train $t_n R_{n-1} r'_n T_{n-1}$.
3. The spike train $t_n R_{n-1} r'_n R_{n-1} r'_n T_{n-1}$, and so on.

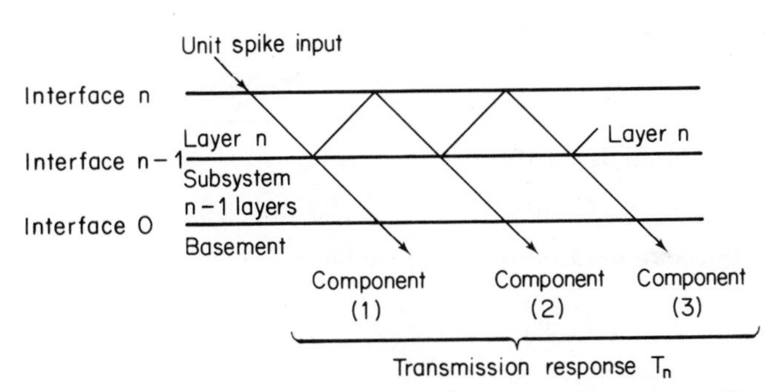

FIGURE 7-6. Schematic make-up of the transmission response T_n.

We choose the time origin of the transmission response as the time of its first break (i.e., at its first nonzero amplitude). Its first break occurs at a delay of $N/2$ (i.e., the one-way travel time) from the time of the surface input spike. With reference to the time origin of the transmission response, spike train 1 occurs with no delay, spike train 2 occurs with a delay of one time unit, spike train 3 occurs with a delay of two time units, and so on. Summing all these contributions, we have

$$\begin{aligned} T_n &= t_n T_{n-1} + t_n R_{n-1} r'_n T_{n-1} z + t_n R_{n-1} r'_n R_{n-1} r'_n z^2 + \cdots \\ &= t_n T_{n-1}[1 + R_{n-1} r'_n z + (R_{n-1} r'_n z)^2 + \cdots] \end{aligned}$$

Summing the geometric series, we have

$$T_n = \frac{t_n T_{n-1}}{1 + r_n R_{n-1} z}$$

We note that the final expressions for R_n and T_n each have the same denominator.

We now wish to define a sequence of polynomials $C_0, C_1, \ldots, C_N$, which we call the feed-forward polynomials, and a sequence of polynomials $D_0, D_1, \ldots, D_N$, which we call the feedback polynomials. In the case of no finite thickness layers, the reflection and transmission responses are

$$R_0 = r_0, \qquad T_0 = t_0$$

These responses may be expressed in terms of polynomials C_0 and D_0 each of zero degree as

$$R_0 = \frac{C_0}{D_0}, \qquad T_0 = \frac{t_0}{D_0}$$

where the polynomials satisfy

$$C_0 = r_0, \qquad D_0 = 1$$

Let us suppose that in the case of $n - 1$ finite thickness layers the reflection and transmission responses may be expressed in terms of polynomials

C_{n-1} and D_{n-1}, each of degree $n - 1$, as

$$R_{n-1} = \frac{C_{n-1}}{D_{n-1}}, \qquad T_{n-1} = \frac{t_{n-1} \ldots t_0}{D_{n-1}}$$

where the polynomials satisfy

$$C_{n-1}(0) = r_{n-1}, \qquad D_{n-1}(0) = 1$$

Using the Einstein addition formula, we have

$$R_n = \frac{r_n + (C_{n-1}/D_{n-1})z}{1 + r_n(C_{n-1}/D_{n-1})z}$$

which is

$$R_n = \frac{r_n D_{n-1} + C_{n-1}z}{D_{n-1} + r_n C_{n-1}z}$$

Let us define the feed-forward polynomial C_n of degree n and the feedback polynomial D_n of degree n by the *recursion formulas*

$$C_n = r_n D_{n-1} + C_{n-1}z, \qquad D_n = D_{n-1} + r_n C_{n-1}z$$

From these recursion formulas, we see that the polynomials satisfy

$$C_n(0) = r_n, \qquad D_n(0) = 1$$

The reflection response R_n is given by the ratio

$$R_n = \frac{C_n}{D_n}$$

Similarly, the transmission response is

$$T_n = \frac{t_n T_{n-1}}{1 + (r_n R_{n-1}z)} = \frac{(t_n t_{n-1} \cdots t_0)/D_{n-1}}{1 + [r_n(C_{n-1}/D_{n-1})z]}$$

which is

$$T_n = \frac{t_n t_{n-1} \ldots t_0}{D_{n-1} + (r_n C_{n-1})z} = \frac{t_n t_{n-1} \ldots t_0}{D_n}$$

Because T_n is the generating function of a stable one-sided time function, it follows that the polynomial D_n is minimum delay. Because the inverse of a minimum-delay function is also minimum-delay, it follows that T_n itself is minimum delay.

Let us call the function $D_n\bar{D}_n$ the spectral function of the feedback polynomial D_n, and the function $C_n\bar{C}_n$ the spectral function of the feed-forward polynomial C_n. Let us now find an expression for the difference of these spectral functions. If we use the recursion formulas given previously, we obtain

$$D_n\bar{D}_n - C_n\bar{C}_n = (1 - r_n^2)(D_{n-1}\bar{D}_{n-1} - C_{n-1}\bar{C}_{n-1})$$

We can now make repeated use of this result to obtain

$$D_n\bar{D}_n - C_n\bar{C}_n = (1 - r_n^2)(1 - r_{n-1}^2) \ldots (1 - r_0^2)$$

The expression on the right is recognized as the product of the two-way transmission factors of the n layers of finite thickness. Let us designate this product by σ_n^2. Thus the difference between the feedback and feed-forward spectral functions is equal to the two-way transmission factor; that is,

$$D_n\bar{D}_n - C_n\bar{C}_n = \sigma_n^2$$

Let $n = N$ in the above expression, and rewrite it as

$$1 - \frac{C_N\bar{C}_N}{D_N\bar{D}_N} = \frac{\sigma_N^2}{D_N\bar{D}_N}$$

which is

$$1 - R_N\bar{R}_N = \frac{\sigma_N^2 T_N\bar{T}_N}{(t_N \ldots t_0)^2}$$

Since

$$\sigma_N^2 = (t_n \ldots t_0)(t'_n \ldots t'_0)$$

we have

$$1 - R_N\bar{R}_N = \frac{t'_N \ldots t'_0}{t_N \ldots t_0} T_N\bar{T}_N \tag{7.1}$$

We now want to show that this equation represents the law of the conservation of energy; that is, the input energy minus the output reflected energy is equal to the output transmitted energy. The instantaneous energy flow of a traveling wave in a layer is proportional to the product of the characteristic impedance of the layer and the square of the amplitude of the wave, where for the present purposes we assume that wave amplitude refers to particle velocity amplitude. Let Z_0 denote the characteristic impedance of the basement, and let Z_{N+1} denote the characteristic impedance of the air. Let the system be at rest initially, and let the input to the layered system be a unit downgoing spike incident on the surface. The output of the system is the reflection response R_N reflected upward into the air and the transmission response T_N transmitted downward into the basement. Because we assume that the sedimentary system is lossless (i.e., that there are not energy degradation effects within the layers), all the input energy must be accounted for by the two outputs R_N and T_N. The energy of the input spike is proportional to Z_{N+1} times unity squared. The energy of the output reflected wave is proportional to $Z_{N+1}R_N\bar{R}_N$, and the energy of the output transmitted wave is proportional to $Z_0T_N\bar{T}_N$. The law of the conservation of energy states that the energy input to the sedimentary system must equal the energy output from the sedimentary system; that is,

$$Z_{N+1} = Z_{N+1}R_N\bar{R}_N + Z_0T_N\bar{T}_N$$

The reflection response R_N is the observed seismic trace (which includes all primary and multiple reflections), and hence is at our disposal. If we bring this known quantity to the left side of the above equation, we obtain

$$1 - R_N\bar{R}_N = \frac{Z_0}{Z_{N+1}} T_N\bar{T}_N \tag{7.2}$$

Let us now find an expression for Z_0/Z_{N+1}. It is well known that the transmission coefficients t_n and t'_n are given in terms of the characteristic impedances Z_{N+1} and Z_n for the layers $n + 1$ and n (where layer $n + 1$ is on top of layer n) by

$$t_n = \frac{2Z_{n+1}}{Z_{n+1} + Z_n}, \qquad t'_n = \frac{2Z_n}{Z_{n+1} + Z_n}$$

It follows that

$$\frac{t'_n}{t_n} = \frac{Z_n}{Z_{n+1}}$$

and, therefore, we have

$$\frac{t'_N \ldots t'_1 t'_0}{t_N \ldots t_1 t_0} = \frac{Z_N}{Z_{N+1}} \cdots \frac{Z_1}{Z_2}\frac{Z_0}{Z_1} = \frac{Z_0}{Z_{N+1}}$$

That is, the ratio Z_0/Z_{N+1} is given by the ratio of the transmission factor $t'_0 t'_1 \ldots t'_N$ upward through the sedimentary system to the transmission factor $t_N \ldots t_1 t_0$ downward through the sedimentary system. Thus Eqs. (7.1) and (7.2) are the same; each represents the law of the conservation of energy.

Let us designate the known left side of Eq. (7.2) by the symbol Φ, which we call the spectral function. That is, the spectral function is defined as

$$\Phi = 1 - R_N \bar{R}_N \tag{7.3}$$

and by the law of conservation of energy, the spectral function is equal to

$$\Phi = \frac{Z_0}{Z_{N+1}} T_N \bar{T}_N$$

Next let us find an expression for T_N. We know that the sedimentary system acts as a pure finite feedback system in producing the transmitted wave, so T_N is proportional to the reciprocal of a polynomial D_N of degree N. If we choose this polynomial so that its leading coefficient is unity, then the proportionality factor is equal to the downward transmission factor $t_N \ldots t_1 t_0$. Thus the transmitted wave has the form

$$T_N = \frac{t_N \ldots t_1 t_0}{D_N}$$

where we have chosen the time origin of the transmitted wave to be at its first break, and where this first break occurs at a delay of $N/2$ time units from the time of the input spike (i.e., at a delay of the one-way travel time through the sedimentary layers). Since T_N is the generating function of a stable one-sided time function, it follows that the denominator polynomial D_N must be minimum delay. It then follows that T_N itself must be minimum delay.

Using the results given above, we see that the spectral function is equal to

$$\Phi = \frac{t'_N \ldots t'_1 t'_0}{t_N \ldots t_1 t_0} \frac{(t_N \ldots t_1 t_0)^2}{D_N \bar{D}_N}$$

If we define σ_N^2 as the two-way transmission factor of the sedimentary system, that is, if we define σ_N^2 as

$$\sigma_N^2 = t'_N t_N \ldots t'_1 t_1 t'_0 t_0 = (1 - r_N^2) \ldots (1 - r_1^2)(1 - r_0^2)$$

then we see that the spectral function is equal to

$$\Phi = \frac{\sigma_N^2}{D_N \bar{D}_N} \tag{7.4}$$

Thus the feedback polynomial D_N and the constant σ_N^2 can be found from the known spectral function Φ by one of the methods of minimum-delay spectral factorization, such as the Fejér–Wold method, the Kolmogorov method, or the normal-equations method (see ref. 2 for a more detailed description of these approaches).

Because

$$D_N T_N = t_N \ldots t_1 t_0 = \text{constant}$$

the polynomial D_N is the generating function of the prediction-error operator that reduces the minimum-delay transmitted wave to a spike.

The sedimentary layers act as a system with both a feed-forward component and a feedback component in producing the output reflected wave R_N (i.e., the observed seismic trace). Moreover, the feedback component is identical to the pure feedback system that produces the output transmitted wave. Thus the generating function of the seismic trace is given by

$$R_N = \frac{C_N}{D_N}$$

where the polynomial C_N of degree N represents the feed-forward component, and the polynomial D_N of degree N represents the feedback component, and is identical to the polynomial D_N appearing in the expression for T_N. We can now deconvolve the observed seismic trace with the prediction-error operator computed above. In terms of generating functions, this deconvolution is given by the multiplication

$$R_N D_N = C_N$$

The result of this deconvolution is the feed-forward component C_N. Now the feed-forward component represents the desired dynamic structure of the sedimentary system (i.e., the reflection coefficients), while the feedback component represents the unwanted, or reverberatory, effects of the sedimentary system. To see that the feed-forward component represents the dynamic structure, let us write down the explicit expression for C_N in terms of the reflection coefficients in the case when $N = 3$. Use of the recursion formulas derived previously yields

$$C_3(z) = r_3 + (r_2 + r_3 r_2 r_1 + r_3 r_1 r_0)z + (r_1 + r_3 r_2 r_0 + r_2 r_1 r_0)z^2 + r_0 z^3$$

Now the magnitude of any reflection can never exceed unity, and in practice the magnitudes of the reflection coefficients will cluster around zero instead of unity. Hence, generally, the product of three or more reflection coefficients will be of a lower order of magnitude than any single reflection coefficient. To this approximation, therefore, the above feed-forward polynomial may be written

$$C_3(z) \approx r_3 + r_2 z + r_1 z^2 + r_0 z^3$$

and, in general, for N sedimentary layers

$$C_N(z) \approx r_N + r_{N-1} z + \cdots + r_1 z^{N-1} + r_0 z^N$$

Since the deconvolution process yields the coefficients of C_N, the process approximately determines the reflection coefficients, which in turn represent the desired dynamic structure of the sedimentary system.

In retrospect, we can write the seismic trace as

$$R_N = C_N \frac{1}{D_N}$$

which in the time domain is

Seismic trace = (dynamic reflection series) * (minimum-delay reverberation)

where the asterisk denotes convolution. The deconvolution process is

(Seismic trace) * (prediction-error operator) = dynamic reflection series

That is, the deconvolution process yields the wanted dynamic structure and destroys the unwanted reverberatory effect. We accordingly call this process the method of *dynamic predictive deconvolution.*

The computational procedure for dynamic predictive deconvolution in terms of the given geophysical model is as follows: let the observed seismic trace (i.e., the reflection response of the sedimentary system to a unit source spike) be the time series

$$x_0, x_1, x_2, x_3, \ldots$$

(*Note:* The generating function of this time series is R_N. Ordinarily, we could use lower case r to denote the coefficients of R_N, but since we have already used r to denote the reflection coefficients, we have used x to denote the coefficients of R_N.)

The first step is to compute the autocorrelation ψ_s of the seismic trace by the formula

$$\psi_s = \sum_{i=0}^{\infty} x_{i+|s|} x_i$$

then compute the source autocorrelation, which under the assumption of a unit spike source is itself a unit spike, and then compute their difference [see Eq. (7.3)], which is an autocorrelation function ϕ given by

$$\phi_0 = 1 - \psi_0$$

$$\phi_s = -\psi_s \qquad \text{for } s \neq 0$$

(*Note:* The generating function of the autocorrelation ϕ_s is seen to be the spectral function Φ.)

The second step is to compute the prediction-error operator (i.e., the deconvolution operator) $d_0 = 1, d_1, d_2, \ldots, d_N$ by solving the system of normal equations,

$$\begin{bmatrix} \phi_0 & \phi_1 & \cdots & \phi_N \\ \phi_1 & \phi_0 & \cdots & \phi_{N-1} \\ & & \cdots & \\ \phi_N & \phi_{N-1} & & \phi_0 \end{bmatrix} \begin{bmatrix} 1 \\ d_1 \\ \cdot \\ \cdot \\ \cdot \\ d_N \end{bmatrix} = \begin{bmatrix} \sigma_N^2 \\ 0 \\ \cdot \\ \cdot \\ \cdot \\ 0 \end{bmatrix}$$

where $d_0 = 1, d_2, \ldots, d_N$ are the coefficients of the feedback polynomial D_N. [*Note:* These normal equations can be derived by writing Eq. (7.4) in the form $D_N\Phi = \sigma_N^2/\bar{D}_N$.]

Since D_N is minimum delay, we have

$$\begin{aligned}(1 + d_1 z + \cdots + d_N z^N)\Phi(z) &= \sigma_N^2(1 + d_1 z^{-1} + \cdots + d_N z^{-N})^{-1} \\ &= \sigma_N^2(1 + \text{terms in negative powers of } z)\end{aligned}$$

and the right side vanishes for positive powers of z. If we equate coefficients in zeroth and positive powers of z on both sides of this relation, we obtain the above normal equations, where in particular the two-way transmission factor σ_N^2 is given by

$$\sigma_N^2 = d_0\Phi_0 + d_1\Phi_1 + \cdots + d_N\Phi_n$$

The third step is to compute the deconvolved seismic trace c_i by the formula

$$c_i = \sum_{s=0}^{N} d_s x_{i-s} \qquad \text{for } i = 0, 1, 2, \ldots, N$$

The deconvolved trace is the set of coefficients of the feed-forward polynomial, and to a first approximation represents the set of reflection coefficients, that is,

$$(c_0, c_1, \ldots, c_N) \approx (r_N, r_{N-1}, \ldots, r_0)$$

For the first and last coefficients, an exact equality sign holds; that is, $c_0 = r_N$ and $c_N = r_0$.

If one does not wish to make the above approximation (i.e., that the reflection coefficients are approximately equal to the coefficients of the feed-forward polynomial), then it is possible to actually decompose the coefficients of the feed-forward and feedback polynomials to yield the reflection coefficients exactly, as we shall now show.

At this point in the computational procedure, we have calculated the coefficients $(c_0, c_1, \ldots, c_N)$ of the feed-forward polynominal C_N and coefficients $(d_0, d_1, \ldots, d_N)$ of the feedback polynomial D_N. To decompose these polynomial coefficients into the reflection coefficients, we must invert the recursion formulas given previously. If we simultaneously solve the recursion formulas for C_{n-1} and D_{n-1} in terms of C_n and D_n, we obtain the *inverse recursion formulas*

$$C_{n-1} = (1 - r_n^2)^{-1}(C_n - r_n D_n)z^{-1}$$
$$D_{n-1} = (1 - r_n^2)^{-1}(D_n - r_n C_n)$$

Also, we know that

$$D_n(0) = 1, \qquad C_n(0) = r_n$$

After the third computational step, we know C_N, D_N, and $r_N = C_N(0)$. We can then use the inverse recursion formulas to obtain C_{N-1}, D_{N-1}, and $r_{N-1} = C_{N-1}(0)$. We can continue using the inverse recursion formulas until we finally obtain all the polynomials and, therefore, all the reflection coefficients.

Finally, let us make the following observations as to the nature of the reflection seismic method. In this section we have seen that the reflection seismic trace may be written as

$$R_N = \frac{C_N}{D_N}$$

where C_N and D_N are polynomials whose coefficients are functions of the reflection coefficients $r_0, r_1, r_2, \ldots, r_N$. For example, if $N = 3$, then

$$C_3 = r_3 + (r_2 + r_3 r_2 r_1 + r_3 r_1 r_0)z + (r_1 + r_3 r_2 r_0 + r_2 r_1 r_0)z^2 + r_0 z^3$$
$$D_3 = 1 + (r_1 r_0 + r_2 r_1 + r_3 r_2)z + (r_2 r_0 + r_3 r_1 + r_3 r_2 r_1 r_0)z^2 + r_3 r_0 z^3$$

In many cases encountered in seismic prospecting, the individual reflection coefficients are small in magnitude. Therefore, as an approximation, if we neglect products of three or more reflection coefficients, then C_3 and D_3 may be written as

$$C_3 \approx r_3 + r_2 z + r_1 z^2 + r_0 z^3$$
$$D_3 \approx 1 + \gamma_1 z + \gamma_2 z^2 + \gamma_3 z^3$$

where the γ's are the autocorrelation coefficients (autoproducts) of the reflection coefficients. In general, to this approximation,

$$C_N \approx r_N + r_{N-1}z + \cdots + r_1 z^{N-1} + r_0 z^N$$
$$D_N \approx 1 + \gamma_1 z + \gamma_2 z^2 + \cdots + \gamma_N z^N$$

where

$$\gamma_j = \sum_{i=0}^{N-j} r_{i+j} r_i, \qquad j = 1, 2, \ldots, N$$

Thus the seismogram trace may be written as

$$R_N \approx \frac{r_N + r_{N-1}z + \cdots + r_0 z^N}{1 + \gamma_1 z + \cdots + \gamma_N z^N}$$

which exhibits the trace as a series of primary reflections $r_N, r_{N-1}, \ldots, r_0$, to each of which is attached a reverberation wavelet, and where the reverberation wavelet is the inverse of the denominator wavelet $(1, \gamma_1, \ldots, \gamma_N)$. In many geologic situations, the sedimentary layers have been laid down in a haphazard sequence, so the reflection coefficients represent a random white sequence; that is, their autocorrelation coefficients are small:

$$\gamma_j \approx 0 \qquad \text{for } j = 1, 2, \ldots, N$$

For such cases the reflection seismogram trace reduces to

$$R_N \approx r_N + r_{N-1}z + r_{N-1}z^2 + \cdots + r_0 z^N$$

which exhibits the trace simply as a series of primary reflections $r_N, r_{N-1}, \ldots, r_0$. (*Note:* According to our convention, N represents the surface and 0 represents the deepest interface.)

Our current civilization may be described as the Age of Petroleum, and most of the petroleum that makes our daily lives possible was discovered by the reflection seismic method. The reflection seismic method was itself discovered empirically in the 1920s. In many areas one could set off an explosion and record a reflection seismogram that exhibited just the primary reflections. Knowledge of these primary reflections made it possible to interpret the raw records by eye, and thereby map the subsurface structure and discover oil. But it was always a mystery why the raw reflection seismic trace only exhibited primary reflections, when in fact there are many more multiple reflections generated by the explosion that never appear on the trace. If these multiples did appear on the trace, the primary reflections would have been lost in the multiples, making visual interpretation of the raw record impossible. This would have meant that little new oil could have been found during the 30-year period from 1930 to 1960. The above result suggests the reason why the empirical seismic method worked: the reflection coefficients were small in magnitude and were random. Since 1960, oil prospecting has extended into areas where these assumptions are no longer valid; the reason why this extension was possible rests in digital-processing methods that can eliminate the nonrandom multiple reflections, water reverberations, and other unwanted signals.

7.4 MINIMUM-DELAY PROPERTY OF THE UNIT-DISTANCE PREDICTION-ERROR OPERATOR

In geophysical data processing it was first discovered empirically that the unit-distance prediction-error operator is always minimum delay. Subsequently, this result was established mathematically. In this section we wish to give a proof for the more general multichannel case, which of course includes the single-channel case. Let

$$x_0, x_1, x_2, \ldots$$

denote a one-sided sequence of rectangular $M \times N$ matrices. Furthermore, let this sequence be stable in the sense that its autocorrelation coefficient for lag 0 is finite; that is,

$$r_0 = \sum_{t=0}^{\infty} x_t x_t^T < \infty$$

where the superscript T denotes the complex-conjugate transpose. The autocorrelation function of this sequence has coefficients

$$r_\tau = \sum_{t=0}^{\infty} x_{t+\tau} x_t^T$$

Each autocorrelation coefficient is an $M \times M$ square matrix. These coefficients satisfy the relationship

$$r_{-\tau} = r_\tau^T$$

The unit-distance prediction operator of length n is one that predicts x_t from $x_{t-1}, x_{t-2}, \ldots, x_{t-n}$. Each coefficient h_i of this operator is an $M \times M$ square matrix. If we let $\hat{x}_t$ denote the predicted value of x_t, then the required prediction operator may be expressed as

$$\hat{x}_t = h_1 x_{t-1} + h_2 x_{t-2} + \cdots + h_n x_{t-n}$$

Let us now formulate the prediction problem in an alternative way. Let us suppose that we wish to predict the matrix X_t made up of the entries x_t, $x_{t-1}, \ldots, x_{t-n+1}$ from the matrix X_{t-1}, which in turn is made up of the entries $x_{t-1}, x_{t-2}, \ldots, x_{t-n}$. The matrices X_t and X_{t-1} are each of order $nM \times N$. In this new formulation the required prediction operation is

$$\begin{bmatrix} \hat{x}_t \\ x_{t-1} \\ x_{t-2} \\ \cdot \\ \cdot \\ \cdot \\ x_{t-n+2} \\ x_{t-n+1} \end{bmatrix} = \begin{bmatrix} h_1 & h_2 & h_3 & \cdots & h_{n-1} & h_n \\ I & 0 & 0 & \cdots & 0 & 0 \\ 0 & I & 0 & \cdots & 0 & 0 \\ & & & & & \\ & \cdots & & & & \\ & & & & & \\ 0 & 0 & 0 & \cdots & 0 & 0 \\ 0 & 0 & 0 & \cdots & I & 0 \end{bmatrix} \begin{bmatrix} x_{t-1} \\ x_{t-2} \\ x_{t-3} \\ \cdot \\ \cdot \\ \cdot \\ x_{t-n+1} \\ x_{t-n} \end{bmatrix}$$

where I is the $M \times M$ identity matrix.

The prediction process can be expressed in this way because all the entries on the left are perfectly predictable except the first. We can write the equation for this new formulation more simply as

$$\hat{X}_t = HX_{t-1}$$

where the matrices $\hat{X}_t$ and X_{t-1} are each $nM \times N$, while the matrix H is square and of order $nM \times nM$. The prediction error is

$$E_t = X_t - HX_{t-1}$$

where the matrix E_t is $nM \times N$.

According to the Gauss method of least squares, the mean-square prediction error is minimized when the error matrix E_t is normal to the past value matrix $X(t-1)$. Thus we have the normal equation

$$\sum E_t X_{t-1}^T = 0$$

where here, and throughout the remainder of this section, all summations are over t from zero to infinity. In this equation, the 0 is an $nM \times nM$ zero matrix.

Let us denote the mean-square prediction-error matrix by σ^2, where σ^2 is an $nM \times nM$ square matrix. We have

$$\begin{aligned}\sigma^2 &= \sum E_t E_t^T \\ &= \sum E_t[X_t^T - X_{t-1}^T H^T]\end{aligned}$$

Making use of the normal equation, we obtain

$$\sigma^2 = \sum E_t X_t^T$$

This equation and the normal equation may be written, respectively, as

$$\sum [X_t - HX_{t-1}]X_t^T = \sigma^2$$

$$\sum [X_t - HX_{t-1}]X_{t-1}^T = 0$$

or

$$R_0 - HR_{-1} = \sigma^2$$

$$R_1 - HR_0 = 0$$

where the autocorrelation coefficients R_{-1}, R_0, R_1 are defined as

$$R_0 = R_0^T = \sum X_t X_t^T, \qquad R_1 = R_{-1}^T = \sum X_t X_{t-1}$$

Each of these autocorrelation coefficients is an $nM \times nM$ matrix. The normal equation gives

$$R_{-1} = R_1^T = R_0 H^T$$

which upon substitution into the equation for σ^2 gives

$$\sigma^2 = R_0 - HR_0H^T$$

Let c be a $1 \times nM$ eigenrow vector of the matrix H; that is, c satisfies

$$cH = c\lambda$$

where λ is the eigenvalue of H corresponding to the eigenvector c. Likewise we have

$$H^T c^T = c^T \lambda^*$$

where the superscript asterisk denotes the complex conjugate. Thus the equation for σ^2 can be written

$$c\sigma^2 c^T = cR_0 c^T - cHR_0 H^T c^T$$

which becomes

$$c\sigma^2 c^T = cR_0c^T - c\lambda R_0\lambda^* c^T$$

or finally

$$c\sigma^2 c^T = (1 - \lambda\lambda^*)cR_0c^T$$

The left side of this equation is a nonnegative number because

$$c\sigma^2 c^T = c \sum E_t E_t^T c^T$$
$$= \sum [cE_t][cE_t]^T \geq 0$$

Similarly, the quantity $c_0 R_0 c^T$ on the right side is a nonnegative number, for

$$cR_0c^T = c \sum X_t X_t^T c^T$$
$$= \sum [cX_t][cX_t]^T \geq 0$$

Hence the other quantity on the right side must also be nonnegative; that is, we have established that

$$1 - \lambda\lambda^* \geq 0$$

or equivalently, that each eigenvalue λ of the matrix H has magnitude less than or equal to 1; that is, $|\lambda| \leq 1$. The eigenvalues λ of the matrix H are the roots of the characteristic polynomial of the matrix H; that is, the eigenvalues λ satisfy the characteristic equation

$$\det(\lambda I - H) = 0$$

where det indicates the determinant of a square matrix. Let us now find a more explicit expression for this determinant. Making use of the expression for the determinant of a partitioned matrix (see ref. 8, p. 344), we can reduce the above characteristic equation to

$$\det[\lambda^n - h_1\lambda^{n-1} - h_2\lambda^{n-2} - \cdots - h_{n-1}\lambda - h_n] = 0$$

Since this equation is merely another expression for the characteristic equation, its roots λ are less than or equal to 1.

If we turn our attention now from the predicted value to the prediction error, we define the prediction-error operator for unit prediction distance as the operator with $M \times M$ matrix coefficients a_i in the form

$$a_0 = I,\ a_1 = -h_1,\ a_2 = -h_2, \ldots, a_n = -h_n$$

Then the prediction error

$$e_t = x_t - \hat{x}_t$$

is given as the convolution of the sequence x_t with the prediction-error operator; that is,

$$e_t = a_0x_t + a_1x_{t-1} + \cdots + a_nx_{t-n}$$

The z transform of the prediction-error operator is defined as the $M \times M$ matrix finite Laurent series

$$A(z) = a_0 + a_1z^{-1} + a_2z^{-2} + \cdots + a_nz^{-n}$$

An operator is *minimum delay* provided that the zeros of the determinant of its z transform $\det A(z)$ are all less than or equal to 1 in magnitude.[4] Thus the unit-distance prediction-error operator is minimum delay, provided that the magnitude of each zero of

$$\det [a_0 + a_1 z^{-1} + a_2 z^{-2} + \cdots + a_n z^{-n}] = 0$$

satisfies $|z| \leq 1$. But the characteristic equation derived above, when put in terms of the prediction-error operator, is

$$\det [a_0 \lambda^n + a_1 \lambda^{n-1} + a_2 \lambda^{n-2} + \cdots + a_n] = 0$$

which can be written,

$$\det \{\lambda^n [a_0 + a_1 \lambda^{-1} + a_2 \lambda^{-2} + \cdots + a_n \lambda^{-n}]\} = 0$$

and we have shown above that $|\lambda| \leq 1$. We immediately see that the characteristic equation is equivalent to the equation $\det A(z) = 0$, provided that we identify λ and z as being the same (i.e., $\lambda = z$). Therefore, all the zeros of $\det A(z) = 0$ satisfy $|z| \leq 1$, and the unit-distance prediction-error operator is minimum delay.

In this section we have shown that the finite-length unit-distance prediction-error operator computed by the method of least squares from any time series has the property that the zeros of the z transform lie within the unit circle, or equivalently that the operator is minimum delay. The equivalence of these two properties was first established by Robinson.[2] Yule[9] and Walker[10] introduced the concept of an autoregressive time series. The general definition of an autoregressive process was given by Wold,[11] according to which the z transform of the associated operator was required to have its zeros within the unit circle (i.e., was required to be an all-pole process). Robinson and Wold[12] proved that the least-squares finite-length unit-distance prediction-error operator computed from any type of time series has the minimum-delay property by showing by actual construction that there exists an autoregressive process associated with the computed operator. This proof was for single-channel time series, but the paper stated that the theorem and its proof extend to the multichannel case.

Subsequently, the link between the z transform of the single-channel autoregressive process and the behavior of polynomials orthogonal on the unit circle became known, and it is a classical fact that the zeros of such orthogonal polynomials lie within the unit circle. Various alternative proofs of this single-channel, minimum-delay property have been published in recent years. In the present section we have given a multichannel proof that reduces the minimum-delay property to its most basic form. In other words, for the matrix prediction equation $\hat{X}(t+1) = HX(t)$, the eigenvalues of the least-squares prediction operator H do not exceed unity. An alternative multichannel proof has been given by Burg.[13]

7.5 WAVEFORM SHAPING FILTERS

At the end of Sec. 7.2 we remarked that the output from the predictive deconvolution filter generally requires smoothing, and that a convenient way to accomplish this goal is to employ a digital waveform shaping filter. Such filters can be designed in either the frequency or time domain, but the most successful geophysical implementations have tended to occur in the time domain. The problem of signal or waveform shaping is so important that we have chosen to deal with this subject in considerable detail. In so doing, we give a novel and simplified mathematical formulation of the design problem. This development has enabled us to shed new light on the detailed relationships between the desired output shape and the associated error measures.

A problem often encountered in digital signal processing may be stated in this way. Given a finite-length input signal, one wishes to find a finite-length operator that shapes the input signal into the form of a given desired finite-length output signal. Except in special circumstances, this shaping operation cannot be perfect, so there will be an error between the actual output and the desired output. The desideratum is to choose that operator which makes this error as small as possible in a least-squares sense.

The key feature of this model is that all the signals and the operator have finite length. From a computer point of view, the concept of infinity is foreign. The inputs and outputs of computers, as well as the operations that computers perform, are finite in every sense. However, many of the mathematical models that are used to describe the physical situation involve the concept of infinity in one form or another, and usually many of the difficulties in analysis can be traced to the problem of reconciling such infinite models with the actual data and computations. For this reason, it is often much simpler and more consistent to replace an infinite-type model with a finite model at the outset.

Let $b(0), b(1), \ldots, b(n)$ represent the given real finite-length *input signal*, and let $d(0), d(1), \ldots, d(m+n)$ represent the real finite-length *desired output signal*.* Here m and n are each nonnegative integers. The problem is to determine the real finite-length *operator* or *filter* $f(0), f(1), \ldots, f(m)$ such that the *actual output signal* $c(0), c(1), \ldots, c(m+n)$ is the least-squares approximation to the desired output. The actual output is equal to the convolution of the input with the operator. This convolution can be written in matrix form as follows. We define the *regressor matrix* B as the $(m+1) \times (m+n+1)$ rectangular matrix whose rows are successively delayed replicas of the input signal:

* We change notation in this section for the sake of simplicity.

$$B = \begin{bmatrix} b(0) & b(1) & \cdots & b(n) & 0 & \cdots & 0 & 0 \\ 0 & b(0) & \cdots & b(n-1) & b(n) & \cdots & 0 & 0 \\ & \cdots & & & & & & \\ 0 & 0 & & & & & b(n) & 0 \\ 0 & 0 & & & & & b(n-1) & b(n) \end{bmatrix}$$

Define the *regression coefficient* f as the $1 \times (m+1)$ row vector of operator coefficients:

$$f = [f(0), f(1), \ldots, f(m)]$$

Define the *regression* c as the $1 \times (m+n+1)$ row vector of the values of the actual output:

$$c = [c(0), c(1), \ldots, c(m+n)]$$

and define the *regressional* d as the $1 \times (m+n+1)$ row vector of the values of the desired output:

$$d = [d(0), d(1), \ldots, d(m+n)]$$

The convolution of the input and the operator can be represented as the matrix multiplication

$$c = fB$$

The error between desired and actual output is $d - c$, and the sum of squared errors is

$$v = (d-c)(d-c)^T$$

where the superscript T denotes matrix transpose. The regression equation can be written as

$$d = fB + (d-c)$$

From the theory of least squares we know that the sum of squared errors v is a minimum if and only if the regressor B is normal to the error $d - c$, that is, if and only if

$$(d-c)B^T = 0$$

This matrix equation represents the scalar normal equations. We may write this *matrix normal equation* as

$$cB^T = dB^T$$

or

$$fBB^T = dB^T$$

The solution of this matrix equation gives the required regression coefficient or, in other words, the required operator,

$$f = dB^T(BB^T)^{-1}$$

Let $r(s)$ denote the autocorrelation of the input signal; that is,

$$r(s) = \sum_{t=0}^{n} b(t+s)b(t)$$

and let $g(s)$ denote the cross correlation of the desired output with the input signal; that is,

$$g(s) = \sum_{t=0}^{n} d(t+s)b(t)$$

Then we see that the matrix BB^T is the $(m+1) \times (m+1)$ autocorrelation matrix R of the input signal; that is,

$$BB^T = R = \begin{bmatrix} r(0) & r(1) & r(2) & \cdots & r(m) \\ r(-1) & r(0) & r(1) & \cdots & r(m-1) \\ r(-2) & r(-1) & r(0) & \cdots & r(m-2) \\ & \cdots & & & \\ r(-m) & r(-m+1) & r(-m+2) & \cdots & r(0) \end{bmatrix}$$

and the matrix dB^T is the $(1) \times (m+1)$ column vector g whose entries are the cross correlation of the desired output with the input signal; that is,

$$dB^T = g = [g(0)\, g(1)\, g(2) \ldots g(m)]$$

In this notation, the normal equations in matrix form are

$$fR = g$$

A special case of the shaping filter is the spiking filter. A spiking filter is a shaping filter for which the desired output is a unit spike. If one looks at the desired output

$$d = [d(0), d(1), \ldots, d(m+n)]$$

one sees that any one of the $m+n+1$ values may represent the spike, while the remaining values are zero. Hence there are $m+n+1$ different spike filters possible in our given model, one spike filter for each of the $m+n+1$ different spike positions. Let

$$a_0 = [a_0(0), a_0(1), \ldots, a_0(m)]$$

be the spiking operator for the zero-delay desired output spike

$$d_0 = [1, 0, 0, \ldots, 0]$$

Similarly, let

$$a_1 = [a_1(0), a_1(1), \ldots, a_1(m)]$$

be the spiking operator for the one-delay desired output spike

$$d_1 = [0, 1, 0, \ldots, 0]$$

and so on, until we let

$$a_{m+n} = [a_{m+n}(0), a_{m+n}(1), \ldots, a_{m+n}(m)]$$

be the spiking operator for the $(m + n)$-delay desired output spike

$$d_{m+n} = [0, 0, 0, \ldots, 1]$$

We see that these successively delayed spikes represent the rows of an $(m + n + 1) \times (m + n + 1)$ identity matrix I. Let A be the $(m + n + 1) \times (m + 1)$ matrix whose rows are the spike operators from zero delay to $(m + n)$ delay; the *spiking operator matrix* is

$$A = \begin{bmatrix} a_0 \\ a_1 \\ \cdot \\ \cdot \\ \cdot \\ a_{m+n} \end{bmatrix}$$

Then the normal equations $aR = dB^T$ for each of the spiking operators can be encompassed in one equation,

$$AR = IB^T$$

or simply

$$AR = B^T$$

Let the $1 \times (m + n + 1)$ row vector

$$c_0 = [c_0(0), c_0(1), \ldots, c_0(m + n)]$$

be the actual output of the zero-delay spiking operator a_0. Let the $1 \times (m + n + 1)$ row vector c_1 be the actual output of the one-delay spiking filter a_1, and so on. Then the $(m + n + 1) \times (m + n + 1)$ square matrix C defined by

$$C = \begin{bmatrix} c_0 \\ c_1 \\ \cdot \\ \cdot \\ \cdot \\ c_{m+n} \end{bmatrix}$$

is called the *spiking actual output matrix* and satisfies the equation

$$AB = C$$

The normal equation $AR = B^T$ may be written as

$$ABB^T = B^T$$

which by postmultiplying by A^T becomes

$$ABB^TA^T = B^TA^T$$

which is

$$CC^T = C^T$$

Let v_i be the sum of squared errors for the spiking filter of delay i. The grand sum of squared errors (i.e., the sum of the sum of squared errors for each spiking operator) is

$$\begin{aligned} V &= v_0 + v_1 + \cdots + v_{m+n} \\ &= \operatorname{tr}\{(I - C)(I - C)^T\} \end{aligned}$$

where tr stands for the trace (i.e., the sum of the diagonal entries) of a square matrix. We have

$$V = \operatorname{tr}(I - C - C^T + CC^T)$$

and because $C^T = CC^T$, we have

$$V = \operatorname{tr}(I - C) = \operatorname{tr} I - \operatorname{tr} C$$

Since I is an $(m + n + 1) \times (m + n + 1)$ identity matrix, $\operatorname{tr} I = m + n + 1$. From this equation we see that $v_i = 1 - c_i(i)$; that is, the sum of squared errors for the spiking filter of delay i is equal to 1 minus the value of the actual output at time instant i.

Let us now find tr C. We have

$$\begin{aligned} \operatorname{tr} C &= \operatorname{tr} AB = \operatorname{tr} B^T R^{-1} B \\ &= \operatorname{tr} B^T (BB^T)^{-1} B \end{aligned}$$

Now the trace of a product of matrices does not depend upon the order of the product, and this result is true even if the matrices are not square; that is,

$$\operatorname{tr} M_1 M_2 = \operatorname{tr} M_2 M_1$$

Thus

$$\operatorname{tr} C = \operatorname{tr}(BB^T)^{-1} BB^T = \operatorname{tr} I$$

where here I is an $(m + 1) \times (m + 1)$ identity matrix. Thus

$$\operatorname{tr} C = m + 1$$

and thus the grand sum of squared errors is

$$V = (m + n + 1) - (m + 1) = n$$

Hence we have come to the conclusion that the grand sum of squared errors for the spiking filters for all possible delays is equal to n, where $n + 1$ is the length of the input wavelet $b(0), b(1), \ldots, b(n)$. Moreover, we see that the grand sum of squared errors V is independent of the filter length $m + 1$. If the sum of squared errors v_i were the same for all possible delays, it would be $v_i = n/(m + n + 1)$ for each spiking filter a_i, where $i = 0, 1, 2, \ldots, m + n$. Generally, the v_i will not be equal for all the spiking filters. The largest possible v_i is unity, because the zero filter $v_i = [0, 0, \ldots, 0]$ would have $v_i = 1$, and any filter computed by least squares could not exceed this sum of squared errors. This maximum error would be obtained, for example, by the zero-delay spiking filter for an input signal whose leading term is zero: $b = [0, b_1, b_2, \ldots, b_n]$; for in this case the right side of the normal equations is zero, and hence the filter is zero and produces maximum error. Consider next the case where all the terms of the input signal are zero except the last: $b = (0, 0, 0, \ldots, 0, b_n)$. In this case the first n spiking filters produce maximum $v_i = 1$ ($i = 0, 1, \ldots, n-1$). The sum of squared errors for these first n spiking filters is therefore n; since this number is identically equal to the grand sum of squared errors, it follows that the last $m+1$ spike filters produce minimum $v_i = 0$ ($i = n, n+1, \ldots, n+m$).

In any case there is some delay i for which the sum of squared errors v_i is a minimum. This minimum need not be unique. The value of i that produces the minimum v_i is called the *optimum delay* or *optimum spike position*, and the corresponding spiking filter a_i is called the *optimum spiking filter* for the given input signal b.

For very short spiking filters it appears that no general rules can be established. However, for sufficiently long spiking filters we observe the following rules:

1. The optimum delay for a minimum-delay input signal is the smallest possible delay, that is, 0.
2. The optimum delay for a maximum-delay input signal is the largest possible delay, that is, $m + n$.
3. The optimum delay for a mixed-delay input signal is intermediate, that is, between the smallest and largest possible delays.

In fact, these rules can actually be used to define the concepts of minimum delay, maximum delay, and mixed delay.

Let us now return to the case of a shaping filter f for an arbitrary desired output d. The matrix normal equation is

$$fR = dB^T$$

But the normal equation for the spiking operator matrix A is $AR = B^T$, so

$$fR = dAR$$

Hence the shaping filter f can be expressed in terms of the desired output d and the spiking operator matrix A as

$$f = dA$$

or

$$f = [d_0, d_1, \ldots, d_{m+n}] \begin{bmatrix} a_0 \\ a_1 \\ \cdot \\ \cdot \\ \cdot \\ a_{m+n} \end{bmatrix}$$

$$= d_0 a_0 + d_1 a_1 + \cdots + d_{m+n} a_{m+n}$$

That is, the shaping filter f is a weighted sum of the spiking filter for all possible delays, the weighting factors being the values of the desired output at times corresponding to the respective delays.

We recall the matrix normal equation for f may be written as

$$(d - c)B^T = 0$$

which when multiplied through by f^T gives

$$(d - c)B^T f^T = 0$$

or

$$(d - c)c^T = 0$$

Hence the sum of squared errors is

$$\begin{aligned} v &= (d - c)(d - c)^T = (d - c)d^T - (d - c)c^T \\ &= (d - c)d^T \\ &= dd^T - cd^T \end{aligned}$$

Since $c = fB$, the sum of squared errors is

$$v = dd^T - fBd^T$$

which, since $g = dB^T$, is

$$v = dd^T - fg^T$$

That is, the sum of squared errors for the shaping filter f is equal to the sum of squared errors of the desired output minus the dot product of the filter with the cross correlation. Because

$$f = dA, \qquad g = dB^T, \qquad C = AB$$

this dot product is

$$fg^T = dABd^T = dCd^T$$

Therefore, the sum of squared shaping filter errors is the quadratic form

$$\begin{aligned} v &= dd^T - dCd^T \\ &= d[I - C]d^T \end{aligned}$$

where the matrix of the quadratic form is $I - C$, that is, the difference between the desired and actual outputs for all possible spiking filters.

This formula gives the minimum squared error v for an arbitrary choice of the desired output d, but the error will not necessarily be small. It is therefore of interest to seek particular choices of the desired output d that will indeed make v small.

It will be convenient to deal with the normalized squared error (NSE) v', which we define in the form

$$v' = \frac{v}{dd^T} = \frac{d(I - C)d^T}{dd^T}$$

Since $v' \geq 0$, the matrix $I - C$ is nonnegative definite. Consider the

$(m + n + 1) \times (m + n + 1)$ spiking actual output matrix C. Since we have seen that

$$CC^T = C^T$$

we may transpose both sides of this relation to obtain

$$CC^T = C$$

We deduce that $C = C^T$ (i.e., C is symmetric), which in turn means that

$$C = CC^T = C^2$$

Now the conditions

$$C = C^T$$

$$C = C^2$$

make C a symmetric idempotent matrix (ref. 14, Definition 12.3.1). Furthermore, the matrix $I - C$ is also symmetric idempotent (ref. 14. Theorem 12.3.5, part [4]), and there follows that

$$\text{rank}\,(I - C) = \text{tr}\,(I - C)$$

(ref. 14, Theorem 9.1.5). But we have already shown that $\text{tr}\,(I - C) = V = n$, and so

$$\text{rank}\,(I - C) = n$$

Therefore, $I - C$ is a symmetric idempotent matrix of rank n, and such a matrix has a total of n nonzero eigenvalues λ_i, each equal to $+1$, while the remaining $m + 1$ eigenvalues are zero (ref. 14, Theorem 12.3.2).

Let us write

$$v' = \frac{d}{dd^T}(\lambda_1 e_1^T e_1 + \lambda_2 e_2^T e_2 + \cdots + \lambda_n e_n^T e_n + \lambda_{n+1} e_{n+1}^T e_{n+1} + \cdots + \lambda_{m+n+1} e_{m+n+1}^T e_{m+n+1})\, d^T$$

where e_i = ith orthonormal $(1) \times (m + n + 1)$ row eigenvector of the matrix $I - C$. The e_i's are orthogonal because $I - C$ is symmetric, and the above expansion of $I - C$ in terms of its eigenvectors and eigenvalues is a direct consequence of the orthogonal transformation

$$I - C = E^T \Lambda E$$

where $E = (m + n + 1) \times (m + n + 1)$ matrix of row eigenvectors e_i, and $\Lambda = \text{diag}\,(\lambda_1, \lambda_2, \ldots, \lambda_{m+n+1})$.

Now let the $(1) \times (m + n + 1)$ desired output signal d be any row eigenvector, say e_j, associated with any of the $m + 1$ zero eigenvalues λ_j,

$$d = e_j, \qquad n + 1 \leq j \leq n + m + 1$$

Then we obtain

$$v' = \frac{e_j\left(\sum_{i=1}^{m+n+1} \lambda_i e_i^T e_i\right)e_j^T}{e_j e_j^T} = \frac{e_j \lambda_j (e_j^T e_j) e_j^T}{e_j e_j^T} = 0$$

because $\lambda_j = 0$, and $e_j e_j^T = \delta_{ij}$, and where

$$\delta_{ij} = \begin{cases} 1 & \text{for } i = j \\ 0 & \text{for } i \neq j \end{cases}$$

This means that we have $m + 1$ possible choices of the desired output signal d that reduce the NSE to its minimum possible value, that is, zero.

It is also interesting to consider what happens when we let d be any row eigenvector, say e_j, associated with any of the n unit eigenvalues λ_j,

$$d = e_j, \qquad 1 \leq j \leq n$$

Then we have

$$v' = \frac{e_j\left(\sum_{i=1}^{n} \lambda_i e_i^T e_i\right) e_j^T}{e_j e_j^T}$$

or

$$v' = \frac{(e_j e_j^T)(e_j e_j^T)}{e_j e_j^T} = 1$$

because $\lambda_j = 1$ and $e_j e_i^T = \delta_{ij}$. This means that we have n possible choices of the desired output signal d that reduce the NSE to its maximum possible value, that is, unity. We remark in passing that our earlier analysis of the error distribution for a set of spiking filters, given an input of the form $b = (0, 0, \ldots, b_n)$, is readily explainable in terms of the above theory.

An important special case of the shaping filter is the prediction filter. Here the desired output is equal to the input signal advanced by a certain time distance α, called the *prediction distance*. The advanced signal is made up of two parts: the irreducible part

$$b(0), b(1), \ldots, b(\alpha - 1)$$

which occurs before time instant zero, and thus is outside the range of the filter, and the reducible part

$$b(\alpha), b(\alpha + 1), \ldots, b(n)$$

which is within the range of the filter. Hence the reducible part represents the desired output of the filter; that is, the desired output is the $1 \times m + n + 1$ row vector d_α given by

$$d_\alpha = [b(\alpha), b(\alpha + 1), \ldots, b(n), 0, \ldots, 0]$$

The prediction filter for prediction distance α is

$$f_\alpha = d_\alpha A = b(\alpha)a_0 + b(\alpha + 1)a_1 + \cdots + b(n)a_{n-\alpha}$$

The sum of squared errors for shaping the input signal into the reducible part is

$$\begin{aligned} v_\alpha &= d_\alpha[I - C]\, d_\alpha^T \\ &= \sum_{i=\alpha}^{n} b^2(i) - \sum_{i=\alpha}^{n} \sum_{j=\alpha}^{n} b(i)c_i(j)b(j) \end{aligned}$$

The prediction error is the sum of the irreducible part, which is all error, and the error between the reducible part and the actual output of the filter. If we let w_α denote the sum of squared prediction errors, we have

$$w_\alpha = \sum_{i=0}^{\alpha-1} b^2(i) + v_\alpha$$

where the first term on the right is the contribution of the irreducible part and the second the reducible part.

For a given input signal b and a given operator length $m + 1$, let us consider the sum of squared prediction errors as a function of the prediction distance; that is, let us consider w_α as a function of α. For sufficiently long operators, w_α is a monotone increasing function of α (where $\alpha = 1, 2, 3, \ldots$); for short operators this is not always so. The irreducible component of w_α is the partial energy of the input signal up to time $\alpha - 1$; thus the irreducible component is a monotonically nondecreasing function of α. No general statement can be made about the reducible component v_α, except that it is zero for α greater than n. As a result, the curve w_α will have a minimum value for one or more values of α; such a value of α is called the *optimum prediction distance.*

It is interesting to note that, whereas the optimum delay (i.e., optimum spike position) of a signal is fundamentally related to the delay properties (i.e., minimum delay, mixed delay, or maximum delay) of the signal, the optimum prediction distance in no way depends on the delay properties of the signal. We can establish this result by considering the normal equation for the determination of the prediction operator. The matrix normal equation is

$$f_\alpha R = d_\alpha B^T$$

The right side is the cross correlation of the reducible part of the signal with the signal; that is, the right side is that portion of the autocorrelation given by

$$d_\alpha B^T = [r(\alpha), r(\alpha + 1), \ldots, r(\alpha + m)]$$

where it is understood that $r(s) = 0$ for $s > n$. Hence the normal equation for the prediction operator f_α involves only the autocorrelation of the input wavelet b, and since the autocorrelation does not depend on the delay properties of the signal, neither does the prediction operator. Thus the optimum prediction distance is independent of the delay properties of the input signal b.

The counterpart of the prediction filter is the hindsight filter. Here the desired output is equal to the input signal delayed by a certain time distance, called the *hindsight distance.* The hindsight distance may be represented as $-\alpha$, where α is intrinsically negative number. That is, the hindsight distance may be thought of as a negative prediction distance. The normal equations for the hindsight filter f_α (where α is intrinsically negative) are given by

$$f_\alpha R = [r(\alpha), r(\alpha + 1), \ldots, r(\alpha + m)]$$

Suppose that $\alpha = -m - 1$, which represents a delay of $m + 1$ time units. Then this hindsight filter satisfies

$$f_{-m-1}R = [r(-m - 1), r(-m), \ldots, r(-1)]$$

where

$$f_{-m-1} = [f_{-m-1}(0), f_{-m-1}(1), \ldots, f_{-m-1}(m)]$$

For real scalar-valued signals, the autocorrelation is symmetric; that is, $r(-k) = r(k)$. Hence the hindsight filter satisfies

$$f_{-m-1}R = [r(m + 1), r(m), \ldots, r(1)]$$

Because R is a symmetrical Toeplitz matrix, we can reverse the coefficients in both row vectors in the above equation to obtain

$$[f_{-m-1}(m), \ldots, f_{-m-1}(1), f_{-m-1}(0)]R = [r(1), \ldots, r(m), r(m + 1)]$$

But we know that the prediction operator f_1 for the prediction distance $\alpha = 1$ satisfies

$$f_1 R = [r(1), \ldots, r(m), r(m + 1)]$$

Hence the reverse of the hindsight operator for hindsight distance $-\alpha = m + 1$ is the same as the prediction operator for prediction distance $\alpha = 1$. More generally, the reverse of the hindsight operator for hindsight distance $-\alpha = m + k$ is the same as the prediction operator for prediction distance $\alpha = k$.

7.6 RECURSIVE SCHEMES FOR NORMAL EQUATIONS INVOLVING TOEPLITZ FORMS

The solution of the least-squares optimum filtering problem involves solving a set of simultaneous equations called the normal equations. In general, there will be one equation for each coefficient in the filter. The requirements for computer time and computer storage space for solving these equations by use of a standard simultaneous equation routine are prohibitive, except in the case of a small number of filter coefficients. This section gives more efficient schemes for arriving at the desired filter coefficients.

These schemes make possible large volumes of seismic computations at reasonable costs; at least 5 million sets of normal equations, many involving up to 100 unknowns or more, are solved each day by the seismic industry.* The approach makes use of the special form of the autocorrelation matrix R, called the Toeplitz form in the scalar case, and the block-Toeplitz form in the

* G. M. Houchins, Amoco Production Company, personal communication, 1976.

matrix case.* This form can be written as

$$R = \begin{bmatrix} r_0 & r_1 & r_2 & \cdots & r_m \\ r_{-1} & r_0 & r_1 & \cdots & r_{m-1} \\ r_{-2} & r_{-1} & r_0 & \cdots & r_{m-2} \\ & \cdots & & & \\ r_{-m} & r_{-m+1} & r_{-m+2} & \cdots & r_0 \end{bmatrix}$$

where the entries are scalars or square matrices; that is, all terms along each diagonal are the same. Thus, given the entries in the left column and the top row, the matrix is fully specified.

The recursive technique involves initially finding a filter of length 1, using this filter to find a filter of length 2, and so on, until the desired length filter is reached. The principal advantages of using the recursive techniques are time and space savings. The standard solution of simultaneous equations requires time proportional to m^3 and space proportional to m^2. The recursive technique reduces these requirements to m^2 and m for time and space, respectively.

An important side benefit of using this scheme is that we can compute the prediction-error variance v at each step of the process. This allows us to formulate a criterion for determining the length of the filter. As the filter becomes longer, the mean-square error will decrease and then level off at some value.

The development of the recursive scheme will be made by using two different notations. The first will be standard algebraic notation; the second will involve the use of a set of vector operators, which we label as *compact notation.*

The recursive forward scheme for the scalar process was first formulated by Levinson.[5] Robinson[15] extended the scalar scheme to the multichannel case. Finally, Wiggins[16] extended the scalar scheme to the multidimensional case. The recursive sideways scheme was proposed by Simpson.[17]

The normal equations for the single-channel case are

$$\sum_{i=0}^{n} f_i r_{j-i} = g_j, \qquad j = 0, 1, \ldots, n$$

where the filter coefficients f_i, the autocorrelation coefficients r_{j-i}, and the right side coefficients g_j are scalars. Associated with these normal equations are the normal equations for the unit-distance prediction-error operator a_i,

$$\sum_{i=0}^{n} a_i r_{j-i} = v\delta_j$$

where $a_0 = 1$, v is the mean-square error, and δ_j is the Kronecker delta function defined as $\delta_j = 1$ for $j = 0$ and $\delta_j = 0$ for $j \neq 0$.

* The matrix case finds application in Sec. 7.7, where we deal with multichannel and multidimensional processes.

The development of the recursion to larger operators given here is a modified version of that given by Levinson. Levinson uses the prediction operator instead of the prediction-error operator. Also, the original Levinson algorithm requires three vector dot products per recursion. One of these dot products is used to calculate the next value of the prediction-error variance v. Since v may become quite small, the accumulation of round-off errors can make the algorithm as proposed by Levinson unstable. The modification of the algorithm as given by Wiggins and Robinson[18] bypasses such a calculation of v, as we shall show later. This modification and its significance have also been discussed by Burg.[13]

The hindsight operator is one that "predicts" past values of a time series from future values. For the scalar case, R is symmetric, and hence the unit-distance hindsight-error operator b_j is just the reverse of the unit-distance prediction-error operator; that is,

$$[b_0, b_1, \ldots, b_n] = [a_n, a_{n-1}, \ldots, a_0]$$

The scheme for extending the unit-distance prediction-error operator $a_0, a_1, \ldots, a_n$ to the new unit-distance prediction-error operator $a'_0, a'_1, \ldots, a'_{n+1}$ involving one more coefficient is first to extend a by adding a zero to the right end:

$$[a_0, \ldots, a_n, 0]\begin{bmatrix} r_0 & \cdots & r_{n+1} \\ r_{-n-1} & \cdots & r_0 \end{bmatrix} = [v, 0, \ldots, 0, u]$$

The quantity u represents the discrepancy

$$u = a_0 r_{n+1} + \cdots + a_n r_1$$

If the discrepancy is zero, the extended operator is the correct one. Generally, the discrepancy will not be zero, so the next step is to modify the coefficients of the extended operator to cancel out the discrepancy. We do this by adding a weighted version of the extended hindsight-error operator to the extended prediction-error operator Thus we obtain

$$[a_0, a_1 + ka_n, \ldots, ka_0]\begin{bmatrix} r_0 & \cdots & r_{n+1} \\ r_{-n-1} & \cdots & r_0 \end{bmatrix} = [v + ku, 0, \ldots, 0, u + kv]$$

The quantity $u + kv$ is set equal to zero in order to determine the value of k,

$$k = -\frac{u}{v}$$

That is, k is equal to the negative of the ratio of the discrepancy u to the prediction-error variance v. The new operator is then

$$a' = [a_0, a_1 + ka_n, \ldots, a_n + ka_1, ka_0]$$

and the new variance is

$$v' = v + ku$$

We now use the new prediction-error operator to extend the length of the filter f. As before, we make the first approximation to f' by adding a zero to the end of f,

$$[f_0, f_1, \ldots, f_n, 0]\begin{bmatrix} r_0 & \cdots & r_{n+1} \\ r_{-n-1} & \cdots & r_0 \end{bmatrix} = [g_0, \ldots, g_n, \gamma_{n+1}]$$

where

$$\gamma_{n+1} = f_0 r_{n+1} + \cdots + f_n r_1$$

If we weight and add the new hindsight-error operator to the above extended filter, we get

$$[f_0 + k_f a'_{n+1}, \ldots, f_n + k_f a'_1, k_f a'_0]\begin{bmatrix} r_0 & \cdots & r_{n+1} \\ r_{-n-1} & \cdots & r_0 \end{bmatrix}$$
$$= [g_0, \ldots, g_n, \gamma_{n+1} + k_f v']$$

Now we choose k_f such that

$$\gamma_{n+1} + k_f v' = g_{n+1}$$

Then the new filter is

$$f' = [f_0 + k_f a'_{n+1}, \ldots, f_n + k_f a'_1, k_f a'_0]$$

Let us now summarize the above results in compact notation, and let us define the following vectors:

$a = [a_0, a_1, \ldots, a_n]$: given prediction-error operator

$a' = [a'_0, a'_1, \ldots, a'_{n+1}]$: new prediction-error operator

$b = [a_n, a_{n-1}, \ldots, a_0]$: given hindsight-error operator

$b' = [a'_{n+1}, a'_n, \ldots, a'_0]$: new hindsight-error operator

$p = [r_{n+1}, r_n, \ldots, r_1]$: segment of autocorrelation

$f = [f_0, f_1, \ldots, f_n]$: given filter

$f' = [f'_0, f'_1, \ldots, f'_n]$: new filter

First we compute the discrepancy by the dot product

$$u = a \cdot p$$

and then k as the negative ratio

$$k = -\frac{u}{v}$$

Next we compute the new prediction-error operator

$$a' = (a, 0) + k(0, b)$$

and the new prediction-error variance

$$v' = v + ku$$

(This computation represents the direct formula, which is the essential departure of the present method from Levinson's, where one computes v' by the dot product $v' = a_0 r_0 + a_1 r_{-1} + \cdots + a_n r_{-n}$.)

The new hindsight-error operator b' is simply the reverse of the new prediction-error operator. Next we compute the dot product

$$\gamma(n + 1) = f \cdot p$$

and then the constant

$$k_f = \frac{[g_{n+1} - \gamma_{n+1}]}{v'}$$

Finally, the new filter is

$$f' = (f, 0) + k_f b'$$

Each step of this scheme requires the computation of two dot products, $a \cdot p$ and $f \cdot p$, instead of three dot products, $a \cdot p$, $f \cdot p$, and $a_0 r_0 + \cdots + a_n r_{-n}$, as required by the Levinson method. Criticisms as to the accuracy of the Levinson method are often due to the errors resulting from computing this third dot product, which is unnecessary in the modified scheme given here.

For a multichannel process with M input channels and L output channels, each autocorrelation coefficient r_i is an $M \times M$ matrix, each filter coefficient as well as each prediction-error operator coefficient and each hindsight-error operator coefficient is an $L \times M$ matrix, and each right-side coefficient is an $L \times M$ matrix. Also, in the multichannel case the hindsight-error operator is not simply the time reverse of the prediction error operator but is a separate operator in its own right,

$$b = [b_n, b_{n-1}, \ldots, b_0]: \quad \text{given hindsight-error operator}$$

$$b' = [b'_{n+1}, b'_n, \ldots, b'_0]: \quad \text{new hindsight-error operator}$$

Moreover the autocorrelation function is not symmetric as in the real scalar case, so now we must make use of two vectors:

$$p = [r_{n+1}, r_n, \ldots, r_1]$$

$$q = [r_{-1}, r_{-2}, \ldots, r_{-n-1}]$$

With these changes, the recursive scheme in the multichannel case proceeds along the same lines as in the scalar case. Thus, in compact notation, the recursive scheme is as follows.

First compute the dot products

$$u_a = a \cdot p$$

$$u_b = b \cdot q$$

In the usual multichannel case, the autocorrelation has the following symmetry:

$$r_{-i} = [r_i]^T$$

where T is the complex-conjugate transpose. In this case only one of the above two dot products has to be computed, as they are related by

$$u_b = u_a^T$$

Next compute

$$k_a = -u_a v_b^{-1}$$
$$k_b = -u_b v_a^{-1}$$

Then we compute the new operators

$$a' = (a, 0) + k_a(0, b)$$
$$b' = (0, b) + k_b(a, 0)$$

and the new variances

$$v_a' = v_a + k_a u_b$$
$$v_b' = v_b + k_b u_a$$

Next we compute the dot product

$$\gamma_{n+1} = f \cdot p$$

and

$$k_f = [g_{n+1} - \gamma_{n+1}] v_b'^{-1}$$

Finally, the new filter is

$$f' = (f, 0) + k_f b'$$

We note that each step in the recursion to increase the length of the filter by 1 involves computing two dot products, $a \cdot p$ and $f \cdot p$. However, in the special case when one desires the prediction-error operator (or equivalently the prediction operator), then with the recognition that a' is the desired result, one can eliminate the computation of the second dot product $f \cdot p$, as well as k_f and f'. This recognition essentially reduces the computation by one half for prediction operators as compared to general filter operators.

7.7 TWO-DIMENSIONAL SHAPING FILTER

Thus far, our discussion has tended to emphasize scalar-valued geophysical applications of signal processing. A growing body of technology does, however, require the treatment of the recordings from an array of geophysical sensors in terms of the more involved vector-valued signal processing theory. These vector-valued processes can sometimes be studied from the multichannel viewpoint, sometimes from the multidimensional viewpoint, and often from a combination of both. Space does not allow us to do justice to this vast subject here; instead, we confine ourselves to the discussion of a design procedure that leads to the two-dimensional least-squares shaping filter. In many instances the implementation of the two-dimensional shaping filter is most conveniently carried out in terms of an appropriate two-dimensional recursive filter.

An exploration seismogram consists of an assemblage of ground motion recordings made as a function of time. These recordings represent the output from individual sensors or the combined output from a group of sensors having specific locations in space. If the actual locations of these sensors do not require explicit consideration, the only independent variable is time. If spatial coordinates do require explicit consideration, we may have to deal with up to four independent variables: time, plus three possible spatial coordinates. Let the dimension of a process by given by the number of its independent variables, and let its order be given by the number of dependent variables representing it at each point in space. For example, the output from a linear array of seismometers measuring ground motion in three different planes would represent a third-order, two-dimensional process. These matters have been treated eloquently by Wiggins.[16]

In an engineering sense, the order of a process is equivalent to the number of channels. An n-trace seismogram with time as the only independent variable is therefore a one-dimensional n-channel process. Thus the multichannel systems discussed earlier in this chapter are all one dimensional. In many geophysical applications the spatial coordinates of the seismometers must be considered explicitly. The same n-channel seismogram recorded with the seismometers positioned along a straight line can also be viewed as a two-dimensional process in the independent variables time (t) and distance (x). Such a duality of viewpoints suggests the possibility of mapping a multidimensional process into an equivalent single-dimensional multichannel process. Wiggins[16] has given the formalism for doing this in the general case, but here we restrict ourselves to the two-dimensional (or planar) problem.

The subject is best treated with a simple example. Consider the planar convolution,

t (time) →, x (distance) ↓

$$\begin{matrix} a_{11} & a_{12} \\ a_{21} & a_{22} \end{matrix} * \begin{matrix} b_{11} & b_{12} \\ b_{21} & b_{22} \end{matrix} = \begin{matrix} c_{11} & c_{12} & c_{13} \\ c_{21} & c_{22} & c_{23} \\ c_{31} & c_{32} & c_{33} \end{matrix}$$

or

$$A*B = C$$

where A and B are (2×2) arrays with elements a_{ij} and b_{ij} and C is a (3×3) array with elements c_{ij}. The symbol (*) denotes convolution. The abcissa and oridnate directions for all arrays represent discrete time (t) and discrete distance (x), respectively. The coefficients c_{ij} can be calculated most simply by performing the two-dimensional polynomial multiplication,

$$A(z, w)B(z, w) = C(z, w)$$

where $A(z, w)$, $B(z, w)$, and $C(z, w)$ are two-dimensional, or planar, generating functions, and where z and w are the unit delay operators along the t and x axes, respectively, so that

$$A(z, w) = a_{11} + a_{12}z + a_{21}w + a_{22}zw$$
$$B(z, w) = b_{11} + b_{12}z + b_{21}w + b_{22}zw$$

Therefore,

$$C(z, w) = a_{11}b_{11} + (a_{11}b_{12} + a_{12}b_{11})z + (a_{11}b_{21} + a_{21}b_{11})w + \cdots + a_{22}b_{22}z^2w^2$$

and

$$c_{11} = a_{11}b_{11}$$
$$c_{12} = a_{11}b_{12} + a_{12}b_{11}$$
$$c_{21} = a_{11}b_{21} + a_{21}b_{11}$$
$$c_{33} = a_{22}b_{22}$$

However, we may also write

$$A(z, w) = (a_{11} + a_{12}z) + (a_{21} + a_{22}z)w = a_1(z) + a_2(z)w$$
$$B(z, w) = (b_{11} + b_{12}z) + (b_{21} + b_{22}z)w = b_1(z) + b_2(z)w$$

which yields

$$C(z, w) = [a_1(z) + a_2(z)w][b_1(z) + b_2(z)w]$$
$$= a_1(z)b_1(z) + [a_1(z)b_2(z) + a_2(z)b_1(z)]w + a_2(z)b_2(z)w^2$$

or

$$C(z, w) = c_1(z) + c_2(z)w + c_3(z)w^2 \tag{7.5}$$

where

$$c_1(z) = a_1(z)b_1(z)$$
$$c_2(z) = a_1(z)b_2(z) + a_2(z)b_1(z) \tag{7.6}$$
$$c_3(z) = a_2(z)b_2(z)$$

Equation (7.5) is reminiscent of the generating function of the convolution between two one-dimensional sequences, except that the coefficients c_i are no longer constants, but rather polynomials in powers of z. This observation suggests writing Eq. (7.6) in the form

$$\begin{bmatrix} b_1(z) & 0 \\ b_2(z) & b_2(z) \\ 0 & b_2(z) \end{bmatrix} \begin{bmatrix} a_1(z) \\ a_2(z) \end{bmatrix} = \begin{bmatrix} c_1(z) \\ c_2(z) \\ c_3(z) \end{bmatrix}$$

or, transposing both sides,

$$[a_1(z) \quad a_2(z)]\begin{bmatrix} b_1(z) & b_2(z) & 0 \\ 0 & b_1(z) & b_2(z) \end{bmatrix} = [c_1(z) \quad c_2(z) \quad c_3(z)] \tag{7.7}$$

We observe that the original two-dimensional filter problem has now been mapped into an equivalent multichannel problem. Let us assume that the (2×2) array A is the filter, while the (2×2) array B is the input. The above multichannel mapping has then been obtained in terms of a (2×3) channel input,

$$\begin{bmatrix} b_1(z) & b_2(z) & 0 \\ 0 & b_1(z) & b_2(z) \end{bmatrix}$$

two of whose channels are null, and in terms of a (1×2) channel filter

$$[a_1(z) \quad a_2(z)]$$

More generally, inductive reasoning tells us that if A is an $(m \times n)$ filter array and B is a $(\mu \times \nu)$ input array, the equivalent multichannel system consists of

1. An $(m \times [m + \mu - 1])$-channel, ν-length input.
2. A $(1 \times m)$-channel, n-length filter.

Several remarks are in order at this point. First, the input channel matrix has the general structure

$$\begin{bmatrix} b_1(z) & b_2(z) & \cdots & b_\mu(z) & 0 & \cdots & 0 \\ 0 & b_1(z) & \cdots & b_{\mu-1}(z) & b_\mu(z) & & 0 \\ & & \cdots & & & & \\ 0 & \cdots & & b_1(z) & & \cdots & b_\mu(z) \end{bmatrix}$$

which is an $(m \times [m + \mu - 1])$ polynomial matrix, only μ of whose polynomial elements are independent, and which has

$$m(m + \mu - 1) - m\mu = m(m - 1)$$

null entries as shown. Second, the above mapping is not the only one possible, for we may write

$$A(z, w) = (a_{11} + a_{21}w) + (a_{12} + a_{22}w)z = \hat{a}_1(w) + \hat{a}_2(w)z$$
$$B(z, w) = (b_{11} + b_{21}w) + (b_{12} + b_{22}w)z = \hat{b}_1(w) + \hat{b}_2(w)z$$

which leads to the equivalent multichannel system

$$[\hat{a}_1(w) \quad \hat{a}_2(w)]\begin{bmatrix} \hat{b}_1(w) & \hat{b}_2(w) & 0 \\ 0 & \hat{b}_1(w) & \hat{b}_2(w) \end{bmatrix} = [\hat{c}_1(w) \quad \hat{c}_2(w) \quad \hat{c}_3(w)] \tag{7.8}$$

where

$$\hat{c}_1(w) = \hat{a}_1(w)\hat{b}_1(w)$$
$$\hat{c}_2(w) = \hat{a}_1(w)\hat{b}_2(w) + \hat{a}_2(w)\hat{b}_1(w)$$
$$\hat{c}_3(w) = \hat{a}_2(w)\hat{b}_2(w)$$

Expressed in this form, if A is $(m \times n)$ and B is $(\mu \times \nu)$, the equivalent multichannel system consists of

1. An $(n \times [n + \nu - 1])$-channel, μ-length input.
2. A $(1 \times n)$-channel, m-length filter.

The input channel matrix now has the general structure,

$$\begin{bmatrix} \hat{b}_1(w) & \hat{b}_2(w) & \cdots & \hat{b}_\nu(w) & 0 & \cdots & 0 \\ 0 & \hat{b}_1(w) & \cdots & \hat{b}_{\nu-1}(w) & \hat{b}_\nu(w) & \cdots & 0 \\ & & & \cdots & & & \\ 0 & \cdots & & \hat{b}_1(w) & \cdots & & \hat{b}_\nu(w) \end{bmatrix}$$

which is an $(n \times [n + \nu - 1])$ polynomial matrix, only ν of whose polynomial elements are independent, and which has

$$n(n + \nu - 1) - n\nu = n(n - 1)$$

null entries as shown.

The discrete two-dimensional least-squares filter problem can be posed and solved by means of either the equivalent multichannel systems of the kind of Eq. (7.7) or (7.8). If we use Eq. (7.7), the problem is

$$\begin{array}{ccc} & \xrightarrow{\text{Desired output}} & [d_1(z) \quad d_2(z) \quad d_3(z)] \\ [a_1(z) \quad a_2(z)] & \begin{bmatrix} b_1(z) & b_2(z) & 0 \\ 0 & b_1(z) & b_2(z) \end{bmatrix} & = [c_1(z) \quad c_2(z) \quad c_3(z)] \\ \text{Filter} & \text{Input} & \text{Actual output} \end{array} \qquad (7.9)$$

where $d_i(z)$, $i = 1, 2, 3$, is the generating function for ith row of some desired (3×3) output array D. Using the least-squares multichannel algorithms described earlier, we can compute the filter $[a_1(z) \ a_2(z)]$ such that the squared error

$$I_1 = \sum_{i=1}^{3} (d_i - c_i)(d_i - c_i)^T$$

is minimized, where d_i and c_i are the ith row vectors of the desired output arrays D and the actual output arrays C, respectively, and where the symbol T denotes transposition.

On the other hand, if we use Eq. (7.8), the problem is

$$\xrightarrow{\text{Desired output}} [\hat{d}_1(w) \quad \hat{d}_2(w) \quad \hat{d}_3(w)]$$

$$\underset{\text{Filter}}{[\hat{a}_1(w) \quad \hat{a}_2(w)]} \underset{\text{Input}}{\begin{bmatrix} \hat{b}_1(w) & \hat{b}_2(w) & 0 \\ 0 & \hat{b}_1(w) & \hat{b}_2(w) \end{bmatrix}} = \underset{\text{Actual output}}{[\hat{c}_1(w) \quad \hat{c}_2(w) \quad \hat{c}_3(w)]} \qquad (7.10)$$

where $\hat{d}_j(w)$, $j = 1, 2, 3$, is the "w"-generating function of the jth column of the same desired (3×3) output array D. Again, the previously described multichannel algorithms can be employed to calculate the filter $[\hat{a}_1(w) \; \hat{a}_2(w)]$ such that the squared error

$$I_2 = \sum_{j=1}^{3} (\hat{d}_j - \hat{c}_j)^T(\hat{d}_j - \hat{c}_j)$$

is minimized, where $\hat{d}_j$ and $\hat{c}_j$ are the jth column vectors of the desired output arrays D and the actual output arrays C, respectively.

If the arrays A and B are square, both formulations (7.9) and (7.10) are equally efficient from the computational viewpoint. In the more general case of rectangular arrays A and B, either one or the other of these formulations may be more efficient, depending on the actual dimensions of the problem and the computer software available.

Let us illustrate the calculation of a two-dimensional least-squares filter with a small numerical example. Let the (2×2) input array be

$$B = \begin{bmatrix} 1 & 2 \\ 3 & 4 \end{bmatrix}$$

We wish to design a (2×2) filter A such that the resulting (3×3) actual output array C approximates a (3×3) desired output array

$$D = \begin{bmatrix} 0 & 0 & 0 \\ 0 & 1 & 0 \\ 0 & 0 & 0 \end{bmatrix}$$

in the least-squares sense. If we use formulation (7.9), we have

$$b_1(z) = 1 + 2z$$

$$b_2(z) = 3 + 4z$$

and

$$d_1(z) = 0$$

$$d_2(z) = 1z$$

$$d_3(z) = 0$$

so that the input polynomial matrix $B(z)$ is

$$B(z) = \begin{bmatrix} 1 + 2z & 3 + 4z & 0 \\ 0 & 1 + 2z & 3 + 4z \end{bmatrix}$$

The input autocorrelation polynomial matrix $R(z)$ is therefore

$$R(z) = B(z)B^T(z^{-1}) = \begin{bmatrix} 14z^{-1} + 30 + 14z & 6z^{-1} + 11 + 4z \\ 4z^{-1} + 11 + 6z & 14z^{-1} + 30 + 14z \end{bmatrix}$$

while the cross-correlation polynomial matrix $G(z)$ is

$$G(z) = D(z)B^T(z^{-1}) = [4 + 3z \quad 2 + z]$$

The required (2×2) autocorrelation coefficients are thus

$$r_0 = \begin{bmatrix} 30 & 11 \\ 11 & 30 \end{bmatrix}, \qquad r_1 = \begin{bmatrix} 14 & 4 \\ 6 & 14 \end{bmatrix}$$

while the required (2×1) cross-correlation coefficients are

$$g_1 = [4 \quad 2], \qquad g_2 = [3 \quad 1]$$

The appropriate normal equations are

$$[(a_{11} \quad a_{12})(a_{12} \quad a_{22})]\begin{bmatrix} r_0 & r_1 \\ r_1^T & r_0 \end{bmatrix} = [g_0 \quad g_1]$$

which can be solved with the multichannel block-Toeplitz algorithm to yield the solution

$$A = \begin{bmatrix} a_{11} & a_{12} \\ a_{21} & a_{22} \end{bmatrix} = \begin{bmatrix} 0.10201 & 0.05138 \\ 0.02374 & -0.01019 \end{bmatrix}$$

The actual output is

$$\begin{bmatrix} 0.10201 & 0.25540 & 0.11276 \\ 0.32977 & 0.59965 & 0.18514 \\ 0.07122 & 0.06439 & -0.04076 \end{bmatrix}$$

with a normalized mean-square error of 0.40035.

The example illustrates the use of the two-dimensional least-squares filter as a two-dimensional shaping operator.

We remark that the normal equations for the discrete two-dimensional least-squares filter can also be obtained directly without recourse to the multichannel mapping approach.[16] The derivations are extremely lengthy however, and the insight into the entire problem afforded by the multichannel mapping approach tends to be lost.

Quite often the convolutional implementation of the least-squares shaping filter becomes computationally laborious. This problem can generally be circumvented by recourse to appropriate two-dimensional recursive (or feedback) filters. The transfer function of a two-dimensional recursive filter can be written in the form

$$A(z, w) = \frac{P(z, w)}{Q(z, w)} = \frac{\sum_{i,j=0}^{k,\ell} p_{ij} z^i w^j}{\sum_{i,j}^{m,n} q_{ij} z^i w^j} \tag{7.11}$$

where $P(z, w)$ and $Q(z, w)$ are numerator and denominator polynomials of appropriate degree. The degree of these polynomials as well as the numerical values of the coefficients p_{ij} and q_{ij} depend on the filter design technique applied to the particular problem at hand. In general, a recursive filter is computationally more efficient than its convolutional counterpart in the sense that for a wide variety of situations the total number of coefficients p_{ij} and q_{ij} is significantly less than the total number of coefficients a_{ij}.

The subject of two-dimensional recursive filter design has received widespread attention in the recent literature (see e.g., refs. 19 and 20), and there appears little reason to repeat details here. However, the two-dimensional least-squares filter discussed above does have bearing on one vital aspect of the two-dimensional recursive filter, its stability. Just as is true in one dimension, an unstable two-dimensional recursive filter produces an unbounded output. Many current design techniques result in filters that require stabilization before they can be implemented. This means that the denominator or feedback polynomial $Q(z, w)$ cannot vanish for $|z|$ and $|w|$ simultaneously less than or equal to 1, which is the condition that $Q(z, w)$ be minimum phase.

A number of techniques have been devised to accomplish this stabilization. One involves use of the two-dimensional complex cepstrum,[21,22] while another is based on the calculation of the two-dimensional Hilbert transform associated with the log-magnitude spectrum of $Q(z, w)$.[23] Both of these approaches are designed to produce a minimum-phase denominator $Q(z, w)$, but both can lead to implementational difficulties.

An alternative method hinges on a conjecture,[24] which appears to hold in many cases, although a counterexample has been found recently:[25] Given an arbitrary real finite array X, its two-dimensional least-squares inverse is (probably!) minimum phase.

To stabilize $Q(z, w)$, we may proceed as follows:

1. Compute a least squares inverse filter array Q' by solving the problem,

$$\overset{\text{Desired output}}{\longrightarrow D}$$

$$\underset{\text{Filter}}{Q'} \quad * \quad \underset{\text{Input}}{Q} \quad = \quad \underset{\text{Actual output}}{C'}$$

where D is the *unit pulse response* array

$$D = \begin{bmatrix} 1 & 0 & 0 & \cdots & 0 \\ 0 & 0 & 0 & \cdots & 0 \\ & & \cdots & & \\ 0 & 0 & 0 & \cdots & 0 \end{bmatrix}$$

The input Q is an $(\mu \times \nu)$ array. Letting the least-squares inverse filter Q' be an $(m \times n)$ array, the desired output D and actual output C must be $([m + \mu - 1] \times [n + \nu - 1])$ arrays. The above conjecture then enables us to state that Q' is minimum phase.

2. Next compute a least-squares inverse filter array $\hat{Q}$ by solving the problem

$$\hat{Q} \quad * \quad Q' \quad = \quad \hat{C} \qquad \xrightarrow{\text{Desired output}} D$$

Filter — Input — Actual output

Now the input Q' is an $(m \times n)$ array. In general we will want the filter $\hat{Q}$ to have the same dimensions as the original recursive filter denominator array Q, that is, $(\mu \times \nu)$. Therefore, the desired output D and the actual output $\hat{C}$ must again be $([m + \mu - 1] \times [n + \nu - 1])$ arrays. The conjecture enables us to state that the array $\hat{Q}$ is minimum phase.

Since Q' is the least-squares inverse of Q, and since $\hat{Q}$ is the least-squares inverse of Q', we may conclude that $\hat{Q}$ is a minimum-phase approximation to the given recursive filter denominator array Q.

Stabilization is accordingly achieved after replacement of Q by its minimum-phase approximation $\hat{Q}$. This method has worked satisfactorily in a wide variety of actual filter design problems, yet the existence of at least one counterexample to the conjecture clearly indicates the need for further work. More details on this question can be found in a forthcoming paper.[26]

In most applied work we will wish the recursive filter to have a zero-phase response. Since the filter $A(z, w)$ of Eq. (7.11) has been defined for the first quadrant only $(k, l, m, n \geq 0)$, a stable zero-phase recursive filter can be synthesized in the form

$$G(z, w) = A_s(z, w)A_s(z^{-1}, w^{-1})$$

where $A_s(z, w)$ is the recursive filter with a minimum-phase denominator $\hat{Q}(z, w)$. Setting $z = e^{-i\omega_t}$ and $w = e^{-i\omega_x}$, where ω_t and ω_x are angular frequencies along the t and x axes, respectively, we obtain

$$G(e^{-i\omega_t}, e^{-i\omega_x}) = A_s(e^{-i\omega_t}, e^{-i\omega_x})A_s(e^{i\omega_t}, e^{i\omega_x})$$

which implies

$$|G(\omega_t, \omega_x)| = |A_s(\omega_t, \omega_x)|^2$$

That is, $G(\omega_t, \omega_x)$ has a phase response of zero and an amplitude response equal to the square of the amplitude response of $A_s(z, w)$. Read et al.[20] describe a number of alternative operations leading to a desired zero-phase response.

7.8 CONCLUSIONS

We have seen how the model of a horizontally stratified earth is used to determine seismic data-processing methods. Even though this model is simplified, it not only provides essential methods for seismic interpretation, but also offers a basis from which more complex models can be constructed.

Because of the large mass of data that must be processed in seismic operations, most seismic models must have a statistical basis. In actual operations in the search for hydrocarbons, data must be sifted and sorted out. To make the correspondence with the field situation, research models must also allow for noisy and uncertain data, which means that the models must incorporate statistical parameters. Statistical data and statistical methods are in everyday use in most or all of the sciences that unravel the world around us, and statistical techniques have proved indispensable in a great many areas of human activity. In geophysics, as in any other branch of science, the scientific method is synonymous with the method of model building. Since the advent of large-scale digital processing of seismic data, the operational uses of the scientific models have become essential; as a result, significant advances in model building have been made. These advances make modern geophysical models useful both in the search for hydrocarbons and in the research for better methods to find them.

We have here tried to describe a number of mathematical models that have found application in applied seismic data processing. We have by no means exhausted the list of such models, but have rather attempted to emphasize those of potentially greatest interest to workers engaged in signal processing in other disciplines. For example, we have omitted a discussion of the method of homomorphic deconvolution[27] because this powerful approach has evolved in the area of speech processing. These matters are dealt with in Chapter 3.

The present chapter is an attempt to describe many of the digital-processing methods in current usage. We have treated multidimensional methods as well as multichannel methods; in fact, one of the most important applications given here is the mapping of multidimensional problems into corresponding multichannel problems, so that already known multichannel methods can be used for their solution.

Although other sources of energy exist, the energy that runs our automobiles, trucks, airplanes, trains, and ships is oil. Liquid oil occurs underground in deep deposits; the most common way to find it is with the reflection seismic method. In geologic areas in which the reflection coefficients are small (in magnitude) and random, the reflection seismogram emphasizes the primary reflections, and as a result such records can be visually interpreted. Such was the case for much of the exploration for oil during the period from 1930 to 1960. However, the increasing demand for petroleum required that geologic areas be explored that did not produce such favorable seismic records. Interpretation of records from these difficult areas was impossible in many cases, because various types of multiple reflections completely masked the desired primary reflections. In particular, the offshore areas were in this category, because the water and mud layers had large reflection coefficients associated with their interfaces; these large reflection coefficients produced a water-confined multiple, or reverberation, called ringing. This phenomenon completely masked the primary reflections. To make such difficult areas amenable to seismic exploration, digital methods were developed. These reduced or eliminated the unwanted multiples. Deconvolution, straight summing of traces, and other multiple elimination methods transformed the raw seismic traces into processed data that could now be interpreted. During the 1960s, the oil exploration industry went through a digital revolution, and today virtually all exploration seismic data are recorded in digital form. They are then processed by a digital computer to yield the output seismogram that can be interpreted to delineate buried geologic structures. As a result, large geologic areas previously veiled are now open for intensive exploration for oil and gas.

REFERENCES

1. L. C. Wood and S. Treitel, "Seismic Signal Processing," *Proc. IEEE*, vol. 63, 1975, pp. 649–661.
2. E. A. Robinson, "Predictive Decomposition of Time Series with Applications to Seismic Exploration," Ph.D. Thesis, Department of Geology and Geophysics, Massachusetts Institute of Technology, Cambridge, Mass., 1954, also *Geophysics*, vol. 32, 1967, pp. 418–484.
3. K. L. Peacock and S. Treitel, "Predictive Deconvolution—Theory and Practice," *Geophysics*, vol. 34, 1969, pp. 155–169.
4. E. A. Robinson, *Random Wavelets and Cybernetic Systems*, Charles Griffin & Co. Ltd., London, 1962.
5. N. Levinson, "The Wiener RMS (Root Mean Square) Error Criterion in Filter Design and Prediction, *J. Math. Phys.*, vol. 25, 1947, pp. 261–278.
6. S. Treitel and E. A. Robinson, "The Design of High-Resolution Digital Filters," *IEEE Trans. Geosci. Electron.*, vol. GE-4, 1966, pp. 25–38.
7. H. A. Lorentz, A. Einstein, H. Minkowski, and H. Weyl, *The Principles of Relativity, a Collection of Original Memoirs*, Methuen, 1923; reprinted by Dover, New York, 1958.
8. T. W. Anderson, *An Introduction to Multivariate Statistical Analysis*, Wiley, New York, 1958.
9. G. U. Yule, "On a Method of Investigating Periodicities in Disturbed Series, with Special Reference to Wolfer's Sunspot Numbers," *Phil. Trans.*, vol. A 226, 1927, pp. 267–298.
10. G. Walker, "On Periodicity in Series of Related Terms," *Proc. Roy. Soc.*, vol. A 131, 1931, pp. 518–532.

11. H. Wold, *A Study in the Analysis of Stationary Time Series*, Almquist and Wiksells, Uppsala, Sweden, 1938.

12. E. A. Robinson and H. Wold, "Minimum-Delay Structure of Least-Squares and Eo-Ipso Predicting Systems for Stationary Stochastic Processes," in *Proceedings of the Symposium on Time Series Analysis*, M. Rosenblatt, ed., Wiley, New York, 1963, pp. 192–196.

13. J. P. Burg, "Maximum Entropy Spectral Analysis," Ph. D. Thesis, Department of Geophysics, Stanford University, Stanford, Calif., 1975.

14. F. A. Graybill, *Introduction to Matrices with Applications in Statistics*, Wadsworth, Belmont, Calif., 1969.

15. E. A. Robinson, "Mathematical Development of Discrete Filters for the Detection of Nuclear Explosions," *J. Geophys. Res.*, vol. 68, 1963, pp. 5559–5567.

16. R. A. Wiggins, "On Factoring the Correlations of Discrete Multivariable Stochastic Processes," Ph. D. Thesis, Department of Geology and Geophysics, Massachusetts Institute of Technology, Cambridge, Mass., 1965.

17. S. M. Simpson, "Recursive Schemes for Normal Equations of Toeplitz Form, Chapter 4 of Scientific Report No. 7 of Contract AF 19(604) 7378, ARPA Project VELA-UNIFORM, MIT, Cambridge, Mass., 1963.

18. R. A. Wiggins and E. A. Robinson, "Recursive Solution to the Multichannel Filtering Problem," *J. Geophys. Res.*, vol. 70, 1965, pp. 1885–1891.

19. R. M. Mersereau and D. E. Dudgeon, "The Representation of Two-dimensional Sequences as One-dimensional Sequences," *IEEE Trans. Acoust. Speech Signal Processing*, vol. ASSP-22, 1974, pp. 320–325.

20. R. R. Read, J. L. Shanks and S. Treitel, "Two-dimensional Recursive Filtering," *Topics in Applied Physics*, vol. 6, Springer, New York, 1975, pp. 131–176.

21. D. E. Dudgeon, "Two-dimensional Recursive Filtering," Ph.D. Thesis, Department of Electrical Engineering, Massachusetts Institute of Technology, Cambridge, Mass., 1974.

22. P. Pistor, "Stability Criterion for Recursive Filters," *IBM J. Res. Develop.*, vol. 18, 1974, pp. 59–71.

23. R. R. Read and S. Treitel, "The Stabilization of Two-dimensional Recursive Filters via the Discrete Hilbert Transform, *IEEE Trans. Geosci. Electron.*, vol. GE-11, 1973, pp. 153–160, 205–207.

24. J. L. Shanks, S. Treitel and J. H. Justice, "Stability and Synthesis of Two-dimensional Recursive Filters," *IEEE Trans. Audio Electroacoust.*, vol. AU-20, 1972, pp. 115–128.

25. Y. Genin and Y. Kamp, "Counter Example in the Least-Squares Inverse Stabilization of 2-D Recursive Filters," *Electron. Letters*, vol. 11, 1975, pp. 330–331.

26. E. I. JURY, V. KOLAVENNU and B. D. O. ANDERSON, "Further Proof of Shanks' Conjecture for Low Degree Polynomial," *Proc. IEEE*, 1977, in press.

27. A. V. OPPENHEIM, R. W. SCHAFER and T. G. STOCKHAM, "Nonlinear Filtering of Multiplied and Convolved Signals," *Proc. IEEE*, vol. 56, 1968, pp. 1264–1291.

INDEX

T

U

V

W